Practical Asymptotics

Practical Asymptotics

Edited by

H.K. KUIKEN

Eindhoven University of Technology,
Eindhoven, The Netherlands

Reprinted from *Journal of Engineering Mathematics*, Volume 39, Nos. 1–4 (March 2001)

Kluwer Academic Publishers
Dordrecht / Boston / London

TABLE OF CONTENTS

Journal of Engineering Mathematics **39:** 1–2, 2001.

Practical asymptotics

H. K. KUIKEN
Department of Mathematics and Computing Science, Eindhoven University of Technology, P.O. Box 513, 5600 MB Eindhoven, The Netherlands

Received 25 July 1998, accepted in revised form 26 September 2000

Abstract. The term *practical asymptotics* is explained. It is argued that many practical problems are amenable to and will benefit from the asymptotic approach. Fifteen papers by authors who are active in a wide range of disciplines demonstrate this. It is argued that the teaching of asymptotic methods should remain an integral part of any sophisticated numerics curriculum, raising it far above the level of mere number crunching.

Key words: asymptotics, asymptotics teaching, non-dimensionalisation, numerics, real-world problems.

A few years ago this journal published a special issue entitled *Large-Scale Numerical Modelling of Problems Involving the Navier-Stokes Equations* (32 (1997) 101–280). That appeared to be a very timely publication. Indeed, these days applied mathematics seems, at first sight anyway, to become more and more dominated by direct numerical simulation. Admittedly, this leads to new insights which, it would seem, could not have been attained by other means. The purpose of that issue was to highlight that particular aspect.

The interest of large sections of the community of applied mathematicians being focussed in the above direction nowadays, the older and broad subject of asymptotics seems to be losing the popularity it once enjoyed and the impact it had. This is partly because many believe that asymptotics deals with exceptional cases which are usually outside the practical domain. However, this is a misconception! To make this clear, we offer the following argument:

Having non-dimensionalised your problem, consider a typical dimensionless group

$$N_D = \alpha_1^{k_1}\alpha_2^{k_2}\alpha_3^{k_3}\dots,$$

where the k_i are real-valued exponents that can be both positive and negative. We know that, within a given set of units, each of the physical parameters α_i can assume values that may vary widely depending upon the problem in hand, *i.e.* $\alpha_i \sim 10^{n_i}$, where the integers n_i can range from large negative to large positive values. Thus,

$$N_D \sim 10^p, \qquad \left(p = \sum_i n_i k_i\right)$$

and we can ask ourselves: what is the probability that $p \sim 0$? Of course, p is most likely to be less than -1 or larger than 1, which means that the problem can be simplified through the application of asymptotic techniques, *without seriously affecting the practical usefulness of the reduced model.* The technique, or rather the craft, by which the reduction is accomplished is called *Practical Asymptotics.* It is a highly intuitive process and is based, to a large extent, on physical reasoning.

Papers which demonstrate the power of practical asymptotics should involve both large-scale computations on the full model and computations based on the asymptotically reduced model. A comparison between the two should then illustrate a considerable reduction in computing effort in terms of cpu time. Further emphasis should be put on the explicit representation of trends, physical relationships, rules, laws, etc., as expressed by mathematical formulae resulting from the asymptotics with numerically generated coefficients. This, then, should contrast with the practice of large-scale numerics which usually results in the presentation of numbers, (colour) plates or contour plots relating to specific examples.

In the present issue fifteen papers have been collected, each seeking to demonstrate the usefulness and practical nature of the asymptotic approach for the discipline from which it arose. These papers address subjects as varied as: the production of glass bottles, semiconductors, surface-tension-driven flows, microwave-assisted joining of tubes, viscous-inviscid interaction, industrial limit cycles, heat exchangers used in the production of foodstuffs, water waves generated by bottom topography, channel flows, chemical-clock reactions, Stokes flows, separated flows, superconductivity, flows over fibrous plates, multiscale problems arising in low-Mach-number flows; a wide range of subjects indeed!

We hope that this issue will demonstrate the continuing usefulness and validity of the asymptotic approach in reducing the complexity of mathematical models to a practical minimum, *i.e.*, without unduly sacrificing their accuracy, thus achieving large reductions in computing effort and increased physical understanding. We believe that the teaching of asymptotic methods should remain an integral part of the academic curriculum as an indispensable tool for the numericist wishing to tackle problems of ever-increasing complexity.

Journal of Engineering Mathematics **39:** 3–24, 2001.

Shear flow over a particulate or fibrous plate

C. POZRIKIDIS
Department of Mechanical and Aerospace Engineering University of California, San Diego, La Jolla, California 92093-0411, U.S.A. (e-mail: cpozrikidis@ucsd.edu, Internet URL: http://stokes.ucsd.edu/c_pozrikidis)

Received: 15 September 1999, accepted in revised form 10 April 2000

Abstract. Simple shear flow over a porous plate consisting of a planar array of particles is studied as a model of flow over a membrane. The main objective is to compute the slip velocity defined with reference to the velocity profile far above the plate, and the drift velocity induced by the shear flow underneath the plate. The difference between these two velocities is shown to be proportional to the thickness of the plate. When the geometry of the particle array is anisotropic, the directions of the slip and drift velocity are generally different from the direction of the overpassing shear flow. An integral formulation is developed to describe flow over a plate consisting of a periodic lattice of particles with arbitrary shape, and integral representations for the velocity and pressure are developed in terms of the doubly-periodic Green's function of three-dimensional Stokes flow. Based on the integral representation, asymptotic expressions for the slip and drift velocity are derived to describe the limit where the particle size is small compared to the inter-particle separation, and numerical results are presented for spherical and spheroidal particles of arbitrary size. The asymptotic results are found to be accurate over an extended range of particle sizes. To study the limit of small plate porosity, the available solution for shear flow over a plane wall with a circular orifice is used to describe flow over a plate with a homogeneous distribution of circular perforations, and expressions for the slip and drift velocity are derived. Corresponding results are presented for axial and transverse shear now over a periodic array of cylinders arranged distributed in a plane. Streamline pattern illustrations confirm that a negative drift velocity is due to the onset of eddies between closely-spaced particles.

Key words: Stokes flow, integral equations, shear flow, porous plate, membrane.

1. Introduction

An important area of research in biomechanics concerns the effect of a shear flow on the equilibrium shapes of, and mass transport through, membranes enclosing vesicles and biological cells [1–3]. The membranes typically consist of lipid bilayers, sometimes resting on a network of proteins comprising the cytoskeleton, separating the vesicle or cytoplasmic fluid from the ambient plasma or buffer fluid. From the point of view of hydrodynamics, a membrane may be regarded as a screen or sieve, and the problem of flow through or over it may be studied within the more general context of flow over a porous or irregular plate separating two semi-infinite regions occupied by the same or different fluids. This more general problem is pertinent to several other areas of biofluiddynamics and mainstream engineering fluid mechanics involving, for example, flow over arrangements of cilia or bundled tubes.

Previous studies of the thermodynamics and hydrodynamics of polymeric and biological membranes have modeled the network of the fundamental molecular units comprising a membrane as a planar lattice of particles over which the smaller molecules of the solute slide [4, 5]. The theory seeks to predict the species diffusivity and to establish the relationship between the pressure drop and the geometrical properties of the membrane expressed in terms of the hydrodynamic resistance for flow normal to the membrane. Ishii [6] studied flow normal to a planar lattice of spherical particles in the asymptotic limit where the particle size is small compared to the particle separation. To the author's knowledge, shear flow over a porous plate or particulate plane has not been addressed explicitly by previous authors, although three related classes of problems have been discussed in detail.

The first class of problems involves flow over an irregular surface with small or large amplitude protuberances possibly of fractal nature [7, 8]. Issues of interest include the establishment of the physical origin of the no-slip boundary condition, the computation of the slip velocity as a function of the morphology of the surface irregularities, and the study of transport properties in terms of effective heat or mass transfer coefficients. The second class of problems involves shear flow over a semi-infinite particulate matrix modeling a porous medium. Reviews and numerical simulations for two- and three-dimensional arrangements were presented by Larson and Higdon [9, 10] and Sangani and Behl [11]. Their results illustrated the dependence of the slip velocity on the geometry of the porous medium microstructure, and helped to establish a theoretical foundation for Brinkman's equation governing the structure of the flow near the surface of a porous medium, also pointing out its limitations. The third class of problems involves shear flow over a wall with a circular hole or side pore, possibly in the presence of suction that drives the fluid through the hole or into the pore, with applications to particle entrainment and particle screening [12–15].

In this paper, we study shear flow over a planar lattice of particles forming a porous surface. When the thickness of the plate is comparable to, or larger than, the gaps between the particles, regions of recirculating flow develop in the intervening spacing, and the flow decays rapidly underneath the array to give a virtually quiescent lower fluid. When, however, the particle size is smaller than the inter-particle separation, the shear flow penetrates the lower fluid, and a uniform drift velocity is established underneath the array. If the particle shape or lattice geometry is anisotropic, the direction of the drift velocity is generally different from that of the overpassing shear flow. Far above the lattice, the flow reduces to simple shear flow with a macroscopic slip velocity similar to that established in shear flow over a porous material. The main goal of the present work is to illustrate the dependence of the slip and drift velocity on the particle shape and lattice geometry.

The asymptotic and numerical studies are based on integral representations of periodic Stokes flow using an appropriate Green's function. In implementing the numerical procedure for solving an integral equation of the first kind for the traction over a particle surface, a general method is developed for removing the eigenfunction of the single-layer Stokes hydrodynamic potential. To complement the results for three-dimensional flow over a planar array, two related configurations are also considered. The first configuration involves shear flow over a flat plate of zero thickness containing a homogeneous distribution of circular perforations of small size, which is relevant to shear flow over a porous plate with a large solid areal fraction. The second configuration involves longitudinal and transverse shear flow over an array of cylinders. Comparisons of the results for three-and two-dimensional configurations illustrate once again the fundamental differences in the nature of the corresponding flows.

2. Shear flow over a planar particle lattice

Consider infinite simple shear flow over a two-dimensional lattice of identical rigid particles positioned at the vertices of a regular lattice that is parallel to the xy plane, as shown in

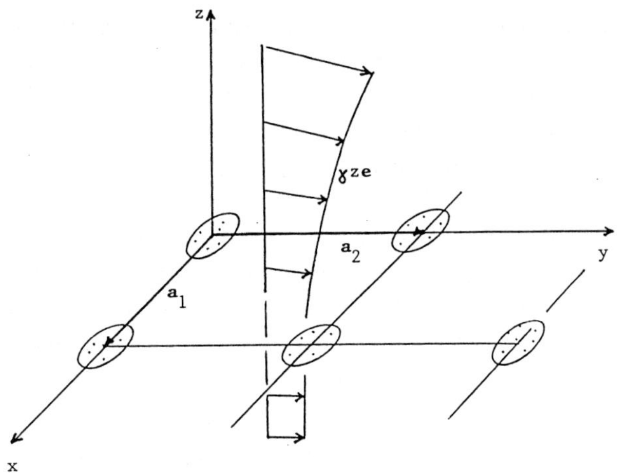

Figure 1. Shear flow over a planar lattice of particles. Far below the lattice, the velocity tends to the uniform drift velocity whose direction and magnitude are determined by the particle shape and size, the lattice geometry, and the direction of the shear flow.

Figure 1. The lattice is described by two base vectors $\mathbf{a}_1$ and $\mathbf{a}_2$, so that the designated centers of two particles labelled n and m are related by

$$\mathbf{X}_n = \mathbf{X}_m + i_1\mathbf{a}_1 + i_2\mathbf{a}_2, \tag{1}$$

where i_1 and i_2 are two integers.

Far above the lattice, as z tends to $+\infty$, the flow reduces to simple shear flow with shear rate γ in the direction of the unit vector $\mathbf{e}$ that is parallel to the xy plane. Thus, the slope of the velocity is required to exhibit the asymptotic behavior

$$\lim_{z\to+\infty} \frac{\mathrm{d}\mathbf{u}}{\mathrm{d}z} \to \gamma\mathbf{e} \tag{2}$$

while the pressure tends to a constant value denoted by $p^{+\infty}$. Far below the lattice, as z tends to $-\infty$, the shear stress is required to vanish,

$$\lim_{z\to-\infty} \frac{\mathrm{d}\mathbf{u}}{\mathrm{d}z} \to 0 \tag{3}$$

and the pressure tends to a constant value denoted by $p^{-\infty}$. Integration of the preceding two equations with respect to z produces the asymptotic forms

$$\lim_{z\to+\infty} \mathbf{u} \to \gamma z\mathbf{e} + \mathbf{U}^{+\infty} + \mathrm{edt} \tag{4}$$

and

$$\lim_{z\to-\infty} \mathbf{u} \to \mathbf{U}^{-\infty} + \mathrm{edt}, \tag{5}$$

where $\mathbf{U}^{+\infty}$ and $\mathbf{U}^{-\infty}$ are two constant velocities parallel to the xy plane, both to be computed as part of the solution, and 'edt' stands for exponentially decaying terms. The magnitude and direction of the slip velocity $\mathbf{U}^{+\infty}$ depend on the definition of the origin of the z axis; in the present case, the origin coincides with the designated particle centers [9, 16]. The magnitude and direction of the uniform drift velocity induced by the shear flow under the lattice, $\mathbf{U}^{-\infty}$, on the other hand, is independent of the origin of the z axis. Neither the slip velocity nor the drift velocity is necessarily oriented in the direction of the shear flow. In practice, the shear flow may be generated by the translation of a flat plate located at $z = z_p$ with velocity $\mathbf{U}_p$ parallel to the particle lattice. Using the asymptotic form (4), we find $\mathbf{U}_p = \gamma z_p \mathbf{e} + \mathbf{U}^{+\infty}$, which provides us with an expression for evaluating $\gamma \mathbf{e}$ and $\mathbf{U}^{+\infty}$, when another linear relationship between them has been established.

Returning to the problem of infinite shear flow, we perform a force balance over a control volume that is confined between (a) four planes that are perpendicular to the xy plane and enclose one lattice cell and thus one particle, (b) two planes that are parallel to the xy plane located far above or below the particle lattice, and (c) the surface of the enclosed particle, to obtain

$$\mathbf{F} \equiv \int_{\text{Particle}} \mathbf{f}(\mathbf{x})\, \mathrm{d}S(\mathbf{x}) = \mu \gamma A \mathbf{e} + A(p^{-\infty} - p^{+\infty})\mathbf{e}_z, \tag{6}$$

where $\mathbf{F}$ is the force exerted on one particle, $\mathbf{f} = \sigma \cdot \mathbf{n}$ is the hydrodynamic traction, σ is the stress tensor, $\mathbf{n}$ is the unit vector normal to the particle pointing into the fluid, μ is the fluid viscosity, A is the area of one lattice cell, and $\mathbf{e}_z$ is the unit vector pointing along the z axis.

The motion of the fluid is governed by the equations of Stokes flow [17]

$$\nabla p = \mu \nabla^2 \mathbf{u}, \qquad \nabla \cdot \mathbf{u} = 0 \tag{7}$$

which are to be solved subject to (a) the far-field conditions expressed by Equations (4) and (5), (b) the no-slip and no-penetration condition $\mathbf{u} = \mathbf{0}$ on the particle surface, and (c) the periodicity condition

$$\mathbf{u}(\mathbf{x}) = \mathbf{u}(\mathbf{x} + i_1 \mathbf{a}_1 + i_2 \mathbf{a}_2) \tag{8}$$

where i_1 and i_2 are two integers. Our main objective is to compute the slip and drift velocities as functions of the particle shape and size and of the lattice geometry, in the limit of vanishing Reynolds number.

2.1. Integral formulation

To prepare the ground for the integral formulation, we introduce the Green's function of the equations of Stokes flow describing the doubly-periodic flow due to a two-dimensional lattice of point forces that is identical to the particle lattice shown in Figure 1. The velocity and pressure field induced by the point forces at the point $\mathbf{x}$ are expressed by

$$u_1(\mathbf{x}) = \frac{1}{8\pi\mu} G_{ij}(\mathbf{x}, \mathbf{x}_0)\, b_j, \quad p(\mathbf{x}) = \frac{1}{8\pi} P_j(\mathbf{x}, \mathbf{x}_0)\, b_j \tag{9}$$

where $\mathbf{G}$ is the velocity Green's function tensor, $\mathbf{P}$ is the pressure Green's function vector, and $\mathbf{b}$ is the strength of a point force; one arbitrarily selected point force is located at $\mathbf{x}_0$. The Green's functions for the velocity and pressure satisfy the periodicity condition shown in Equation (8). Moreover, as $z - z_0$ tends to $+\infty$, we require the asymptotic behavior

$$\lim_{z \to +\infty} \mathbf{G}(\mathbf{x}, \mathbf{x}_0) \to -\frac{8\pi}{A}(z - z_0)\, \mathbf{J} + \text{edt}, \qquad \lim_{z \to +\infty} \mathbf{P}(\mathbf{x}, \mathbf{x}_0) \to \frac{8\pi}{A}\mathbf{e}_z + \text{edt}, \tag{10}$$

where $\mathbf{J}$ is the identity matrix but with the third diagonal component corresponding to the z axis set equal to zero. As $z - z_0$ tends to $-\infty$, all components of the Green's functions are required to vanish. When the point forces are parallel to the lattice, the pressure field decays far from the lattice. When the point forces are perpendicular to the lattice, a pressure difference is established across the lattice to balance the point force exerted on the volume of fluid confined within each periodic cell.

The computation of the Green's functions in terms of Fourier series or Ewald sums has been discussed by Ishii [6], Sangani and Behl [11], and Pozrikidis [18]. As the observation point $\mathbf{x}$ approaches a point force, the expressions in terms of Fourier series converge extremely slowly and are inadequate for the purposes of numerical computation. As an alternative, we evaluate the Green's function using the fast summation method developed in [18], based on the expression

$$\begin{aligned}\mathbf{G}(\mathbf{x}, \mathbf{x}_0) &= \sum_n \left(\mathbf{I} \frac{1 - 3\xi r_n + \xi^2 r_n^2}{r_n} + (\mathbf{x} - \mathbf{x}_n)(\mathbf{x} - \mathbf{x}_n) \frac{1 + \xi r_n - \xi^2 r_n^2}{r_n^3} \right) \mathrm{e}^{-\xi r_n} + \\ &+ \frac{4\pi}{A} \sum_{\substack{m \\ |\mathbf{l}_m| \neq 0}} (-\mathbf{I}\nabla^2 + \nabla\nabla) \\ &\frac{1}{2} \left(\frac{1 + \rho_m}{|\mathbf{l}_m|^3} \mathrm{e}^{-\rho} + \frac{(2 - \zeta_m^2)(1 + \delta\sqrt{1 + \zeta_m^2}) + \delta^2(1 + \zeta_m^2)}{\xi^3 (1 + \zeta_m^2)^{5/2}} \right) \cos(\mathbf{l}_m \cdot \mathbf{x}) + \\ &+ \frac{4\pi}{A} (-\mathbf{I}\nabla^2 + \nabla\nabla) \frac{2 + 2\delta + \delta^2}{2\xi^3} \mathrm{e}^{-\delta} \cos(\mathbf{l}_m \cdot \mathbf{x})\end{aligned} \tag{11}$$

where ξ is an arbitrary splitting parameter, $r_n = |\mathbf{x} - \mathbf{x}_n|$, $\mathbf{x}_n$ is the location of the nth point force, the first sum on the right-hand side of (11) runs over all point forces, and the second sum runs over the nodes of the reciprocal planar wave number lattice with base vectors $\mathbf{b}_1 = (2\pi/A)\mathbf{a}_2 \times \mathbf{e}_z$ and $\mathbf{b}_2 = (2\pi/A)\mathbf{e}_z \times \mathbf{a}_1$; $\rho_m = |\mathbf{l}_m|(z - z_0)$, $\zeta_m = |_m|/\xi$, and $\delta = \xi(z - z_0)$. The computations reported in later sections were conducted using $\xi = 2\pi/\sqrt{A}$. Figure 2(a) shows the streamline pattern in the xz plane induced by a square lattice of point forces directed along the x axis. The unit lattice vectors are given by $\mathbf{a}_1 = (L, 0, 0)$, $\mathbf{a}_2 = (0, L, 0)$, and one of the point forces is located at the origin, The onset of regions of recirculating flow above and below the lattice is a dominant feature of the motion.

It can be shown that, subject to the previously stated conditions, the Green's function satisfies the properties

$$\mathbf{G}(\mathbf{x}, \mathbf{x}_0) = \mathbf{G}(\mathbf{x}_0, \mathbf{x}) - \frac{8\pi}{A}(z - z_0)\, \mathbf{J} \tag{12}$$

and

$$\mathbf{P}(\mathbf{x}, \mathbf{x}_0) = \mathbf{P}(\mathbf{x}_0, \mathbf{x}) - \frac{8\pi}{A}\mathbf{e}_z \tag{13}$$

which allow us to switch the location of the singular point and field point, provided that we also include a complementary shear flow or decrease the pressure by an appropriate amount.

Considering now the control volume confined between (a) four planes that are perpendicular to the xy plane and enclose one lattice cell and thus one particle, (b) two planes that are parallel to the xy plane and are located far above or below the particle lattice, and (c) the surface of the enclosed particle, we use the integral representation for Stokes flow to express

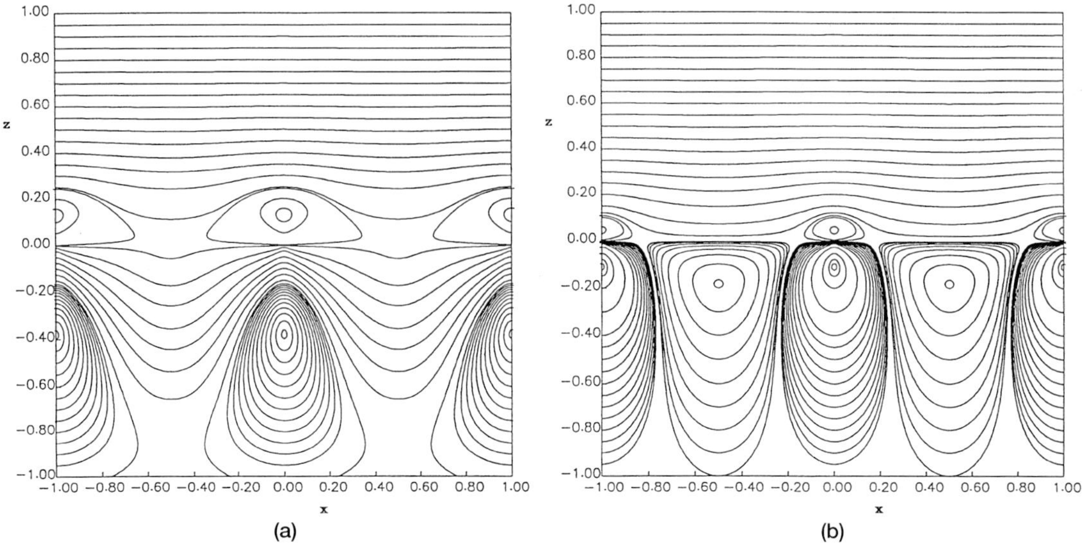

Figure 2. (a) Streamline pattern in the xz plane induced by a square lattice of three-dimensional point forces directed along the x axis. (b) Streamline pattern induced by an array of two-dimensional point forces directed along the x axis.

the flow variables in terms of boundary integrals [17]. Exploiting the stipulated periodicity of the flow and the conforming periodicity of the Green's function, shifting the bottom boundary of the control volume to negative infinity, and noting that far below the lattice the velocity of the shear flow tends to a constant value whereas the Green's function for the velocity vanishes, we derive the following representation for the velocity at a point $\mathbf{x}_0$:

$$u_j(\mathbf{x}_0) = -\frac{1}{8\pi\mu}\int_{\text{Particle}} G_{ij}(\mathbf{x},\mathbf{x}_0) f_i(\mathbf{x})\,\mathrm{d}S(\mathbf{x}) + \\ +\frac{1}{8\pi\mu}\int_{\text{Top}} G_{ij}(\mathbf{x},\mathbf{x}_0) f_i(\mathbf{x})\,\mathrm{d}S(\mathbf{x}) - \frac{1}{8\pi}\int_{\text{Top}} n_i(\mathbf{x}) T_{ijk}(\mathbf{x},\mathbf{x}_0) u_k(\mathbf{x})\,\mathrm{d}S(\mathbf{x}) \tag{14}$$

where $\mathbf{T}$ is the stress tensor associated with the Green's function, and "Top" denotes the top of the control volume. Over the particle surface, the unit normal vector $\mathbf{n}$ points into the fluid, whereas over the top of the control volume, $\mathbf{n}$ points toward the positive z axis. Observing the limit as the top of the control volume moves to infinity, and noting the asymptotic behavior of the flow under consideration and of the periodic Green's function, we obtain a simplified representation in terms of the single-layer Stokes potential and the unknown slip velocity,

$$u_j(\mathbf{x}_0) = -\frac{1}{8\pi\mu}\int_{\text{Particle}} G_{ij}(\mathbf{x},\mathbf{x}_0) f_i(\mathbf{x})\,\mathrm{d}S(\mathbf{x}) + \gamma z_0\, e_j + U_j^{+\infty}. \tag{15}$$

Taking now the limit as the field point $\mathbf{x}_0$ moves far below the particle array, and using the asymptotic behavior of the Green's function described earlier, we find

$$U_j^{-\infty} \to -\frac{1}{8\pi\mu}\left(-\frac{8\pi}{A}\right)\int_{\text{Particle}} (z - z_0) J_{ij} f_i(\mathbf{x})\,\mathrm{d}S(\mathbf{x}) + \gamma z_0 e_j + U_j^{+\infty}. \tag{16}$$

Breaking up the integral into two parts, and using the integral force balance expressed by equation (6), we obtain a relationship between the slip and drift velocities,

$$U_j^{-\infty} = U_j^{+\infty} + J_{ij}\frac{1}{\mu A}\int_{\text{Particle}} z f_i(\mathbf{x})\,\mathrm{d}S(\mathbf{x}) \tag{17}$$

which shows that their difference is on the order of the particle size normal to the lattice. For discoidal particles of zero thickness, the difference will vanish.

It is both useful and physically appealing to recast the representation (15) into an alternative form expressing the flow induced by a distribution of point forces over the particle surfaces. For this purpose, we use identity (12) to switch the arguments of the Green's function, finding

$$u_j(\mathbf{x}_0) = -\frac{1}{8\pi\mu}\int_{\text{Particle}} G_{ji}(\mathbf{x}_0,\mathbf{x}) f_i(\mathbf{x})\,\mathrm{d}S(\mathbf{x}) + J_{ij}\int_{\text{Particle}} (z-z_0) f_i(\mathbf{x})\,\mathrm{d}S(\mathbf{x}) + \gamma z_0 e_j + U_j^{+\infty}. \tag{18}$$

The first integral on the right-hand side of (18) represents the requisite distribution of point forces. To simplify the remaining terms, we use an integrated form of the reciprocal relation for Stokes flow over the boundaries of the control volume previously defined, written for the flow under consideration and for a test shear flow with velocity $\mathbf{u}^{\text{test}} = \gamma^{\text{test}}(z-z_0)\mathbf{e}^{\text{test}}$, where γ^{test} is an arbitrary shear rate and $\mathbf{e}^{\text{test}}$ is an arbitrary unit vector parallel to the xy plane, obtaining

$$\int_{\text{Particle, Bottom, Top}} \gamma^{\text{test}}(z-z_0)e_i^{\text{test}} f_i(\mathbf{x})\,\mathrm{d}S(\mathbf{x}) = \int_{\text{Particle, Bottom, Top}} u_i(\mathbf{x}) f_i^{\text{test}}(\mathbf{x})\,\mathrm{d}S(\mathbf{x}). \tag{19}$$

Requiring the boundary condition at the particle surface, shifting the top and bottom boundaries of the control volume far above or below the particle lattice, and taking into consideration the asymptotic behavior of the flow, we obtain the identity

$$\int_{\text{Particle}} (z-z_0) f_j(\mathbf{x})\,\mathrm{d}S(\mathbf{x}) = -\mu\gamma A z_0 + \mu\gamma^{\text{test}} A (U_i^{-\infty} - U_i^{+\infty}). \tag{20}$$

Finally, we combine equations (18–20) and derive the desired representation

$$u_j(\mathbf{x}_0) = -\frac{1}{8\pi\mu}\int_{\text{Particle}} G_{ji}(\mathbf{x}_0,\mathbf{x}) f_i(\mathbf{x})\,\mathrm{d}S(\mathbf{x}) + U_j^{-\infty}. \tag{21}$$

On the basis of this representation, we find that pressure difference across the lattice is given by

$$p^{-\infty} - p^{+\infty} = \frac{1}{A}\mathbf{F}\cdot\mathbf{e}_z \tag{22}$$

in agreement with (6).

Evaluating equation (15) or (21) at the particle surface, and requiring the boundary condition $\mathbf{u} = \mathbf{0}$, we obtain a Fredholm integral equation of the first kind for the distribution of the traction $\mathbf{f}$. Equation (6) provides us with a supplementary condition that we may use to compute the slip velocity $\mathbf{U}^{+\infty}$ or drift velocity $\mathbf{U}^{-\infty}$. Having computed one, the other follows from Equation (17).

2.2. MULTIPLE ARRAYS

The formulation for shear flow over a two-dimensional particle array developed in this section may be extended readily to shear flow over multiple arrays or a semi-infinite lattice modeling an ordered porous medium, considered by Sangani and Behl [11]. In the case of flow over a semi- infinite lattice, the fast decay of the velocity down the lattice causes the drift velocity to vanish, yielding the counterpart of the representation (21)

$$u_j(\mathbf{x}_0) = -\frac{1}{8\pi\mu}\sum_{l=1}^{\infty}\int_{P_l} G_{ji}(\mathbf{x}_0,\mathbf{x}) f_i(\mathbf{x})\,\mathrm{d}S(\mathbf{x}) \tag{23}$$

where P_l denotes a particle that belongs to the lth array. Evaluating this representation at a point located on a particles surface, and requiring the boundary condition $\mathbf{u} = \mathbf{0}$, we obtain a homogeneous Fredholm integral equation of the first kind for the distribution of the traction $\mathbf{f}$, which is to be solved subject to a constraint imposed by the x and y components of the integral force balance

$$\sum_{l=1}^{\infty}\int_{P_l} \mathbf{f}(\mathbf{x})\,\mathrm{d}S(\mathbf{x}) = \mu\gamma A\mathbf{e} + A(p^{-\infty} - p^{+\infty})\mathbf{e}_z. \tag{24}$$

The pressure drop across the lattice arises as part of the solution in terms of the z component of the force exerted on the individual particles.

3. Asymptotics for small particles

When the typical particle size a is small compared to the characteristic dimension of the particle lattice L, an approximate solution of the problem posed in Section 2 may be found using the method of matched asymptotic expansions. Considering the integral representation (21), we find that, to leading order with respect to the small parameter $\varepsilon = a/L$, the velocity induced at the designated center of one particle by all other particles is given by

$$u_j(\mathbf{x}_p) \cong \frac{1}{8\pi\mu} D_{ji} F_i + U_j^{-\infty}, \tag{25}$$

where $\mathbf{D}$ is the difference between the free-space Green's function or Stokeslet, denoted by $\mathbf{S}$, and the periodic Green's function, evaluated at the particle center $\mathbf{x}p$,

$$D_{ji} \equiv (S_{ji} - G_{ji})(\mathbf{x}_p, \mathbf{x}_p). \tag{26}$$

Using the principles of matched asymptotic expansions, we find that to leading order in ε, the force exerted on the particle is given by

$$F_k \cong 6\pi\ \mu\ a\ R_{kj}\Big(\frac{1}{8\pi\mu}\ D_{ji} F_i + U_j^{-\infty}\Big), \tag{27}$$

where $\mathbf{R}$ is a dimensionless resistance tensor relating the force exerted on a stationary solitary particle to the velocity of an infinite uniform incident flow. For a spherical particle of radius a, $\mathbf{R}$ is the identity matrix $\mathbf{I}$. Solving Equation (27) for $\mathbf{U}^{-\infty}$, and using expression (6) to evaluate the force, we obtain

$$\mathbf{U}^{-\infty} \cong \frac{\gamma A}{6\pi a}(\mathbf{R}^{-1} - \tfrac{3}{4}a\ \mathbf{D})\cdot\mathbf{e} \tag{28}$$

accurate to first order with respect to ε. An alternative form of (28) is

$$\mathbf{U}^{-\infty} \cong \frac{1}{6}\gamma a \frac{1}{\phi}\left(\mathbf{R}^{-1} - \tfrac{3}{4}\phi^{1/2}\left(\frac{A}{\pi}\right)^{1/2}\mathbf{D}\right)\cdot\mathbf{e}, \tag{29}$$

Table 1. Diagonal components of the matrix **D** for a rectangular lattice of side ratio δ.

δ	LD_{xx}	LD_{yy}	LD_{zz}
0·1	−13·02	33·95	6·98
0·2	−13·44	3·11	−3·44
0·3	−11·67	− 3·33	−5·00
0·4	−10·19	− 5·38	−5·18
0·5	− 9·04	− 6·09	−5·04
0·6	− 8·14	− 6·29	−4·81
0·7	− 7·41	− 6·27	−4·56
0·8	− 6·81	− 6·14	−4·32
0·9	− 6·30	− 6·01	−4·10
1·0	− 5·85	− 5·85	−3·90

where $\phi = \pi a^2/A$ is the effective areal fraction of the planar array occupied by the particles.

Expression (28) illustrates the singular behavior of the drift velocity, and thus of the slip velocity, in the limit as the particle size becomes much smaller than the particle separation. Physically, the small surface area of the particles requires a large incident velocity to produce a force that is able to balance the shear stress exerted at the top of the lattice. This singular behavior is analogous to that exhibited by the velocity of a non-deforming three-dimensional particle lattice upon which a constant force is exerted due, for example, to gravity [19]. Moreover, expression (28) illustrates that when, either the resistance tensor for the force, or the Green's function tensor **D**, or both, are anisotropic, the drift velocity, and thus the slip velocity, is not necessarily directed parallel to the shear flow; that is, a shear flow in a certain direction may cause a drift or slip velocity in another direction.

For a rectangular lattice with base vectors $\mathbf{a}_1 = L(1,0,0), \mathbf{a}_2 = L(0,\delta,0)$ parallel to the x or y axis, where δ is the dimensionless aspect ratio, the matrix $\mathbf{D}(\mathbf{x}_p, \mathbf{x}_p)$ is diagonal. Numerical computation produces the values of the diagonal components listed in Table 1 in terms δ. The values of D_{zz} are implicit in a graph presented by Ishii ([6] Figure 2), and agree with the present results within plotting resolution. It is interesting to note the inversion in the sign of D_{yy} and D_{zz} for small lattice aspect ratios; the physical interpretation of this behavior is not clear. For a hexagonal lattice corresponding to $\mathbf{a}_1 = L(1,0,0), \mathbf{a}_2 = L(0{\cdot}5, 3^{1/2}/2, 0)$, the tensor $\mathbf{D}(\mathbf{x}_p, \mathbf{x}_p)$ is diagonal with components $D_{xx} = D_{yy} = 6{\cdot}32/L$, and $D_{zz} = 4{\cdot}21/L$.

3.1. Shear flow over a semi-infinite lattice

The preceding asymptotic solution may be extended readily to shear flow over multiple particle arrays. The extension to a semi-infinite lattice, considered previously by Sangani and Behl [11] using a somewhat different method, requires modification due to the vanishing of the drift velocity. For simplicity, we assume that all particle layers are identical, and the layer separations are constant and equal to H.

As the particle size becomes smaller than the inter-particle separation, the integral representation (23) evaluated at the center of a particle located at the kth layer, yields

$$u_i(\mathbf{x}^{(k)}) \cong \frac{1}{8\pi\mu} D_{ij}(\mathbf{x}^{(k)}, \mathbf{x}^{(k)}) F_j^{(k)} - \frac{1}{8\pi\mu} \sum_{\substack{l=1 \\ l \neq k}}^{\infty} G_{ij}(\mathbf{x}^{(k)}, \mathbf{x}^{(l)}) F_j^{(l)}, \tag{30}$$

where $\mathbf{F}^{(m)}$ is the force exerted on a particle the belongs to the mth layer, and the tensor $\mathbf{D}$ is defined in (26). A global force balance requires

$$\sum_{l=1}^{\infty} \mathbf{F}^{(l)} = \mu\gamma A\, \mathbf{e} + A\,(p^{-\infty} - p^{+\infty})\, \mathbf{e}_z. \tag{31}$$

Using the method of asymptotic expansions, we find

$$\mathbf{F}_m^{(k)} = 6\pi\mu a\, \mathbf{R}_{mi} \left[\frac{1}{8\pi\mu} D_{ij}(\mathbf{x}^{(k)}, \mathbf{x}^{(k)}) F_j^{(k)} - \frac{1}{8\pi\mu} \sum_{\substack{l=1 \\ l \neq k}}^{\infty} G_{ij}(\mathbf{x}^{(k)}, \mathbf{x}^{(l)}) F_j^{(l)} \right], \tag{32}$$

where $\mathbf{R}$ is the resistance tensor for the force exerted on a particle in infinite uniform flow. Rearranging Equation (32), we obtain the homogeneous linear system

$$\left(\mathbf{I} - \frac{3}{4} a\, \mathbf{R} \cdot \mathbf{D}(\mathbf{x}^{(k)}, \mathbf{x}^{(k)})\right) \cdot \mathbf{F}^{(k)} + \frac{3}{4} a\, \mathbf{R} \cdot \sum_{\substack{l=1 \\ l \neq k}}^{\infty} \mathbf{G}(\mathbf{x}^{(k)}, \mathbf{x}^{(l)}) \cdot \mathbf{F}^{(l)} = \mathbf{0} \tag{33}$$

which is to be solved subject to the global constraint imposed by (31). Recalling the linear dependence of the Green's function for the velocity on the vertical coordinate z far above a plane of the point forces, we derive the approximate form

$$\left(\mathbf{I} - \frac{3}{4} a\, \mathbf{R} \cdot \mathbf{D}(\mathbf{x}^{(k)}, \mathbf{x}^{(k)})\right) \cdot \mathbf{F}^{(k)} - \frac{6\pi a}{A} \mathbf{R} \cdot \sum_{l=k+1}^{\infty} (z^{(k)} - z^{(l)}) \mathbf{J} \cdot \mathbf{F}^{(l)} = \mathbf{0} \tag{34}$$

which is a generalization of the asymptotic form of Equation (59) derived by Sangani and Behl [11] for spheres.

To establish the asymptotic behavior far below the top layer, we note that the solution of the discrete problem expressed by Equation (34) bears some resemblance to the solution of the integral equation

$$\left(\mathbf{I} - \tfrac{3}{4} a\, \mathbf{R} \cdot \mathbf{D}(\mathbf{x}^{(k)}, \mathbf{x}^{(k)})\right) \mathbf{F}(w) - \frac{6\pi a}{A\, H} \mathbf{R} \cdot \int_{-\infty}^{w} (w - \nu) \cdot \mathbf{J} \cdot \mathbf{F}(\nu)\, \mathrm{d}\nu = \mathbf{0}, \tag{35}$$

where $w = -z$ is the positive distance from the top layer down the array, and $\mathbf{F}$ is a distributed body force. The integral constraint (31) requires

$$\frac{1}{H} \int_0^{\infty} \mathbf{F}(w)\, \mathrm{d}w = \mu\gamma A\, e + A(p^{-\infty} - p^{+\infty})\, \mathbf{e}_z. \tag{36}$$

Differentiating both sides of (35) twice with respect to w, we derive the differential equation

$$\left(\mathbf{I} - \tfrac{3}{4} a \mathbf{R} \cdot \mathbf{D}(\mathbf{x}_k, \mathbf{x}_k)\right) \frac{\mathrm{d}^2 \mathbf{F}(w)}{\mathrm{d}w^2} - \frac{6\pi a}{A\, H} \mathbf{R} \cdot \mathbf{J} \cdot \mathbf{F}(w) = \mathbf{0}. \tag{37}$$

As the size of the particles becomes smaller, the second term in the first set of parentheses on the left-hand side of (37) becomes negligible, and a solution of the resulting simplified equation that decays to zero as w tends to infinity is given by

$$\mathbf{J} \cdot \mathbf{F}(w) = \mathbf{c} \cdot \exp(-\sigma w) \tag{38}$$

where $\mathbf{c}$ is a vectorial constant whose first two components must be evaluated by requiring the satisfaction of the x and y components of the force balance (36), and $\boldsymbol{\sigma}$ is a tensorial rate-of-decay given by

$$\sigma^2 = \frac{6\pi a}{AH}\mathbf{R}. \tag{39}$$

In the case of spherical particles, corresponding to $\mathbf{R} = \mathbf{I}$, this expression is in agreement with that given in Equation (62) of Sangani and Behl [11].

Treating the field (38) as a distributed body force, we now describe the average flow in the semi-infinite matrix in terms of an effective velocity, designated by angular brackets, by means of the equation

$$\mu\mathbf{J} \cdot \frac{\mathrm{d}^2\langle\mathbf{u}\rangle}{\mathrm{d}w^2} + \mathbf{c} \cdot \exp(-\sigma w) = \mathbf{0}. \tag{40}$$

Solving for $\mathbf{J} \cdot \langle\mathbf{u}\rangle$, and expressing the exponential term on the right-hand side in terms of the solution, we obtain the standard form of Brinkman equation for flow in an anisotropic medium

$$\mathbf{J} \cdot \left(\frac{\mathrm{d}^2\langle\mathbf{u}\rangle}{\mathrm{d}w^2} - \sigma^2 \cdot \langle\mathbf{u}\rangle\right) = \mathbf{0}. \tag{41}$$

4. Numerical results for arbitrary particle sizes

To study flow over particles with arbitrary shape and size, we solve the integral Equation (15) using a boundary element method. In the numerical procedure, the surface of a particle is discretized into a collection of six-node curved triangles generated by successively subdividing an octahedron into four descendant elements. The three components of the traction are approximated with constant functions over each element, and the integral equation is evaluated at collocation points located at the centroid of each triangle to produce a system of linear algebraic equations. The Green's function is integrated using the seven-point triangle quadrature over the nonsingular elements [20].

To compute the integral over a singular triangle, we subdivide the triangle into four flat triangles defined by the vertex and mid-nodes, and then integrate individually over each flat triangle in local polar coordinates centered at the singular point. The vector of unknowns includes the three Cartesian components of the traction over each element, and the two components of the slip velocity. Two additional equations arise by appending to the linear system the discrete forms of the x and y components of the integral force balance expressed by the global constraint (6).

It is well known that the integral Equation (15) has an infinite number of solutions: conservation of mass for the flow induced by a lattice of point forces requires

$$\int_{\text{Particle}} n_j(\mathbf{x}) G_{ji}(\mathbf{x}, \mathbf{x}_0)\, \mathrm{d}S(\mathbf{x}) = 0. \tag{42}$$

Because of this identity, any particular solution of the integral Equation (15) may be enhanced with an arbitrary multiple of the normal vector. Combining identity (42) with (12), and using the divergence theorem, we find the reciprocal identity

$$\int_{\text{Particle}} \left(\gamma z e_j + U_j^{+\infty}\right) n_j(\mathbf{x})\, \mathrm{d}S(\mathbf{x}) = 0. \tag{43}$$

Projecting both sides of (15) onto the normal vector, integrating over the particle surface, and using identity (43), we derive the solvability condition

$$\int_{\text{Particle}} \left(\gamma z e_j + U_j^{+\infty}\right) n_j(\mathbf{x})\, \mathrm{d}S(\mathbf{x}) = 0 \tag{44}$$

whose satisfaction can be demonstrated readily by use of the divergence theorem.

The non-uniqueness of solution of the integral equation causes the numerical solution of the discretized system to be sensitive to the numerical or round-off error: the more accurate the numerical solution, the higher the condition number of the matrix resulting from the boundary-element discretization. To remove this sensitivity, we consider the discrete form of the integral identity (42) corresponding to the discretization of the integral equation, and express it in the symbolic form

$$\mathbf{A} \cdot \boldsymbol{w} = \mathbf{0}. \tag{45}$$

The discrete approximate eigenvector $\boldsymbol{w}$ contains the components of the normal vector evaluated at the collocation points. If the boundary elements were flat triangles, in which case the discretized particle surface would be polyhedral, the normal vector over each element would be constant, and condition (45) would be satisfied up to the numerical error associated with the numerical computation of the influence matrix $\mathbf{A}$. More generally, $\boldsymbol{w}$ tends to become an exact eigenvector as the discretization error is reduced by increasing the number of the elements.

To the author's knowledge, with a recent exception [21], in all previous numerical solutions of the present or similar integral equations, the nearly-singular behavior of the linear system was ignored, and the numerical method relied on the discretization error, the round-off error, or an inherent geometrical symmetry to produced a sensible solution. General methods for regularizing nearly singular algebraic systems by spectrum deflation have been discussed by several authors in a more general framework [22, 23]. In the present case, the availability of the adjoint eigenvectors of the single-layer potential corresponding to the null eigenvalue allows us to perform regularization with the least amount of perturbation, as follows.

The discrete form of the solvability condition (44) corresponding to the discretization underlying the system (15) is

$$\sum_k (\gamma\, z\, e_i + U_i^{+\infty})_k\, n_i^{(k)} h_k = 0, \tag{46}$$

where summation is implied over $i = 1, 2, 3$, the index k runs over the collocation points, and h_k is an integration weight with respect to surface area playing the role of a discrete metric. In the present implementation, h_k is the element surface area.

To regularize system (45), we multiply both sides by the preconditioning matrix $\mathbf{P}$ defined as

$$\mathbf{P} = \mathbf{I} - \boldsymbol{v}^T \boldsymbol{v}, \tag{47}$$

where the length of the vector $\boldsymbol{v}$ is equal to three times the total number of collocation points, and the entries of $\boldsymbol{v}$ host the discrete adjoint eigenvector $\alpha n_i^{(k)} h_k$; α is a scaling coefficient adjusted so that the length of $\boldsymbol{v}$ is equal to unity. It is evident that $\boldsymbol{v}$ is an exact eigenvector of the transpose of the preconditioned singular matrix corresponding to the zero eigenvalue, and the solvability condition is also satisfied to machine accuracy. Thus, the projection guarantees the satisfaction of both the integral identity (43) and the solvability condition (45), while introducing the mildest possible perturbation of the right-hand side: when (43) is fulfilled, the projection has no effect on the right-hand side.

To obtain one solution of the singular preconditioned system, we set the value of the z component of the traction at the last collocation point equal to an arbitrary value that was selected to be zero. Discarding the corresponding equation, we obtain a reduced system with a unique solution that may be computed using any dense-system linear solver, for example, by the method of Gauss elimination.

All computations reported in this section were carried out with 128 boundary elements corresponding to the second level of discretization of the regular octahedron. The sums in real and reciprocal wave number space involved in the computation of the Green's function were truncated at the value of 10. Each solution of the integral equation required several hours of CPU time on a SUN SPARCstation 20, with the majority of the computational effort devoted to evaluating the doubly-periodic Green's function. To estimate the magnitude of the numerical error, the numerical procedure was used to compute flow past a triply periodic array of spheres, by replacing the doubly-periodic Green's function with the triply-periodic Green's function computed in terms of Ewald sums [18]. Comparison with results published by Zick and Homsy [24] for uniform flow through a triply-periodic lattice of spheres showed that the numerical error is less than 0.5% for all reported cases.

In Figure 3(a), we plot with circles and squares, respectively, the slip and drift velocity for shear flow over a square lattice of spherical particles of radius a, where the flow is directed along the x or y axis. The solid line represents the predictions of the asymptotic expression (28), yielding

$$\mathbf{U}^{+\infty} = \mathbf{U}^{-\infty} \cong \gamma L \frac{1}{6\pi} \frac{L}{a} (\alpha - 4{\cdot}39 \, \frac{a}{L}) \mathbf{e}, \tag{48}$$

where L is the length of the base vectors, and $\alpha = 1$. The agreement between the asymptotic and numerical results is excellent up to $a/L = 0{\cdot}15$, at which point the magnitudes of the slip and the drift velocity have become very small. The slip velocity becomes negative when the particle radius exceeds a certain threshold, whereas the drift velocity remains positive for all particle sizes considered. Inversion in the sign of the slip velocity was reported previously by Larson and Higdon [9, 10] and Sangani and Behl [11], respectively, for shear flow over a semi-infinite lattice of cylinders or spheres.

In Figure 3(b), we present corresponding results for shear flow over a hexagonal lattice of spherical particles corresponding to the base vectors $\mathbf{a}_1 = L(1, 0, 0)$, $\mathbf{a}_2 = L\,(0{\cdot}5, 3^{1/2}/2, 0)$. The solid line represents the predictions of the asymptotic expression (28), yielding

$$\mathbf{U}^{+\infty} = \mathbf{U}^{-\infty} \cong \gamma L \frac{0.866}{6\pi} \frac{L}{a} (1 - 4{\cdot}74 \, \frac{a}{L}) \, \mathbf{e}. \tag{49}$$

As in the case of the square lattice, the agreement between the numerical and asymptotic results is surprisingly good even when the particles are not small. In spite of some quantitative differences, the results for the square and the hexagonal lattice are generally similar.

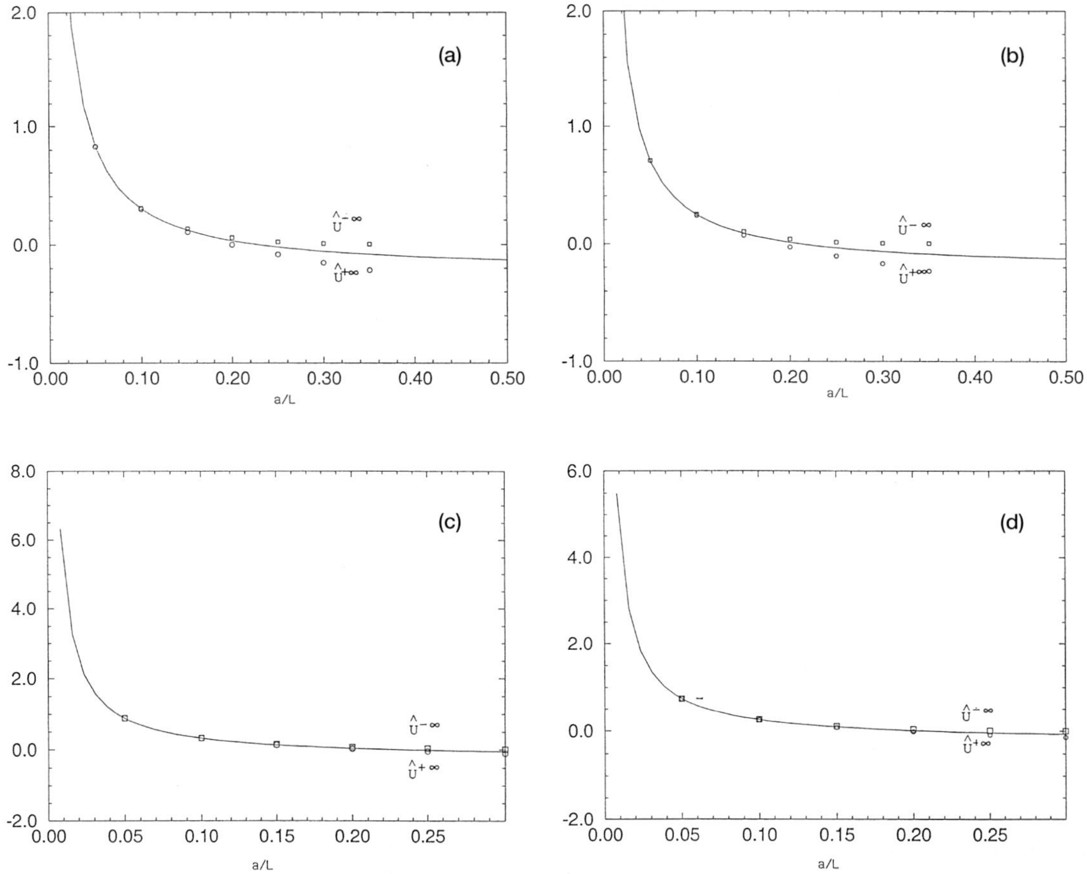

Figure 3. Slip and drift velocity plotted, respectively, with circles and squares for (a) a square lattice of spherical particles of radius a, (b) a hexagonal lattice of spherical particles of radius a, and (c, d) a square lattice of spheroidal particles of aspect ratio 2:1 and equivalent radius a. Panel (c) shows results for flow along the major axis, and panel (d) results for flow along the minor axis. The solid lines represent the predictions of the asymptotic theory for small particle sizes.

In Figure 3(c, d), we present results for shear flow over a square lattice of spheroidal particles with axes ratio 2:1 and equivalent volume radius a, positioned with the major axis parallel to the x axis. The solid lines represent the predictions of the asymptotic expression (28) yielding Equation (48) with $\alpha = 1{\cdot}046$ in the case of flow along the major axis, and $\alpha = 0{\cdot}914$ in the case of flow along the minor axis. The agreement between the asymptotic and numerical results is excellent even for small particle separations. The numerical results show once again that the slip velocity becomes negative when the equivalent radius a crosses a threshold, whereas the drift velocity remains positive for all cases considered. Similar results were obtained for spheroids with different aspect ratios.

As the size of the particles is increased, the tips of the spheroids touch yielding an array of corrugated cylinders. The structure of the flow in this limit may be studied in the context of flow over a periodic array of cylinders, to be discussed in Section 6. The results will show that a negative slip velocity may be established for highly elongated and closely spaced particles due to the onset of regions of recirculating flow in the intervening spaces.

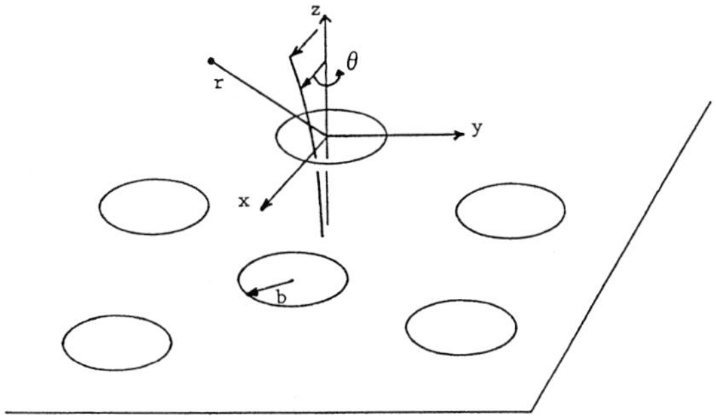

Figure 4. Schematic illustration of shear flow over a flat plate of zero thickness containing a homogeneous distribution of circular perforations.

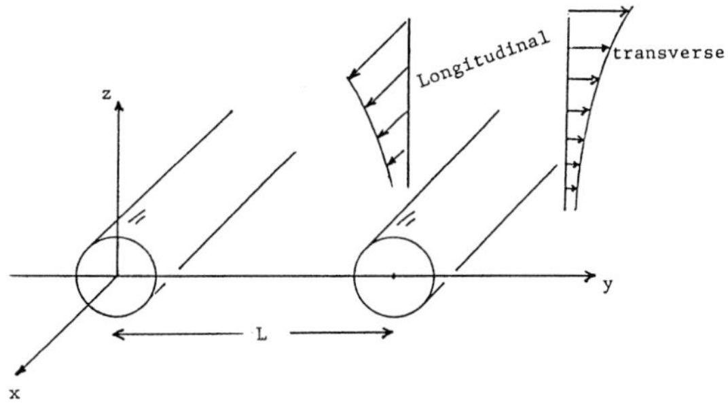

Figure 5. Schematic illustration of longitudinal or transverse shear flow over an array of evenly-spaced cylinders.

Overall, we found that the asymptotic expansions produce sufficiently accurate results in the regimes where the slip and drift velocities are not negligible compared to γa. In practical terms, the data presented in Figure 3 provide us with a basis for computing the size of particles comprising a sieve or membrane in terms of measured values of the drift velocity.

5. Flow over a perforated plate

In the preceding sections, we have discussed flow over a planar lattice of particles of small and moderate size. When the particle size is large, the lattice resembles a plane wall with a periodic distribution of protrusions separated by gaps or perforations. To study the structure of the flow in this limit, we consider a model configuration consisting of an infinite plate of zero thickness with a homogeneous distribution of identical circular perforations separating two semi-infinite domains, and evaluate the drift velocity as a function of the areal fraction of the perforations. The results of Section 2 ensure that the slip and drift velocity will be identical.

When the radius of the perforations is small compared to the distance between any two neighboring perforations, the flow may be expressed as the sum of the incident shear flow and the disturbance flows due to the individual perforations. The building block for generating the latter is the simple shear flow over a circular orifice on a plane wall derived by Smith [12] and Davis [13]. In the Cartesian and accompanying polar coordinates defined in Figure 4, with $\theta = 0$ corresponding to the first quadrant of the zx plane, the velocity components for shear flow along the x axis over a circular orifice of radius b are given by

$$u_x = \gamma\left(f_1 \cos 2\theta + f_2 + \frac{1}{2}(1 + \frac{z}{|z|})z\right), \quad u_y = \gamma f_1 \sin 2\theta, \quad u_z = \gamma f_3 \cos\theta \tag{50}$$

where γ is the shear rate of the overpassing shear flow. We have introduced the auxiliary functions

$$
\begin{aligned}
f_1 &= \frac{1}{3\pi}\frac{\lambda^2}{b}\zeta\frac{1-\zeta}{(\xi^2+1)(\xi^2+\zeta^2)},\\
f_2 &= \frac{1}{\pi}b\zeta\left(1-\frac{|\lambda|}{b}\arctan(\frac{b}{|\lambda|}-\frac{1}{3}\frac{\zeta^2}{\xi^2+\zeta^2}\right),\\
f_3 &= \frac{2}{3\pi}\lambda\zeta\frac{\zeta^2}{\xi^2+\zeta^2}\frac{1-\zeta^2}{1+\xi^2}
\end{aligned}
\tag{51}
$$

and we have defined

$$
\lambda^2=\frac{1}{2}\left[r^2-b^2+\left((r^2-b^2)^2+4z^2b^2\right)^{1/2}\right],\quad \lambda=\frac{z}{|z|}|\lambda|,\,\zeta=\frac{z}{\lambda},\quad \xi=\frac{\lambda}{b}. \tag{52}
$$

The flow in the lower half-plane is the mirror image of the disturbance flow in the upper half-plane. As the distance from the center of the orifice r tends to infinity, the functions f_1, f_2, f_3 tend to vanish. In this limit, $|\lambda|$ tends to r, and the function f_2 assumes the asymptotic form

$$
f_2=\frac{1}{\pi}b\frac{z}{r}\left(1-\frac{r}{b}\arctan\left(\frac{b}{r}\right)-\frac{1}{3}\left(1+\frac{r^4}{z^2b^2}\right)^{-1}\right). \tag{53}
$$

As z tends to positive or negative infinity, the infinite sum of the variables expressing the superposition of the individual components of the disturbance velocity reduces to an area integral. Inspecting the expressions given in (50), we find that the assumption of a homogeneous distribution of perforations requires that all integral vanish, with the exception of the integrals of the function f_2 corresponding to the direction of the shear flow. Considering the limit as z tends to positive or negative infinity, we use the asymptotic form (53) and obtain

$$
\begin{aligned}
u_x \to u_x^{\pm\infty} = \gamma\frac{1-\phi}{b^2}\lim_{z\to\infty}2\pi\int_0^\infty \frac{1}{\pi}b\frac{z}{(z^2+\rho^2)^{1/2}}\Bigg(1-\frac{(z^2+\rho^2)^{1/2}}{b}\\
\arctan\left(\frac{b}{(z^2+\rho^2)^{1/2}}\right)\Bigg)-\frac{1}{3}\left(1+\frac{(z^2+\rho^2)^2}{z^2b^2}\right)\rho\,\mathrm{d}\rho+\gamma\frac{1}{2}\left(1+\frac{z}{|z|}\right)z,
\end{aligned}
\tag{54}
$$

where $1-\phi$ is the areal fraction of the perforations. Taking the limit as z tends to infinity while the ratio ρ/z remains finite, we find the asymptotic expression

$$
u_x^{\pm\infty}-\gamma\frac{1}{2}(1+\frac{z}{|z|})=\frac{2}{3}(1-\phi)\gamma b\int_0^{\pi/2}\sin^3\omega\,\mathrm{d}\omega=\frac{4}{9}(1-\phi)\gamma b=\frac{4\pi}{9}n\gamma b^3 \tag{55}
$$

which is valid when the solid areal fraction ϕ of the plate is sufficiently close to unity and the perforations are separated by a great distance; n on the right-hand side is the number density of the perforations defined as the number of perforations per unit area of the plate.

To compare expression (55) with the diametrically opposite expansion (29) corresponding to the limit of small solid fractions, we set $b=a((1-\phi)/\phi)^{1/2}$, and find

$$
u_x^{\pm\infty}-\gamma\frac{1}{2}(1+\frac{z}{|z|})z=U_x^{+\infty}=\frac{4}{9}\gamma a\frac{(1-\phi)^{3/2}}{\phi^{1/2}}. \tag{56}
$$

It is evident that different functional forms arise in the asymptotic limit of ϕ tending to zero or to unity.

Previous numerical solutions have shown that the structure of shear flow over a solitary orifice with non-zero thickness is sensitive to the thickness of the plate hosting the orifice, and this suggests that the drift and slip velocity will be strong functions of the plate thickness. Indeed, the numerical solutions presented in Section 4 revealed that the slip velocity becomes negative for sufficiently large values of ϕ, whereas (56) predicts positive values.

6. Shear flow over an array of cylinders

The formulation of the preceding sections may be extended in a straightforward fashion to describe shear flow over a periodic array of cylinders separated by the distance L, modeling, for example, a membrane consisting of elongated fibers, as depicted in Figure 5. A shear flow in an arbitrary direction may be decomposed into a unidirectional longitudinal flow along the x axis parallel to the generators, and a transverse flow along the y axis normal to the generators. Both the longitudinal and the transverse flow over a semi-infinite regular lattice were discussed by Larson and Higdon [9, 10]. Their results illustrated the structure of the streamline patterns and established the dependence of the slip velocity on the cylinder shape and areal fraction.

6.1. Integral formulation

Considering first the case of longitudinal flow, we work as in Section 2 and derive the integral representation

$$u_x(\mathbf{X}_0) = -\frac{1}{\mu}\int_{\text{Cylinder}} G^L(\mathbf{x}, \mathbf{x}_0) f_x(\mathbf{x})\, \mathrm{d}l(\mathbf{x}) + \gamma z_0 + U_x^{+\infty}, \tag{57}$$

where the integral is over the contour of one cylinder in the yz plane, l is the arc length, and f_x is the x-component of the traction. The kernel G^L is the upward biased singly-periodic Green's function of Laplace's equation given by

$$G^L(\mathbf{x}, \mathbf{x}_0) = \frac{1}{2\pi} H(\mathbf{x}, \mathbf{x}_0) - \frac{1}{2}\frac{z - z_0}{L}, \tag{58}$$

where

$$H(\mathbf{x}, \mathbf{x}_0) = -\frac{1}{2}\log\left[2\cosh\left(\frac{2\pi}{L}(z - z_0)\right) - 2\cos\left(\frac{2\pi}{L}(y - y_0)\right)\right] \tag{59}$$

representing the harmonic potential at the point $\mathbf{x}$ due to a periodic array of two-dimensional point sinks of unit strength, where one of the point sinks is located at the point $\mathbf{x}_0$. Far above the array of point sinks, as $z - z_0$ tends to $+\infty$, the Green's function exhibits the asymptotic behavior $G^L \to -(z - z_0)/L + \text{edt}$; far below the array, as $z - z_0$ tends to $-\infty$, the Green's function vanishes at an exponential rate. Using these conditions, we find that the slip and drift velocity are related by

$$U_x^{-\infty} = U_x^{+\infty} + \frac{1}{\mu L}\int_{\text{Cylinder}} z\, f_x(\mathbf{x})\, \mathrm{d}l(\mathbf{x}). \tag{60}$$

Considering next the case of transverse flow along the y axis, we work as in Section 2 and derive the integral representation

$$u_j(\mathbf{x}_0) = -\frac{1}{4\pi\mu}\int_{\text{Cylinder}} G_{ij}^{2D-1P}(\mathbf{x},\mathbf{x}_0)\, f_i(\mathbf{x})\, \mathrm{d}l(\mathbf{x}) + \delta_{jy}\,\gamma\, z_0 + \delta_{jy} U_y^{+\infty} \tag{61}$$

where the indices i and j range over y and z, and G_{ij}^{2D-1P} is the singly-periodic Green's function of Stokes flow with components

$$\begin{aligned} G_{yy}^{2D-1P}(\mathbf{x},\mathbf{x}_0) &= H(\mathbf{x},\mathbf{x}_0) + (z-z_0)\frac{\partial H(\mathbf{x},\mathbf{x}_0)}{\partial z} - 2\pi\frac{z-z_0}{L}, \\ G_{yz}^{2D-1P}(\mathbf{x},\mathbf{x}_0) &= G_{zy}^{2D-1P}(\mathbf{x},\mathbf{x}_0) = -(z-z_0)\frac{\partial H(\mathbf{x},\mathbf{x}_0)}{\partial y}, \\ G_{zz}^{2D-1P}(\mathbf{x},\mathbf{x}_0) &= H(\mathbf{x},\mathbf{x}_0) + (z-z_0)\frac{\partial H(\mathbf{x},\mathbf{x}_0)}{\partial z} \end{aligned} \tag{62}$$

representing the velocity due to an array of two-dimensional point forces. The streamline pattern established when the point forces are directed parallel to the array is depicted in Figure 2(b). Far above the array of the point forces, as $z - z_0$ tends to $+\infty$, $G_{xx}^{2D-1P} \to -4\pi(z-z_0)/L +$edt, while the other components decay at an exponential rate. Far below the array, as $z - z_0$ tends to $-\infty$, all components of G_{ij}^{2D-1P} decay at an exponential rate. Using these conditions, we find that the slip and drift velocities are related by

$$U_y^{-\infty} = U_y^{+\infty} + \frac{1}{\mu L}\int_{Particle} z\, f_y(\mathbf{x})\, \mathrm{d}l(\mathbf{x}). \tag{63}$$

As in the case of three-dimensional flow, the difference between these two velocities is on the order of the cylinder size normal to the array.

6.2. Asymptotics for Small Circular Cylinders

The ill-posedness of infinite two-dimensional Stokes flow prevents us from using the method of matched asymptotic expansions to derive expressions for the slip and drift velocity in the limit of small cylinder sizes. Hasimoto [25] modified a method pioneered by Burgers to study uniform flow through a doubly-periodic array of circular cylinders of small size. The method involves writing the integral representation (57) or (61) at a point on the surface of a cylinder, expanding the Green's function into a Laurent series whose leading term is the free-space Laplace Green's function or the two- dimensional Stokeslet, retaining only the singular and constant terms, and then requiring that the integral of the velocity over the contour of the cylinder vanish. Hasimoto [25], Sangani and Acrivos [26], and Drummond and Tahir [27] confirmed the validity of this approach by recomputing and extending the asymptotic expansion using different methods.

Applying Burgers' method to the present problem, we find

$$U^{\pm\infty} \cong \frac{\gamma L}{2\pi}\left(-\log\left(\frac{a}{L}\right) - 1{\cdot}879\right) \tag{64}$$

for longitudinal flow, and

$$U^{\pm\infty} \cong \frac{\gamma L}{4\pi}\left(-\log\left(\frac{a}{L}\right) - 2{\cdot}3379\right) \tag{65}$$

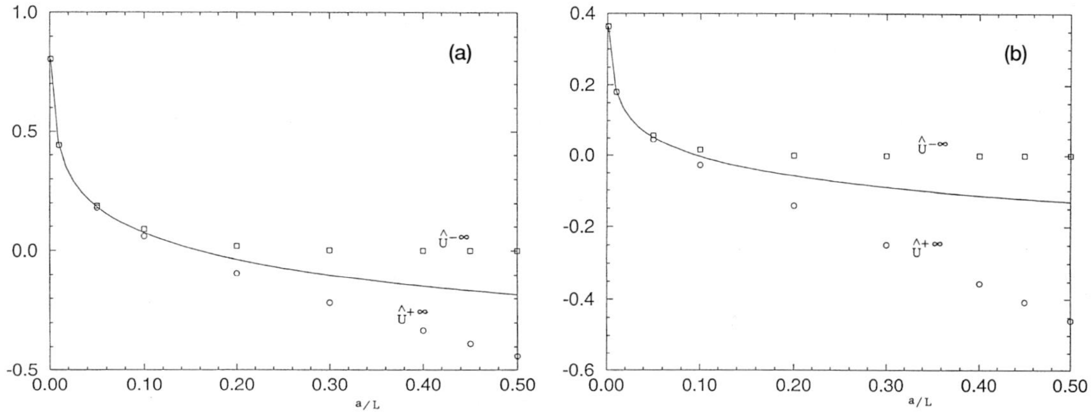

Figure 6 Slip and drift velocity plotted, respectively, with circles and squares for (a) longitudinal, and (b) transverse flow past an array of cylinders of radius a separated by the distance L.

for transverse flow. The logarithmic dependence may be contrasted with the inverse first power behavior encountered previously for three-dimensional flow, underscoring once more the distinct nature of two- and three-dimensional Stokes flow.

6.3. Numerical results for circular cylinders

The integral equations arising from the integral representations (57) and (61) were solved using the counterpart of the numerical method discussed in Section 4. In Figure 6(a, b), we plot, respectively, with circles and squares the slip and drift velocity for longitudinal and transverse flow past an array of cylinders of radius a separated by the distance L; the caret indicates nondimensionalization with respect to γL. The solid lines represent the predictions of the asymptotic expressions (64) and (65). In both cases, the numerical results confirm that as a/L tends to vanish, both velocities diverge at a logarithmic rate. Significant differences between the asymptotic predictions and the numerical results arise when a/L becomes approximately greater than 0·10.

For small cylinder sizes, the slip and drift velocities are positive for both longitudinal and transverse flow. As the cylinders become larger, the slip velocity becomes negative at a certain cylinder radius. In the limit as a/L tends to 0·5, corresponding to touching cylinders, the slip velocity tends to a limiting value that is consistent with a value extrapolated from graphs presented by Larson and Higdon [9, 10] for shear flow over a semi-infinite lattice. To make this comparison, we recomputed the slip velocity in terms of the flow rate above the array, as discussed by the previous authors. In the case of longitudinal flow, the drift velocity is positive for any cylinder radius, in agreement with physical intuition. In the case of transverse flow, however, as the cylinders become larger, the drift velocity changes sign at a particular cylinder radius and then it shows weak fluctuations.

To explain the physical reason for a negative drift velocity, in Figure 7(a, b) we present streamline patterns for $a/L = 0{\cdot}10$ and 0·30. In the first case, regions of recirculating flow do not develop, and the fluid moves in the direction of the shear flow above and below the array. To demonstrate the effect of the cylinder geometry, in Figure 7(c) we present the streamline pattern for flow over a plane wall with a slit, computed using the exact solution derived by Smith [12]. The comparison reveals topologically similar features near the opening, but significant differences below the plate due to the establishment of the drift velocity in the case

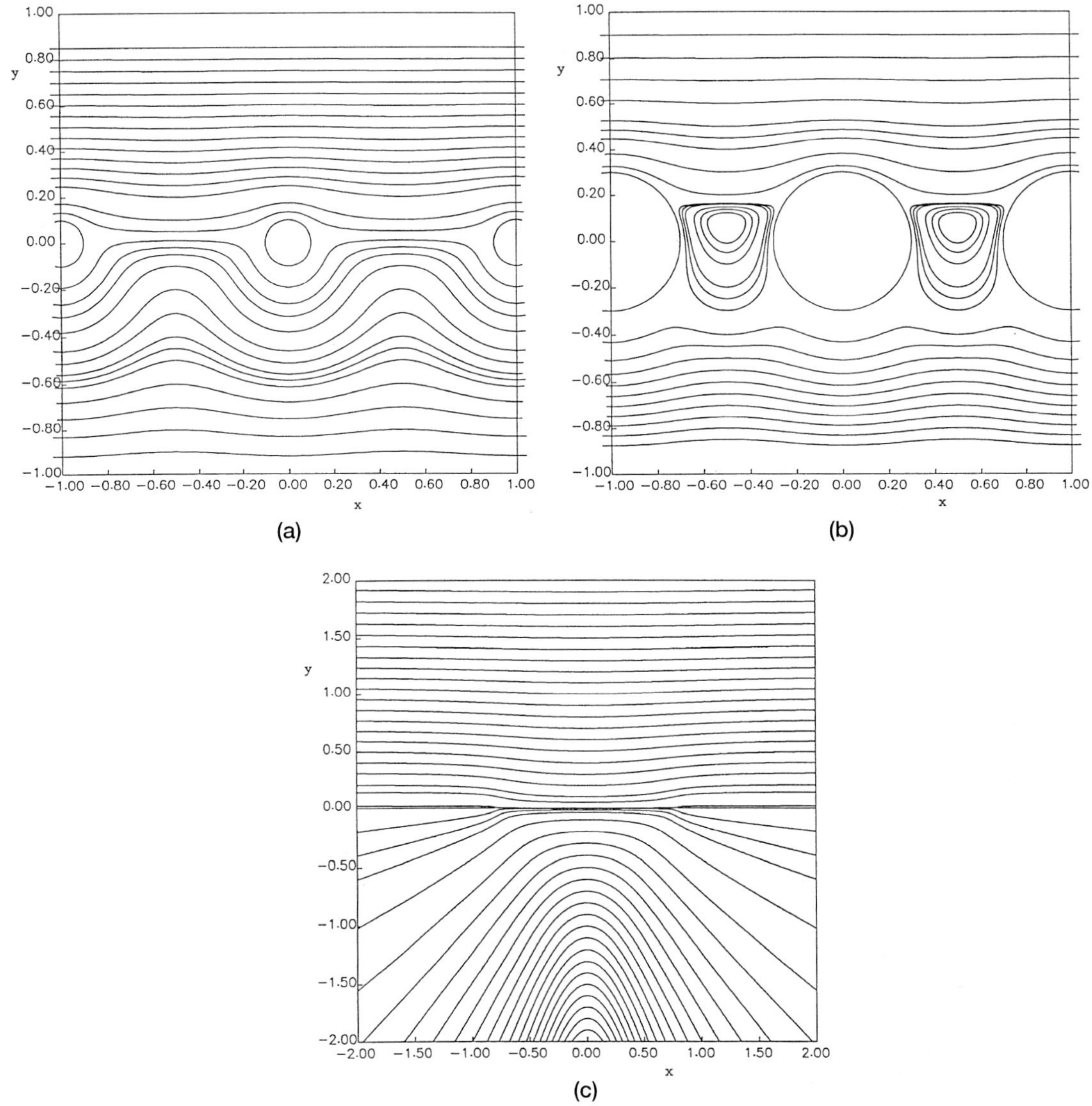

Figure 7 Streamline patterns for shear flow over an array for circular cylinders for (a) $a/L = 0{\cdot}10$ and (b) $0{\cdot}30$. Streamline patterns for shear flow over a plane wall with a slit.

of flow over the cylinders. Figure 7(b) reveals that when $a/L = 0{\cdot}30$, regions of recirculating flow develop between the cylinders. The backward motion of the fluid at the bottom of the eddies drives a weak flow with a negative drift velocity.

7. Concluding remarks

We have studied shear flow over a planar lattice of particles by asymptotic and numerical methods. The asymptotic results were found to generate accurate predictions over an extended range of particle sizes in the case of three-dimensional flow, and over a moderate range of particle sizes in the less realistic case two-dimensional flow. We have also derived asymptotic expressions for the slip and drift velocity of shear flow over a plate with a homogeneous distribution of perforations, in the limit of large solid fractions. These expressions are suitable for deriving modified boundary conditions that relate the local jump in the shear stress to the jump in the velocity across a membrane subjected to an ambient flow.

The model problems considered in this study involve flow over regular arrays. In reality, the particles are likely to be distributed randomly over the plane of the array. The hydrodynamical laws of flow through random particulate media and over random porous surfaces have been shown to be fundamentally different from those for ordered media, at least in the limit of small volume fractions [28–30]. Accordingly, the slip and drift velocities of flow over regular arrays are expected to obey laws that are different than those for random arrays in the limit of small particle sizes. Whether or not the dichotomy persists at higher solid fractions needs to be addressed in future analysis.

Acknowledgement

This research has been supported by a grant provided by the National Science Foundation.

References

1. R. Lipowsky, The conformation of membranes. *Nature* 349 (1991) 475–481.
2. U. Seifert, Configurations of fluid membranes and vesicles. *Adv. in Phys.* 46 (1997) 13–137.
3. R. Lipowsky and E. Sackmann (Eds.), *Structure and Dynamics of Membranes* (Vol I). Amsterdam: Elsevier (1995) 1020 p.
4. H. Yasuda, G.E. Lamaze and A. Peterlin, Diffusive and hydraulic permeabilities of water in water-swollen polymer membranes. *J. Polym. Sci. A* 2 (1971) 1117–1131.
5. G. H. Malone, T. E. Hutchinson and S. Prager, Molecular models for permeation through thin membranes: the effect of hydrodynamic interaction on permeability. *J. Fluid Mech.* 65 (1974) 753–767.
6. K. Ishii, Viscous flow past multiple planar arrays of small spheres. *J. Phys. Soc. Japan* 46 (1979) 675–680.
7. E. O. Tuck and A. Kouzoubov, A laminar roughness boundary condition. *J. Fluid Mech.* 300 (1995) 59–70.
8. K. Sarkar and A. Prosperetti, Effective boundary conditions for Stokes flow over a rough surface. *J. Fluid Mech.* 312 (1996) 1–19.
9. R. E. Larson and J. J. L. Higdon, Microscopic flow near the surface of two-dimensional porous media. Part 1. Axial flow. *J. Fluid Mech.* 166 (1986) 449–472.
10. R. E. Larson and J. J. L. Higdon, Microscopic flow near the surface of two-dimensional porous media. Part 2. Transverse flow. *J. Fluid Mech.* 178 (1987) 119–136.
11. A. S. Sangani and S. Behl, The planar singular solutions of Stokes and Laplace equations and their application to transport processes near porous surfaces. *Phys. Fluids A* 1 (1989) 21–37.
12. S. H. Smith, Stokes flow past slits and holes. *Int. J. Multiphase Flow* 13 (1987) 219–231.
13. A. M. J. Davis, Shear flow disturbance due to a hole in the plane. *Phys. Fluids* A3 (1991) 478–480.
14. Z.-Y. Yan, A. Acrivos and S. Weinbaum, Fluid skimming and particle entrainment into a small circular side pore. *J. Fluid Mech.* 229 (1991) 1–27.
15. W.-Y. Wu, S. Weinbaum and A. Acrivos, Shear flow over a wall with suction and its application to particle screening. *J. Fluid Mech.* 243 (1992) 489–518.
16. 24 G. S. Beavers and D. D. Joseph, Boundary conditions at a naturally permeable wall. *J. Fluid Mech.* 30 (1967), 197–207.
17. C. Pozrikidis, *Boundary Integral and Singularity Methods for Linearized Viscous Flow.* Cambridge: Cambridge University Press (1992) 25 p.
18. C. Pozrikidis, Computation of periodic Green's functions of Stokes flow. *J. Eng. Math.* 30 (1996) 79–96.
19. P. G. Saffman, On the settling speed of free and fixed suspensions. *Stud. Appl. Math.* 52 (1973) 115–127.
20. C. Pozrikidis, *Numerical Computation in Science and Engineering.* New York: Oxford University Press, (1998) 62 p.
21. C. Pozrikidis, A spectral-element method for particulate Stokes flow. *J. Comp. Phys.* 156 (1999) 360–381.

22. T. F. Chan, Deflated decomposition of solutions of nearly singular systems. *SIAM J. Numer. Anal.* 21 (1984) 738–745.
23. P. C. Hansen, *Rank-deficient and Discrete Ill-posed Problems*. Philadelphia: *SIAM* (1997) 24 p.
24. A. A. Zick and G. M. Homsy, Stokes flow through periodic arrays of spheres. *J. Fluid Mech.* 115 (1982) 13–26.
25. H. Hasimoto, On the periodic fundamental solutions of the Stokes equations and their application to viscous flow past a cubic array of spheres. *J. Fluid Mech.* 5 (1959) 317–328.
26. A. S. Sangani and A. Acrivos, Slow flow through a periodic array of spheres. *Int. J. Multiphase Flow* 8 (1982) 343–360.
27. J. E. Drummond and M. I. Tahir , Laminar viscous flow through regular arrays of parallel solid cylinders. *Int. J. Multiphase Flow* 10 (1984) 515–540.
28. P. G. Saffman, On the boundary condition at the surface of a porous medium. *Stud. Appl. Math.* 50 (1971) 93–101.
29. I. D. Howells, Drag due to the motion of a Newtonian fluid through a sparse random array of small fixed rigid objects. *J. Fluid Mech.* 64 (1974) 449–475.
30. A. S. Sangani and C. Yao, Transport processes in random arrays of cylinders. II Viscous flow. *Phys. Fluids* 31 (1988) 2426–2434.

Journal of Engineering Mathematics **39:** 25–46, 2001.

Current-voltage characteristics from an asymptotic analysis of the MOSFET equations

ELLIS CUMBERBATCH, HENOK ABEBE and HEDLEY MORRIS[1]
Claremont Graduate University, Claremont, CA 91711, U.S.A.
[1]*San Jose State University, San Jose, CA 95192, U.S.A.*

Received 17 September 1999; accepted in revised form 21 August 2000

Abstract. Phenomenological formulae for the current-voltage characteristics of a MOSFET are derived by extending the asymptotic method of Ward. Practical methods for combining the perturbation approximations and numerically implementing the Ward equations are developed. A detailed comparison with real MOSFET data is presented and the model is shown to be effective over a range of device geometries.

Key words: semiconductor modeling, asymptotic analysis, MOSFET current/voltage formulae, comparison with data.

1. Introduction

A semiconductor chip is made by fabricating up to 10^6 transistors and their inter-connections on the surface of a silicon wafer. The transistors are usually MOSFET's (metal-oxide-silicon-field-effect-transistors), drawn from a small set of designs (n-type, p-type) and of various sizes (lengths and widths). Each chip has circuit functions, and in order to design the circuits using the transistor building blocks, the VLSI industry uses a program called SPICE. SPICE is the most common program for the simulation of electronic circuits. The original SPICE was developed in 1971 at U.C. Berkeley, and updated versions exist in many forms both commercial and public domain. See [1] for more information. For SPICE to produce a design, the software must access the current/voltage relation for *each* transistor a number of times. In order that this be effected in a reasonable time, the current/voltage relationships for the transistors must have simple form.

The classical approximation governing the flow of electrons and holes in a semi-conductor relevant to the low field-strength operation of a transistor has been in use for over forty years [2, 3]. It results in a set of nonlinear partial differential equations for the electron density, n, the hole density, p, and the electric field, $\mathbf{E}$. Even in the time-independent case, which will be considered here and which is sufficient for many applications, the nonlinear character and the non-trivial geometry of the MOSFET preclude approaches that would generate exact solutions, least of all any that lead to simple current/voltage relationships. Hence there has resulted a variety of approximate solutions. A typical MOSFET geometry is shown in Figure 1.

The first approximation, which allows a reduction from a PDE system to an ODE system has the designation 'long channel' or 'quasi-one-dimensional' (called 1-D in the sequel). As indicated, this approximation has validity when the channel length, L, is relatively large. (See Figure 1 for the coordinate system, for identification of various regions on the device, etc. Also refer to Appendices A, B for variable definitions.) When voltages are applied at the gate and/or drain, electrons and holes are attracted or repelled from these locations. The depletion region, bereft of either electrons or holes, plays an important role; a measure of its size is $L_D(\log\lambda/\lambda)^{\frac{1}{2}}$ where L_D is the Debye length and λ is a measure of the doping level relative to the intrinsic level. (λ is large, typically 10^6–10^7.) The ratio of these lengths, which plays the role of the aspect ratio of the device, is

$$\epsilon = \frac{L_D}{L}\sqrt{\frac{\log\lambda}{\lambda}}. \tag{1}$$

When ϵ is small most of the region under the gate is uniform in the x_2 -direction and the 1-D approximation is exact for the fields there; the approximation breaks down in the source-gate and drain-gate corners where the fields lose their 1-D character. When ϵ is large the corner regions influence the solution across the whole gate region and a two-dimensional solution is required. Ward [4], made the 1-D approximation into an explicit asymptotic expansion by scaling x_2, the variable along the channel, with L, and scaling x_1, the perpendicular variable, with L_D. Series expansions in powers of ϵ^2 give the 1-D approximation as the leading term of these series. In a subsequent paper [5], Ward obtained some solutions for the corner regions and matched them to the 1-D approximation in the central region under the gate. This enhanced solution would be expected to be more accurate down to smaller channel lengths, but the solution obtained is restricted to small drain voltages. For a typical device $L_D = 33$ microns, $\lambda = 10^6$, and for $L = 10$ microns, $\epsilon = 0{\cdot}012$. For ϵ this small we expect the 1-D approximation to be accurate; this is borne out by the data. However we show that by adjustments of some of the parameters we can make the 1-D approximation match the data at much reduced channel lengths. At the time of writing, the standard industry channel length is 0·25 micron, and the SPICE software, based on the 1-D approximation (but improved empirically), is still being used.

The 1-D approximation has the descriptor 'quasi' because the ODE in the x_1 direction contains x_2 as a parameter via the quasi-Fermi potential, φ. The electric potential, ψ, satisfies the second-order nonlinear ODE in x_1, which contains φ, and therefore its solution depends on x_2 parametrically. The form of $\varphi(x_2)$ is determined subsequently when the current flowing across the device is calculated. The ODE allows one integration in the case of constant doping. The current can then be expressed as a double integral called the Pao-Sah integral, see [6]. An approximation allows reduction to a single integral [7]. Further approximations valid over different regimes of the applied voltages V_{ds}, V_{gs} then yield explicit formulae for $I_{ds}(V_{ds}, V_{gs})$. These latter approximations are motivated by neglect of various physical effects in one or other regimes of operation. A comprehensive survey of these approximations and their inter-relations is found in [3] and [8]. These separate formulae, valid over distinct ranges of voltages, were found to be accurate for long channel use. As the need for denser packing of transistors on chips has intensified, and the technology has been able to satisfy those needs by reducing lengths and widths of transistors, these formulae have been adapted in various ways for use at the reduced sizes. Often these adaptations are little more than 'fudge factors', which have parameters valid only over limited ranges of size and voltage regimes. Hence these parameters are changed regime to regime ('binning'). Current technology uses of the order of 150–200 parameters for $I_{ds}(V_{ds}, V_{gs})$ formulae for use in SPICE. Most of these parameters are not universal constants, and they must be 'identified' for each fabrication batch. This means considerable testing and data collection, followed by numerical optimization for parameter identification. All of this is expensive, both in time and money. There are pressures to improve the models so that they are less cumbersome and based more on physics. The work reported here, done by the Claremont Graduate University Mathematics Clinic for the Information Sciences Institute MOSIS Program, has these more efficient models as its goal. Related goals were (a) comparison of the asymptotic results with data to determine the domain of validity of the asymptotic approximations, and (b) improvement of solution 'blending' by which solutions valid over different voltage regimes are patched together.

The work presented in here generates explicit $I_{ds}(V_{ds}, V_{gs})$ formulae based on the 1-D approximation. It takes further the asymptotic results for large λ derived by Ward in [4, 5]. These results in turn were based on the asymptotic scheme originated by Please, [9], for the $p - n$ junction. This scheme, based on the matched asymptotic expansion concept [10], obtains solutions valid in different regions and matches them across overlapping regions to provide a composite expansion valid over the whole region (to the order computed). In the 1-D MOSFET approximation there are three main regions: the inversion layer, a thin boundary layer containing a profusion of holes or electrons under some voltage regimes, the depletion layer, and the substrate which is determined by conditions far from the channel. The separate solutions in these regions are, by themselves, not difficult to obtain. However, matching them, and satisfying the boundary condition at the gate, are non-trivial exercises. Ward successfully accomplished these, in [4] for constant doping and for variable doping of constant signature, and in [5] for variable doping of opposite signature. Reference [11] presents the Ward analysis, and refers to other problems treated both asymptotically and numerically among a broad introduction to an advanced treatment of the semi-conductor equations.

The asymptotic solution achieved by Ward requires numerical work (to solve non-linear transcendental equations) and the $I_{ds}(V_{ds}, V_{gs})$ formula remains an integral to be computed numerically. Hence its use in the semi-conductor industry (via SPICE) is precluded. We briefly outline Ward's asymptotic results, and in Section 3 we show how the Ward analysis may be taken further (in the constant doping case) to achieve an *explicit* $I_{ds}(V_{ds}, V_{gs})$ formula. Part of our analysis is justifiable as a continuation of the asymptotics and it agrees with the numerical solution of the exact ODE/BC system. A second approximation is not justifiable asymptotically. It approximates by a constant a function that is varying inside an integral; with this, though, the integral can be evaluated analytically, and the two approximations combined to produce a formula for the current.

2. Equations, boundary conditions, and the quasi-1-D asymptotic approximation

Voltages may be applied at the source, drain, gate and substrate contacts (S, D, G, B in Figure 1); we shall assume that the substrate and source are at the same voltage. The voltages drain-to-source and gate-to-source are denoted by V_{ds}, V_{gs}, respectively. We shall treat an n-channel device (in inversion the channel is profuse in electrons). Here the silicon has been doped with acceptor atoms (typically the number ionized is $N_A^- \sim 10^{16} - 10^{17}$). Source, drain regions have been overlaid with donor atoms ($N_D^+ \sim 10^{19}$). An insulated layer (silicon dioxide) separates the gate contact from the silicon. As the gate voltage is raised, holes are repelled from the region under the gate, creating a depleted region. Further increase in V_{gs} attracts electrons to the insulator-substrate interface, creating the 'channel'. Voltages at the drain then cause a current to flow from source to drain. The aim of analysis of the device is to obtain the current created by this flow of electrons, I_{ds}, as a function of the applied voltages V_{ds} and V_{gs}. In addition, I_{ds} also depends on the parameters of the device: its size, the doping levels, the mobility of electrons, etc.

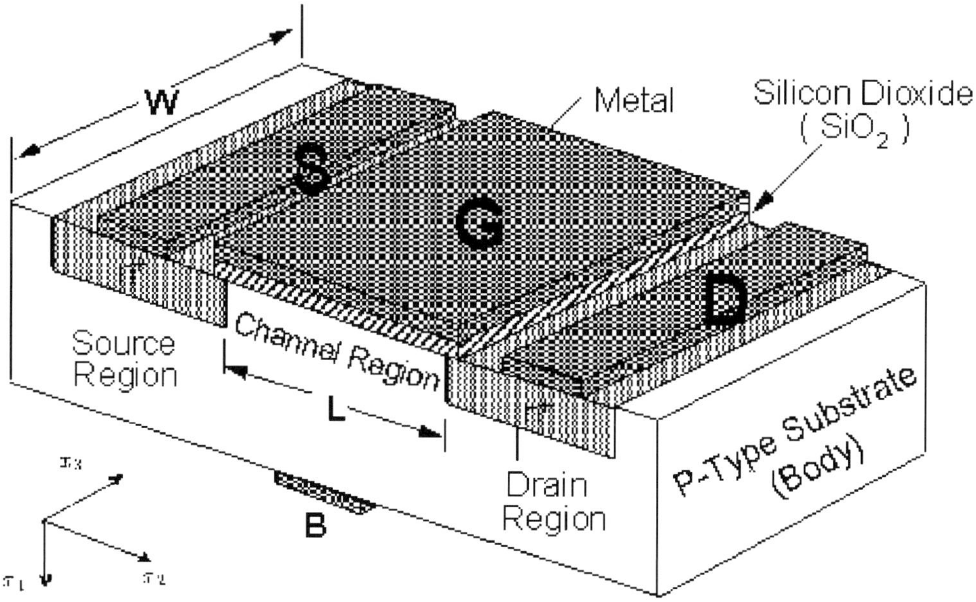

Figure 1. A schematic diagram of a MOSFET.

The standard model for the MOSFET comprises Gauss' equation for charge

$$\epsilon_s \nabla \cdot \mathbf{E} = -\rho = q(p - n + N), \tag{2}$$

together with the drift-diffusion model for the motion of electrons

$$\mathbf{J}_n = q\mu_n \left(\frac{kT}{q} \nabla n + n\mathbf{E} \right) \tag{3}$$

and the static conservation equation with no recombination/generation for the flux of electrons

$$\nabla \cdot \mathbf{J}_n = 0. \tag{4}$$

The assumption that the holes remain in thermal equilibrium allows the hole density to be obtained as

$$p = n_i \mathrm{e}^{-q\frac{\psi}{kT}}. \tag{5}$$

It is convenient to write

$$n = n_i \mathrm{e}^{q\frac{\psi - \varphi_n}{kT}} \tag{6}$$

with φ_n replacing n as a primary variable; it is called the quasi-Fermi potential. For other notation and definitions see Appendix A.

The boundary conditions comprise

(i) specification of the electric potential at the gate contact:

$$\psi = V_{gs} - V_{fb} \text{ at } x_1 = -t, \tag{7}$$

where V_{fb} is the 'flat-band voltage', which is a voltage that results from terminal contacts with the substrate [2].

(ii) continuity of electric potential and electric displacement at the oxide-substrate interface:

$$\psi^- = \psi^+ \quad \text{and} \quad \epsilon_i \frac{\partial \psi^-}{\partial x_1} = \epsilon_s \frac{\partial \psi^+}{\partial x_1} \text{ at } x_1 = 0. \tag{8}$$

(iii) specification of the electric and quasi-Fermi potentials at the source-substrate and drain-substrate boundaries. These are given in [4], and will not be written out here. They are required for the specification of ψ, φ_n at these boundaries, and they will be introduced when needed subsequently.

(iv) current flow occurs only across the source-substrate and drain-substrate boundaries. A consequence of this is that that the total current flowing across a plane $x_2 =$ const does not vary with x_2 for $0 \leq x_2 \leq L$.

The independent and dependent variables are scaled, see Appendix B. With this scaling, the governing equations, (2–4), become

$$\left.\begin{aligned} &\widetilde{\nabla}^2 w = \frac{1}{\lambda}\left[\exp((w-\varphi)\log\lambda) - \exp(-w\log\lambda)\right] + d(x) \\ &\widetilde{\nabla}^2\varphi + \widetilde{\nabla}\varphi \cdot \left(\frac{1}{\mu}\widetilde{\nabla}\mu + \log\lambda\widetilde{\nabla}(w-\varphi)\right) = 0 \\ &\text{where } \widetilde{\nabla} \equiv \left(\frac{\partial}{\partial x}, \epsilon\frac{\partial}{\partial y}\right) \end{aligned}\right\}. \tag{9}$$

The differential scaling of the independent variables introduces the parameter ϵ, Equation (1), into the field equations, (2) and (4). The quasi-one-dimensional approximation is obtained for the solution in the mid-channel region (away from the source and drain corners) by the asymptotic expansion for the scaled electric and quasi-Fermi potentials

$$w(x, y) = w^0(x, y) + \epsilon^2 w^1(x, y) + \cdots, \tag{10}$$

$$\varphi(x, y) = \varphi^0(x, y) + \epsilon^2 \varphi^1(x, y) + \cdots. \tag{11}$$

The electron mobility in (3) is not necessarily constant: it may vary spatially and with field strength. The Ward assumption is

$$\mu_n = \mu_n(x, \frac{\partial \varphi^0}{\partial y}) > 0. \tag{12}$$

To leading order

$$\frac{\partial^2 w^0}{\partial x^2} = \frac{1}{\lambda}(\exp((w^0 - \varphi^0)\log(\lambda)) - \exp(-w^0\log(\lambda)) + d(x), \tag{13}$$

$$\frac{\partial^2 \varphi^0}{\partial x^2} = -\left(\frac{1}{\mu_n}\frac{\partial \mu_n}{\partial x} + \log(\lambda)\left(\frac{\partial w^0}{\partial x} - \frac{\partial \varphi^0}{\partial x}\right)\right)\frac{\partial \varphi^0}{\partial x}. \tag{14}$$

The equation for φ^0 can be integrated, and with the no-flux boundary condition it follows that

$$\frac{\partial \varphi^0}{\partial x} = 0 \quad \text{and} \quad \varphi^0 = \varphi(y), \tag{15}$$

where we have dropped the zero superscript for convenience.

The order ϵ^2 terms give for φ^1

$$\frac{\partial^2\varphi^1}{\partial x^2}+\left(\frac{1}{\mu_n}\frac{\partial\mu_n}{\partial x}+\log(\lambda)\frac{\partial w^0}{\partial x}\right)\frac{\partial\varphi^1}{\partial x}=-\left(\varphi''+\left(\frac{1}{\mu_n}\frac{\partial\mu_n}{\partial y}+\log(\lambda)\left(\frac{\partial w^0}{\partial y}-\varphi'\right)\right)\varphi'\right), \tag{16}$$

where primes represent derivatives with respect to y. By means of an integrating factor

$$\begin{aligned}\frac{\partial}{\partial x}\left(\mu_n\exp(\log(\lambda)w^0)\frac{\partial\varphi^1}{\partial x}\right)&=\\ =-\mu_n\exp(\log(\lambda)w^0)&\left(\varphi''+\left(\frac{1}{\mu_n}\frac{\partial\mu_n}{\partial y}+\log(\lambda)\left(\frac{\partial w^0}{\partial y}-\varphi'\right)\right)\varphi'\right).\end{aligned} \tag{17}$$

If we integrate each side of (17) from $x=x^*$ (x^* is a point in the substrate where there is no current flux across $x=x^*$) we obtain for the left-hand side

$$\left[\left(\mu_n\exp(\log(\lambda)w^0)\frac{\partial\varphi^1}{\partial x}\right)\right]_0^{x^*}=0, \tag{18}$$

as $\partial\varphi^1/\partial x=0$ when $x=0$ and $x=x^*$. It follows that we must have

$$\varphi''+\frac{\partial}{\partial y}\log\left(\int_0^{x^*}\mu_n\exp(\log(\lambda)(w^0-\varphi))\,\mathrm{d}x\right)\varphi'=0. \tag{19}$$

The notation

$$\sigma_m=\int_0^{x^*}\mu_n\exp(\log(\lambda)(w^0-\varphi))\,\mathrm{d}x \tag{20}$$

allows the equation determining the y dependence of φ to be integrated once to give

$$\varphi'\sigma_m\log(\lambda)=C. \tag{21}$$

3. Ward's $\lambda \gg 1$ asymptotic expansion

We copy the 1-D scaled charge conservation equation, (13), written in the form

$$\begin{array}{llll}w_{xx}= & \exp((w-\varphi-1)\log\lambda) & -\exp(-(w+1)\log\lambda) & +1 \quad \text{for } x>0.\\ (i) & (ii) & (iii) & (iv)\end{array} \tag{22}$$

The second line provides notation for the identification of terms: terms (i–iv) on the RHS represent charge densities due to electrons, holes and the static impurity doping, respectively. The impurity doping is taken to be constant: for an n-channel device and with the scaling adopted in Appendix B, it has the value unity. The electric potential, w, asymptotes to the value $w=-1$ in the substrate (terms (iii) and (iv) balance) and this provides the boundary condition at large x. There are no charges in the oxide layer, $-t<x_1<0$, giving a potential linear in x. This together with the boundary condition that the potential is $V_{gs}-V_{fb}$ at $x_1=-t$ and the continuity of electric displacement at $x=0$ means that the latter can be written

$$\left.\frac{\partial w}{\partial x}\right|_s=c_{ox}\sqrt{\frac{\log\lambda}{\lambda}}(w_s-V_{gs}^*), \tag{23}$$

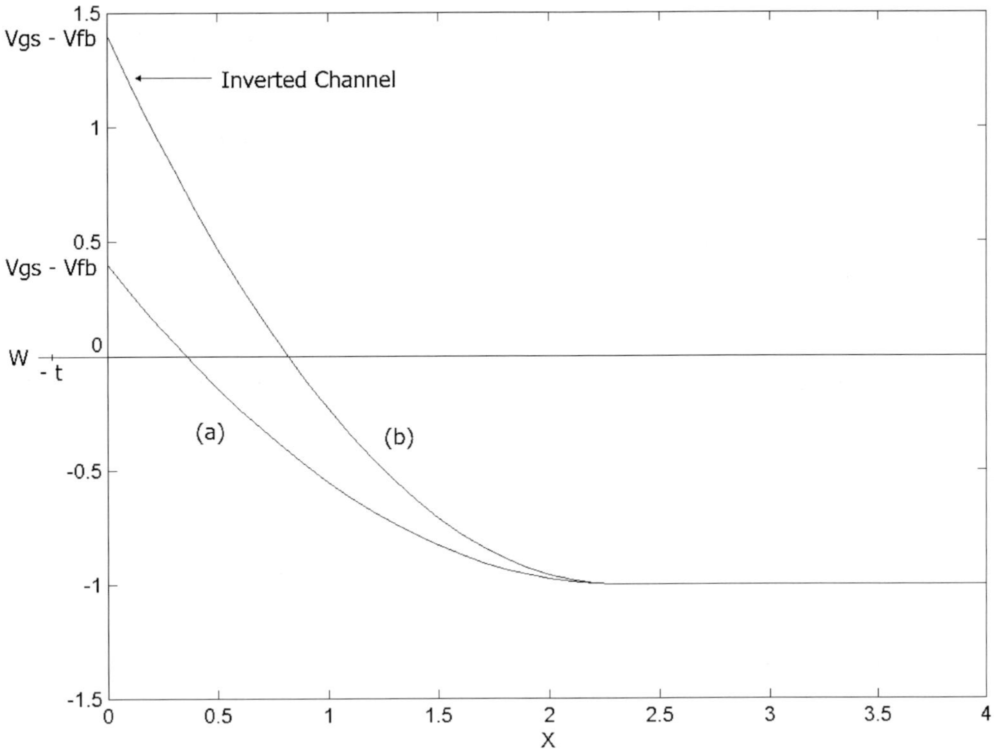

Figure 2. The function $w(x)$ in (a) depletion and (b) inversion.

where s refers to values taken at $x = 0^+$. Figure 2 indicates $w(x)$. (See also Figure 2 in [4].)

In Equation (22) φ is a function only of y which is determined subsequently. Equation (22) allows one integration which together with the boundary condition in the substrate yields

$$\tfrac{1}{2}w_x^2 = \left\{\exp\left[w - \varphi - 1)\log\lambda\right] + \exp\left[-(w+1)\log\lambda\right]\right\} / \log\lambda + w + 1 + 0(\lambda^{-2}). \quad (24)$$

Evaluating this at $x = 0^+$ relates w_s, w_{xs} and φ while (23) relates w_s, w_{xs} and V_{gs}^*. Elimination of w_{xs} yields a relation between w_s, φ and V_{gs}^*. The formula for the current requires $w_s(\varphi)$ in an integration. Hence this may be accomplished numerically.

An alternative to the procedure just outlined is to solve (22) by the method of matched asymptotic expansions. In this, the spatial domain is separated into λ-dependent regions in which various terms in (22) may be neglected. More amenable solutions are achieved, then suitably 'matched'. There are three regimes of operation depending on whether term (ii) is totally absent, present, or present only over part of the y domains. Since $\varphi(0) = 0, \varphi(1) = V_{ds}/V_{th}\log\lambda$, and $\varphi(y)$ is monotonic, the first two regimes categorize easily:

3.1. Weak inversion – depletion:

$$-1 < w_s < 1 + \frac{V_{ds}}{V_{th}\log\lambda}, \quad \text{for all} \quad y, \quad (25)$$

3.2. Strong inversion:

$$w_s > 1 + \frac{V_{ds}}{V_{th}\log\lambda}, \quad \text{for all} \quad y. \quad (26)$$

The last regime has part of the device (adjacent to the source) in strong inversion and part (adjacent to the drain) in weak inversion-depletion and the change-over depends on $\varphi(y)$. This is called pinch-off. When the pinch-off location is designated by the subscript, *po*, the third regime is:

3.3. Pinch off: Weak inversion – depletion

$$-1 < w_s < 1 + \varphi_{po}, \quad \text{for} \quad y_{po} < y < 1. \tag{27}$$

In the weak inversion-depletion regime there is an x-domain adjacent to $x = 0$ in which the dominant terms in (22) are (i) and (iv). This solution must be matched to the solution in the substrate via a transition layer (in which (i) = (iii) + (iv)). Note that φ plays no role in this solution. In the strong inversion regime there is an x-domain adjacent to $x = 0$ in which the dominant terms in (22) are (i) and (ii). (The latter term represents proliferation of electrons in a narrow layer near $x = 0$ – the channel – and it is flow of these electrons along the channel that gives strong current flow). This inverted region must be matched to a depletion solution and at larger x to the substrate solution. All these solutions and matchings are obtained in Ward; only the results will be quoted here.

3.4. Strong inversion regime: $w_s > 1 + \varphi$

From the solution given in [4], Equation (6.1),

$$\frac{\partial w}{\partial x}|_s = -\sqrt{2}\alpha_0 \coth \gamma, \tag{28}$$

where

$$\alpha = \sinh \gamma, \quad \alpha = \alpha_0 (\log \lambda)^{1/2} \lambda^{(1-w_s+\varphi)/2}, \tag{29}$$

$$\alpha_0^2 = 2 + \varphi + K - 1/\log \lambda, \tag{30}$$

$$K = \frac{\log(\log \lambda)}{\log \lambda} + \frac{2}{\log \lambda}\left[\log(2\alpha_0) - \gamma\right]. \tag{31}$$

3.5. Pinch-off: Asymptotic solution

In the pinch-off regime a portion of the device is in weak inversion-depletion. The asymptotic solution ([4, Section 4.1]) is substituted in the boundary condition (23) giving

$$(w_s)_{po} = V_{gs}^* + \frac{1}{c^2 \log \lambda} - \frac{1}{c^2 \log \lambda}\left\{2c^2 \log \lambda \left(V_{gs}^* + 1 - \frac{1}{\log \lambda}\right) + 1\right\}^{1/2}. \tag{32}$$

Since the device becomes inverted for $w_s > 1 + \varphi$ (term (ii) on the right-hand side of (22) becomes dominant) the pinch-off boundary is at φ_{po} where

$$\varphi_{po} = (w_s)_{po} - 1. \tag{33}$$

The asymptotic solution (28) is a replacement for the exact relations between w_s,w_{xs} and φ obtained from (24). The elimination of w_{xs} from (23) and (28) to generate $w_s(\varphi)$ remains a numerical task; Sarvas and Spanier, [12], obtained results by Newton iteration. In the following section we take the asymptotics a little further so that $w_s(\varphi)$ is obtained in analytic form.

4. Further asymptotics

If we combine the expressions for α_0 and K in (29–31) we obtain

$$\alpha_0^2 = 2 + \varphi + \frac{\log(\log\lambda)}{\log(\lambda)} + 2\frac{\log(2\alpha_0) - \gamma - \frac{1}{2}}{\log(\lambda)}. \tag{34}$$

Using the definition of γ, we can write this as

$$\alpha_0^2 = 2 + \varphi + \frac{\log(\log\lambda)}{\log(\lambda)} + \frac{2}{\log(\lambda)}\log(2\frac{z\sqrt{\log(\lambda)}}{1+\sqrt{\frac{z^2}{\alpha_0^2}\log(\lambda)+1}}) - \frac{1}{\log(\lambda)}, \tag{35}$$

where

$$z = \alpha_0/\alpha\sqrt{\log\lambda}. \tag{36}$$

Also we write

$$c_{ox} = c\sqrt{\lambda}. \tag{37}$$

In Ward $c_{ox} = 450$, and in the data supplied by ISI $c_{ox} = 1244$; for $\lambda \backsim 10^6$ this gives c to be an $O(1)$ quantity.

Elimination of w_{xs} between (23) and (28), and use of (37) yields

$$\left\{2\left(1+\alpha_0^2\log\lambda\ \lambda^{1-w_s+\varphi}\right)\right\}^{1/2}\lambda^{(w_s-1-\varphi)/2} = c\log\lambda(V_{gs}^* - w_s). \tag{38}$$

This complicated relationship between w_s and φ can be solved by iteration. It is convenient to introduce the notation:

$$r = w_s - 1 - \varphi,\ s = V_{gs}^* - 1 - \varphi,\ z = \lambda^{r/2}/\log\lambda. \tag{39}$$

We have noted above that $z = \alpha_0/\alpha\sqrt{\log\lambda}$, but it helps to keep the notation just introduced.

With the substitutions (39), Equation (38) reads

$$\sqrt{2(z^2 + \alpha_0^2/\log\lambda)} = c\left(s - 2\frac{\log\log\lambda}{\log\lambda} - \frac{2\log z}{\log\lambda}\right). \tag{40}$$

Since $\log\lambda \gg 1$, an iterative solution to (35) and (40) can be constructed in the following way. Define the sequences $\{\beta_j\}_{j=0}^{\infty}$ and $\{z_j\}_{j=0}^{\infty}$ by the recurrence formulae

$$\beta_{j+1}^2 = 2 + \varphi + \frac{\log(\log\lambda)}{\log(\lambda)} + \frac{2}{\log(\lambda)}\log(2\frac{z_j\sqrt{\log(\lambda)}}{1+\sqrt{\frac{z_j^2}{\beta_j^2}\log(\lambda)+1}}) - \frac{1}{\log(\lambda)}, \tag{41}$$

$$z_{j+1} = \frac{c}{\sqrt{2}}(s - 2\frac{\log(\log\lambda)}{\log(\lambda)} - 2\frac{\log(z_j)}{\log(\lambda)})(1 + \frac{\beta_j^2}{z_j^2\log(\lambda)})^{-\frac{1}{2}} \tag{42}$$

for $j \geq 0$, and start the sequences with the values

$$\beta_0^2 = 2 + \varphi + \frac{\log(\log\lambda)}{\log(\lambda)}, \tag{43}$$

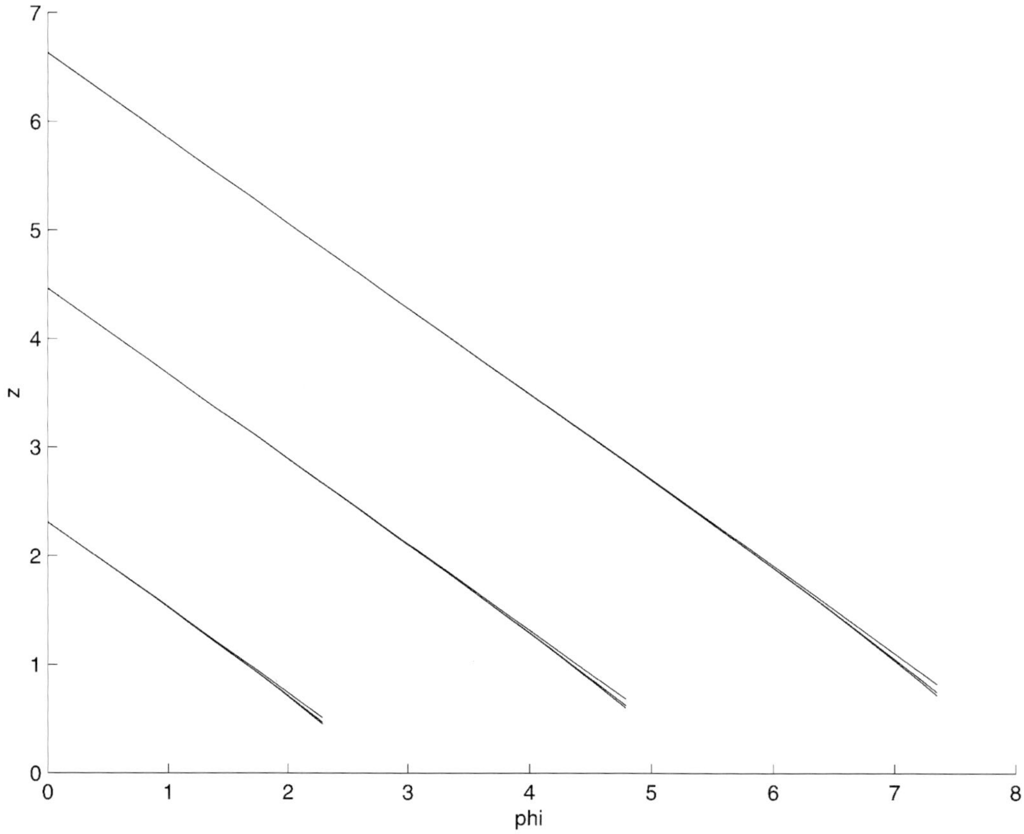

Figure 3. The first three iterations for (left to right) $V_{gs} = 1, 2, 3$ V.

$$z_0 = \frac{c}{\sqrt{2}}(s - 2\frac{\log(\log\lambda)}{\log(\lambda)}). \tag{44}$$

Provided these sequences are convergent we have

$$z_j \to z \text{ and } \beta_j \to \alpha_0 \text{ as } j \to \infty. \tag{45}$$

These values of z give r and hence w_s, in terms of s, and hence φ, with V^*_{gs} as a parameter.

Figure 3 shows the first three iterations (z_1 to z_3) given by the iterative scheme for (41–42), and shows excellent convergence to the exact solution. The latter (not shown) was obtained from Sarvas and Spanier, [12], who solved (23), (24) numerically (and provided us with the MATLAB programs to reproduce their results). The scheme has effectively converged at z_3. These calculations relate only to the inverted region: that is, Figure 3 shows calculations terminated at φ_{po} given by (33).

The fast iteration shown in Figure 3 enables us to generate explicit formulae for the solution. For instance, the first iterate, z_1, is a reasonable approximation to the exact solution, and it can be written in a formula taking up only a couple of lines.

Using the iterative solution for z the relation $w_s(\varphi)$ may be obtained from the relations (39) in the inverted region, and $(w_s)_{po}$ from (32) extends it for $\varphi > \varphi_{po}$. These results are shown in Figure 4 for $\overline{V}_{gs} = 70, 80, 90$, respectively.

The approximations pre-and post-pinch-off have been blended by use of the blending Y, see Appendix D. Also shown (circles) are the exact results obtained in [12], which indicate

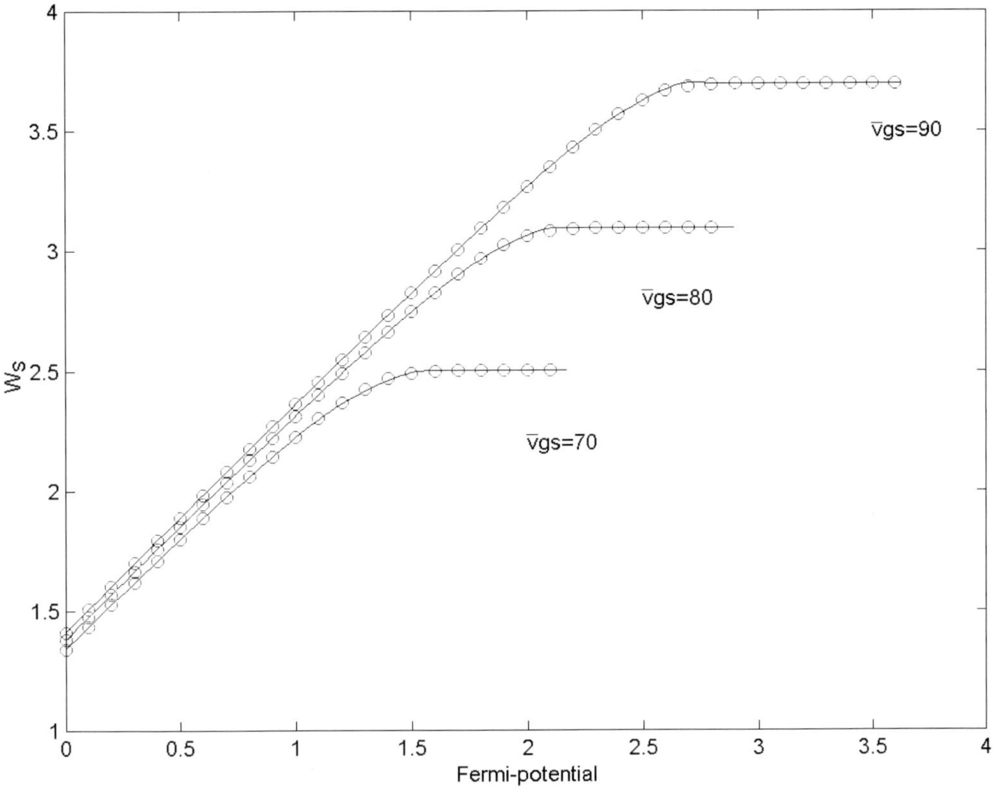

Figure 4. $w_s(\varphi)$ for $\overline{V}_{gs} = 70, 80, 90$.

that a very high level agreement between the asymptotic and exact numerical approaches has been achieved.

5. Current

The definition of the quasi-Fermi potential φ_n, (6), allows the current flux $\mathbf{J}_n$, (3), to be expressed as

$$\mathbf{J}_n = q\mu_n \frac{kT}{q} n \nabla \varphi_n. \tag{46}$$

The current flowing from source to drain, I_{ds}, is obtained:

(a) by first integrating the x_2 - component of $\mathbf{J}_n$ in the x_1-direction to obtain the current flowing across a plane x_2 = a constant. We may evaluate this integral, using the asymptotic representation of the solution, [4].

(b) by performing a second integration in the x_2-direction from source to drain.

Since, as has been noted already, the current across a plane x_2 = constant does not vary with x_2, this integration of I_{ds} yields on the left-hand side LI_{ds}, whilst on the right-hand side the conjunction $\frac{\mathrm{d}\varphi_n}{\mathrm{d}x_2} dx_2 = \mathrm{d}\varphi_n$ yields an integration with respect to φ_n. Since the solution is known in terms of φ_n, this artifice avoids determining φ_n as a function of x_2.

We obtain expressions for I_{ds} in the regimes when the whole device is inverted, and when it is in pinch-off (partially inverted, partially depleted).

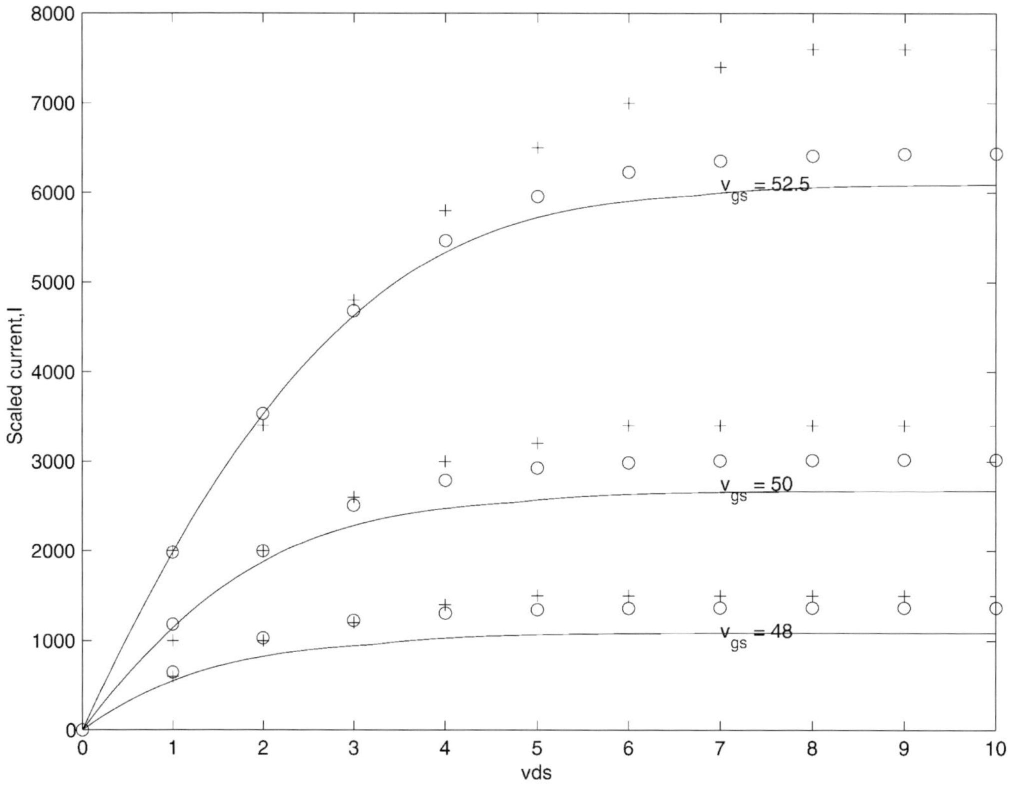

Figure 5. The scaled current for $\overline{V}_{gs} = 48, 50, 52{\cdot}5$ (For $\lambda = 10^6$).

5.1. Linear Regime

When the whole device is inverted the current rises approximately linearly with V_{ds} : hence the title. For constant mobility the current is found from [4, Equations (3.6,7), (6.5,6)] as

$$\frac{I_{ds}}{\mu I_c} = \sqrt{2}\log\lambda \int_0^{\overline{V}_{ds}/\log\lambda} \exp\left[\frac{1}{2}(w_s - \varphi)\log\lambda - \gamma\right] \mathrm{d}\varphi. \tag{47}$$

The substitutions (39) allow us write the current integral in the form

$$\frac{I}{\mu I_c} = \sqrt{2\lambda}(\log(\lambda))^2 \int_{z_d}^{z_s} H(z, \alpha_0^2(z))\mathrm{d}z, \tag{48}$$

where

$$H(z, \alpha_0^2(z)) = z\frac{\mathrm{d}s}{\mathrm{d}z}\mathrm{e}^{-\gamma}. \tag{49}$$

Equation (40) yields

$$\frac{\mathrm{d}s}{\mathrm{d}z} = \left\{\frac{\sqrt{2}}{c}\frac{z}{\sqrt{z^2 + \alpha_0^2/\log\lambda}} + \frac{2}{z\log\lambda}\right\}\left\{1 + \frac{1}{c\log\lambda(2z^2 + 2\alpha_0^2/\log\lambda)^{1/2}}\right\}^{-1}. \tag{50}$$

Using the relations (28–31) we can express

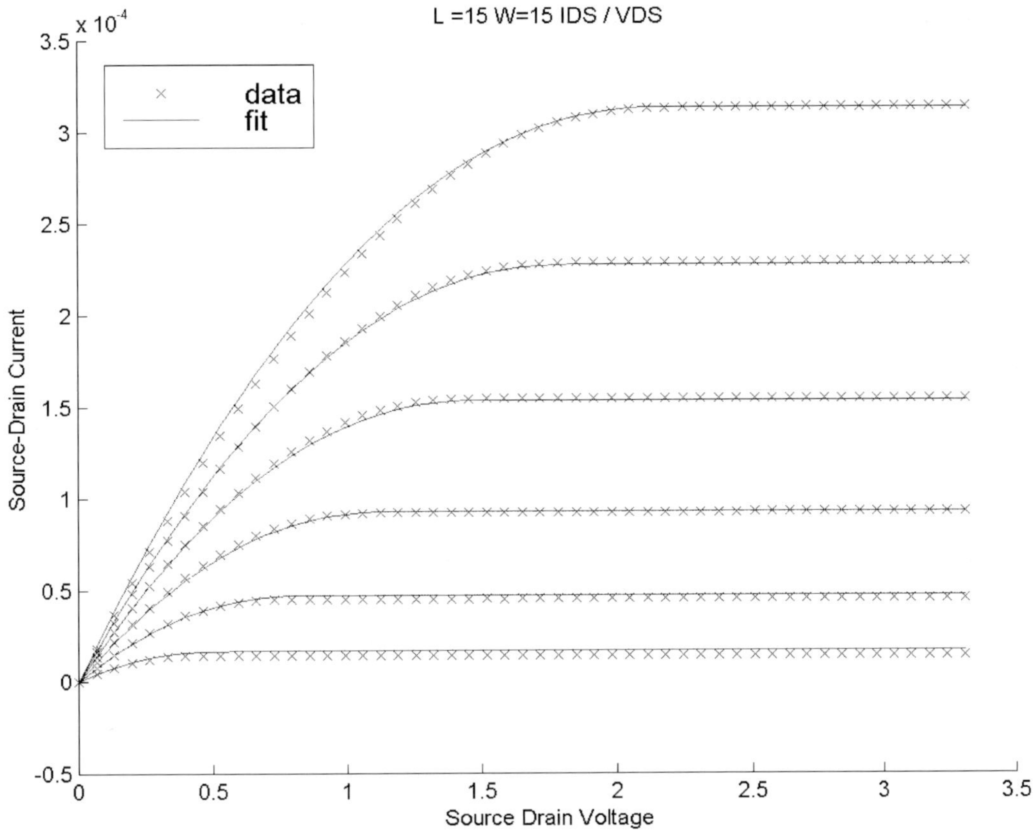

Figure 6. A long $L = 15$ and width $W = 15$ device ($\lambda = 10^7$).

$$e^{-\gamma} = (1 + \alpha_0^2/z^2 \log \lambda)^{1/2} - \alpha_0/z\sqrt{\log \lambda}. \tag{51}$$

The functions ds/dz and $e^{-\gamma}$ can be seen to be decreasing functions of α_0^2 for fixed z by direct observation of their functional forms in (50) and (51). Consequently, the current integrand $H(z, \alpha_0^2(z))$ is monotonically decreasing as a function of α_0^2 for fixed z. The function $\alpha_0^2(z)$ is similarly monotonically decreasing and bounded on $[z_d, z_s]$

$$m \le \alpha_0^2(z) \le M \quad \text{for} \quad z \in [z_d, z_s]. \tag{52}$$

Define the function

$$W(t) = \int_{z_d}^{z_s} H(z, t)dz \quad \text{for} \quad t \in [m, M]. \tag{53}$$

This is monotonic in t and lies in the interval $[W(M), W(m)]$. As $H(z, \alpha_0^2(z))$ is monotonically decreasing as a function of α_0^2 for fixed z we can write

$$W(M) \le \int_{z_d}^{z_s} H(z, \alpha_0^2(z))dz \le W(m). \tag{54}$$

By the intermediate-value theorem there is a value of t for which $W(t)$ takes any value in $[W(M), W(m)]$. In particular there is a value $t = t^* \in [m, M]$ for which

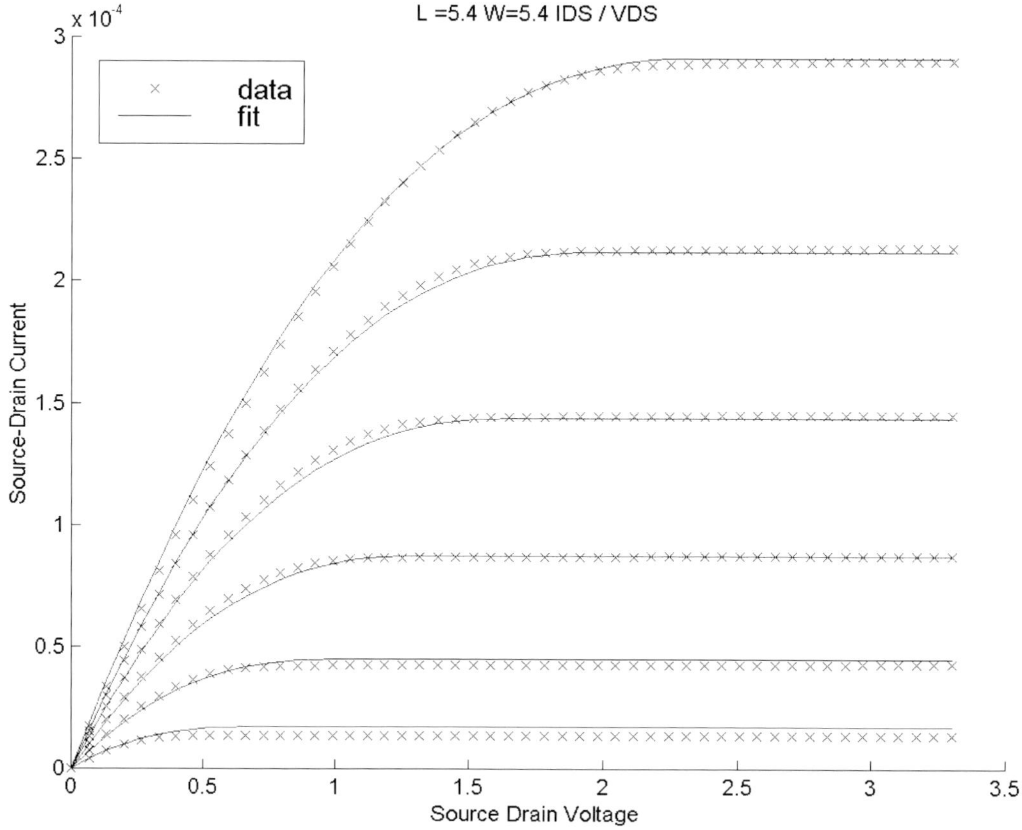

Figure 7. A $L = 5{\cdot}4$ and width $W = 5{\cdot}4$ device ($\lambda = 10^7$).

$$\int_{z_d}^{z_s} H(z, \alpha_0^2(z))\,\mathrm{d}z = \int_{z_d}^{z_s} H(z, t^*)\,\mathrm{d}z. \tag{55}$$

As $\alpha_0^2(z)$ is monotonic decreasing on $[z_d, z_s]$ there exists a $z^* \in [z_d, z_s]$ for which

$$\alpha_0^2(z^*) = t^* = \overline{\alpha}_0^2 \tag{56}$$

From (35) the equation for $\overline{\alpha}_0^2$ is

$$\begin{aligned} \overline{\alpha}_0^2 = {} & 1 + V_{gs}^* - \frac{\sqrt{2}}{c}\sqrt{(z^*)^2 + \frac{\overline{\alpha}_0^2}{\log(\lambda)}} - \frac{\log(\log\lambda)}{\log(\lambda)} + \\ & + \frac{2}{\log(\lambda)} \log\left(\frac{2}{\sqrt{e}} \frac{\sqrt{\log(\lambda)}}{1 + \sqrt{\frac{(z^*)^2}{\overline{\alpha}_0^2}\log(\lambda) + 1}} \right) \end{aligned} \tag{57}$$

which can be written in the form

$$\overline{\alpha}_0^2 = P - \frac{\sqrt{2}}{c}\sqrt{(z^*)^2 + \frac{\overline{\alpha}_0^2}{\log(\lambda)}}, \tag{58}$$

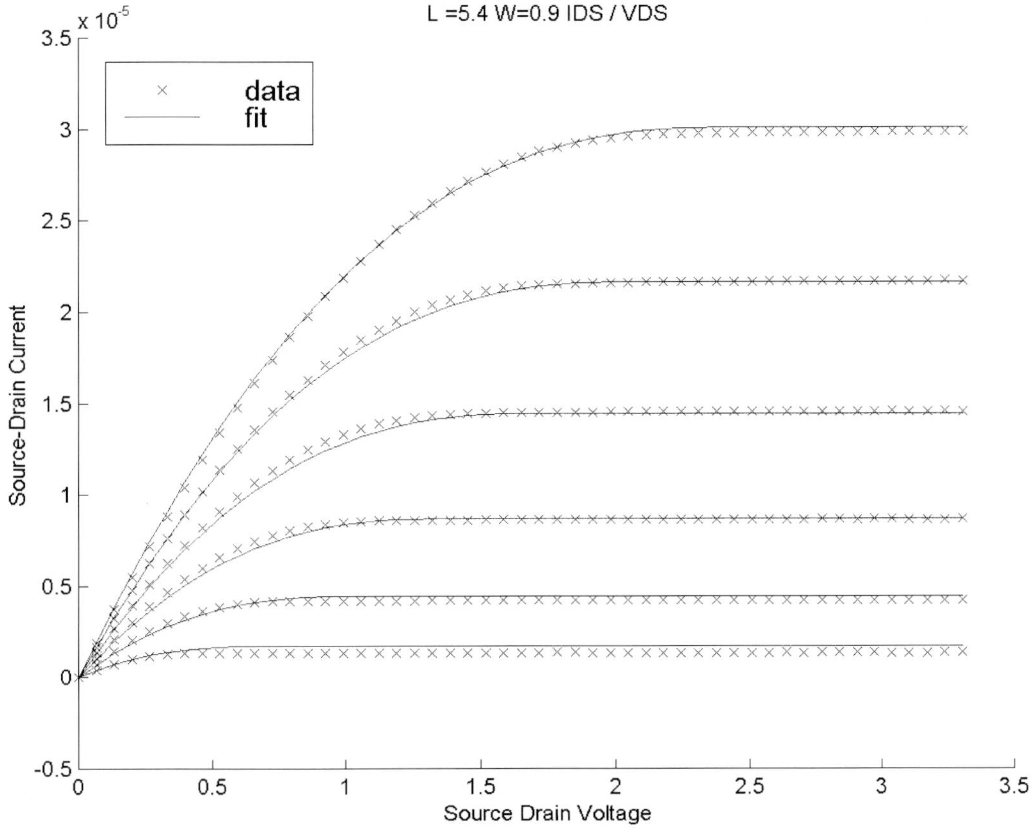

Figure 8. A $L = 5{\cdot}4$ and width $W = 0{\cdot}9$ device ($\lambda = 10^7$).

involving

$$P = 1 + V_{gs}^* - \frac{\log(\log\lambda)}{\log(\lambda)} + \frac{\kappa}{\log(\lambda)} \tag{59}$$

and

$$\kappa = 2\log\left(\frac{2}{\sqrt{e}}\frac{\sqrt{\log(\lambda)}}{1+\sqrt{\dfrac{(z^*)^2}{\overline{\alpha}_0^2}\log(\lambda)+1}}\right). \tag{60}$$

Assuming that z^* is small, we can write the solution for $\overline{\alpha}_0$ in the form

$$\overline{\alpha}_0^2 \backsim P - \frac{\sqrt{2}}{c}\sqrt{\frac{P}{\log(\lambda)}} \tag{61}$$

with P given by (59). The unknown value z^* has a dependence upon V_{gs}^* but this is expected to be weak. In the fitting process (see Section 6) κ is treated as a free parameter and determined by the least-squares optimization process. As α_0 can be treated as a constant $\overline{\alpha_0}$ in (48) the integral can be evaluated explicitly. Having done this, we obtain the final result for the current in the linear regime, which is

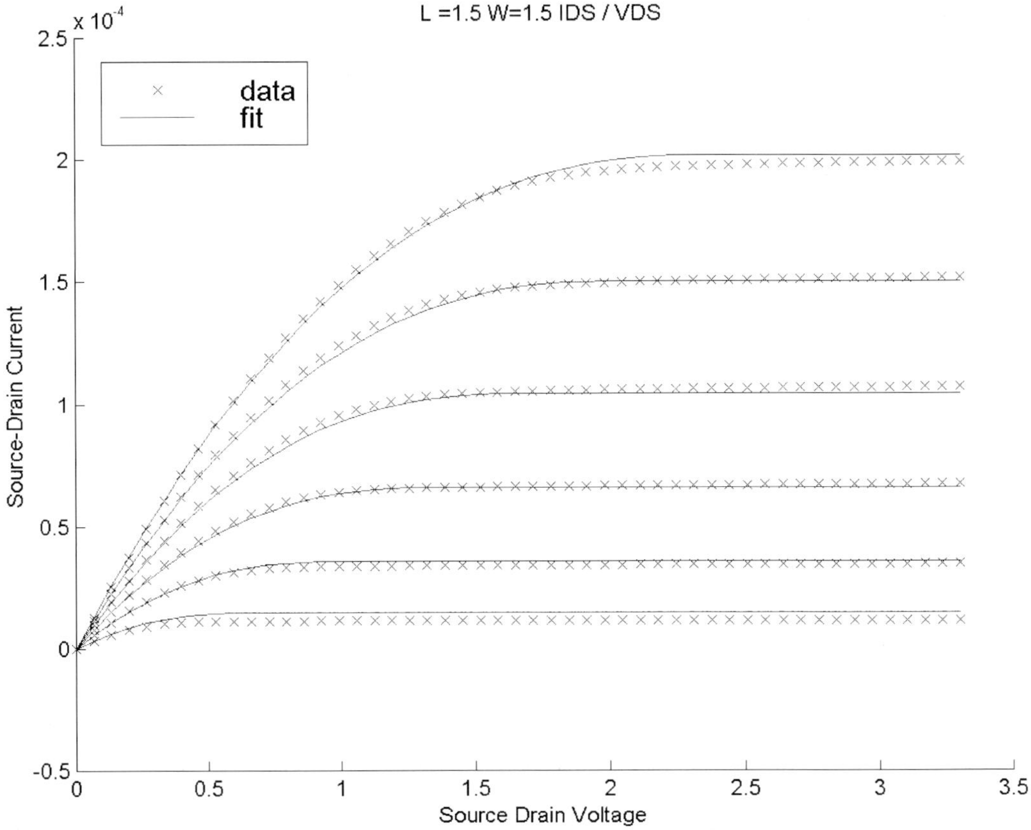

Figure 9. A $L = 1{\cdot}5$ and width $W = 1{\cdot}5$ device ($\lambda = 10^7$).

$$\frac{I_{ds}}{\mu I_c} = \left\{2\lambda(\log\lambda)^3\right\}^{1/2}\overline{\alpha_0}\left\{I\left(\frac{z_s\sqrt{\log\lambda}}{\overline{\alpha_0}}\right) - I\left(\frac{z_d\sqrt{\log\lambda}}{\overline{\alpha_0}}\right)\right\}, \tag{62}$$

where $I(x)$ is defined by

$$\begin{aligned} I(x) &= \frac{\overline{\alpha_0}}{c}\sqrt{\frac{2}{\log\lambda}}\left(\frac{1}{2}y^2 - \beta^2 y + \beta^4\log(y+\beta^2)\right) + \\ &\quad + \frac{2}{\log\lambda}\left(y + \frac{\log(y-1)}{2(1+\beta^2)} - \frac{\log(y+1)}{2(1-\beta^2)} + \frac{\beta^6}{1-\beta^4}\log(y+\beta^2)\right) - \\ &\quad - \frac{\overline{\alpha_0}}{c}\sqrt{\frac{2}{\log\lambda}}\left(y - \beta^2\log(y+\beta^2)\right) - \\ &\quad - \frac{2}{\log\lambda}\left(\frac{\log(y-1)}{2(1+\beta^2)} + \frac{\log(y+1)}{2(1-\beta^2}) + \frac{\beta^4}{\beta^4-1}\log(y+\beta^2)\right) \end{aligned} \tag{63}$$

where $y = \sqrt{1+x^2}$ and $\beta^2 = 1/c\overline{\alpha_0}\sqrt{2\log\lambda}$.

5.2. Saturated regime

When V_{ds} is large enough a region near the drain changes from inverted to depleted. In this region the elevated values of φ eliminate from importance term (ii) in Equation (22) and the results used for inverted $w_s(\varphi)$ are no longer valid. The integral for current is separated into two parts: (1) that valid in the inverted region and available from the results of the previous section, merely substituting z_{po} for z_d in the upper limit in (48), and (2) that valid in the depleted region for which we can use [4, Equations (4.6), (4.7)]. This gives for the current in the saturated regime the formula

$$\frac{I_{ds}}{\mu I_c} = \left\{2\lambda(\log\lambda)^3\right\}^{1/2}\overline{\alpha}_0\left\{I\left(\frac{z_s\sqrt{\log\lambda}}{\overline{\alpha}_0}\right) - I\left(\frac{z_{po}\sqrt{\log\lambda}}{\overline{\alpha}_0}\right)\right\} + Q_c\left(\exp\left(-\overline{V}_{ds}\right) - \exp\left(-\varphi_{po}\log\lambda\right)\right), \tag{64}$$

where $I(x)$ is given by (63), and φ_{po} is given by (33), and Q_c is given by (see [4, Equation (4.7)]:

$$Q_c = -e^{w_s\log\lambda}(2\lambda\log\lambda)^{-\frac{1}{2}}(w_s + 1 - 1/\log\lambda)^{-\frac{1}{2}}\left(1 + \frac{1}{2(w_s + 1 - 1/\log\lambda)\log\lambda}\right) \tag{65}$$

with w_s taking the value w_{spo}.

Figure 5 shows the scaled drain current I_{ds} as a function of $\overline{V}_{ds}$ for values of $\overline{V}_{gs} = 48$, 50, 52·5. The + are Ward results, 0 are Spanier and Sarvas numerical results [12], and – are the results of the calculations reported on here.These curves were obtained by blending the current in the linear region, (62), with that in the saturated region (65). The three approaches have the same shape as experimental data, but provide slightly different amplitudes. This is attributed to the following:

Comment 1: Since the agreement between our results and Sarvas and Spanier has been excellent so far, the difference in the results for current is attributed to the approximations made in generating the explicit formula (62) from the integral (47).

Comment 2: Divergence between the Sarvas and Spanier and Ward current results was also shown in [12]. The results generated in this paper show the possible source of the differences. Ward states 'γ is transcendentally small' and 'α is transcendentally small in $\log\lambda$' in the linear regime, and bases some approximations on these assertions. However the $w_s(\varphi)$ relation (see Figures 4 and 5) give $w_s - 1 - \varphi \cong 0{\cdot}4$ and this gives

$$\alpha = \alpha_0\sqrt{\log\lambda}\ \lambda^{-0\cdot2}. \tag{66}$$

There is a negative power of λ in α, but it is small. Since $\alpha = \alpha_0/z\sqrt{\log\lambda}$ and z is $0(1)$ over a substantial portion of the linear regime (see Figure 3), $\alpha = 0(1/\sqrt{\log\lambda})$ there, and may not be neglected without some penalty.

6. Optimized fitting

A number of assumptions have been made in constructing the model of the MOSFET that has been used. It has been supposed that the mobility μ is a constant in the sense that it is independent of z. It does, however, have a dependence upon V_{gs}^* that must be taken into account if we are to compare our formulae to real data. A common form used in SPICE is given by

$$\mu = \frac{1}{1 + \theta(V_{gs} - V_T)} \tag{67}$$

and we adopt this also. The mobility scaling factor μ_0 is also a parameter that has to be provided in SPICE and our model. These parameters can be extracted from experimental data but it is best to regard such extracted values as estimates. We shall take both these physical parameters, θ and μ_0, as fitting parameters to be determined by the optimization process. In Section 5 we were able to make a partial evaluation of the integrals by invoking the intermediate value theorem. This required the introduction of an additional non-physical parameter z^*. Although this cannot be extracted from the data we do have a sense of its value as it must lie in $[z_d, z_s]$ which we can estimate. Thus we have three variables and approximate values from which to start an optimization process to obtain the best fit to the data. The method of least squares implemented in the form of the Marquardt algorithm, [13], is used to determine the values of these three parameters. We should remark that a similar parameter optimization technique is used in SPICE. None of the physical parameters is known well enough for typical fabricated devices to warrant any alternative approach.

Using 0·35 micron technology data provided by ISI we have applied our method to a number of devices of different channel length and width. Each device was separately fitted. We present the results as a series of plots. Figure 6 is a long and wide device for which we would expect the best agreement.

The fit is excellent. This continues to be true as the channel length and width are reduced. Figure 7 shows a 5·4 micron length device with width 5·4 microns.
As figure 8 shows we can even reduce the width to 0·9 microns still retain an acceptable fit. The fits remain good down to about 1·5 microns as shown by Figure 9.
At shorter lengths the experimental I-V curves develop a non-zero asymptotic slope and are outside the range of validity of our assumptions.

7. Concluding remarks

Explicit formulae for the dependence of current flow on applied voltages and device geometry, doping, etc., have been obtained for the standard MOSFET by asymptotic analysis. The analysis is valid for long channel devices (ϵ, the aspect ratio, Equation (1), small), and for large λ, where λ is the ratio of the doping level to its intrinsic level. However, comparison of the results with experimental data shows that accuracy is maintained for a more extended range of values of ϵ.

The results obtained here extend the work of Ward [4] and generate *explicit* formulae for the current. These formulae are derived directly from the physics of the drift-diffusion model. Ad-hoc assumptions are at a minimum. (Only the mobility has an empirically based dependence). However, some (three) parameters are given values determined by optimally fitting the derived formule with the data; this, though, is standard in the industry. The formulae, though complicated, provide a much simpler model than the ones currently used in industry which require up to 200 fitting parameters.

The semiconductor industry continues to minaturise the device–channel lengths of $0{\cdot}18\mu$ are being fabricated commercially. At this value the aspect ratio is $\epsilon = 2/3$. Hence a two-dimensional theory would be valuable. An asymptotic 2-D scheme along the lines suggested by Ward [5] and realised only in the zero-current case, seems to be the only choice if explicit results are required. This, in the case of non-zero current flow, is a challenging problem.

Acknowledgements

We wish to acknowledge Cesar Piña and Vance Tyree of the University of Southern California Information Sciences Institute for technical discussions and for provision of the MOSFET data that was used in this study.

Appendix

A. Primary Notation and Symbols

x_1, x_2	co-ordinates along and perpendicular to channel (See Figure 1)	ρ	is the space charge density
L, W	are channel length and width	n_i	is the intrinsic carrier concentration
t	is oxide thickness	q	is the charge of a proton
n, p	are the electron and hole concentrations	k	is the Boltzmann's constant
N	is the concentration of selected impurities or dopants	T	is the lattice temperature (assumed constant)
$\mathbf{J}_n$	is the electron current density	$V_{th} = kT/q$	is called the thermal voltage
$\mathbf{E}, \psi$	are the electric field and electrostatic potential	V_T	is the threshold voltage, a measurable characteristic of the device
ϕ_n	is the quasi-Fermi potential for electrons	$L_D = \left(\frac{kT\varepsilon_s}{n_i q^2}\right)^{\frac{1}{2}}$	is called the Debye length
μ_n, μ_0	are the electron mobility, typical mobility magnitude	V_{ds}	is drain-to-source voltage
ε_s	is the dielectric constant of the semiconducting material	V_{gs}	is gate-to-source voltage
ε_i	is the dielectric constant of the oxide insulator	V_{fb}	is flat-band voltage
I_{ds}	is the source-to-drain current		

B. Scaled Variables

$$\lambda = \max |N(x_1)/n_i| \qquad c_{ox} = \varepsilon_i L_D/\varepsilon_s t$$

$$x_1 = L_d \left(\frac{\log\lambda}{\lambda}\right)^{1/2} x \qquad x_2 = Ly \qquad c = c_{ox}/\sqrt{\lambda}$$

$$d(x) = N(L_D(\tfrac{\log\lambda}{\lambda})^{1/2}x)/n_i \qquad V_{gs} - V_{fb} = V_{th}(\overline{V}_{gs} - \log\lambda) = V_{th}\log\lambda \;\; V^*_{gs}$$

$$(\psi, \phi_n) = V_{th}\log\lambda \;\; (w, \phi) \qquad V_{ds} = V_{th}\overline{V}_{ds}$$

C. Numerical Values

$L_D = 33$ microns $\quad V_{fb} = -0{\cdot}6$ volt

$V_{th} = 0{\cdot}0259$ volt $\quad \mu_0 = 0{\cdot}0407\ \mathrm{m}^2\ \mathrm{volt}^{-1}\ \mathrm{s}^{-1}$

The table below shows the corresponding magnitudes of $\overline{V}_{gs}$ and V_{gs} for $\lambda = 10^6$

$\overline{V}_{gs}$	48	50	52·5	70·0	80·0	90·0
V_{gs} volt	0·2854	0·3372	0·4019	0·8552	1·1142	1·3732

If the value of λ is increased the values of $\overline{V}_{gs}$ are larger. For example if $\lambda = 10^7$ then $\overline{V}_{gs} = 155$ corresponds to $V_{gs} = 2{\cdot}997$ volt.

D. Blending

In the theory of matched asymptotic expansions the construction of two approximate solutions each valid in one of two domains that are contiguous and overlapping is a typical situation. The construction of a composite expansion valid over the union of the domains is then indicated, and there are techniques available prescribing how to accomplish this. The composite expansion (to some order) will be accurate (to that order) uniformly over the whole domain and will have no reduction in smoothness from the component solutions. The desirability of this procedure is evident, and it transcends the domain of asymptotic analysis: one may have separate solutions valid in separate domains that one wishes to join or blend smoothly. These solutions may in some sense be asymptotic – and so complicated that one shrinks at the chore of finding the composite – or not. This section explores some ideas on how to blend such solutions in an empirical, ad-hoc way.

This topic was taken up by Wollkind, [14], and the analysis concerned problems in which (in the language of matched asymptotic expansions) the two component solutions consist of an inner solution valid near a boundary of the domain (a boundary layer) and an outer solution valid elsewhere. We have used the ideas presented by Wollkind to form a blended solution from the asymptotic formulae for separate 'linear' and 'saturated' components.

Let us start the discussion with a simple example. Suppose we have approximations (in $x > 0$)

$$y = \begin{cases} \alpha x & \text{close to } x = 0 \\ 1 & \text{away from } x = 0 \end{cases}. \tag{D1}$$

Following Wollkind, a blended solution is

$$y = (\alpha x - 1)\mathrm{e}^{-x/\epsilon} + 1, \tag{D2}$$

since it has the behavior of (D1), for $x \gg \epsilon$ and for $x \ll \epsilon$. (The Wollkind technique would find ϵ by least squares fitting (D2) to (D1) (αx joined to 1 at $x = 1/\alpha$)). Clearly this procedure gets more accurate the more closely this example resembles a boundary layer problem ($\alpha \gg 1$).

Let us now extend the discussion beyond the boundary layer context. Suppose we have approximations

$$y = L(x),\ R(x) \tag{D3}$$

on the left and right, respectively, of $x = a$. The task is to blend L, R in a smooth formula to provide a blended approximation, $y = Y(x)$.

A first attempt at a blend is given by

$$Y(x) = b_L(x)L(x) + b_R(x)R(x), \tag{D4}$$

$$b_R, b_L = \frac{1}{2}\left\{1 \pm \tanh \frac{x-a}{\epsilon}\right\}.$$

The blending functions b_R, b_L are, for small ϵ, essentially unity in $x \gtrless a$ and zero in the complement, and use of them assures that Y gives (D3) outside a transition region around $x = a$. Since Y is a convex linear combination of L, R at each x it lies between L and R and provides a blend.

Comments

Choice of ϵ: In problems in which asymptotic series have been developed for $L(x), R(x)$ there will be a small parameter which may be identified with ϵ, in which case L, R have been blended through a suitably chosen transition layer. This is the case in the applications in this paper. In other circumstances, ϵ may have to be determined numerically, say by a least-squares fitting of the blended function to data.

Choice of a: In the theoretical development above, the value of a was determined as the intersection point for L, R. Where L, R do not intersect, or where a may not be found accurately, experimentation or extra numerical procedures may be required in order to determine a in order to arrive at a good blending function.

References

1. A. Massobrio and P. Antognetti, *Semiconductor Device Modeling with SPICE*. New York: McGraw-Hill (1993) 479 pp.
2. S. Sze, *Physics of Semiconductor Devices*. New York: John Wiley (1969) 868 pp.
3. Y. Tsividis, *Operation and Modeling of the MOS Transistor*. New York: McGraw-Hill (1987) 505 pp.
4. M. J. Ward, F. M. Odeh, and D. S. Cohen, Asymptotic methods for metal oxide semiconductor field effect transistor modeling. *SIAM J. Appl. Math* 50 (1990) 1099–1125.
5. M. J. Ward, Singular perturbations and a free boundary problem in the modeling of field-effect transistors. *SIAM J. Appl. Math.* 52 (1992) 112–139.
6. H. Pao and C. Sah, Effects of diffusion current on characteristics of metal-oxide (insulator)-semiconductor transistors. *Solid State Electr.* 10 (1966) 927–937.
7. R. Pierret and J. Shields, Simplified long channel MOSFET theory. *Solid State Electr.* 26 (1983) 143–147.

8. M. Martelli, E. Cumberbatch, A. Klafter, S. Pilorz, R.L. Polosa, G. Ruddock, and A. Smith (for Jet Propulsion Laboratory) Modelling short channel MOSFETS for use in VLSI. Claremont Graduate School Final Clinic Report (May 1986).
9. C. Please, An analysis of semiconductor P-N junctions. *IMA J. Appl. Math.* 28 (1982) 301–318.
10. J. Kevorkian and J. D. Cole, *Multiple Scale and Singular Perturbation Methods*. New York: Springer (1996) 632 pp.
11. P.A. Markowich, C.A. Ringhofer, and C. Schmeiser, *Semiconductor Equations*. Wien, New York: Springer-Verlag (1990) 248 pp.
12. J. Sarvas and J. Spanier, A direct approach to solving the drift-diffusion model equations for use in certain MOSFET devices. *Mathl. Comput. Modelling* 22 (1995) 17–31.
13. D. Marquardt, An algorithm for least-squares estimation of nonlinear parameters. *SIAM J. Appl. Math.* 11 (1963) 431–441.
14. D. J. Wollkind, Singular perturbation techniques: a comparison of the method of matched asymptotic expansions with that of multiple scales. *SIAM Rev.* 19 (1977) 502–516.

Journal of Engineering Mathematics **39:** 47–62, 2001.

Separating shear flow past a surface-mounted blunt obstacle

S. BHATTACHARYYA[1], S.C.R. DENNIS[2] and F.T. SMITH[3]
[1]*Department of Mathematics, Indian Institute of Technology, Kharagpur, 721302, West Bengal, India*
[2]*Dept. of Applied Mathematics, University of Western Ontario, London, Ontario, Canada, N6A 5B7*
[3]*Department of Mathematics, University College London, Gower Street, London WC1E 6BT, UK*

Received 28 October 1999; accepted in revised form 25 May 2000

Abstract. The planar flow of incompressible fluid past a blunt obstacle mounted on a flat (horizontal) fixed solid surface of infinite extent is examined in the presence of an incident linear velocity profile, modelling the fluid behaviour close to a small surface roughness for instance. The motion is taken to be steady and laminar. The obstacle is blunt in the sense that its typical surface slopes are not small, a feature which here always induces flow separation both upstream and downstream of the obstacle. Computations and nonlinear theory are applied, together with comparisons. The direct computations of the Navier-Stokes equations, using for example a higher order upwind-difference scheme, deal with a moderate range of Reynolds numbers up to 200, based on the obstacle height and the incident uniform shear. In addition the accuracy is necessarily limited as the Reynolds number increases. The theory is for large Reynolds numbers and is based on viscous-inviscid reasoning, back-pressure effects from the obstacle and slender-layer separation locally, among other influences. The comparisons nevertheless yield encouragingly close agreement, for the present computed cases of a vertical flap or a rectangular block. This is both quantitatively, in terms of the upstream separation and downstream reattachment positions in particular, and generally, in terms of the separating flow structure, even at the notably moderate Reynolds numbers covered accurately by the computations.

Key words: blunt roughness, computations, separation, theory

1. Introduction

Configurations involving shear flow over a blunt obstacle, mounted on an otherwise flat or nearly flat surface, are manyfold and have much physical and technological interest. A blunt or bluff obstacle here is one whose typical surface slopes are of order unity or greater, such as a flap standing normal to the flat surface, a rectangular block, a forward-facing step or a hemispherical object. Configurations of practical interest include boundary-layer flow past roughnesses on airfoils and turbine blades, atmospheric boundary layer motion over hills, wind over water, the use of trip wires to generate transition to turbulent flow downstream, the flow past excrescences in pipes or at artery walls, and the Gurney-flap device placed near an airfoil or blade trailing edge. Some of the theoretical and/or computational studies relevant to the above configurations and especially to the present research are by Dennis and Smith [1], Mei and Plotkin [2], Durst and Loy [3] on internal flows, Savin *et al.* [4] on transition, Smith and Walton [5] on roughness flows, and Giguère *et al.* [6], Smith [7] on the advantages of a flap device being buried within the trailing-edge boundary layer; but see also the many references therein on computations, theory and experiments.

Our specific concern is with the case of a blunt obstacle so small and close to the flat surface that the surrounding fluid motion, *i.e.* the farfield flow as far as the scale of the obstacle is concerned, may be viewed as uniform shear flow. The whole motion is taken to be laminar, planar and steady, and the fluid to be incompressible and Newtonian, yielding an idealised setting of course but one which has correspondingly clearer findings, it is hoped, and which is fundamental to all the practical configurations of interest. Computation and theory are applied for the determination of the major flow properties and for their basic understanding. The theory, which is for large Reynolds numbers, is meant to provide extra insight of course but it also gives particular predictions, for example for the positions of separation upstream and reattachment downstream of the obstacle, and it helps to guide the all-important numerical gridding, especially at increasing Reynolds numbers. The quite large-scale computation is needed at almost all moderate Reynolds numbers and it helps to indicate the range of practical validity of the theory, how much the flow solution is independent of the precise shape of obstacle, and so on. The joint approach using computation and theory is very much a hand-in-hand one. A not dissimilar approach has also been used successfully on occasion in studies of external flow past a blunt body, without surface mounting, as in many papers on the subject by Dennis, Sychev, Fornberg, Smith, Chernyshenko and co-workers: *e.g.* see review in [8, Chapters 23–25].

Section 2 below describes the flow set-up, for the surface-mounted obstacle in a uniform-shear flow, at finite Reynolds number based on the typical obstacle height and on the slope of the incident velocity profile at the flat surface. Computational methodology is presented in detail in Section 3; compare the local leading-edge and trailing-edge studies by Van de Vooren, Dijkstra, Veldman in the 1970s (*e.g.* see [8, Chapters 23–25]). The present computations are designed especially for obstacles which are vertical flaps or rectangular blocks, although the accuracy is of course limited. Theory for large Reynolds numbers is given in Section 4, this being quite distinct from the small-disturbance linear analyses of Stewartson [9], Smith [10] because in the current setting the physical slopes present are not small and so they demand nonlinear reasoning. While intuition may at first suggest that, with increasing Reynolds number, the recirculating eddy upstream of the obstacle should shrink due to inertial effects, the subsequent action of viscous effects near the upstream surface implies that the eddy must elongate, for overall smoothness of the separating flow. This is reflected in the prediction of the separation position (far) upstream, and likewise in the predicted reattachment station (far) downstream. Moreover, the arguments used generalize, for example to any obstacle shape, and they simplify certain related ones used previously, enabling the upstream separation point for instance to be obtained from a single, rapid, forward numerical march. Section 5 shows results and comparisons between computation and theory. Section 6 provides further comments.

2. The flow configuration

The blunt shaped obstacle of concern is on a fixed surface, is deep inside the surface boundary layer or other shear flow and is so localized that the surface appears flat and indefinitely long (the x axis) and the fluid appears to be of semi-infinite extent normally, with uniform incident shear flow. This implies that the only relevant geometric length scales are those of the obstacle itself. We take the typical height of the obstacle as the flow's characteristic length scale ℓ^*, which is often comparable with the typical obstacle length anyway for a blunt shape. Further, as there is no velocity scale u^* directly we take the prescribed slope λ^* of the incident velocity profile at the surface multiplied by ℓ^*, leaving $u^* = \lambda^*\ell^*$ as the velocity scale based on the obstacle height. Nondimensional quantities are then used, so that in Cartesian coordi-

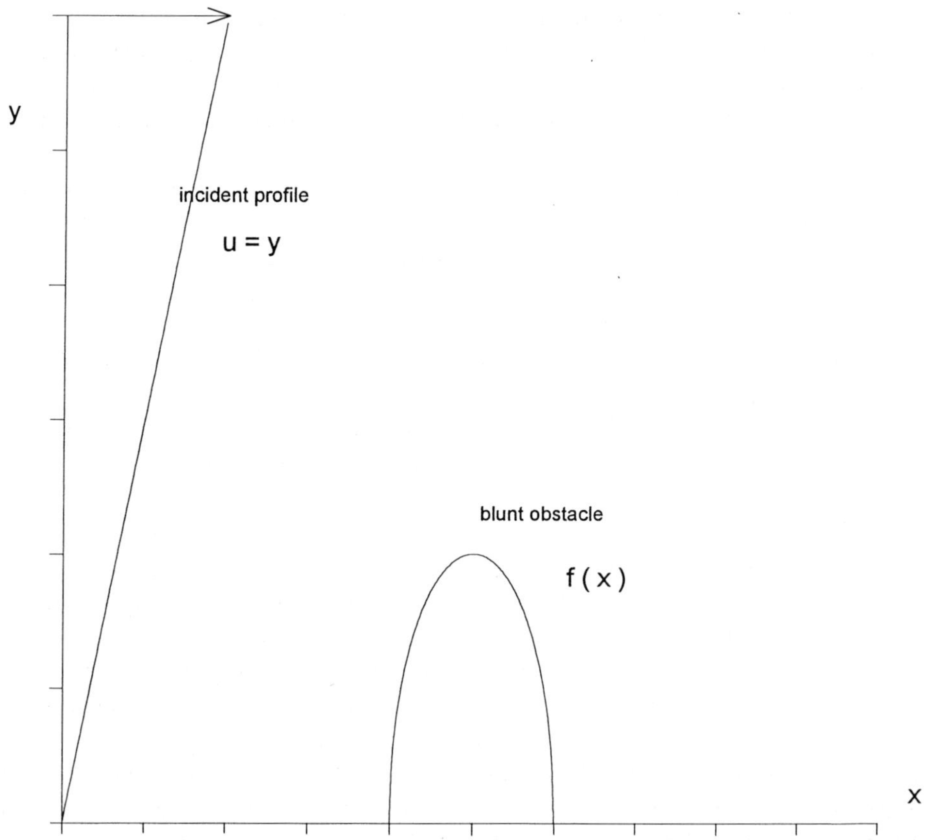

Figure 1. Sketch of the flow configuration, in nondimensional terms, for a blunt obstacle (smooth or non-smooth) with $|f'|$ typically of order unity.

nates $\ell^*(x, y)$ which are streamwise and normal respectively the corresponding fluid velocity $\lambda^*\ell^*(u, v)$ and pressure $\rho^*\lambda^{*2}\ell^{*2}p$ satisfy the continuity and Navier-Stokes equations

$$\frac{\partial u}{\partial x}+\frac{\partial v}{\partial y}=0, \qquad u\frac{\partial u}{\partial x}+v\frac{\partial u}{\partial y}=-\frac{\partial p}{\partial x}+\mathrm{Re}^{-1}\nabla^2 u, \tag{2.1a,b}$$

$$u\frac{\partial v}{\partial x}+v\frac{\partial v}{\partial y}=-\frac{\partial p}{\partial y}+\mathrm{Re}^{-1}\nabla^2 v. \tag{2.1c}$$

Here the shear-based Reynolds number $\mathrm{Re}\equiv\lambda^*\ell^{*2}/\nu^*$ has characteristic values of order unity, the constants ρ^*, ν^* denote in turn the density and kinematic viscosity of the incompressible fluid, and ∇^2 is the Laplacian $\partial^2/\partial x^2+\partial^2/\partial y^2$. The governing Equations (2.1a–c) hold above the obstacle in the upper half plane. They are subject to the boundary conditions

$$u=v=0 \text{ at } y=f(x), \qquad u\sim y,\ v\to 0 \text{ as } x^2+y^2\to\infty, \tag{2.1d,e}$$

for no slip at the given nondimensional surface $f(x)$ of the obstacle or flat plate and for the approach to the specified uniform shear flow in the farfield, respectively. See Figure 1.

The task is to solve (2.1a-e), then. An alternative to (2.1a) is to use a stream function ψ defined by $u = \partial\psi/\partial y,\ v = -\partial\psi/\partial x$ with, say, ψ zero on the obstacle surface, and of course (2.1b,c) can be combined into the vorticity equation

$$u\frac{\partial\zeta}{\partial x} + v\frac{\partial\zeta}{\partial y} = \mathrm{Re}^{-1}\nabla^2\zeta, \tag{2.2a}$$

with

$$\zeta = -\nabla^2\psi \tag{2.2b}$$

being the unknown scaled vorticity. Again, polar coordinates such that $(x, y) = r(\cos\theta, \sin\theta)$ are also used later. In the absence of any obstacle $f(x)$ is zero and we suppose then that the uniform shear flow $(\psi, u, v, \zeta, p) = (\frac{1}{2}y^2, y, 0, -1, 0)$ holds everywhere for positive y. With $f(x)$ present, on the other hand, computations are required in general because of the nonlinearity associated with blunt shaped obstacles, in contrast with the previous linear analyses mentioned in the introduction for non-blunt shapes. Computations are considered in Section 3, followed by theory for large Re values in Section 4.

3. Computational methods

Two separate computational approaches were used on (2.1a–e) and/or (2.2a,b) as described below. They were applied to the blunt non-smooth obstacles formed by a vertical thin flap ($x = 0$ for $0 \leqslant y \leqslant 1$) standing normal to the flat horizontal surface $y = 0$ or by a rectangular block placed on the horizontal surface. It has to be acknowledged immediately that the numerical problem is difficult, due partly to the corners on the obstacles and partly to increasing Re and its accompanying multiple scales, which limits the general accuracy and resolution. For that reason and for the sake of comparison we eventually decided to apply the two distinct computational approaches mentioned. One gave interesting results sooner and this is described first below in some detail, followed by the second.

In one approach the numerical scheme used consists of the alternating-direction implicit (ADI) method and a fourth-order accurate compact difference scheme. For this steady problem the equivalent time coordinate in the differenced Equation (2.2a) for the vorticity is treated as fictitious with each time step considered equivalent to an iteration. In the ADI scheme each time step is split into two halves, resulting in two finite-difference equations each of duration $\delta t/2$ where δt is the time step, used in turn over successive time steps. The first equation is implicit in the x-direction while the second is implicit in the y-direction.

Flows that are dominated by convection as here suffer from numerical instability at high values of the Reynolds number. To suppress such instability upwind differencing in the convection terms could be employed. The artificial dissipation introduced by the upwind difference scheme stabilizes the numerical solution at high values of the Reynolds number. The following quasilinearization is used for the convective terms,

$$\left[gu_x\right]^{n+1} = [g]^n[u_x]^{n+1}. \tag{3.1}$$

The superscript n represents the time step. The spatial derivatives in (3.1) are approximated as

$$\left[gu_x\right]_{i,j} = g_{i,j}\left\{u_{i+2,j} - 2u_{i+1,j} + 9u_{i,j} - 10u_{i-1,j} + 2u_{i-2,j}\right\}/(6\delta x) \tag{3.2a}$$

for $g_{i,j}$ positive, and

$$g_{i,j}\left\{-2u_{i+2,j}+10u_{i+1,j}-9u_{i,j}+2u_{i-1,j}-u_{i-2,j}\right\}/(6\delta x) \tag{3.2b}$$

for $g_{i,j}$ negative. The truncation error of this scheme is given by

$$\frac{1}{4}(\delta x)^3\left[g\frac{\partial^4 u}{\partial x^4}\right]. \tag{3.2c}$$

The upwinding produces a truncation error proportional to the fourth derivative of the vorticity. This numerical dissipation stabilizes computations at high Reynolds numbers and in addition does not significantly affect the physical dissipation. The velocities in the convective terms are calculated using the fourth order accurate Hermitian relations and are given by

$$u_{i,j-1}+4u_{i,j}+u_{i,j+1}=3(\psi_{i,j+1}-\psi_{i,j-1})/\delta y, \tag{3.3a}$$

$$v_{i-1,j}+4v_{i,j}+v_{i+1,j}=-3(\psi_{i+1,j}-\psi_{i-1,j})/\delta x, \tag{3.3b}$$

respectively.

The vorticity boundary condition on the surface is approximated by a second order accurate cubic polynomial approximation and is given by

$$\zeta_{i,0}=-(8\psi_{i,1}-\psi_{i,2})/(2(\delta y)^2). \tag{3.4}$$

The boundary condition on ζ along the obstacle is approximated in a similar manner. It may be noted that no boundary condition could or should be prescribed directly on the vorticity at the outer (top) edge of the flap or on the corners of the square obstacle. Following Dennis and Smith [1], in order to avoid these points, we rotate the axes through an angle $\pi/4$ which leaves the vorticity transport equation (2.2a) unaltered and then we approximate the ζ derivatives as before.

At every fictitious time step the Poisson equation (2.2b) for the stream function is solved by a fourth-order accurate compact difference scheme. This fourth-order method considers as unknowns at each discretized point not only the value of the function $g(i)$ itself but also its first $g'(i)$ and second derivatives $g''(i)$. The system is closed by considering the following relationships between the function and its derivatives at three successive discretization points:

$$g'_{i-1}+4g'_i+g'_{i+1}=3(g_{i+1}-g_{i-1})/h+O(h^4), \tag{3.5a}$$

$$g''_{i-1}+10g''_i+g''_{i+1}=12(g_{i+1}-2g_i+g_{i-1})/h^2+O(h^4), \tag{3.5b}$$

where h represents the spatial step of the discretization. The second-order derivatives $g''(i)$ can be expressed as

$$g''_i=-(g'_{i+1}-g'_{i-1})/(2h)+2(g_{i+1}-2g_i+g_{i-1})/h^2. \tag{3.5c}$$

This expression is fourth-order accurate. In order to reduce the number of unknowns, the second-order derivatives can be eliminated from the governing equations through the relation (3.5c). Besides higher accuracy, another advantage of this method is that it takes into account the boundary conditions on both ψ and its first-order partial derivatives which are data of the physical problem. This method is described in detail by Loc and Bouard [11].

The computational procedure starts by assuming initial values for $\zeta_{i,j}$ and $\psi_{i,j}$ at all mesh points. Using the available values of the stream function, the vorticity transport equation is solved for $\zeta_{i,j}$. Then utilizing the values of $\zeta_{i,j}$, the Poisson equation for the stream function is solved to update $\psi_{i,j}$. At the end of each time step new estimates of $\zeta_{i,j}$ and $\psi_{i,j}$ are obtained by this process. The procedure is repeated as the computational scheme progresses in time until a certain relative error criterion, preselected as a condition for convergence, is satisfied.

As rapid changes in the flow take place around the obstacle, we included more densely packed grid points around the obstacle. Further away from the obstacle grid stretching is applied, defined by

$$x_{i+1} = x_i + \beta(x_i - x_{i-1}), \tag{3.6}$$

where $\beta = \frac{5}{4}$. Similar operations are made for $y(j)$. To check on how much the numerical solution is dependent on the grid size, the grid sizes around the obstacle were made to range between 0·001 × 0·001 and 0·01 × 0·01 at selected Reynolds numbers. The effects of grid sizes on the solution are found to be minimal (see later figure). We found that a grid size 0·005 × 0·005 near the obstacle produces the optimal solution. The fictitious time step was taken as 0·001 originally but was then increased at subsequent times t.

It may be noted that the fourth order accurate compact difference scheme cannot be used where variable grid sizes are considered. Away from the obstacle, where variable grid sizes are used, the Poisson equation for stream function is solved through a central difference scheme along with the successive over relaxation technique.

A second computational approach has also been used on the flow problem. Only preliminary results from this were available at the time of writing and we hope to report fully on this separate approach in the near future. For now we note that this approach is based on Dennis and Hudson's [12], it is a fourth order scheme and it proves to be both accurate and stable: see the comparisons in [12] itself. While the results from the second approach are still to be regarded as tentative so far, and are for the vertical thin flap only, subsequent comparisons in Section 5 with results from the first approach above are found to be affirmative.

Results are described in Section 5.

4. Theory

The theory presented in this section concerns the flow properties at large Re specifically, although without doubt small Re is also of interest. It is predicted, and later found, that for large Re values, and for *any* blunt obstacle, the length of upstream influence is markedly large, as is the upstream separation distance, and the downstream influence length and reattachment distance are even larger.

The obstacle shape is taken as suitably smooth for now. Non-smooth shapes are addressed at the end of this section. Some guidance for the theory comes from Smith [13], Smith and Walton [5], suggesting a beginning on $O(1)$ scaled streamwise lengths.

First, on the actual obstacle $f(x)$ a thin attached boundary layer is produced, with pressure variations of order unity, followed downstream by a thin separated free shear layer. The latter adjusts its position ($F(x)$ say) to make the pressure variation almost negligible inside the accompanying long eddy of recirculating fluid, broadly, over length scales x of order one, and in consequence the height $F(x)$ of the free shear layer tends to a (positive) constant further downstream, so that $F(\infty) > 0$. The corresponding pressure and the stream-function perturbation both tend to $-\frac{1}{2}F^2(\infty)$.

Second, in the outer majority of the flow inviscid forces dominate. So, sufficiently far away, on a long x scale where the disturbance to the surrounding uniform-shear motion is relatively small, $u - y$ and v satisfy Laplace's equation (whereas p satisfies $y\nabla^2 p = 2\partial p/\partial y$). The equation is subject to farfield decay which is to be found and to matching the disturbance of the stream function, $\psi - \frac{1}{2}y^2$, with $-\frac{1}{2}F^2(\infty)$ as y decreases towards zero in the downstream portion of the flow and with zero in the upstream portion. Hence the fundamental decay is algebraic, being given simply by

$$u - y \sim \Gamma r^{-1} \cos\theta, \tag{4.1}$$

for large distances r, with the constant $\Gamma \equiv F^2(\infty)/(2\pi)$ and with the corresponding stream-function disturbance being $\Gamma(\theta - \pi)$. The above result coincides in fact with the source-like effect of the obstacle shape combined with the free-shear-layer shape/positioning, which yields an overall jump in the mass flux over a long streamwise scale, coupled with the Bernoulli relation for the pressure. In particular, the slip induced from (4.1) near the flat surface decays as the inverse of distance.

Third, close to the flat surface in the upstream portion of the flow a viscous wall layer must be provoked. There the velocity u is $y + 0(|x|^{-1})$, from (4.1), but the viscous y scale has to be of order $\mathrm{Re}^{-1/3}|x|^{1/3}$ of course in a uniform shear motion, from the inertial-viscous balance of $y\partial/\partial x$ against $\mathrm{Re}^{-1}\partial^2/\partial y^2$. Therefore we equate $\mathrm{Re}^{-1/3}|x|^{1/3}$ with $|x|^{-1}$, in terms of orders of magnitude, implying the long streamwise length scale $|x| \sim \mathrm{Re}^{1/4}$ for the viscous-inviscid interaction that is necessary for nonlinear upstream influence and in particular the possibility of separation upstream. It follows that in the viscous wall layer the velocity components and pressure are $\mathrm{Re}^{-1/4}U$, $\mathrm{Re}^{-3/4}V$, $\mathrm{Re}^{-1/2}P$ to leading order, with x, y scaled as $\mathrm{Re}^{1/4}X$, $\mathrm{Re}^{-1/4}Y$ in turn, and the controlling equations from (2.1a–c) become the interactive boundary-layer system

$$\frac{\partial U}{\partial X} + \frac{\partial V}{\partial Y} = 0, \qquad U\frac{\partial U}{\partial X} + V\frac{\partial U}{\partial Y} = -P'(X) + \frac{\partial^2 U}{\partial Y^2}, \tag{4.2a,b}$$

for the order-one quantities $U(X, Y)$, $V(X, Y)$, $P(X)$. There is negligible normal pressure gradient because of (2.1c). The boundary conditions require no slip along $Y = 0$ for all negative X values and

$$U \sim Y + \Gamma X^{-1} \,\mathrm{as}\, Y \to \infty, \tag{4.2c}$$

in view of (4.1) holding in the motion outside. Here the constant Γ is defined just after (4.1) and is positive but the imposed displacement in (4.2c) is upward, *i.e.* $U - Y$ is negative, as X is negative, thus indicating an increasing tendency towards flow separation as X increases. The solution of the reduced problem (4.2a–c) starts as a small perturbation from the incident profile $U = Y$ far upstream at large negative X and thereafter can be obtained from a single forward march in X, at least up to separation (flow reversal). Fortunately this reduced problem can be normalized to one solved numerically by Smith and Walton [5], even though the present work is a generalization of theirs to allow for the finite ratio of the length of the obstacle to its width in the present configurations and for the variable eddy height $F(\infty)$ downstream, as well as being a simplification in respect of the source effect (4.1). Separation is found to occur at $X = -0{\cdot}142\,F(\infty)^{3/2}$. Beyond this station a FLARE (after Flügge-Lotz and Reyhner) approximation in which the first term in (4.2b) is neglected wherever U is negative seems to work fairly well as usual numerically, permitting continued forward marching. The flow solution continues towards $X = 0-$, that is, towards order-one length scales around the obstacle, developing there an asymptotic grossly separated form in line with the displacement effect in (4.2c). Thus whereas the outer inviscid response exhibits quite simple decay lasting over great distances, in (4.1), the near-surface viscous response is rather more intricate, its perhaps most significant feature being the predicted behaviour

$$x_1 = -0.142(F(\infty))^{3/2}\mathrm{Re}^{1/4} \tag{4.3}$$

of the upstream separation point at large Reynolds numbers.

Fourth, after the increasingly separated motion beyond the station (4.3) joins to the thinner boundary layer on the obstacle itself, as described previously, this thin boundary layer proceeds downstream through a second smooth separation, on/from the obstacle, to form the thin free shear layer also mentioned previously. In particular if the height of the real obstacle is negligible downstream so that $f(\infty)$ is zero then that of the effective obstacle, $F(\infty)$, is nonzero as the free shear layer emerges almost horizontally atop the long eddy far downstream on the above length scales. Closure of that downstream eddy and a flow-reattachment process to the surface then take place on the much longer length scale where x is of order Re, with the characteristic y scale being of order unity. Accordingly the boundary-layer equations (as in (4.2a,b)) again describe the process there, given that u, v, p are expected to scale with the orders 1, Re^{-1}, 1 in turn, owing to the boundary conditions in (2.1e) (applied as in [14]) and the outer decay in (4.1). The starting profile, however, for the downstream reattachment process is discontinuous, at $y = F(\infty)$, or has discontinuous derivatives, in order to merge with the incident separated-motion properties further upstream. The flow solution in normalized form can be deduced from Smith and Daniels [14], fortunately, and in particular it gives the prediction

$$x_2 = 0{\cdot}076(F(\infty))^3\mathrm{Re} \tag{4.4}$$

for the downstream reattachment point at large Reynolds numbers. This, while subject to certain assumptions about the starting profile noted above, actually confirms the earlier arguments leading to (4.1)–(4.3) on the shorter length scales.

The above points are the major points, and more analytical detail is as in [5],[13] essentially, but the most relevant brief comments here are the following. The argument supposes that the effective height parameter $F(\infty)$ is typically of order unity, whether due to the actual obstacle height $f(\infty)$ being nonzero or to the long eddy shape downstream. So the reasoning extends to non-smooth obstacles also, such as a normal (vertical) thin flap or a rectangular block as in the numerical cases of Sections 3,5, although the theoretical separation from the obstacle is different then and indeed may be multiple. The scales, by the way, in (4.3),(4.4), are such that if Re, x, y are based on the eddy height instead of obstacle height then the factors $F(\infty)$ are replaced by unity. Further, the quite long, algebraic, upstream-influence scale exemplified by (4.3) contrasts with those in Smith [13], Dennis and Smith [1] which are only logarithmic, the distinction stemming from the slow algebraic decay to the state of uniform shear as in (4.1). Finally here, similar reasoning may well apply also in three-dimensional motions under uniform near-surface shear at large Re values.

5. Results and comparisons

The flow past a variety of obstacle shapes has been studied numerically by means of the methodology described in Section 3. The results shown here are all either for a vertical thin

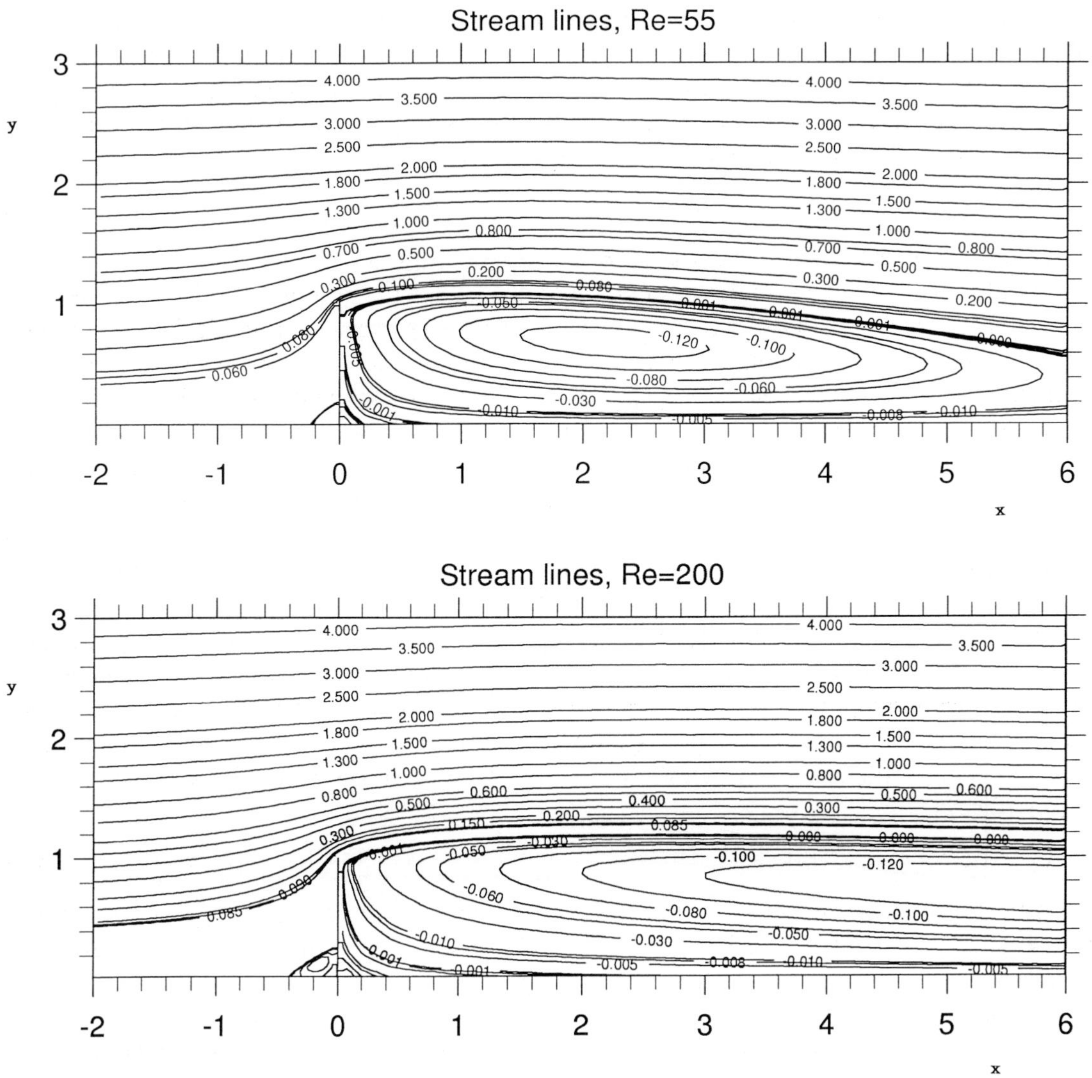

Figure 2. Computational results for the flow streamlines at Re $=$ 55, 200, for (a,b) the vertical-flap shape and (c,d) the square-block shape.

flap of nondimensional height 1 ($= f(0)$) or for a square block of nondimensional height 1, mounted on the horizontal surface.

The main computational results are presented in Figures 2–6, showing the induced pressure gradients, surface shears, streamlines and other properties. Thus Figure 2 gives the streamline plots determined at the particular representative values 55 and 200 of the Reynolds number Re, for the vertical flap and the square block, while Figure 3 gives vorticity curves and velocity vectors, again for two Re values. In Figure 3(a) the main contour values of the vorticity are marked except for the most negative values which are -12, -8, -7, -6, -5, and similarly the most negative vorticity values shown in Figure 3(c) are -10, -5. Figure 4 then shows the pressure gradients produced along the horizontal surface $y = 0$ at Re equal to 100, and Figure 5 presents the computed results at Re of 100 for the horizontal surface vorticity,

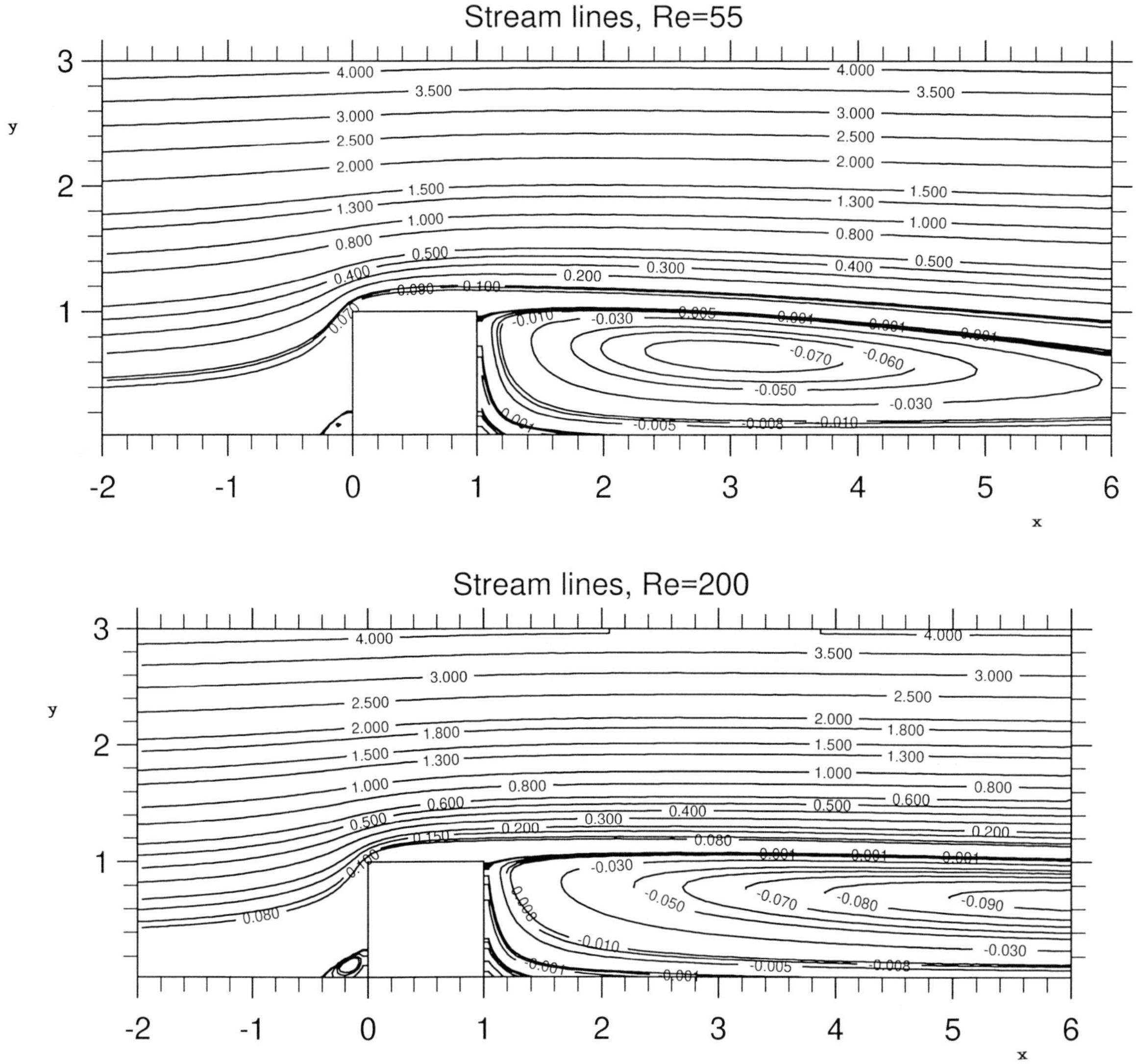

Figure 2. Continued.

including a study of grid-size effects which suggests that the results are of quite high accuracy up to that Reynolds number.

Figure 6(a,b) provides plots of the upstream separation and downstream reattachment positions x_1, x_2, versus Re, which are reasonably sensitive measures of the upstream and downstream eddy features. The results shown are as deduced from the direct computations of the surface vorticity and from the theory summarized by (4.3), (4.4), for both the vertical-flap case and the square-block case. Any slight "wiggles" in the computational results in Figure 6 are due to the interpolation used to estimate the values of x_1, x_2 numerically; while in the theory the quantity $F(\infty)$ is taken mostly as the obstacle height. We should add that most of the results shown are from the first computational method described in detail in section 3, but comparisons between results from both computational approaches used are also presented in Figures 5(a), 6(a), 6(b) for the case of the flap. Given the clear extrapolation effects in Figure 5(a), and the fact that only the results obtained from the smallest grid used so far in

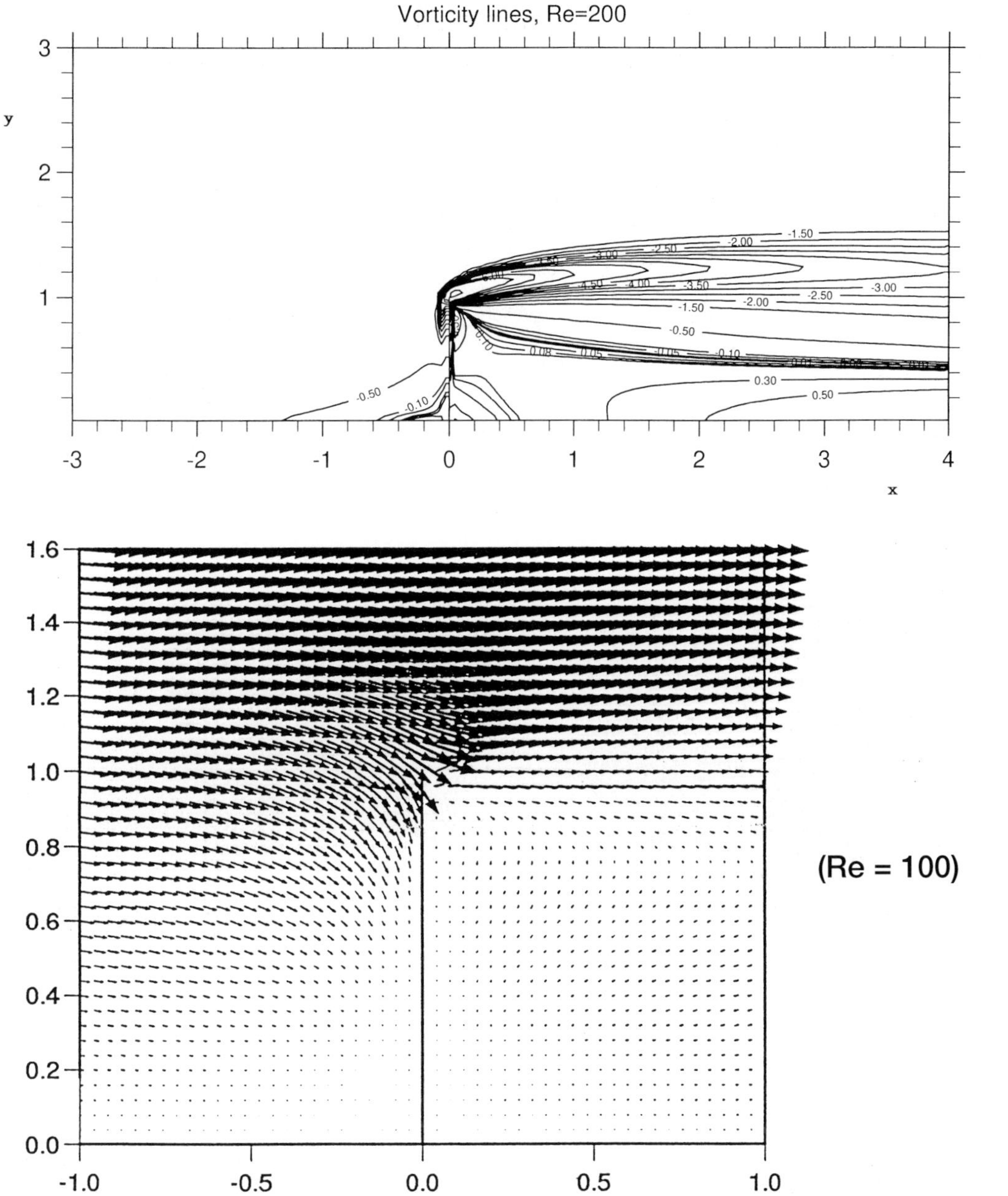

Figure 3. Computed vorticity contours and velocity vectors, at various Re, for (a,b) vertical flap at Re = 200, 100 respectively and (c,d) square block at Re = 200.

the second computational method are plotted in Figures 6(a), 6(b), the sets of results would appear to be consistent and accurate.

The above comparison between the computations and the theory in the determination of x_1, x_2 in Figure 6(a,b) indicates that there is fairly good quantitative agreement on the separation and reattachment positions, from lowish Re values, despite there being only one term, the leading term, in (4.3), (4.4). Earlier comparisons on separation and reattachment points for other shapes addressed experimentally by Klebanoff and Tidstrom [15] and numerically

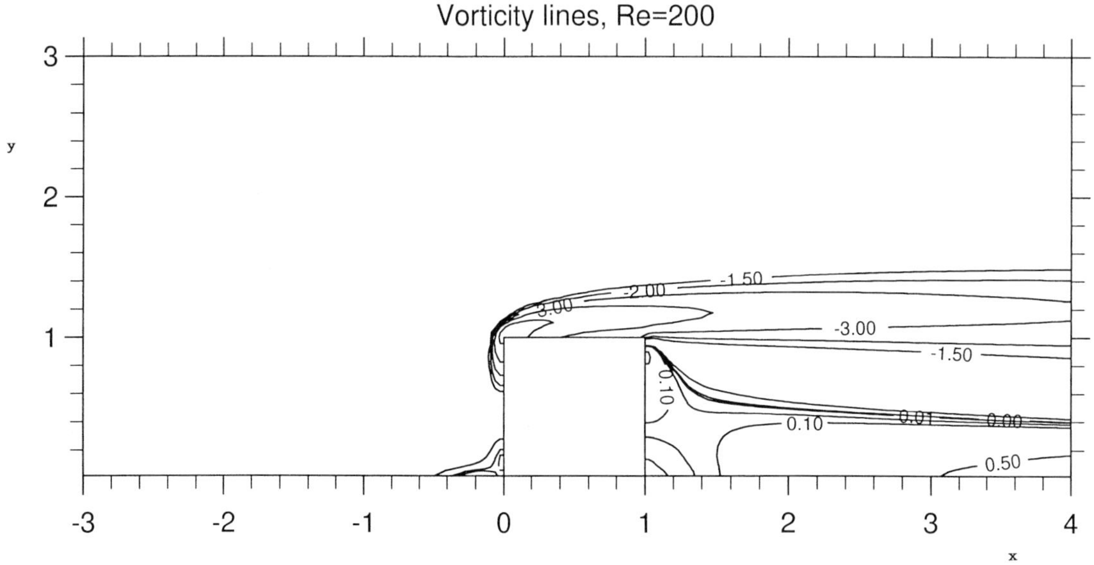

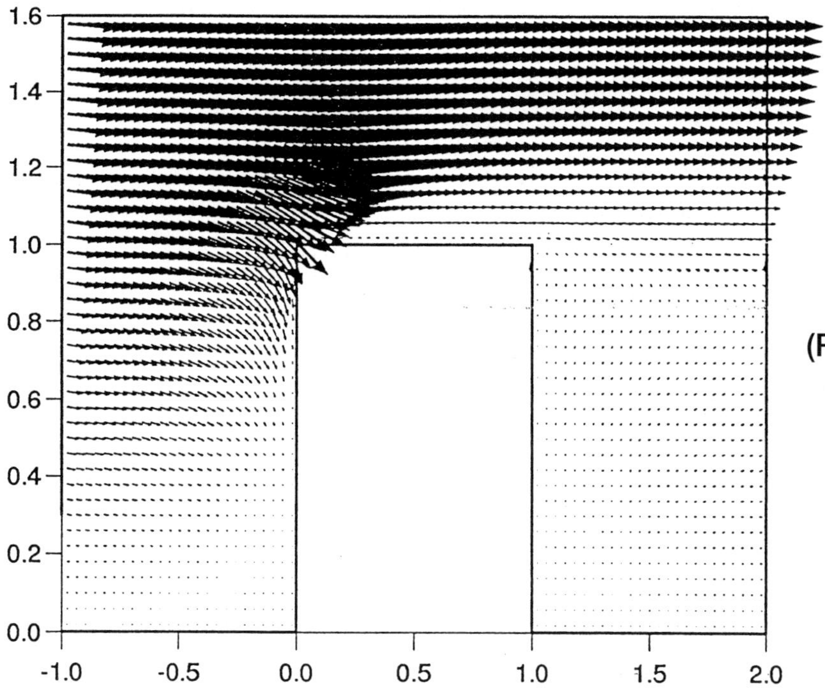

Figure 3. Continued.

by Bogolepov [16] are noted by Smith and Walton [5], Smith [7]. Even more recent computational results by Ngo Boum *et al.* [17] also agree at least qualitatively with the present ones as regards the trend of the downstream reattachment position x_2 with increasing Re, although this last paper puts forward an (unjustified) formula different from the present (justified) one (4.4) and it also overlooks the smaller separation arising upstream, possibly due to lack of numerical resolution.

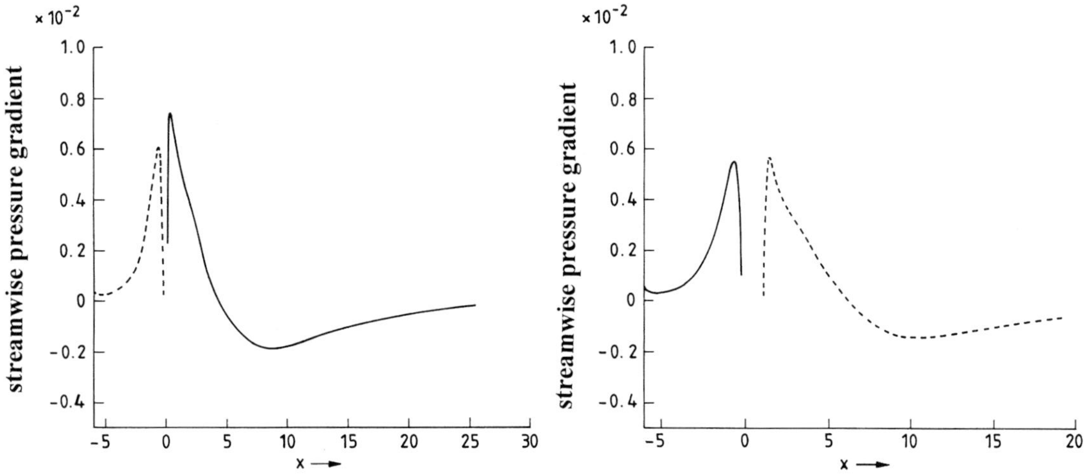

Figure 4. Computed pressure gradients $\partial p/\partial x$ on the horizontal surface $y = 0$ at Re = 100; (a) flap, (b) block.

Returning to the present computations, we note that another point of comparison concerns the position of maximum surface pressure gradient upstream. The theory according to Figure 4 of [5] predicts the above position to be approximately twice as far upstream as the separation position x_1, and the computational results in our Figures 4,5 are reasonably in line with this at the shown Re values. The whole trend of the surface pressure gradient inferred from Figure 4 of [5] is also in line with, although somewhat higher than, the results at Re of 100 in our Figure 4(a,b). There is in addition good qualitative agreement, between the present computations and the present theory, on the overall flow structure, in particular the presence of attached or detached thin layers as Re increases. Indeed, an increase in the value of $F(\infty)$ suggested perhaps by the streamline plots of Figure 2 would bring the theoretical and computational values even closer together; see Figure 6(b). Moreover the behaviour of the upstream separation point x_1 is clearly in response to the pressure feedback influence from the presence of the blunt obstacle, which leads in virtually all cases to a streamwise rise in the horizontal surface pressure ahead of the obstacle, that is, a slight adverse pressure gradient there. Hence on inviscid reasoning the slight, induced, effective slip velocity u_w then falls in the streamwise direction but this effect is on top of the original incident linear velocity profile of (2.1e) and so acts near the wall as an increasing *displacement* response [that is, of the form $y + u_w(x)$ with $u_w(x)$ monotonically decreasing with x]. It is this last response that provokes the upstream separation and hence specifically the result (4.3). Again theory and computation appear to be in agreement on the main physics of the separating flow structure here.

6. Further comments

The flow past a blunt obstacle mounted on an otherwise flat surface is intricate as well as fascinating even in the steady laminar two-dimensional regime. Its accurate determination seems to be aided considerably by the combination of theory and direct computing, the former emphasizing the reduced equations of motion that capture the major physics at large Reynolds numbers whereas the latter address the full equations at reduced Reynolds numbers. The fairly close agreement between the results of the two as investigated in the previous section would

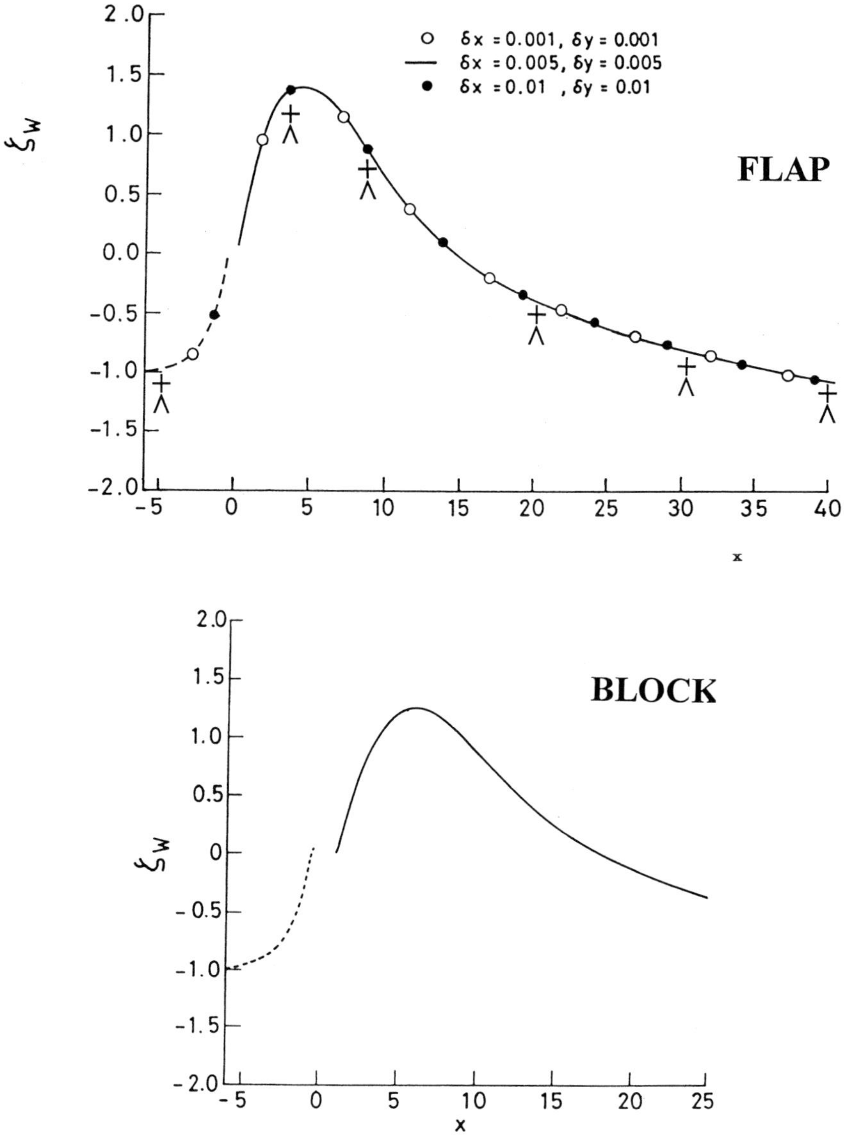

Figure 5. Computational results for the horizontal surface shear ζ at $y = 0$ (for Re $= 100$), and a study on the influences of the grid sizes. (a) Flap. (b) Block. Results from the separate computational method are also shown in (a), for grid spacings 0·05, 0·01 ($\wedge$, $+$, respectively).

appear to enhance the value of both. This is especially so in view of the moderate values of Reynolds number involved in all the comparisons.

Extensions of the work would be interesting on both the direct computational and the theoretical sides. An extension to higher blunt obstacles could allow for the effects of interference from the full velocity profile, acting across the boundary layer, internal flow or other more global motion. Interesting experimental investigations have been made by Giguère, Dumas and Lemay [6] on the Gurney flap and its scaling concerning lift-to-drag ratio of an airfoil. This flap is typically a tiny fence standing normal to the airfoil surface, near the trailing edge, as in their Figure 1. For a particular airfoil they note that Liebeck earlier found increased lift

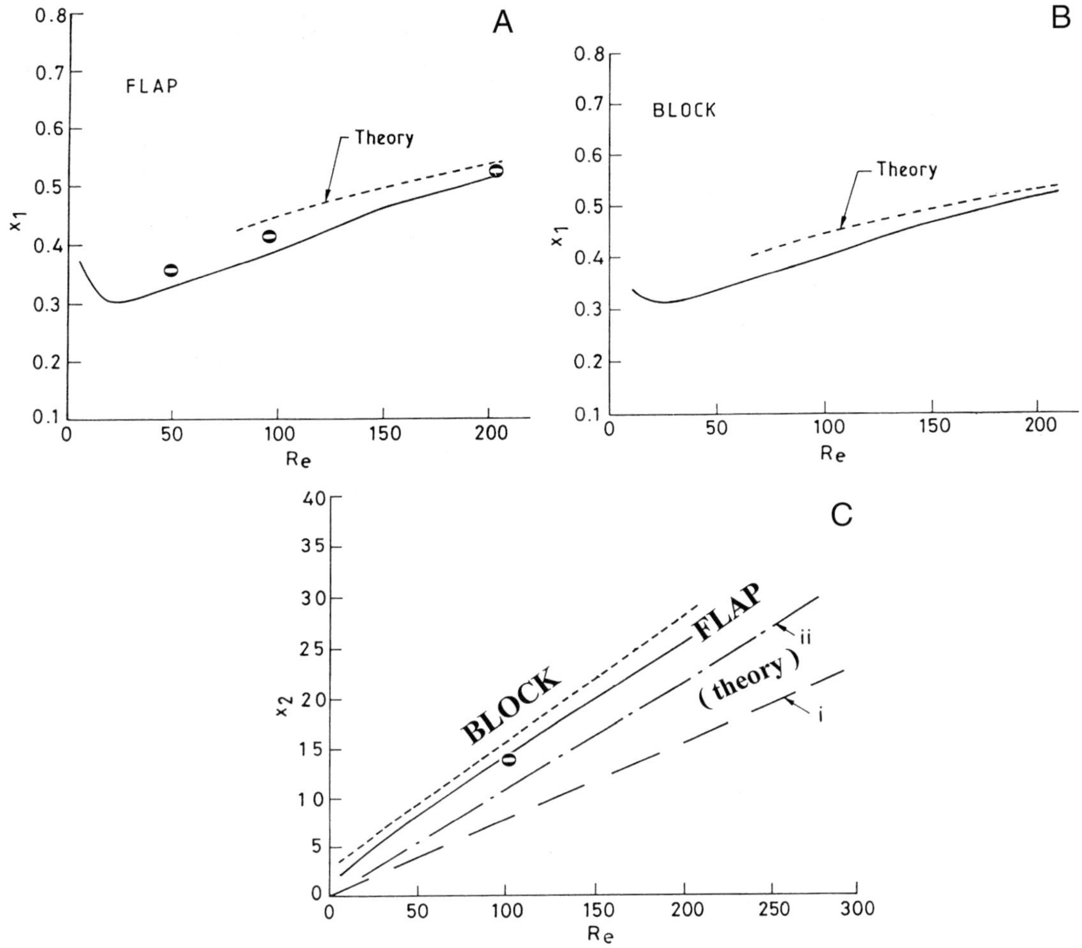

Figure 6. Comparisons, for the vertical flap and the square block. (a, b) Theory and computations for the upstream separation and downstream reattachment positions x_1, x_2 versus Re. (Results from the separate computational method are included as open circles for the flap case, for the smallest grid used.) In (b) curve (i) is from (4.4) with $F(\infty)$ taken as the obstacle height $f(\infty)$, while (ii) has $F(\infty)$ taken as $1{\cdot}115 f(\infty)$.

and reduced drag for high lift coefficients from the addition of a flap of height 1.25% chord, and the benefits of the device were maximized with heights between 1% and 2%. Results in broadly the same vein are given in their Figures 1,2. Although they give the opinion that the physical mechanism associated with this device is still an open question, the present view is that it is the pressure-feedback mechanism of Section 4 extended to higher obstacles exactly as above. Another extension, to unsteady motions, seems desirable in the context of arterial blood flows for example and likewise concerning the possibility of flow transition. Extending the work to three-dimensional flows can also be done in principle, some theoretical studies by Professor S.N. Brown, Mr. N.C. Ovenden and F.T.S. being in progress.

References

1. S.C.R. Dennis and F.T. Smith, Steady flow though a channel with a symmetrical constriction in the form of a step. *Proc. R. Soc. London* A 372 (1980) 393–414.

2. R.W. Mei and A. Plotkin, Navier Stokes solutions for laminar incompressible flows in forward-facing step geometries. *A.I.A.A. J.* 24 (1986) 1106–1111.
3. F. Durst and T. Loy, Investigations of laminar flow in a pipe with sudden contractions of cross sectional area. *Comp. Fluids* 13 (1985) 15–36.
4. D.J. Savin, F.T. Smith and T. Allen, Transition of free disturbances in inflectional flow over an isolated roughness. *Proc. R. Soc. London* A 455 (1999) 491–541.
5. F.T. Smith and A.G. Walton, Flow past a two- or three-dimensional steep-edged roughness. *Proc. R. Soc. London* A 454 (1998) 31–69.
6. P. Giguère, G. Dumes and J. Lemay, Gurney-flap scaling for optimum lift-to-drag ratio. *A.I.A.A. J.* 35 (1997) 1888–1890.
7. F.T. Smith, On physical mechanisms in two- and three-dimensional separations. Theme Issue, *Phil. Trans. R. Soc. London* A (2000) in press.
8. R.W. Johnson (ed.), *The Handbook of Fluid Dynamics*. Boston; CRC Press, Boston (1998).
9. K. Stewartson, On the flow near the trailing edge of a flat plate, II. *Mathematika* 16 (1969) 106–121.
10. F.T. Smith, Laminar flow over a small hump on a flat plate. *J. Fluid Mech.* 57 (1973) 803–824.
11. T.P. Loc and R. Bouard, Numerical solution of the early stage of the unsteady viscous flow around a cylinder: a comparison with experimental visualisation and measurements. *J. Fluid Mech.* 160 (1985) 93–177.
12. S.C.R. Dennis and J.D. Hudson, Compact h^4 finite-difference approximations to operators of Navier-Stokes type. *J. Comp. Phys.* 85 (1989) 390–416.
13. F.T. Smith, The separating flow through a severely constricted symmetric tube. *J. Fluid Mech.* 90 (1979) 725–754.
14. F.T. Smith and P.G. Daniels, Removal of Goldstein's singularity at separation in flow past obstacles in wall layers. *J. Fluid Mech.* 110 (1981) 1–37.
15. P.S. Klebanoff and K.D. Tidstrom, Mechanism by which a two-dimensional roughness element induces boundary layer transition. *Phys. Fluids* 15 (1972) 1173–1188.
16. V.V. Bogolepov, Separated flows near an upward-facing step. Presentation at Euromech Colloquium, Manchester, July 6–9, 1998, and private communication.
17. G.B. Ngo Boum, S. Martemianov and A. Alemany, Computational study of laminar flow and mass transfer around a surface-mounted obstacle. *Int. J. Heat Mass Transfer* 42 (1999) 2849–2861.

Journal of Engineering Mathematics **39:** 63–78, 2001.

Microwave joining of two long hollow tubes: an asymptotic theory and numerical simulations

GREGORY A. KRIEGSMANN and JONATHAN H.C. LUKE
Department of Mathematical Sciences, Center for Applied Mathematics and Statistics, New Jersey Institute of Technology, University Heights, Newark, NJ 07102, U.S.A.

Received 25 November 1999; accepted in revised form 2 May 2000

Abstract. A nonlinear heat equation which models the microwave assisted joining of two large SiC tubes is analyzed. By exploiting the small fineness ratio of the structure and disparate time scales an asymptotic theory for this problem is systematically deduced. Specifically, a one-dimensional nonlinear heat equation is described which governs the temperature in the outer region. This is a numerically well posed problem and it is efficiently solved using standard methods. This solution is not valid in the inner region which includes the microwave source. An inner asymptotic approximation is derived to describe the temperature in this region. This approximation yields two unknown functions which are determined from matching to the outer solution. The results of the asymptotic theory are compared to calculations done on the full problem. Since the full problem is numerically ill conditioned, the asymptotic theory yields enormous savings in computational time and effort.

Key words: microwaves, matched asymptotic methods, ceramics, joining, heat transfer

1. Introduction

The use of microwaves to join ceramics has received a considerable amount of study over the past several years [1–3]. Problems of practical and theoretical interest occur on both the microscopic and macroscopic levels. The microstructure and integrity of the bonding region and the mechanical strength of ceramic joints are important questions that must be resolved on the microscopic level. However, problems on the macroscopic level that include localizing the electric field in the joint region and determining, *a priori*, the temperature distribution at the joint and the surrounding material have a profound influence on the micro-processes that occur during joining. It is this macroscopic description that we address here for the specific application of joining two hollow cylinders of silicon carbide, such as might be used as radiant burner tubes for heat treating of metals.

The mathematical model that we build here focuses only on the heat transfer aspects of the joining process. Specifically we assume that the electric field present in the microwave cavity is completely specified by the lowest mode and is independent of the material being heated. This assumption is a good first order approximation for thin walled tubes in a nominally resonant cavity. The mathematical statement of the problem is then a nonlinear initial value problem for the heat equation with a source that incorporates the modal structure of the electric field. The nonlinearity arises from radiative heat transport on the surface of the hollow cylinders which is important at joining temperatures.

Although the initial boundary value problem is fairly straightforward to solve numerically, even with the nonlinear losses, there are several features of this problem which mitigate against a frontal numerical attack. First, the aspect ratio of the hollow tube is very small. For the radiant tube mentioned above the outer radius is 2in. and its length is 4ft giving an aspect ratio of $\frac{1}{24}$. Thus, the initial boundary value problem must be solved in a long skinny domain which manifests itself as a small parameter that multiplies the axial diffusion term. This makes the problem nearly singular and numerically ill-posed. Secondly, there are three time scales in the problem, two of the same order and the other much smaller. Specifically for the present application the radial diffusion time is much smaller than the axial diffusion time which is of the same order as the convective time. This later time is a measure of how quickly the tube will cool in the absence of the electric source. The disparity of the time scales causes another computational problem. To capture the thermal dynamics on the shortest time scale, *i.e.*, radial diffusion, numerical integration must continue for a very long time until the physical effects on the long time scale become dominant.

We present here an analysis of this problem which is based on the method of matched asymptotic expansions [4] which exploits the small fineness ratio and disparate time scales. In the inner region, within and near the cavity, we derive an expansion which captures the physics of the heating process. This expansion becomes invalid outside the cavity and it contains undetermined functions which depend upon time. We then construct an outer expansion valid away from the cavity. The leading order term satisfies a one-dimensional axial heat equation with a loss term that models radiative and convective heat transfer. The first order correction satisfies a linear version of this equation. In both cases boundary conditions are prescribed at one end of the cylinder, but not at the other which corresponds to the cavity. Matching is then applied to determine these boundary conditions. The first order correction is proved to be identically zero and the leading order equation must be solved numerically. Once this is done, the unknown functions of time in the inner region are uniquely determined.

The asymptotic scheme we present here is closely related to others that have appeared in the literature. Specifically our exploitation of the disparity between the radial diffusion and convective time scales (*i.e.*, a small Biot number) has been done in the context of FZ silicon processing [5]. In fact the matching performed in this paper is similar to ours with the solid-liquid inner region corresponding to our cavity region. Perhaps our method is more closely related to the schemes developed to study thermal-solutal diffusion in float zones [6] and Ohno continuous casting of cored rods [7]. These asymptotic schemes, and ours, exploit both the smallness of the fineness ratio and Biot number. However, our problem is time dependent and these previous works were not. This additional feature is important in ceramic joining where the time history of the temperature plays an important role in determining the microstructure of the joint [3].

Returning to the results of our paper, the one-dimensional nonlinear heat equation that describes the leading order term in the outer expansion is numerically well posed. Thus, an accurate approximation can easily be constructed in this region. This, coupled with the analytic information contained in the inner expansion, yields an efficient and accurate description of the heating process. We also perform a direct numerical attack on this problem for comparison. Because of the singular nature of the equations this direct numerical approach takes much longer to solve than the one-dimensional heat equation, about three orders of magnitude. The results of the direct numerical calculations and those of our asymptotic method agree almost exactly with errors that are in concert with our asymptotic remainders.

In summary the problem we present here is a simplified model of a real physical experiment. Although certain effects are neglected we believe that it does capture many of the important physical features of the real problem. Moreover, it provides an excellent example

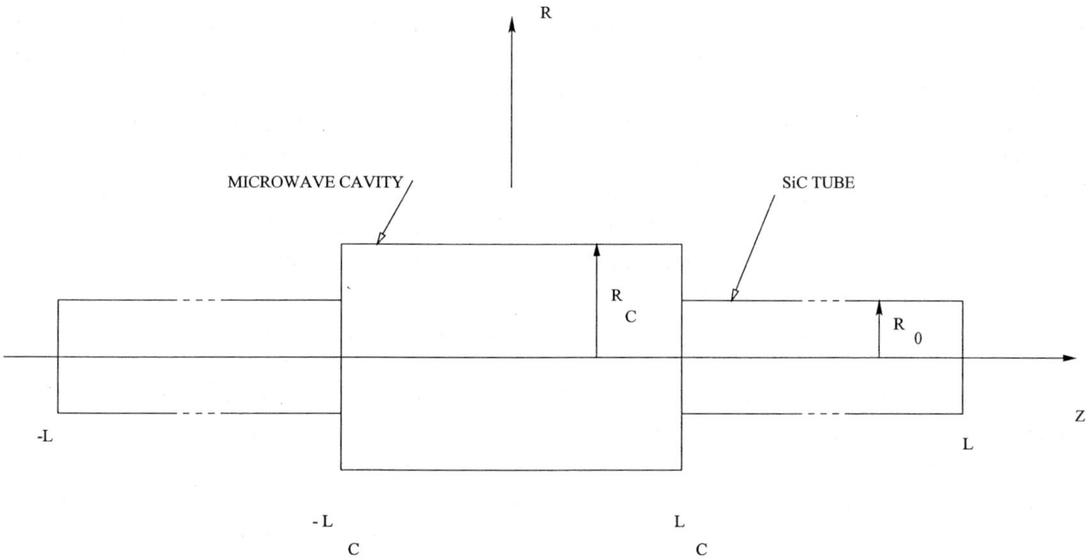

Figure 1. Geometrical sketch.

of how asymptotic analysis can reduce a complicated problem to a simpler one which is amenable to numerical simulation.

2. Problem formulation

Two hollow identical SiC tubes of length L are brought together in perfect thermal contact at $Z = 0$. The inner radius of each tube is R_i, the outer, R_0. A cylindrial cavity of radius $R_C > R_0$, centered at $Z = 0$, encloses the portion of the joined pipes between $\pm L_C$. Microwaves are introduced into the cavity by external circuitry and are contained therein. The setup is shown schematically in Figure 1.

The temperature in the pipes is governed by the heat equation

$$\rho C_P \frac{\partial}{\partial T} = K\nabla^2 T + \frac{\sigma}{2}|\mathbf{E}|^2, \quad |Z| < L, \quad R_i < R < R_0, \tag{1a}$$

$$K\frac{\partial}{\partial R}T = -S(T), \quad R = R_0; \quad S(T) \approx h(T - T_A) + s\epsilon_0(T^4 - T_A^4), \tag{1b}$$

$$K\frac{\partial}{\partial R}T = 0, \quad R = R_i, \tag{1c}$$

$$T = T_A, \quad Z = \pm L, \tag{1d}$$

$$T(X, Y, Z, 0) = T_A, \tag{1e}$$

where ρ is the density, C_P the thermal capacity, K the thermal conductivity, and σ the effective electrical conductivity of the tube, respectively. The parameter h is the effective heat transfer coefficient, s is the Boltzmann constant, ϵ_0 is the emissivity of the tube, T_A is the temperature of the ambient air in which the joined pipes are situated, and $|\mathbf{E}|$ is the amplitude of the electric field.

There are several points that must be made at this junction. First, the thermal parameters and the effective electrical conductivity are taken to be independent of temperature. This is a reasonable assumption for highly conductive materials, such as SiC. However, for other materials this temperature dependence is important and this point will be revisited in Section 6. Secondly, the thermal boundary condition at $R = R_0$ takes into account both convective and radiative heat loss, the later being dominant at the joining temperature T_J. Finally, the thermal boundary condition at $R = R_i$ assumes that neither radiative nor convective losses are important at this surface. This assumption will also be revisited in Section 6 this paper.

To finish the mathematical description of the problem, we must make assumptions regarding the electric field in the cavity. The determination of the field from first principles requires the solution of Maxwell's equations in a complicated cavity composed of the metallic walls, the SiC cylinders, and feeding antennae which couple the microwave power into the cavity. This problem is quite involved and requires a numerical resolution. Once the electric field is known throughout the cavity, then (1) is solved for the temperature distribution. It is the later problem that we focus on in this paper; accordingly we shall simply specify the source term $|\mathbf{E}|^2$ as

$$|\mathbf{E}|^2 = E_0^2 S(\pi Z/2L_C) P(\omega_1 R/R_C), \tag{2a}$$

$$S(\pi Z/2L_C) = \cos^2(\pi Z/2L_C) H(-|Z| + L_C), \quad R_i < R < R_0, \tag{2b}$$

$$P(\omega_1 R/R_C) = J_1^2(\omega_1 R/R_C), \quad |Z| < L_C, \tag{2c}$$

where E_0^2 is a measure of the microwave power, J_1 is the Bessel function of order 1 and ω_1 its first zero, and H is the Heaviside function. The cosine and Bessel function in (2) model the lowest mode in the cavity (TE_{101}) and the Heaviside function crudely models the other aspect of the cavity which confines the electric field to its interior. This neglects the chokes that are appended to the cavity (through which the pipes must pass) and their effect of exponentially attenuating the electric field. Again, this simple choice of $|\mathbf{E}|^2$ allows us to focus on the thermal aspects of the joining problem.

Before closing this section we will rewrite (1–2) in a dimensionless form. This is a necessary first step in both our asymptotic and numerical analyses that will follow. We begin by introducing the dimensionless variables

$$r = R/R_0, \quad z = Z/L, \quad \eta = t/\theta, \quad u = (T - T_A)/T_A \tag{3a}$$

and by defining the dimensionless parameters

$$\begin{gathered}\theta_C = \rho C_P R_0/h, \quad \theta = \rho C_P L^2/K, \quad \alpha = \theta/\theta_C, \quad \theta_R = \rho C_P R_0^2/K,\\ \epsilon = R_0/L, \quad \gamma = \omega_1 R_0/R_C, \quad \chi = \frac{\pi}{2}\frac{R_0}{L_C}, \quad r_1 = \frac{R_i}{R_0},\\ \beta = s\epsilon_0 T_A^3/h, \quad q = [\sigma E_0^2/2][R_0 L/K T_A], \quad L(u) = u + \beta[(u+1)^4 - 1].\end{gathered} \tag{3b}$$

Here θ_C is a time scale associated with convective losses, θ is a time scale associated with axial diffusion and θ_R with radial diffusion. $L(u)$ is the dimensionless power lost by radiation and convection, β is the ratio of radiative to convective power lost at the surface, and ϵ is the fineness ratio of the cylinder. For a typical SiC joining experiment we take $L = 4\text{ft}$, $R_0 = 2\text{in}$, $R_i = 1{\cdot}75\text{in}$, $L_C = 6\text{in}$, $R_C = 3{\cdot}9\text{in}$, $h = 10\text{W/m}^{2\circ}\text{K}$, $K = 100\text{W/m}^\circ\text{K}$, $\epsilon_0 = {\cdot}75$, and $s = 5{\cdot}67 \times 10^{-9}\text{W/m}^{2\circ}\text{K}$. This implies that all the parameters defined above are $O(1)$ except $\epsilon \sim 0{\cdot}04$. In particular the scaling $\alpha = O(1)$ states the physical fact that $\theta_R/\theta_C = 0(\epsilon^2)$, *i.e.*, the radical and axial diffusive time scales are disparate. Finally, we have chosen $R_C = 3{\cdot}9\text{in}$ so that the maximum of J_1^2 occurs at $R = 1{\cdot}875\text{in}$, *i.e.*, at the midpoint of the tube wall.

Using the definitions defined in (3) we find that (1–2) take the dimensionless form

$$\epsilon^2 \frac{\partial}{\partial \eta} u = \frac{\partial^2}{\partial r^2} u + \frac{1}{r}\frac{\partial}{\partial r} u + \epsilon^2 \frac{\partial^2}{\partial z^2} u + q\epsilon S(\chi \frac{z}{\epsilon}) P(\gamma r), \tag{4a}$$

$$\frac{\partial}{\partial r} u + \alpha \epsilon^2 L(u) = 0, \quad r = 1, \quad \frac{\partial}{\partial r} u = 0, \quad r = r_1 \tag{4b}$$

$$u = 0, \quad z = 1, \tag{4c}$$

$$u = 0, \quad \eta = 0. \tag{4d}$$

The appearance of the small parameter ϵ^2 in front of the time derivative and the second order spatial derivative implies that the solution is singular in the limit as $\epsilon \to 0$. This is further suggested by the singular character of S in this limit. It is precisely this characteristic which makes the initial, boundary-value problem (4) ill conditioned for direct numerical simulations.

3. Asymptotic analysis

In this section we will construct an asymptotic approximation of the solution to (4) in the limit as $\epsilon \to 0$ with all the other parameters in the problem fixed and $O(1)$. This will be done in several stages. First, we shall construct an 'inner' approximation which is valid near the cavity where $z \sim O(\epsilon)$ or equivalently where $Z \sim L_C$. In the second stage we will derive an 'outer approximation' which is valid away from the cavity. Finally, we will use matching to determine the various unknown terms in the inner and outer expansions and thus obtain a complete description of the temperature distribution in the cylinders.

We note here that time has been scaled so that (4) describes the evolution of the temperature on the longest time scale where convection and axial diffusion are the dominant heat transfer mechanisms. This occurs when $\eta \sim 0(1)$ or equivalent, $t \sim \theta$. Since the temperature will evolve to a steady state on this time scale, it is the most important for a joining application. However, there is a possibility that features on the shortest time scale, where radial diffusion dominates, will not be captured by (4). This would occur when $t \sim \theta_R$ or equivalently, when $\eta \sim 0(\epsilon^2)$. The inability of (4) to describe the temperature evolution on this short time scale will mathematically manifest itself as an initial time layer of thickness $O(\epsilon^2)$ where the asymptotic approximation deduced from (4) will become invalid. This nonuniformity will be addressed and a remedy will be described in Section 3.4.

3.1. The inner expansion

We begin by defining the inner axial variable $\bar{z} = z/\epsilon$ which we take as $O(1)$. This is equivalent to $Z \sim L_C$. Inserting this change of variables into (4) gives

$$\epsilon^2 \frac{\partial}{\partial \eta} u = \frac{\partial^2}{\partial r^2} u + \frac{1}{r}\frac{\partial}{\partial r} u + \frac{\partial^2}{\partial \bar{z}^2} u + q\epsilon S(\chi \bar{z}) P(\gamma r), \quad r_1 < r < 1, \quad 0 < \bar{z} < \infty, \tag{5a}$$

$$\frac{\partial}{\partial r} u + \alpha \epsilon^2 L(u) = 0, \quad r = 1, \quad \frac{\partial}{\partial r} u = 0, \quad r = r_1, \tag{5b}$$

$$\frac{\partial}{\partial \bar{z}} u = 0, \quad \bar{z} = 0, \tag{5c}$$

$$u = 0, \quad \eta = 0. \tag{5d}$$

We observe that the source function S, the Equation (5a), boundary conditions (5b), and initial condition (5d) are symmetric in $\bar{z}$ so that we need only solve the problem on the half interval $\bar{z} > 0$. This symmetric is reflected in the Neumann boundary condition (5c). We next expand u in the asymptotic series

$$u = \sum_{n=0}^{\infty} u_n(r, \bar{z}, \eta)\epsilon^n. \tag{6}$$

Inserting this expansion into (5), expanding the nonlinear term using a Taylor series expansion, and equating to zero the coefficients of the powers of ϵ yields an infinite set of equations. These sequentially determine $u_n(r, \bar{z}, \eta)$. We shall just list the first three which are sufficient to deduce u to $O(\epsilon^2)$. They are:

$$\bar{\nabla}^2 u_0 = 0, \quad r_1 < r < 1, \quad 0 < \bar{z} < \infty, \tag{7a}$$

$$\frac{\partial}{\partial r} u_0 = 0, \quad r = r_1, 1, \tag{7b}$$

$$\frac{\partial}{\partial \bar{z}} u_0 = 0, \quad \bar{z} = 0; \quad u_0 = 0, \eta = 0, \tag{7c}$$

$$\bar{\nabla}^2 u_1 = -qS(\chi \bar{z}) P(\gamma r), \quad r_1 < r < 1, \quad 0 < \bar{z} < \infty, \tag{8a}$$

$$\frac{\partial}{\partial r} u_1 = 0, \quad r = r_1, 1, \tag{8b}$$

$$\frac{\partial}{\partial \bar{z}} u_1 = 0, \quad \bar{z} = 0; \quad u_1 = 0, \eta = 0, \tag{8c}$$

$$\bar{\nabla}^2 u_2 = \frac{\partial}{\partial \eta} u_0, \quad r_1 < r < 1, \quad 0 < \bar{z} < \infty, \tag{9a}$$

$$\frac{\partial}{\partial r}u_2 = 0, \quad r = r_1, \frac{\partial}{\partial r}u_2 = -\alpha L(u_0), \tag{9b}$$

$$\frac{\partial}{\partial \bar{z}}u_2 = 0, \quad \bar{z} = 0; \quad u_2 = 0, \eta = 0, \tag{9c}$$

where $\bar{\nabla}^2$ is the Laplacian operator in r and $\bar{z}$.

The solution to the boundary value problem (7) is just $u_0 = a_0(\eta)$. This function is unknown at this stage of our analysis. The boundary value problem (8) is more involved. Its solution is given by the expansion

$$u_1 = b_0\psi_0 + \sum_{m=1}^{\infty} b_m(\bar{z}, \eta)\psi_m(r), \tag{10a}$$

where the eigenfunctions ψ_m satisfy

$$\frac{d^2\psi_m}{dr^2} + \frac{1}{r}\frac{d\psi_m}{dr} + \lambda_m\psi_m = 0, \tag{10b}$$

$$\frac{d\psi_m}{dr} = 0, \quad r = r_1, 1. \tag{10c}$$

This is a regular Sturm Liouville problem; accordingly the eigenfunctions ψ_m form an orthonormal set and their corresponding eigenvalues are real and ordered, $\lambda_0 < \lambda_1$..... In fact the lowest eigenvalue $\lambda_0 = 0$ and the corresponding eigenfunction $\psi_0 = \sqrt{\frac{2}{1-r_1^2}}$, *i.e.*, a constant. We find from (10) and (8) that the functions b_m satisfy the ordinary differential equations

$$\frac{d^2 b_m}{d\bar{z}^2} - \lambda_m b_m = -qS(\chi\bar{z})P_m, \tag{11a}$$

$$\frac{db_m}{d\bar{z}} = 0, \quad \bar{z} = 0, \tag{11b}$$

$$P_m = \int_{r_1}^{1} \psi_m P(\gamma r) r dr \equiv \langle \psi_m, P \rangle \tag{11c}$$

The solutions to these simple boundary value problems are

$$b_0 = a_1(\eta) - qP_0 \int_0^{\bar{z}} (\bar{z} - p)S(\chi p)dp, \tag{12a}$$

$$b_m = \gamma_m \cosh\beta_m\bar{z} - \frac{qP_m}{\beta_m}\int_0^{\bar{z}} \sinh\beta_m(\bar{z} - p)S(\chi p)dp, \tag{12b}$$

where $\beta_m = \sqrt{\lambda_m}, \quad \bar{z}_c = Z_C/R_0$, and

$$\gamma_m = \frac{qP_m}{\beta_m}\int_0^{\bar{z}_C} S(\chi p)e^{-\beta_m p}dp. \tag{12c}$$

The constant γ_m is chosen so that b_m decays exponentially as $\bar{z} \to \infty$, a condition that is later required for matching, and the function $a_1(\eta)$ is unknown at this stage of the analysis.

The solution to the boundary value problem (9) is determined along similar lines. Omitting the details of the analysis we simply state the result

$$u_2 = ar + \frac{b}{2}r^2 + [a_2(\eta) + c_0]\psi_0 + \sum_{m=1}^{\infty} c_m(\bar{z}, \eta)\psi_m(r), \tag{13a}$$

$$a = \frac{\alpha r_1}{1 - r_1} L(u_0), \quad b = \frac{-\alpha}{1 - r_1} L(u_0), \tag{13b}$$

$$c_0\psi_0 = \frac{\bar{z}^2}{2}\Delta_0; \quad \Delta_0 = \frac{\mathrm{d}u_0}{\mathrm{d}\eta} - 2b - a\psi_0^2(1 - r_1), \tag{13c}$$

$$c_m = \frac{\Delta_m}{\lambda_m}; \quad \Delta_m = -a\int_{r_1}^{1} \psi_m \mathrm{d}r, \quad m \geq 1, \tag{13d}$$

where $a_2(\eta)$ is unknown at this stage of the analysis.

Combining (10a) and 13a) with (6) gives the inner expansion

$$\begin{aligned} u = \; & a_0(\eta) + \epsilon\left[\sum_{m=0}^{\infty} b_m\psi_m\right] \\ & + \epsilon^2\left[ar + \frac{b}{2}r^2 + a_2(\eta)\psi_0 + c_0\psi_0 + \sum_{m=1}^{\infty} c_m(\bar{z}, \eta)\psi_m(r)\right] + O(\epsilon^3), \end{aligned} \tag{14}$$

where $O(\epsilon^3)$ represents the contributions of the higher order terms. We shall need the far field expansion, *i.e.*, $\bar{z} >> 1$, of this expansion shortly. It is easily deduced to be

$$\begin{aligned} u = \; & a_0(\eta) + \epsilon[a_1(\eta) + qP_0(d_1 - d_0\bar{z})]\psi_0 \\ & + \epsilon^2\left[ar + \frac{b}{2}r^2 + a_2(\eta)\psi_0 + \frac{\bar{z}^2}{2}\Delta_0 + \sum_{m=1}^{\infty}\frac{\Delta_m}{\lambda_m}\psi_m\right], \end{aligned} \tag{15a}$$

$$\mathrm{d}_0 = \int_0^{\bar{z}_c} S(\chi P)\mathrm{d}p, \quad \mathrm{d}_1 = \int_0^{\bar{z}_c} pS(\chi P)\mathrm{d}p, \tag{15b}$$

where exponentially small terms have been neglected. The constant parameter $qP_0\mathrm{d}_0$ is related to the total power supplied by the microwave source and $qP_0\mathrm{d}_1$ to its first moment. It is clear from the linear and quadratic terms in this expression that (14) is not a uniform asymptotic expansion. Another representation must be found and this is the subject of the next section.

3.2. The Outer Expansion

We now seek an asymptotic expansion of the original problem (4) when $z = 0(1)$. The source term in (4a) is zero when z is restricted in this faction. We again expand u in an asymptotic expansion

$$u = \sum_{m=0}^{\infty} w_m(r, z, \eta)\epsilon^m, \tag{16}$$

where the coefficients w_m now depend on the outer variable z. Inserting this expansion into (4), expanding the nonlinear term using a Taylor series expansion, and equating to zero the coefficients of the powers of ϵ yields an infinite set of equations. These sequentially determine $w_m(r, z, \eta)$. We shall just list the first four which are sufficient to deduce u to $O(\epsilon^2)$. They are:

$$\frac{\partial^2}{\partial r^2}w_j + \frac{1}{r}\frac{\partial}{\partial r}w_j = 0, \quad r_1 < r < 1, \quad 0 < z < 1, \tag{17a}$$

$$\frac{\partial}{\partial r}w_j = 0, \quad r = r_1, 1, \tag{17b}$$

$$w_j = 0, \quad z = 1; \quad w_j = 0, \quad \eta = 0 \tag{17c}$$

for $j = 0, 1$,

$$\frac{\partial^2}{\partial r^2}w_2 + \frac{1}{r}\frac{\partial}{\partial r}w_2 = \frac{\partial}{\partial \eta}w_0 - \frac{\partial^2}{\partial z^2}w_0, \tag{18a}$$

$$\frac{\partial}{\partial r}w_2 = 0, \quad r = r_1; \quad \frac{\partial}{\partial r}w_2 = -\alpha L(w_0), \quad r = 1, \tag{18b}$$

$$w_2 = 0, \quad z = 1; \quad w_2 = 0, \quad \eta = 0, \tag{18c}$$

$$\frac{\partial^2}{\partial r^2}w_3 + \frac{1}{r}\frac{\partial}{\partial r}w_3 = \frac{\partial}{\partial \eta}w_1 - \frac{\partial^2}{\partial z^2}w_1, \tag{19a}$$

$$\frac{\partial}{\partial r}w_3 = 0, \quad r = r_1; \quad \frac{\partial}{\partial r}w_3 = -\alpha \dot{L}(w_0)w_1, \quad r = 1, \tag{19b}$$

$$w_3 = 0, \quad z = 1; \quad w_3 = 0, \quad \eta = 0, \tag{19c}$$

where $\dot{L}$ is the derivative of L with respect to its argument.

The solutions of (17) are $w_0(z, \eta)$ and $w_1(z, \eta)$ for $j = 1$ and $j = 2$ respectively. Both of these functions are unknown at this stage of our analysis. Multiplying (18a) by r, integrating the result from $r = r_1$ to $r = 1$, and applying the boundary conditions (18b) we find that

$$\frac{\partial}{\partial \eta}w_0 = \frac{\partial^2}{\partial z^2}w_0 - \alpha\psi_0^2 L(w_0). \tag{20a}$$

Using this information we find the solution of (18) is

$$w_2 = f_2(z, \eta) - \frac{\alpha}{2}\psi_0^2 L(w_0)\left[\frac{r^2}{2} - r_1^2\log(r)\right], \tag{20b}$$

where f_2 is unknown at this stage. Similar considerations of (19) give the result

$$\frac{\partial}{\partial \eta}w_1 = \frac{\partial^2}{\partial z^2}w_1 - \alpha\psi_0^2 \dot{L}(w_0)w_1. \tag{21}$$

Summarizing the results to this point our outer expansion is given by

$$u = w_0(z,\eta) + \epsilon w_1(z,\eta) + \epsilon^2\left\{f_2(z,\eta) - \frac{\alpha}{2}\psi_0^2 L(w_0)\left[\frac{r^2}{2} - r_1^2\log(r)\right]\right\} + O(\epsilon^3), \quad (22)$$

where w_0 and w_1 satisfy Equations (20a) and (21), respectively.

3.3. Matching

At this stage of our analysis there are several undetermined quantities which are necessary for the description of the temperature in both the inner and outer regions. In the inner region the functions $a_0(\eta)$ and $a_1(\eta)$ are unknown. These are required to obtain a local approximation to $O(\epsilon)$ there. In the outer region we know that w_0 and w_1 satisfy the one-dimensional heat equations (20a) and (21), respectively. Both of these functions satisfy (17c), *i.e.*, they are zero initially and both satisfy a homogenous Dirichlet condition at $z = 1$. To complete the boundary value problems that each satisfies we need to know their behavior at $z = 0$.

The determination of these unknown quantities is obtained by matching. According to this procedure we expand the inner solution for $\bar{z} >> 1$ and then rewrite the result in terms of the outer variable z. Then we expand the outer solution as $z \to 0$ and compare the two results. The principal of Matched Asymptotics [4] states that these two results must be the same. This equality then determines the unknown quantities.

From (15) we find that the inner expansion for $\bar{z} >> 1$ can be written as

$$\begin{aligned} u = {} & a_0(\eta) - qP_0 d_0\psi_0 z + \frac{\Delta_0}{2}z^2 + \epsilon[a_1(\eta) + qP_0 d_1]\psi_0 + \\ & + \epsilon^2\left[ar + \frac{1}{2}br^2 + a_2(\eta)\psi_0 + \sum_{m=1}^{\infty}\frac{\Delta_m}{\lambda_m}\psi_m\right] + O(\epsilon^3). \end{aligned} \quad (23a)$$

Now the outer result (22) must be expanded as $z \to 0$. Using the Taylor series expansions of both w_0 and w_1 we obtain

$$\begin{aligned} u = {} & w_0(0,\eta) + z\frac{\partial}{\partial z}w_0(0,\eta) + \frac{z^2}{2}\frac{\partial^2}{\partial z^2}w_0(0,\eta) + \cdots + \\ & + \epsilon\left[w_1(0,\eta) + z\frac{\partial}{\partial z}w_1(0,\eta) + \cdots\right] + O(\epsilon^2). \end{aligned} \quad (23b)$$

Comparing these two expansions we find that

$$w_0(0,\eta) = a_0(\eta), \qquad \frac{\partial}{\partial z}w_0(0,\eta) = -qd_0P_0\psi_0, \qquad \frac{\partial^2}{\partial z^2}w_0 = \psi_0\Delta_0, \quad (24a,b,c)$$

$$w_1(0,\eta) = [qd_1P_0 + a_1(\eta)]\psi_0, \qquad \frac{\partial}{\partial z}w_1(0,\eta) = 0. \quad (24d,e)$$

Now when (24b) is used in conjunction with (17c) and (20a) we obtain a well posed initial boundary value problem for w_0. Similarly, when (24e) is used along with (17c) and (21) we obtain a well posed initial boundary value problem for w_1. In fact the solution of this problem is $w_1 = 0$. This follows because the initial condition and boundary conditions for w_1 are zero and there is no source term in the equation. The matching condition (24d) then yields

$$a_1 = -qd_1P_0. \quad (25)$$

Now the initial boundary values problem for w_0 must be numerically solved in general. Once this is done $w_0(0, \eta)$ is determined and $a_0(\eta)$ is obtained from (24a). Inserting this information and (25) into (14) determines the temperature in the inner region to $O(\epsilon)$.

The last thing to check is the condition (24c). By using (20a) evaluated at $z = 0$ and replacing w_0 by u_0 there we find, using the definition of Δ_0 in (13b), that (24c) is an identity. We can push this analysis further and match higher order terms, but we do not do so here.

3.4. THE ASYMPTOTIC MODEL

We briefly restate here the results of our asymptotic analyses for reference and later comparison. Since w_1 is identically zero the temperature in the outer region is given by

$$u = w_0(z, \eta) + O(\epsilon^2), \tag{26}$$

where the $O(\epsilon^2)$ represents the error at this level of approximation. The leading order term satisfies the initial boundary value problem

$$\frac{\partial}{\partial \eta} w_0 = \frac{\partial^2}{\partial z^2} w_0 - \alpha \psi_0^2 L(w_0), \quad 0 < z < 1, \quad \eta > 0, \tag{27a}$$

$$\frac{\partial}{\partial z} w_0 = -q P_0 \mathrm{d}_0 \psi_0, \quad z = 0; \quad w_0 = 0, \quad z = 1, \tag{27b}$$

$$w_0 = 0, \quad \eta = 0, \tag{27c}$$

which must be resolved numerically. We note here that the term driving w_0 is the nonzero Neumann condition (27b) where the right hand side of this boundary condition is the projection of the total microwave power onto the eigenfunction ψ_0. Equation (27) is a one-dimensional heat equation which is clearly easier to solve numerically than the original initial boundary value problem (5). The gain from our asymptotic analysis is not just a reduction in dimension; the original problem is numerically ill-conditioned when $\epsilon << 1$ and (27) is not.

Once w_0 is numerically determined its value at the origin is known. Using this information and (25) we find from (14) that the temperature in the inner region is given by

$$\begin{aligned} u = \; & w_0(0, \eta) - \epsilon q P_0 \left[\int_0^{\bar{z}_c} p S(\chi p) \mathrm{d}p + \int_0^{\bar{z}} (\bar{z} - p) S(\chi p) \mathrm{d}_p \right] \psi_0 + \\ & + \epsilon \sum_{m=1}^{\infty} \frac{q P_m}{\beta_m} \psi_m \left[\cosh \beta_m \bar{z} \int_0^{\bar{z}_c} S(\chi p) \mathrm{e}^{-\beta_m p} \mathrm{d}p - \right. \\ & \left. \int^{\bar{z}_0} \sinh \beta_m (\bar{z} - p) S(\chi p) \mathrm{d}p \right] + \epsilon^2 u_2 + O(\epsilon^3), \end{aligned} \tag{28}$$

where u_2 is given by (13). Thus the determination of the temperature is completely described to $O(\epsilon)$ on the entire physical interval $0 < Z < L$ when (27) is solved numerically.

We now revisit a point that was made in the opening of this section. There we speculated that there might be a boundary layer ($\eta \sim O(\epsilon^2)$) where the inner and outer expansions are not valid. Clearly the outer expansion (26) satisfies the initial condition at $\eta = 0$ by design through (27c). However, the inner expansion only vanishes to $O(\epsilon)$.

We shall briefly describe the method which reconciles this discrepancy and omit the details for brevity. First, we introduce a new short time variable $\bar{t} = \eta/\epsilon^2 \equiv t/\theta_R$ into (4) and seek

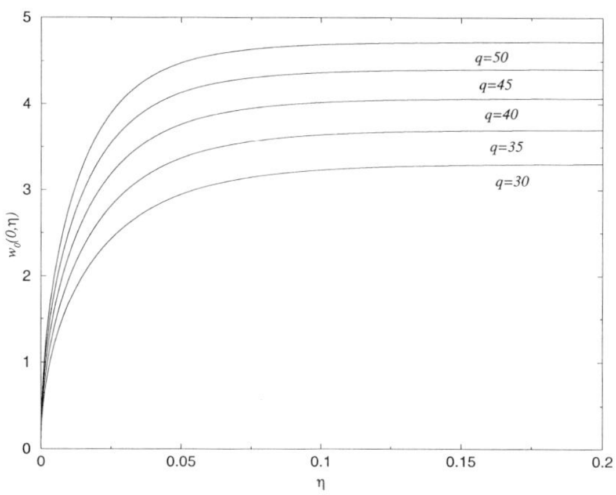

Figure 2. Temperature evolution at the joint. Leading-order asymptotic estimate: $w_0(0, \eta)$.

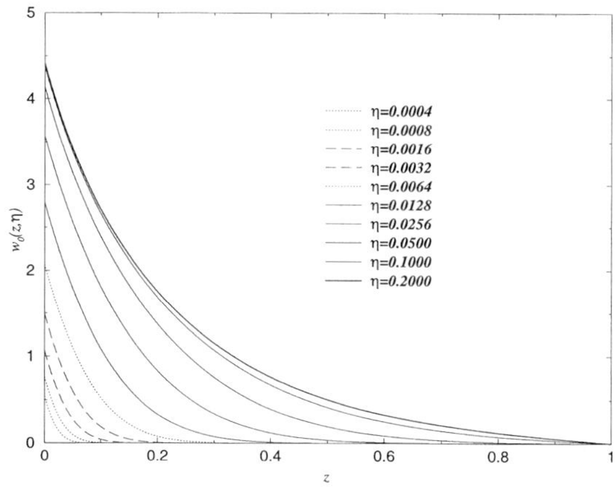

Figure 3. Temperature distribution at various times. Leading-order asymptotic estimate: $w_0(z, \eta)$ with $q = 45$.

another asymptotic expansion of the temperature. This would proceed in exactly the same way as the development of the inner expansion in Section 3.1, except for two differences. First, the expansion would begin with an order ϵ term since this is the size of the source term. More importantly, the sequential boundary value problems are all parabolic and they describe diffusion in the radial and axial directions on the short time scale. Then by expanding this new asymptotic expansion for a fixed $\bar{t}$ and $\bar{z} \to \infty$ we can show that $u \to 0$. This result then matches, as $\bar{t} \to \infty$, into the outer expansion (26) as $\eta \to 0$. Thus, the homogeneous initial condition (27c) is valid. We can also show that this new expansion matches, as $\bar{t} \to \infty$ (28) as $\eta \to 0$. Thus, the initial conditions are transformed by the radial and axial diffusion on the short time scale into the proper function on the long time scale. Finally, we can deduce from this matching that $a_0(\eta) = w_0(0, \eta) \sim 2P_0 d_0 \sqrt{\eta/\pi}$ as $\eta \to 0$, and this is observed in our numerical experiments described below.

4. Numerical example

Using the parameter values stated in Section 2 we find that $\epsilon = 0{\cdot}04$, $\gamma = 1{\cdot}95$, $\chi = 0{\cdot}52$, $\alpha = 1{\cdot}74$, $\beta = 0{\cdot}01$, $r_1 = 0{\cdot}875$, $\psi_0 = 2{\cdot}92$, $\bar{z}_c = 3$, $\theta_C = 3\ hrs$, and $\theta_R = 7$ mins. From the definition of d_0 and d_1, equation (15b), straight forward integration gives $d_0 = \bar{z}/2 = 1{\cdot}5$, and $d_1 = (\bar{z}_c^2 - 1/\chi^2)/4 = 1{\cdot}33$. The constant $P_0 = \int_{r_1}^1 r\psi_0 J_1^2(\gamma r)dr$ must be solved numerically, its value is $P_0 = 0{\cdot}115$. The only free parameter at this stage is q which from (3b) is seen to be proportional to the microwave power; the flux in (27b) is $-0.5q$.

We have used a finite-difference scheme based on Crank-Nicolson differencing to solve (27) numerically. The nonlinear loss term, $L(w_0)$, is handled by linearizing locally at each time step. We have taken very fine numerical grids to assure that the solution is fully resolved. Figure 2 shows a time history for $w_0(0, \eta)$ for $q = 30, 35, 40, 45, 50$. From (28) this is the lowest order behavior of the inner solution. As expected, the temperature at the origin behaves like $\sqrt{\eta}$ and rises rapidly for small values of η, indeed at an infinite rate initially, due to heat flux imposed at the origin. As convective and radiative losses balance this flux,

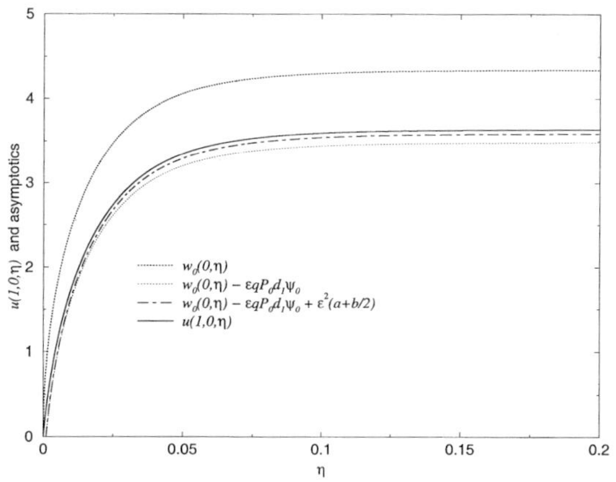

Figure 4. Comparison of $u(1, 0, \eta)$ and asymptotics.

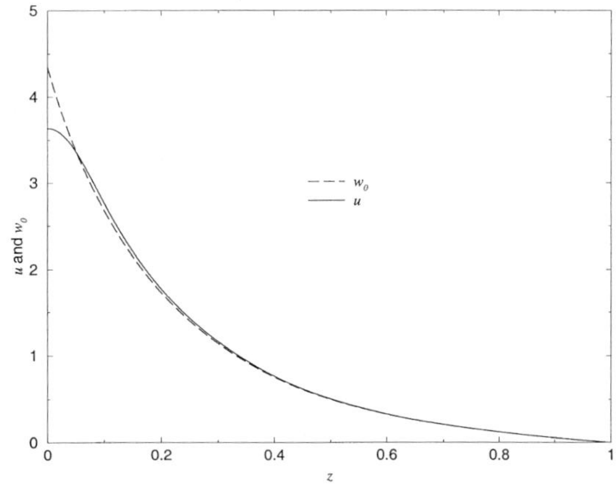

Figure 5. Steady-state surface temperature. Leading-order asymptotics and full simultation.

the temperature at the origin saturates at a steady value which is a monotonically increasing function of q. Figure 3 shows $w_0(z, \eta)$ for various values of η and $q = 45$. Clearly, as η becomes large w_0 goes to a steady state with its maximum temperature at the origin. We note that the steady steate temperature is established more rapidly at $z = 0$ than at points with $z > 0$.

An important question that can be answered at this point is to estimate the power required to obtain a prescribed temperature at the joint, $Z = 0$. For example if the joining temperature is $T_J = 1100\,°\mathrm{C}$, then this corresponds to a dimensionless temperature of $u_J = 3{\cdot}666$ for $T_A = 30\,°\mathrm{C}$. From Figure 2 we see that selecting $q \approx 35$ gives a temperature of about u_J. However, using the higher-order corrections to w_0 given in (28) and (14) can refine this estimate considerably. In applying these corrections, we use the result that the P_m/λ_m's for $m \geq 1$ and $a_2(\eta)$ are negligibly small in our case. Applying the $O(\epsilon)$ correction gives $q \approx 46$; with the $O(\epsilon^2)$ correction, we estimate $q_J = 45$. This result compares favorably with the value of $q_J = 44$ estimated from the direct simulations of (4) discussed in the next section. In dimensional terms the value of $q = 44$ corresponds to about 1040 W.

5. A full numerical simulation

In this section we briefly describe numerical experiments on the full system (4). We treat (4) using an alternating-directions splitting. Diffusion in the r and z directions is treated with Crank-Nicolson differencing. Nonlinear terms are linearized locally. Because of the very large diffusion coefficient in the radial direction extremely small time steps are required. The computational resources, measured in total number of arithmetic operations, required to accurately solve the full system (4) exceed those needed to obtain accurate solutions of the asymptotic model (27) by at least a factor of a thousand. We take $q = 44$ in the following.

In Figure 4 we compare the time histories of the surface temperature at the origin for the full scale simulation and asymptotic analysis. The surface temperature at the origin for the simulation based on (4) is denoted $u(1, 0, \eta)$. The leading order asymptotics $w_0(0, \eta)$ are well above $u(1, 0, \eta)$ for $\eta > 0$. The relative error in the prediction of the steady state temperature is about 20%. The first order correction $w(0, \eta) - \epsilon q P_0 d_1 \psi_0$ lies below and much closer to $u(1, 0, \eta)$. We note that since $w(0, \eta) = 0$ the first order correction gives a negative temperature for very small values of η. Furthermore, we note that the first order correction is insensitive to the nonlinearity in the problem which is measured by β. The relative error in the prediction of the steady state temperature is about 4%. The nonlinearity affects the $O(\epsilon^2)$ correction which gives very good agreement with $u(1, 0, \eta)$. The relative error in the prediction of the steady state temperature is about 2%.

In Figure 5 we show the steady state surface temperature obtained from numerical solution of the full system (4) as the solid curve and the steady state temperature distribution from numerical solution of the leading order asymptotic system (27) as the dotted curve. We judge the solutions to have effectively reached steady state when $\eta = 0{\cdot}2$. The two are almost identical for $z \geq 3$ and begin to differ when z is reduced. The difference is at most 2% for $z_c = 1.5 \leq z < 3$, *i.e.*, up to the cavity and the inner region. This excellent agreement between the outer expansion and the numerical simulation there is almost unexpected. Moreover, the outer expansion continues to give an excellent approximation well inside the cavity until the two curves intersect. At this point the inner expansion really must be employed. Because the radiative loss is important here we need the inner expansion to $O(\epsilon^2)$. This is a very complex expression which we will not evaluate at this time for arbitrary $\bar{z}$. The value of this expansion at $\bar{z} = 0$ has been described in the previous section and Figure 4 demonstrates its effectiveness at this point.

6. Conclusion

We have analyzed a nonlinear heat equation which models the microwave assisted joining two large SiC tubes. By exploiting the small fineness ratio of the structure and the disparate time scales θ_R and θ_C we deduced an asymptotic theory for this problem. Specifically, we found that the temperature in the outer region was governed by a one-dimensional nonlinear heat equation. This is a numerically well posed problem and we efficiently solve it using a standard method. This solution is not valid in the inner region which includes the microwave source. We have derived an asymptotic approximation to the temperature in this region. This approximation yielded two unknown functions which were determined from matching and were obtained from the outer solution. We have compared the results of our asymptotic theory to calculations done on the full problem. Since the full problem is numerically ill conditioned, our asymptotic theory yields enormous savings in time and computational effort.

We close this section by making two observations. First, all of our analysis can be generalized to handle the case of temperature dependent thermal and electrical conductivities. If we set $k(u) = K(T)/K(T_A)$ and $f(u) = \sigma(T)/\sigma(T_A)$, then the initial boundary value problem (27) is replaced by

$$\frac{\partial}{\partial \eta} w_0 = \frac{\partial}{\partial z}\left(k(w_0)\frac{\partial}{\partial z} w_0\right) - \alpha\psi_0^2 L(w_0), \quad 0 < z < 1, \quad \eta > 0, \tag{29a}$$

$$k(w_0)\frac{\partial}{\partial z} w_0 = -q f(w_0) P_0 d_0 \psi_0, \quad z = 0; \quad w_0 = 0, \quad z = 1, \tag{29b}$$

$$w_0 = 0, \quad \eta = 0. \tag{29c}$$

The inner result (28) is modified in a similar fashion. Thus, we again obtain a one-dimensional problem. All of the remarks made above apply to this case. The second observation is that the homogeneous boundary $(\partial/\partial_R)T = 0$ at $R = R_i$ is a very good approximation. In Appendix A we show, under the assumptions that the pipe is a black body and that no air is forced through the inside of the tube, that the asymptotic results presented in this paper are valid to at least $O(\epsilon^2)$.

Acknowledgements

This work was supported by the Air force Office of Scientific Research Under Grant No. AFOSR F49620-94-1-0338, the Department of Energy Under Grant No. DE-FG02-94ER25196, and the National Science Foundation Under Grant No. DMS 9305828. The authors wish to thank Dr. Richard Silberglitt for his encouragement to undertake this work, and for his many helpful and insightful suggestions in its completion.

Appendix A

The dimensionless radiative thermal balance is

$$\frac{\partial}{\partial r}u + \alpha\beta\epsilon^2\left[(u(r,z,\eta)+1)^4 - \int_{-1}^{1}(u(r,z',\eta)+1)^4\frac{1}{\epsilon}K\left(\frac{z-z'}{\epsilon}\mathrm{d}\right)z'\right] = 0, \tag{A1a}$$

$$r = r_1$$

where the Kernel is an even function of its argument and satisfies

$$\int_{-1}^{1}\frac{1}{\epsilon}K\left(\frac{z-z'}{\epsilon}\right)\mathrm{d}z' = 1 \tag{A1b}$$

for all ϵ. The kernel is an involved trigonometric integral obtained from a view angle analysis [8] and we do not produce it here. The integral in (A1a) represents the incident heat flux coming from all the interior points of the surface $r = r_1$. If we rewrite (A1a) in terms of the inner variable $\bar{z}$, then we find

$$dis\frac{\partial}{\partial r}u + \alpha\beta\epsilon^2\left[(u+1)^4 - \int_{-\infty}^{\infty}(u+1)^4K(\bar{z}-p)\mathrm{d}p\right] = 0, \quad r = r_1. \tag{A2}$$

This will enter the analysis at the $O(\epsilon^2)$ stage by forcing u_2, *i.e.*,

$$\frac{\partial}{\partial r}u_2 + \alpha\beta\epsilon^2\left[(a_0+1)^4 - \int_{-\infty}^{\infty}(1+a_0)^4K(\bar{z}-p)\mathrm{d}p\right] = 0, \quad r = r_1. \tag{A3}$$

But a_0 is just a function of η and so by virtue of (A1b) $(\partial/\partial_r)u_2 = 0$ at $r = r_1$. Thus the inner expansion remains the same at least to $O(\epsilon^2)$.

In the outer region (A1a) holds. We observe that $\frac{1}{\epsilon}K((z-z')/\epsilon)$ behaves like a delta function as $\epsilon \to 0$. Thus, the integral picks off $(u+1)^4$ at $z' = z$ and produces a correction term which is proportional to $(\partial^2/\partial z^2)(u+1)^4$ at $z' = z$. The result is

$$\frac{\partial}{\partial r}u + \alpha\beta\epsilon^4 c_1\frac{\partial}{\partial z}\left[4(u+1)^3\frac{\partial}{\partial z}u\right] = 0, \quad r = r_1, \tag{A4}$$

where c_1 is an $O(1)$ constant. This implies that the outer expansion will not be affected until the $O(\epsilon^4)$ term. Thus the outer expansion remains the same at least to $O(\epsilon^3)$.

References

1. R. Silberglitt, I. Ahmad, W.M. Black, and J.D. Katz. Recent developments in microwave joining. *Material Res. Soc. Bull.* 18 (1993) 47–50.
2. J.G.P. Binner, P.A. Davis, T.E. Cross, and J.A. Fernie. Microwave joining of engineering ceramics. *Ceram. Trans.* 59 (1995) 335–346.
3. G.A. Danko, R. Silberglitt, and P.S. Day. Joining of Sintered SiC Tubes to SiC/SiC Composites Plates Using Preceramic Polymer with Microwave Hybrid Heating. Indianapolis: 101st Annual Meeting & Exposition of the American Ceramic Society (1999) Paper BS4-016-99.
4. J. Kevorkian and J.D. Cole. *Perturbation Methods and Applied Mathematics*. New York City: Springer-Verlag (1981) 558 pp.
5. H.K. Kuiken and P.J. Roksnoer. Analysis of the temperature in FZ silicon crystals. J. Cryst. Growth 47 (1979) 20–42.
6. G.W. Young and Chait. Steady-state thermal-solutal diffusion in a float zone. *J. Cryst. Growth* 96 (1989) 65–95.
7. S.A. Morman and G.W. Young. An asymptotic approach to the mathematical modeling of Ohno continuous casting of cored rods. *J. Eng. Math.* 38 (2000) 51–76.
8. E.M. Sparrow and R.D. Cess, *Radiation Heat Transfer*. London: Hemisphere Publishing (1978) 366 pp.

Journal of Engineering Mathematics **39:** 79–86, 2001.

Fast computation of limit cycles in an industrial application

SHAY GUERON and NADAV LIRON[1]
Department of Mathematics, University of Haifa, Haifa, 31905, Israel
[1]*Department of Mathematics, Technion – I.I.T, Haifa, 32000, Israel*

Received 6 December 1999; accepted in revised form 4 July 2000

Abstract. A model problem is described that requires the study of a system of the form $\dot{\mathbf{v}}(t) = \varepsilon \mathbf{F_P}(\mathbf{v(t)}, \mathbf{t})$ which depends on a set of parameters **P**, and where $\varepsilon \ll 1$. The problem comes from an industrial application where it is a kernel of an optimization procedure. The optimization depends on computing the limit cycle, and the problem needs to be solved repeatedly. Short computation time is therefore essential. The naive approach is to integrate the equation forward in time, starting from an arbitrary initial condition, until the transients disappear and the limit cycle is approximated within a given tolerance. This approach is too slow and thus impractical in the context of the optimization procedure. The problem involves two types of asymptotic considerations: long-time asymptotics and small-parameter asymptotics. Here a simple approach is demonstrated, based on implementing the averaging method. This reduces the solution time to the point that the optimization procedure becomes feasible.

Key words: limit cycle, dynamical systems, Poincaré-Bendixson

1. Motivation

Our study is motivated by an industrial engineering problem of designing a piezoelectric motor. The design stage of a certain component of the product required the solution of its model which is represented by a system of the form $\dot{\mathbf{v}}(t) = \varepsilon \mathbf{F_P}(\mathbf{v(t)}, \mathbf{t})$. This was the kernel of an optimization procedure. Here, **v** is an unknown m-dimensional vector modeling velocity, **F** is the given m-dimensional right-hand side (RHS) of the system, modeling acceleration. The notation $\mathbf{F_P}$ indicates that the problem depends on a set of parameters denoted by **P**. The RHS was continuous and piecewise smooth, typically bounded between two linear functions, and also periodic in t. The periodicity accounts for a periodic forcing component in the system, and the nonsmoothness accounts for the changes in the direction of this force with respect to other forces such as friction. The small parameter ε appears, due to the fast oscillations in the system, by rescaling the time ($\omega\tau \to t$).

Based on practical arguments, the physical system as well as the solution of its model were assumed to approach some periodic behavior for large enough t. In mathematical terms, this implies that there exists a stable (attracting) limit cycle. The design stages required the approximation of the limit cycle and some of its functionals such as for example, the average over one period. An analytic solution was not always available and a numerical solution was attempted. The approach adopted by the R&D team in that industry was to choose zero initial conditions for the velocity (as in a real experimental situation), to solve the system numerically until all transients have disappeared, to obtain the limit cycle numerically, and to compute the required functionals. The problem with this approach was (a) the large number of parameters (over ten in this case) on which the system depended required a great deal of runs, (b) the long time before the transients disappeared for each run slowed down the study of the model so much so as to hinder the time-table set by the industry for the development of the project. The speed of approach to the limit cycle depends crucially on the initial guess.

The approach we describe here takes advantage of asymptotics in two ways.

(i) We use 'small ε' perturbation to obtain an initial guess for the entire computation

(ii) We consider the solution at time intervals of the period of the limit cycle as values in a fixed-point iteration process and use acceleration techniques to rapidly approach a point on the limit cycle.

The combination of these two techniques indeed solved the problem practically.

The method we describe was successfully applied to systems with dimensionality $m > 10$. In this note we demonstrate the study of a one dimensional ($m = 1$) system, where we can easily prove that the solution converges to a limit cycle, and the study of a two dimensional ($m = 2$) system.

2. A case study

Consider the one dimensional ordinary differential equation (ODE)

$$\dot{v}(t) = \varepsilon F_{\mathbf{P}}(v(t), t), \tag{1}$$

where $(\dot{})$ denotes differentiation with respect to time t and ε is a small parameter. The notation $F_{\mathbf{P}}$ indicates that the right-hand side of Equation (1) depends on a set of parameters, namely $\mathbf{P}$. For simplicity, we omit the explicit notation of this parameter dependence, and hereafter write $F(v, t) = F_{\mathbf{P}}$. The function $F(v, t)$ is continuous and piecewise smooth. Suppose also that $F(v, t)$ is periodic in the variable t, and denote its period by T.

Equations and systems of the type (1) with smooth and nonsmooth right-hand side are an important class of dynamical systems that have various applications in different fields. In modeling and design applications Equation (1) is studied with different parameter values, and must therefore be repeatedly solved many times (see, for example, [1], [2]).

We assume here that, if the initial condition is chosen within some known interval, the resulting solution of (1) is bounded in $t \in [0, \infty)$. This is a natural assumption in the context of the real applications, where $v(t)$ is interpreted as the velocity of a modeled object, and (1) describes its acceleration. As we explain below, this guarantees the existence of an attracting limit cycle, that is, that $v(t)$ approaches a certain periodic solution (with period T) as $t \to \infty$. We are interested in the following computational problems:

PROBLEM 1. *Compute the limit cycle to which the solution $v(t)$ approaches as $t \to \infty$.*

PROBLEM 2. *Compute the average of the limit cycle solution $v(t)$ over one period,*

$$\frac{1}{T}\int_t^{t+T} v(s)\mathrm{d}s, \qquad \text{for } t \to \infty. \tag{2}$$

An analytic solution to (1) is often not available, or cumbersome to write down, manipulate and get insight from, especially for a function of many parameters. Thus, a numerical approach is required. When these computations need to be repeated, reducing the computational effort of this numerical task is imperative. To that end, we try to minimize the required number of evaluations of $F(v, t)$. This is equivalent to attempting to decrease the number of times that we integrate (1) numerically, over a time interval of length T.

2.1. THE STRAIGHTFORWARD NUMERICAL APPROACH

The straightforward approach for solving problems 1 and 2 is to solve (1) numerically. To that end, we start with some initial condition $v(0) = v_0$ and propagate the numerical trajectory $(t, v(t))$ in time, by applying some numerical ODE integrator (for example a Runge-Kutta scheme). We integrate the ODE for a long enough time interval, until the transients disappear, obtaining a discrete set of points (t_i, v_i), $1 \leq i \leq L$, (L is the number of discretization points) covering one cycle.

A practical way to check that transients have indeed disappeared is to compare the numerical trajectory of the n-th cycle to the $n+1$st cycle, and to stop the procedure when the differences are sufficiently small in some norm, for example, when the RMS $= \sqrt{\frac{1}{L}\sum_1^L (v_j^{n+1} - v_j^n)^2}$ falls below a certain prefixed value. This requires the storing of an appropriate number ($2L$) of trajectory points. For Problem 2 we need to store only the estimated average over the cycles.

We measure the computational cost of this procedure in terms of the number of times that (1) is integrated over a time interval of length T. Obviously, this cost can be reduced by a good guess for v_0.

2.2. THE FIXED-POINT APPROACH

We now consider our computational problem from a different point of view. Suppose that at some time t_1 we have computed the numerical trajectory point (t_1, v_1), which is a point on the limit cycle. Since $F(v, t)$ is periodic in t with period T, this implies that a) The period of the limit cycle is also T; b) If we choose the initial condition (t_1, v_1), and propagate (1) in time up to $t_2 = t_1 + T$, the computed trajectory point (t_2, v_2) must be equal to the point (t_1, v_1). This motivates the following stability assumption: if point (t_1, v_1), is close to a point on the limit cycle, and we propagate (1) to $t_2 = t_1 + T$, using initial condition (t_1, v_1), then the computed trajectory point (t_2, v_2) is close to the point (t_1, v_1).

Formally, we now define the following map $M : R \to R$; $M = M_t(u^*)$ is the value of the solution of (1) at $t + T$, *i.e.*, $v(t+T)$, given initial condition $v(t) = u^*$.

In these terms, our problem is restructured to the problem of computing the fixed point of the map M. Indeed, once the fixed point is computed (estimated), we can solve problems 1 and 2 at the cost of one additional integration of (1) over one cycle of length T.

Clearly, any numerical ODE integration can be interpreted as an iterative method for computing the fixed point of M, namely:

$$v_{n+1} = M(v_n), \qquad n = 0, 1, \ldots. \tag{3}$$

Since t is arbitrary, we can take any large t and shift it to, say $t = 0$. Thus, the explicit dependence of Equation (3) on the chosen time t is omitted.

We now make the following observations that lead to a practical solution of the problem:

1. The computational cost of one iteration of (3) is the cost of integrating (1) along a time interval of length T. Thus, we can solve the long-time asymptotic problem by using (3) with some ODE integration method. To that end, we need only to store a few values of $M(v_n)$.
2. For any choice of an ODE integrator to compute M, the iterative scheme (3) can be replaced by any other and possibly faster iterative scheme that does not require the use of the derivative of M, for example the Secant Method. Note that we do not use the derivative

of M because this would require the differentiation of an ODE with respect to its initial condition, which we cannot perform here.

3. For any choice of an ODE integrator to compute M in (3), we can accelerate the convergence by applying some acceleration technique to the iterations. One possible example is applying Aitken's acceleration method,

$$\hat{M}_n = M_n - \frac{(M_{n+1} - M_n)^2}{M_{n+2} - 2M_{n+1} + M_n}.$$

2.3. Initial choice for $0 < \varepsilon \ll 1$

A crucial ingredient in speeding up the convergence of the iterations, is a good choice of the initial guess v_0. Since $\varepsilon \ll 1$, we can use small-parameter asymptotics and argue that the oscillations of the limit cycle have a small magnitude, and apply some asymptotic technique in order to obtain a good estimate for v_0. The simplest method to obtain a good estimate for v_0 or to obtain a good estimate for the average of v over a cycle is to write $v(t) = \bar{v} + \varepsilon\nu(t) + \ldots$ for some unknown parameter $\bar{v}$ and unknown function $\nu(t)$. Since the limit cycle is periodic with period T, and satisfies (1), the equation for the zero order approximated equation for unknown $\bar{v}$ is

$$\int_0^T F(t, \bar{v})\mathrm{d}t = 0. \tag{4}$$

Equation (4) can be solved numerically.

In the next sections we demonstrate the application of the above points.

3. Proving the existence of a limit cycle

A comprehensive study on the existence and stability of limit cycles for problems of our type can be found for example in [3, Chapters 5 and 7]. However, for dimensionality $m = 1$, and with the specific form of the RHS in mind, we can carry out a simple analysis. If the solution of (1) is uniformly bounded for any initial condition, then the Poincaré–Bendixson theorem guarantees that this solution converges either to a unique limit cycle or to a singular point. Since the RHS of (1) is periodic, there are no singular points. Therefore, the iterations (3) converge to a fixed point of M. We now apply this to our case and state the following:

Suppose that $F(v, t)$ is periodic in the variable t with period T, and that $F(v, t)$ is bounded by two linear functions of v (*e.g.*, as in [2]), that is,

$$-a_1 v + b_1 \le F(t, v) \le -a_2 v + b_2 \tag{5}$$

for some positive constants a_1, a_2, b_1, b_2. Then, there exists a value v_0 for which the solution $v(t)$ of (1) with the initial condition $v(0) = v_0$ satisfies $v(T) = v_0$. Since $F(v, t)$ is periodic with period T, this implies that the solution $v(t)$ is the periodic solution of (1).

Proof:

Denote the solution of (1) with $v(0) = v_0$ by $v(t)$. For any t we have

$$\frac{b_1}{a_1} + \left(\frac{b_1}{a_1} + v_0\right)\mathrm{e}^{-a_1 t} \le v(t) \le \frac{b_2}{a_2} + \left(\frac{b_2}{a_2} + v_0\right)\mathrm{e}^{-a_2 t} \tag{6}$$

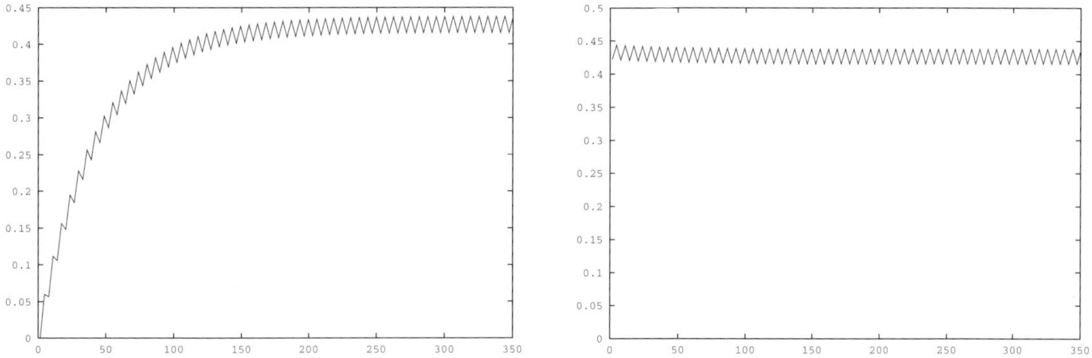

a) b)

Figure 1. The numerical trajectory for Equation (1). Application of the Runge–Kutta method, $dt = 0{\cdot}01256$, (a) initial condition $v_0 = 0$; (b) initial condition $v_0 = \bar{v} = 0{\cdot}43333$.

Therefore, $e^{-a_1 T} < \lim_{v_0 \to \infty} \dfrac{v(T)}{v_0} \le e^{-a_2 T} < 1$.

This implies that there exists a sufficiently large value $A_0 > 0$ such that the map M maps the interval $[-A_0, A_0]$ *into itself*. It then follows that M has a fixed point, which is exactly the required value of v_0.

4. Examples

We first illustrate the method with a simple test case

$$\dot{v}(t) = \varepsilon(\text{sign}\,(v(t) - \delta\cos(\omega t)) - \alpha v(t) + b); \tag{7}$$

We choose the parameter values $\varepsilon = 0{\cdot}01$, $\delta = 1$, $\alpha = 3$, $b = 1$, and $\omega = 1$ (*i.e.*, $T = 2\pi$).

Since the solution v is interpreted as the velocity of an object that starts at rest, the natural choice for initial condition is $v_0 = 0$. We used a Runge-Kutta method of order 4, with 30 points along one interval (*i.e.*, $dt \approx 0{\cdot}01256$) and obtained the numerical trajectory displayed in Figure 1, Panel a.

Clearly, integration over many *cycles* (more than 50) is required before the transients disappear and the trajectory settles down at the limit cycle. Note the small oscillations of the limit cycle. Computations using the numerical trajectory show that the average of $v(t)$ over one cycle is $\bar{v} \approx 0{\cdot}42664$.

We show that it is advantageous to first solve (4) numerically. Using the trapezoidal rule with 50 nodes on $[0, T]$, we obtained the zero order approximation $\bar{v} \approx 0{\cdot}43333$, which is a good approximation. If we are only interested in this average, we are already done. To compute the limit cycle itself we start the same iterations with the initial condition $v_0 = \bar{v} \approx 0{\cdot}43333$. The convergence is much faster than in the previous case. Figure 1, Panel b displays the resulting numerical trajectory.

Recall now that the approach is to consider the problem as an iterative procedure, Therefore, instead of storing the complete information on the numerical trajectory, we store only the values of the map $M_n = M(v_n)$, for $n = 0, 1, \ldots$ where n is the number of the cycle. Figure 2 Panel a shows M_n as a function of n obtained for the computations with, $v_0 = 0$ and with $v_0 = \bar{v} = 0{\cdot}43333$. Note that starting with $v_0 = \bar{v}$, we are practically there after ≈ 10 cycles, as compared with starting with $v_0 = 0$ which requires some 50 cycles for convergence.

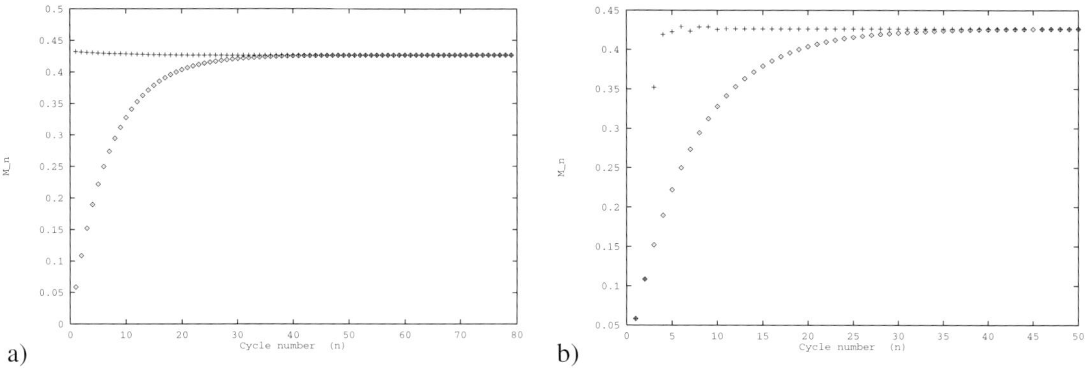

Figure 2. The iterations M_n, $n = 0, 1, \ldots$ where n is the number of the cycle. (a) The $\diamondsuit$ and the $+$ symbols indicate the solution for $v_0 = 0$ and $v_0 = \bar{v} = 0{\cdot}43333$, respectively. (b) The $\diamondsuit$ symbol indicates the solution with $v_0 = 0$. The $+$ symbol indicates the accelerated solution with $v_0 = 0$.

Accelerating the convergence by applying Aitken's method to the iterations can reduce the number of required cycles. To illustrate, Figure 2, Panel b shows the values of M_n as a function of n obtained for the computations with $v_0 = 0$ with and without acceleration. With acceleration, the required number of cycles is reduced to ≈ 10. Applying Aitken's method to the solution with the improved guess $v_0 = \bar{v}$ reduces the number of required cycles to ≈ 6.

4.1. An example with $m = 2$

For $m = 2$, we demonstrate our method with the system

$$\begin{aligned} \dot{v}(t) &= \varepsilon(\text{sign}\,(u(t) - \delta\cos(\omega t)) - \alpha_1 u(t) + b_1); \\ \dot{u}(t) &= \varepsilon(\text{sign}\,(v(t) - \delta\cos(\omega t)) - \alpha_2 v(t) + b_2). \end{aligned} \tag{8}$$

The parameter values used here are $\varepsilon = 0{\cdot}01$, $\delta = 1$, $\alpha_1 = 3$, $\alpha_2 = 4$, $b_1 = b_2 = 1$, and $\omega = 1$ (*i.e.*, $T = 2\pi$).

Solving the system of equations for the zero order approximation, namely $\int_0^T F_{1,2}(\bar{u}, \bar{v}, t)\,\mathrm{d}t = 0$ gives the solution $(\bar{u}, \bar{v}) = (0{\cdot}3250, 0{\cdot}4333)$. Figure 3, Panel a displays the numerical trajectory $(u(t), v(t))$ as a function of t, obtained with the initial conditions $v_0 = u_0 = 0$ and with the improved initial conditions $u_0 = 0{\cdot}3250$, $v_0 = 0{\cdot}4333$. Figure 3, Panel b displays the two dimensional iterations M_n^u, M_n^v, as a function of the number of the cycle, n. The advantage of starting with a better initial guess, by use of small-ε asymptotics, is evident. Panel c shows that applying an acceleration procedure also improves the convergence.

4.2. Extending the method to cases with an unknown period

The basic averaging idea is simple, and can be extended in several ways. For a system of equations $m \geq 2$, as we illustrated in the second example for $m = 2$, the increase in time saving can be substantial. Further, an extended version of the our technique can be applied to nonlinear (autonomous) problems where the period is not known apriori.

One example which we briefly illustrate here is the Van der Pol (VDP) oscillator $\ddot{x} + x = \varepsilon(1 - x^2)\dot{x}$ where $\varepsilon > 0$ is a parameter (see [3, Chapter 11] for the application of the averaging method to this equation). We can apply our numerical procedure to find the underlying periodic solution of the VDP equation, although the period T is not known *a priori*.

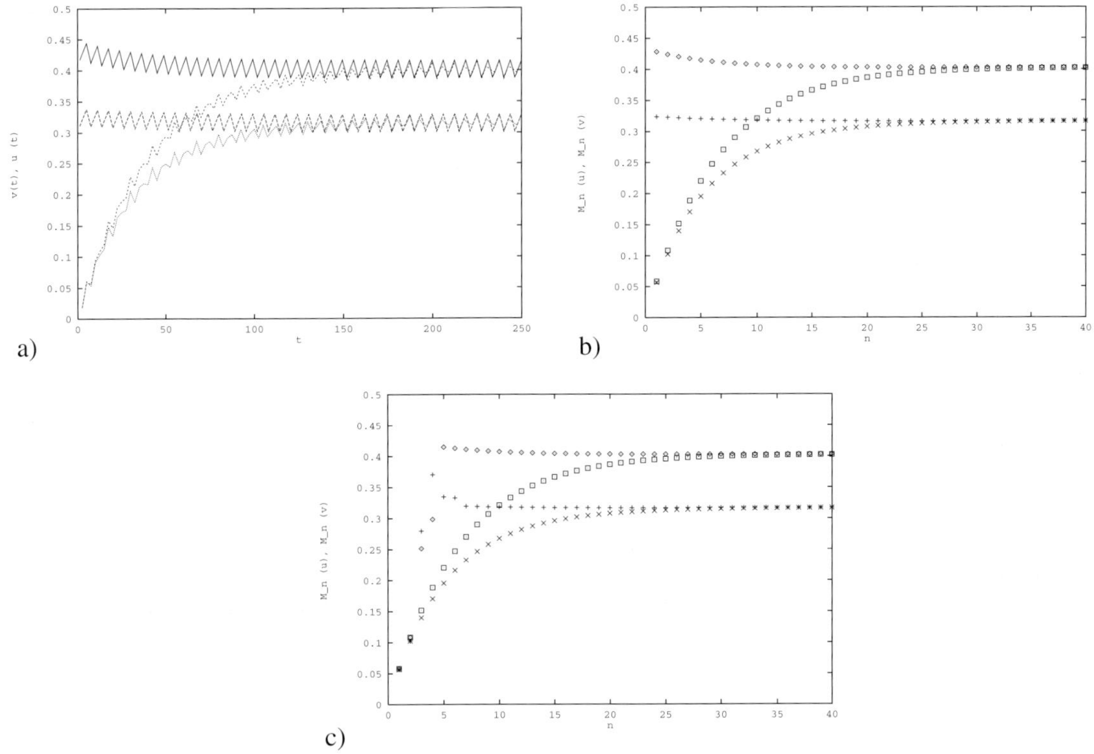

Figure 3. The solution of the system (8) using the Runge–Kutta method, $dt = 0{\cdot}01256$. (a) The numerical trajectory $u(t)$ and $v(t)$ (+) for initial conditions $v_0 = u_0 = 0$ and for $u_0 = 0{\cdot}3250$, $v_0 = 0{\cdot}4333$. (b) The iterations M_n^u, M_n^v, $n = 0, 1, \ldots$ where n is the number of the cycle. The square and $\times$ symbols indicate the iterations for $v_0 = u_0 = 0$. The $\diamondsuit$ and the + symbols indicate the iterations for $u_0 = 0{\cdot}3250$, $v_0 = 0{\cdot}4333$. (c) The accelerated solution with $v_0 = u_0 = 0$.

This can be done by considering T as an additional unknown and adding the requirement $\dot{x}(t_0) = \dot{x}(t_0 + T)$ to the procedure. The implementation uses a numerical minimization of $\mathcal{F}(x_0, \dot{x_0}, T) = (x(0) - x(T))^2 + (\dot{x}(0) - \dot{x}(T))^2$ where ε is a given parameter and $x(0) = x_0$, $\dot{x}(0) = \dot{x_0}$, T are the initial guess.

Computing a value of $\mathcal{F}$ requires the numerical integration of the VDP trajectory $x(s)$ over one period, starting from x_0 and $\dot{x_0}$. The numerical minimization procedure attempts to locate the limit cycle, by an appropriate choice of x_0 and $\dot{x_0}$. At the limit cycle, $\mathcal{F}$ is ideally zero. In practice, this is approximated by the values of x_0 $\dot{x_0}$, for which $\mathcal{F}$ is numerically minimized. Note that $\mathcal{F} = 0$ implies that $\int_0^T \dot{x}(s)ds = 0$ and $\int_0^T \ddot{x}(s)ds = 0$. We use these auxiliary conditions for validating the optimization.

For $\varepsilon = 0{\cdot}1$ and the initial guess $x(0) = 0$, $\dot{x}(0) = 1$, $T = 5$, this procedure produced the estimate $T \approx 6{\cdot}286$ to the unknown period, and enabled integrating the limit cycle. For comparison, the asymptotic expansion using the averaging method gives $T = 2\pi(1 + \frac{1}{16}\varepsilon^2) + O(\varepsilon^4) \approx 6{\cdot}28711$ ([3, Chapter 10]).

5. Discussion

We have demonstrated here the use of 'practical asymptotics' to a problem arising in industry (piezoelectric motors in our case). For a known qualitative limiting behavior of a limit cycle with a known period, we showed how finding the limit cycle, as a long time behavior, can be accelerated by a numerical acceleration technique, and also by a parameter expansion for the case of small oscillations. This idea, called also 'the averaging method', appears in the literature of dynamic systems in different contexts and applications. A comprehensive study can be found in [3, Chapters 4, 5, 11].

Acknowledgements

This research was supported by the U.S. - Israel Binational Science Foundation Grant number 97-00400, by a grant from the Israeli Ministry of Science, by the Technion Vice Provost for Research Funds - The David and Miraim Mondrey Research Fund, by The B. and G. Greenberg Research Fund (Ottawa), and by the Fund for the Promotion of Research of the University of Haifa.

References

1. B. Brogliato, *Nonsmooth Mechanics* (2nd ed.). Springer-Verlag: New York (1999) 400 pp.
2. O. Y. Zharii. Modeling of a mode conversion ultrasonic motor in the regime of slip. *IEEE Trasn. Ultrasonic, Ferroelectric Freq. Control* 40 (1993) 411–416.
3. J. K. Hale, *Ordinary Differential Equations* (2nd edition). Pure and Applied Mathematics series. New York: Wiley-Interscience (1980) 421 pp.
4. F. Verhulst, *Nonlinear Differential Equations and Dynamical Systems*. Springer-Verlag: New York (1996) 303 pp.

Journal of Engineering Mathematics **39:** 87–130, 2001.

Asymptotic analysis of the steady-state and time-dependent Berman problem

J. R. KING and S. M. COX
School of Mathematical Sciences, University of Nottingham, University Park, Nottingham NG7 2RD, UK (e-mail: john.king@nottingham.ac.uk; stephen.cox@nottingham.ac.uk)

Received 14 December 1999; accepted in revised form 14 September 2000

Abstract. The Berman problem for two-dimensional flow of a viscous fluid through an infinite channel is studied. Fluid motion is driven by uniform suction (or injection) of fluid through the upper channel wall, and is characterised by a Reynolds number R; the lower wall is impermeable. A similarity solution in which the streamfunction takes the form $\psi = -x\mathcal{F}(y, t)$ is examined, where x and y are coordinates parallel to and normal to the channel walls, respectively. The function $\mathcal{F}$ satisfies the Riabouchinsky–Proudman–Johnson equation, a partial differential equation in y and t; steady flows satisfy an ordinary differential equation in y. The steady states are computed numerically and the asymptotics of these solutions described in the limits of small wall suction or injection, large wall injection and large wall suction, the last of these being given more concisely and more accurately than in previous treatments. In the time-dependent problem, the solution appears to be attracted to a limit cycle when $R \gg 1$ (large wall suction). This solution has been computed numerically for $\epsilon = 1/R$ down to 0·011, but the structure of the solution makes further numerical progress currently infeasible. The limit cycle consists of several phases, some with slow and others with very rapid evolution. During one of the rapid phases, the solution achieves a large amplitude, and this feature of the solution lies behind the practical difficulties encountered in numerical simulations. The profile of the solution is plotted during the various phases and corresponding asymptotic descriptions are given. An exact solution to the Riabouchinsky–Proudman–Johnson equation covers most of the phases, although separate discussion is required of the boundary layers near the two walls and an interior layer near a zero of $\mathcal{F}$. Particular consideration is required when this zero approaches the upper channel wall.

Key words: similarity solution, Navier–Stokes equations, Berman problem, asymptotics, channel flow.

1. Introduction

We consider a similarity solution of the Navier–Stokes equations for plane flow of a viscous fluid confined between parallel walls. The flow is driven by uniform withdrawal (or injection) of the fluid through the upper channel wall, the lower wall being impermeable. The problem is characterised by a Reynolds number R, based on the speed at which fluid is withdrawn from the channel, and our particular interest is in the asymptotic structure of the flow when this Reynolds number is large.

The governing Navier–Stokes equations and boundary conditions permit a flow whose streamfunction ψ takes the form $-x\mathcal{F}(y, t)$, where x is the coordinate parallel to the channel walls, y is the normal coordinate and t is time. The governing partial differential equation for the similarity function $\mathcal{F}$ is due to Riabouchinsky [1] – see also Proudman and Johnson [2]; as a special case, steady flows (those for which $\mathcal{F} = f(y)$) satisfy an ordinary differential equation [3] studied by Berman [4] in the context of channel flow. We shall refer to the problem of determining $\mathcal{F}(y, t)$ for various Reynolds numbers as the Berman problem.

The large Reynolds number asymptotic behaviour of the steady solutions has proved remarkably subtle, and was one of the first applications of exponential asymptotics in fluid mechanics (by Terrill [5]; see also [6] and [7]). It is this large Reynolds number behaviour, of both steady and unsteady solutions, that we primarily treat below.

The Berman problem was originally investigated in the case where both channel walls are equally permeable (see, for example, [4, 8, 9]). The extension to a more general case, where the fluid is withdrawn from the two channel walls at different rates, has also been carried out [10–15].

The limiting case of 'complete' asymmetry, with one permeable and one impermeable wall, has previously been investigated [7, 15, 16], and it is this problem that is addressed in this paper. In this case, there is a unique steady solution for all values of R, except in the range $7{\cdot}05 \approx R_1 < R < R_2 \approx 7{\cdot}31$, where there are three solutions [15, 16]. After formulating the problem in Section 2, in Section 3 we compute the steady solutions numerically and offer asymptotic solutions in the three limits of large wall injection, large wall suction, and small wall suction or injection. Numerical solution of the ordinary differential two-point boundary-value problem is simplified by means of a rescaling due to Terrill [8] which converts it into an initial-value problem. This rescaling allows calculation of numerical solutions at much larger Reynolds numbers than would otherwise be possible by direct calculation, for example by shooting from the channel walls and matching.

It is possible to compute the stability of the steady flows to disturbances which are also of the similarity form $\psi = -x\mathcal{F}(y,t)$. It is then found [15] that the steady state becomes unstable in a Hopf bifurcation at $R = R_3$, where $R_3 \approx 12{\cdot}755$. Numerical evidence suggests that this bifurcation is supercritical, and that a limit cycle exists for $R > R_3$. This limit cycle seems to be stable (within the context of the Riabouchinsky–Proudman–Johnson equation, henceforth the RPJ equation) for all values of the Reynolds number at which we have carried out numerical simulations of the initial-value problem (see Section 4). As the Reynolds number is increased, the limit cycle separates into slow and fast phases of evolution. During one of the fast phases, the solution gains large amplitude, and this combination of rapid evolution and large amplitude makes reliable numerical simulation difficult. Indeed with our (admittedly rather unsophisticated) numerical scheme we have been unable to compute the limit cycle reliably beyond $R \approx 100$ without the run times becoming prohibitively long. The asymptotic structure of the limit cycle, described in Sections 5, 6 and 7, makes clear why such numerical difficulties are encountered. We note that there have heretofore been very few asymptotic treatments of time-dependent solutions to the RPJ equation, a notable exception being the blow-up study of Grundy and McLaughlin [17].

We discuss our results in Section 8, and make our concluding remarks in Section 9. Appendix A outlines the general solution to the inviscid RPJ equation in the case of zero pressure and also analyses its similarity solutions, these playing a crucial role in Sections 6 and 7. Appendix B discusses symmetry (and related) properties of the full (viscous) RPJ equation. We note that the notation of Section 3 sometimes departs from that of the remainder of the paper.

2. The Berman problem

We consider two-dimensional flow of a Newtonian fluid with kinematic viscosity ν in a channel of half-width h. The flow is driven by uniform withdrawal of the fluid through the upper channel wall with speed V; the lower channel wall is impermeable. After adopting the scales h and h/V for length and time, respectively, we note that a particular class of solutions to the governing (dimensionless) Navier–Stokes equations may be written in terms of a stream-function $\psi(x, y, t)$ via $\psi = -x\mathcal{F}(y, t)$, where x and y are coordinates parallel and normal to the channel walls, respectively. The velocity field is then given by $u = \partial\psi/\partial y = -x\mathcal{F}_y$, $v = -\partial\psi/\partial x = \mathcal{F}$. After elimination of the pressure from the Navier–Stokes equations, we find, for the similarity flow of interest, that the function $\mathcal{F}$ satisfies the RPJ equation [9]

$$\mathcal{F}_{yyt} = \epsilon\mathcal{F}_{yyyy} + \mathcal{F}_y\mathcal{F}_{yy} - \mathcal{F}\mathcal{F}_{yyy}, \tag{1}$$

where $R = hV/\nu$ is the Reynolds number and $\epsilon = R^{-1}$, subject to the boundary conditions

$$\mathcal{F}(-1, t) = \mathcal{F}_y(-1, t) = \mathcal{F}_y(1, t) = 0, \qquad \mathcal{F}(1, t) = 1, \tag{2}$$

corresponding to normal suction (for $\epsilon > 0$) or injection (for $\epsilon < 0$) through the upper wall at $y = 1$, and an impermeable no-slip lower wall at $y = -1$. For numerical simulation of (1) it is useful to note that this equation may be integrated once in y to give [1, 2]

$$\mathcal{F}_{yt} = \epsilon\mathcal{F}_{yyy} + \mathcal{F}_y^2 - \mathcal{F}\mathcal{F}_{yy} + p(t), \tag{3}$$

where $p(t)$ is a function of integration, with the fluid pressure being $\hat{p}(y, t) + x^2 p(t)/2$, where $\hat{p}$ is given by

$$\mathcal{F}_t = \epsilon\mathcal{F}_{yy} - \mathcal{F}\mathcal{F}_y - \hat{p}_y; \tag{4}$$

$p(t)$ is to be determined by imposing the four boundary conditions on $\mathcal{F}$. The vorticity is given by

$$\omega = v_x - u_y = x\mathcal{F}_{yy}, \tag{5}$$

the quantity $\mathcal{F}_{yy}$ playing an important role in what follows.

3. Steady flows

3.1. Numerical results

Steady flows, for which $\mathcal{F}(y, t) = f(y)$, satisfy

$$\epsilon f'''' + f'f'' - ff''' = 0, \tag{6}$$

subject to the boundary conditions

$$f(-1) = f'(-1) = f'(1) = 0, \qquad f(1) = 1. \tag{7}$$

For values of ϵ that are not too small, the system (6–7) may readily be solved numerically. The resulting bifurcation diagram is given in Figure 1, showing $-f''(1)$, which is proportional to the shear stress at the upper channel wall, as a function of the Reynolds number R. A corresponding solution $f(y)$ for small ϵ (*i.e.* large R) is shown in Figure 2.

For the purposes of calculating numerical solutions to (6–7), it is useful to apply the following technique, due to Terrill [8], for converting the boundary-value problem (6–7) into an initial-value problem. (The technique also allows one to develop an alternative asymptotic

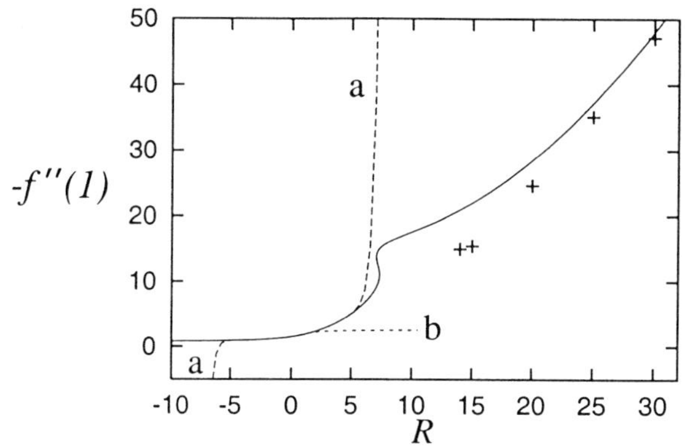

Figure 1. Solid line is a plot of $-f''(1)$ against Reynolds number $R = \epsilon^{-1}$ for numerical solutions to (6–7). Note that the solution is unique, except for a range of Reynolds numbers around $R = 7$. Where there are multiple solutions, the upper and lower branches represent temporally stable solutions to (1–2); the middle branch is unstable. Elsewhere, where the solution is unique, it is stable for $R < R_3 \approx 12{\cdot}755$, but becomes unstable in a Hopf bifurcation at $R = R_3$. The dashed line (a) shows the small-R asymptotic approximation in (11) (the plot includes terms up to R^4). When this series is reverted to give R in powers of $(f''(1) + \frac{3}{2})$, the dashed line (b) is obtained (the plot includes terms up to $(f''(1) + \frac{3}{2})^9$). Crosses show the large-R asymptotic approximation from Section 3.4, optimally truncated. For R less than around 14 it is not possible to apply this large-R approximation consistently since Equations (28) and (30) have no real solutions; it is noteworthy that this occurs fairly close to the fold.

description of f to the one given below [6, 7].) First f and y are rescaled by introducing ϕ and η through

$$f(y) = \tfrac{1}{2}\epsilon b\phi(\eta), \qquad \eta = \tfrac{1}{2}b(y+1), \tag{8}$$

where the constant b is arbitrary at this stage. The parameter ϵ is then absent from the equation for $\phi(\eta)$, which is

$$\phi_{\eta\eta\eta\eta} + \phi_\eta\phi_{\eta\eta} - \phi\phi_{\eta\eta\eta} = 0. \tag{9}$$

This equation is integrated numerically, subject to the initial conditions

$$\phi(0) = 0, \qquad \phi'(0) = 0, \qquad \phi''(0) = A, \qquad \phi'''(0) = B, \tag{10}$$

where A and B are arbitrary, until a zero of $\phi'(\eta)$, at $\eta = \bar{\eta}$, say, is obtained from the numerical calculation. By taking $b = \bar{\eta}$ we obtain from (8) a solution to the boundary-value problem (6–7); the corresponding value of ϵ is $2/(b\phi(b))$. By a suitable rescaling of ϕ and η, we may set $B = 1$. Solutions may then be calculated in the limit as $\epsilon \to 0^+$ by letting $A \to A_c^+$ [18], where $A_c \approx -1{\cdot}232$ [19].

This reformulation of the problem allows us to compute $f(y)$ for much larger Reynolds numbers than is possible by solving the full boundary-value problem (6–7) directly, for example by shooting. Since they arise as a result of the calculation, the Reynolds numbers for which we are able to compute solutions are not necessarily round numbers (this explains the 'odd' values quoted in this paper). To illustrate the practical utility of the reformulation we note that we have been able to compute $f(y)$ from (6–7) at Reynolds numbers up to around 42 by shooting from each wall and matching at an interior point, whereas we have computed f up to Reynolds numbers in the region of 10^4 using (9–10). We have checked a sample of our solutions obtained by Terrill's rescaling against direct solutions to the boundary-value

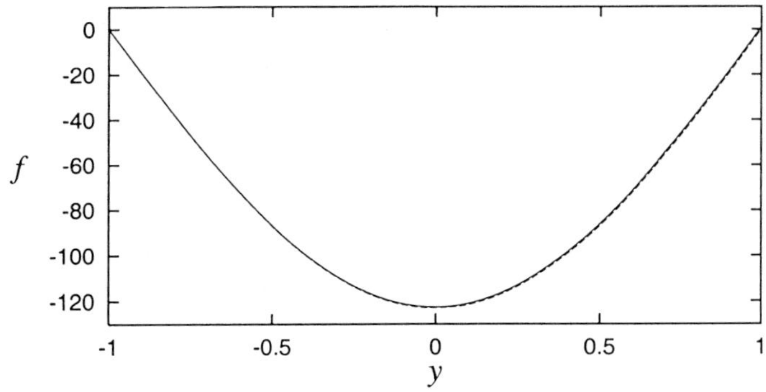

Figure 2. The solution to (6–7) for $R = 12256{\cdot}893$ (solid line), together with the leading-order outer asymptotic approximation from Section 3.4 that $f \sim -122{\cdot}909 \cos \frac{1}{2}\pi y$ (dashed line). The two lines are almost indistinguishable.

problem (6–7), where these are feasible, and have found excellent agreement. All numerical calculations were performed using integration routines from the NAG library in a double precision FORTRAN program.

In the subsections that follow we describe the asymptotic behaviour of the steady solution in the limits of small R, large wall injection ($-R \gg 1$), and large wall suction ($R \gg 1$).

3.2. Steady flows in the limit $R \to 0$

For small Reynolds numbers, the solution to (6) may be written as a power series

$$f(y) = \sum_{n=0}^{\infty} R^n f_n(y). \tag{11}$$

The first couple of terms in this series are readily found to be $f_0(y) = \frac{1}{4}(1+y)^2(2-y)$ and $f_1(y) = -\frac{1}{1120}(1-y^2)^2(y^3+2y+35)$. From these expressions, a small-R expansion of $f''(1)$, plotted in Figure 1, is found to be $f''(1) = -\frac{3}{2} - \frac{19}{70}R + O(R^2)$. This series may readily be reverted to give R in powers of $(f''(1) + \frac{3}{2})$; the result is plotted in Figure 1. The pressure term

$$p = -\epsilon f''' - f'^2 + ff'', \tag{12}$$

has the corresponding expansion

$$Rp = \sum_{n=0}^{\infty} R^n p_n, \tag{13}$$

where the first few terms are $p_0 = 3/2$, $p_1 = -81/140$ and $p_2 = 2929/107800$.

The radius of convergence of the small-R series for p may be estimated using the extension by Mercer and Roberts [20] of the method due to Domb and Sykes [21]. We first compute a large number of terms in (13); we have computed up to $n = 36$ using the computer algebra package Maple. We then compute the quantities

$$B_n = \frac{p_{n+1}p_{n-1} - p_n^2}{p_n p_{n-2} - p_{n-1}^2} \quad \text{and} \quad \cos\theta_n = \frac{1}{2}\left[\frac{p_{n-1}B_n}{p_n} + \frac{p_{n+1}}{p_n B_n}\right]. \tag{14}$$

If the series in (13) has finite radius of convergence r, with convergence-limiting singularities of order q at $r\exp(\pm i\theta)$, then for large n,

$$B_n = \frac{1}{r} - \frac{1+q}{rn} + O(n^{-2}) \qquad \text{and} \qquad \cos\theta_n = \cos\theta + O(n^{-2}). \tag{15}$$

We estimate using the coefficients p_n available to us that $r \approx 5{\cdot}5$, $q \approx 0{\cdot}37$ and $\theta \approx 1{\cdot}46$, so the convergence-limiting singularities appear to lie close to the imaginary axis. Curves (a) and (b) in Figure 1 indicate the range of validity that can be obtained from the series.

3.3. Steady flows in the limit $R \to -\infty$

In the limit $R \to -\infty$, the solution to (6–7), away from a boundary layer at $y = 1$, takes the form [16]

$$f(y) \sim \cos\tfrac{\pi}{4}(1-y) - \tfrac{1}{2}(-R)^{-1/2} z_0 (\pi/4)^{1/2}(1-y)\sin\tfrac{\pi}{4}(1-y) + O((-R)^{-1}), \tag{16}$$

where the constant z_0 is to be determined by matching with the boundary-layer solution.

Near $y = -1$, f takes the form

$$f(y) \sim (-R)^{-1/2}\left\{\theta_0(\mathcal{Y}) + O((-R)^{-1/2})\right\}, \tag{17}$$

where $\mathcal{Y} = (-R)^{1/2}(y+1)$ and θ_0 satisfies

$$\theta_0'''' + \theta_0\theta_0''' - \theta_0'\theta_0'' = 0 \tag{18}$$

subject to $\theta_0(0) = \theta_0'(0) = 0$ and the matching condition $\theta_0 \sim \pi\mathcal{Y}/4$ as $\mathcal{Y} \to \infty$. Thus if we write $\theta_0(\mathcal{Y}) = (\pi/4)^{1/2}\rho(z)$, with $z = (\pi/4)^{1/2}\mathcal{Y}$, it follows that $\rho(z)$ satisfies the Falkner–Skan equation ([22], p. 316)

$$\rho''' + \rho\rho'' - \rho'^2 = -1 \tag{19}$$

(this is the case $m = 1$ in the usual notation), subject to $\rho(0) = \rho'(0) = 0$ and $\rho'(z) \sim 1$ as $z \to \infty$. Numerical evaluation of $\rho(z)$, by shooting, reveals that as $z \to \infty$, $p(z) \sim z - z_0$, where $z_0 \approx 0{\cdot}647900$ [7], and this provides the constant required in (16).

3.4. Steady flows in the limit $R \to \infty$

3.4.1. *Interior- and boundary-layer analysis*

We consider the behaviour close to $y = 1$ of solutions to (6) which satisfy

$$\text{at } y = 1 \qquad f = 1, \quad f' = 0. \tag{20}$$

We start by writing

$$y = 1 + \epsilon Y, \qquad Y < 0,$$

and define $\delta(\epsilon) \ll 1$ by taking the location of the zero of f closest to $y = 1$ to be $Y = -1/\delta$. Moreover, we define $\alpha(\epsilon) = O(1)$ and $\nu(\epsilon) \ll 1$ such that

$$\text{at } Y = -1/\delta \qquad f = 0, \qquad f_Y = \delta\alpha, \qquad f_{YYY} = -\delta^2\alpha^2\nu, \tag{21}$$

asymptotic expressions for each of the unknowns δ, α and ν being obtained below.

3.4.2. *Interior layer*
In view of (21), we scale according to

$$Y = -\frac{1}{\delta} + \frac{\hat{Y}}{(\alpha\delta)^{1/2}}, \qquad f = (\alpha\delta)^{1/2}\hat{f}$$

to give

$$\begin{aligned} &\hat{f}_{\hat{Y}\hat{Y}\hat{Y}\hat{Y}} + \hat{f}_{\hat{Y}}\hat{f}_{\hat{Y}\hat{Y}} - \hat{f}\hat{f}_{\hat{Y}\hat{Y}\hat{Y}} = 0, \\ &\text{at } \hat{Y} = 0 \qquad \hat{f} = 0, \quad \hat{f}_{\hat{Y}} = 1, \quad \hat{f}_{\hat{Y}\hat{Y}\hat{Y}} = -\nu; \end{aligned} \tag{22}$$

the final condition here motivates the inclusion of the factor α^2 in the definition of ν in (21). The only parameter appearing in (22) is ν, which will prove to be exponentially small in δ (see (31) below). The first three terms in the required $\nu \to 0$ limit of (22), in which exponential growth is suppressed as $\hat{Y} \to -\infty$ (this is needed to match into the outer region), are given by

$$\hat{f} \sim \hat{Y} - \tfrac{1}{6}\nu\hat{Y}^3 + \nu^2\hat{F}, \tag{23}$$

where, writing $\hat{G} = \hat{F}_{\hat{Y}\hat{Y}\hat{Y}\hat{Y}}$, we have

$$\hat{G} \sim \hat{Y} - \sqrt{\frac{\pi}{2}}\, e^{\hat{Y}^2/2}\left(1 + \operatorname{erf}(\hat{Y}/\sqrt{2})\right); \tag{24}$$

since

$$\hat{G} - \hat{Y}\hat{F}_{\hat{Y}\hat{Y}\hat{Y}} + \hat{F}_{\hat{Y}\hat{Y}} = -\tfrac{1}{3}\hat{Y}^3,$$

$\hat{F}$ is completely specified by integrating (24) four times using

$$\text{at } \hat{Y} = 0 \qquad \hat{F} = 0, \quad \hat{F}_{\hat{Y}} = 0, \quad \hat{F}_{\hat{Y}\hat{Y}} = \sqrt{\frac{\pi}{2}}, \quad \hat{F}_{\hat{Y}\hat{Y}\hat{Y}} = 0.$$

3.4.3. *Boundary layer* $Y = O(1)$
Writing, as $\delta \to 0$,

$$f = 1 + \delta F, \quad \alpha \sim \sum_{n=0}^{N} \delta^n \alpha_n, \quad F \sim \sum_{n=0}^{N-1} \delta^n F_n, \tag{25}$$

we have

$$\begin{aligned} &F_{nYYYY} - F_{nYYY} = \sum_{m=0}^{n-1}\left(F_m F_{(n-1-m)YYY} - F_{mY}F_{(n-1-m)YY}\right), \\ &\text{at } Y = 0 \qquad F_n = F_{nY} = 0, \\ &\text{as } Y \to -\infty \qquad F_n \sim \alpha_{n+1} + \alpha_n Y, \end{aligned} \tag{26}$$

where $\alpha_0 = 1$; here we have matched with the first term in (23), which is valid to all powers of δ. The boundary-value problem (26) determines both F_n and α_{n+1} and is readily iterated forward in n to give closed form solutions to any desired order. The first few terms are $\alpha_1 = 1$, $\alpha_2 = 4$, $\alpha_3 = 129/4$, and $F_0(Y) = 1 + Y - \exp Y$, $F_1(Y) = 4 + Y - (\frac{1}{2}Y^2 - 3Y + 4)\exp Y$.

We note that we have been able to scale in order to formulate the above problems such that the only parameter appearing in (22) is ν and the only one in (25–26) is δ. This makes our approach both more concise and more generally applicable than previous attempts and facilitates the construction of the solution to any order in δ. At this stage of the analysis, the fact that $\nu(\epsilon)$ and $\delta(\epsilon)$ remain to be determined is not a difficulty; in solving (22) we need only to know that $\nu \ll 1$ and (25–26) is implied by $\delta \ll 1$ on the assumption (justified below) that ν is exponentially small in δ. Subsequent matching will determine δ and ν as functions of ϵ; the first stage of matching, which we now describe, relates δ to ν.

3.4.4. *Matching between interior and boundary layers*

Here we must match terms from the boundary layer which are exponentially small as $Y \to -\infty$. The important quantity in the matching is the term $-\beta_n e^Y$ in $G_n = F_{nYYYY}$, the constant β_n also being determined by solving (26); the first few terms are $\beta_0 = 1$, $\beta_1 = -2$, $\beta_2 = -\frac{9}{2}$ and $G_0(Y) = -\exp Y$, $G_1(Y) = (-\frac{1}{2}Y^2 - Y + 2)\exp Y$. Since (24) implies

$$\hat{G} \sim -\sqrt{2\pi}\, e^{\hat{Y}^2/2} \qquad \text{as } \hat{Y} \to +\infty,$$

writing

$$\beta \sim \sum_{n=0}^{N} \delta^n \beta_n \tag{27}$$

and matching requires, to any order N, that

$$\delta\beta(\delta) \sim \nu^2(\alpha(\delta)\delta)^{5/2}\sqrt{2\pi}\, e^{\alpha(\delta)/2\delta}; \tag{28}$$

hence, as promised, ν is exponentially small in δ.

3.4.5. *The outer solution*

The leading-order outer solution is simply, since the ϵ term in (6) is negligible,

$$f \sim -\frac{2\delta\alpha}{\epsilon\pi}\cos(\tfrac{1}{2}\pi y), \tag{29}$$

where we have required that to leading order (since $\delta/\epsilon \gg 1$) $f = 0$ at $y = \pm 1$ and $f' = \delta\alpha$ at $y = 1$ (to match with (23)). Matching with the second term in (23) then requires

$$\nu \sim \frac{\pi^2\epsilon^2}{4\alpha\delta}, \tag{30}$$

and (28) and (30) are the two equations determining $\delta(\epsilon)$ and $\nu(\epsilon)$, the functions $\alpha(\delta)$ and $\beta(\delta)$ being known from the analysis above. To leading order we have

$$\delta \sim \tfrac{1}{8}\log(1/\epsilon), \qquad \nu \sim 2\pi^2\epsilon^2\log(1/\epsilon) \qquad \text{as } \epsilon \to 0 \tag{31}$$

and, since the expansions proceed in powers of δ and are divergent, the use of optimal truncation methods when determining α and β (with the full balances in (28) and (30) being solved for ν and δ) typically provides the most effective way to obtain an accurate solution. The results of such a calculation are summarised in Figure 1 and Table 1.

Table 1. Comparison between the 'exact' (numerical) value of $M \equiv \max|f(y)|$, and the leading-order asymptotic approximation that $M_a = 2\alpha\delta/\pi\epsilon$. Terms up to $\alpha_{N+1}\delta^{N+1}$ and $\beta_N\delta^N$ are included in the sums used to compute α and β, where N is either 19 (the largest value of N for which we have computed α_{N+1} and β_N), or some smaller value of N, indicated in the table, which corresponds to truncating the series for β after its smallest term. For the examples given, this truncation coincides with the optimal truncation for α.

N	R	δ	α	M_a	M	Error
5	20·0	0·07331	1·1290	1·0538	1·2148	13%
5	30·0	0·05858	1·0867	1·2157	1·3404	9·3%
14	203·4503	0·03101	1·0365	4·1612	3·9662	4·9%
19	711·8385	0·02375	1·0266	11·0502	10·7866	2·4%
19	2462·234	0·01928	1·0211	30·853	30·565	0·95%
19	3572·189	0·01825	1·0198	42·314	42·018	0·71%
19	8457·171	0·01623	1·01746	88·915	88·579	0·38%
19	12256·89	0·01549	1·01660	122·909	122·495	0·34%

3.4.6. *Boundary layer at $y = -1$*

Because (29) does not satisfy $f' = 0$ at $y = -1$, a boundary layer is also needed at the left-hand (lower) boundary, but this is passive as far as the matching is concerned, the behaviour near $y = 1$ being what is crucial in governing the leading-order solution. The inner scalings are

$$y = -1 + \epsilon\bar{Y}/(\alpha\delta)^{1/2}, \qquad f = (\alpha\delta)^{1/2}\bar{F},$$

leading, on matching with (29), to the Falkner–Skan problem

$$\begin{aligned} &\bar{F}_0''' + \bar{F}_0'^2 - \bar{F}_0\bar{F}_0'' = 1, \\ &\text{at } \bar{Y} = 0 \qquad \bar{F}_0 = \bar{F}_0' = 0, \\ &\text{as } \bar{Y} \to +\infty \qquad \bar{F}_0 \sim -\bar{Y}, \end{aligned}$$

so that $\bar{F}_0 = -\rho(\bar{Y})$ with ρ given by (19).

3.5. SUMMARY

As evidenced by Figure 1 and by the preceding discussion, asymptotic methods provide very effective means for constructing accurate analytic approximations over most of the range of Reynolds numbers. Our asymptotic analysis here has been fairly complete – we have applied regular perturbation methods for small R, together with appropriate manipulations of the resulting series (we note that treatment of the series by Padé approximants (*cf.* Drazin and Tourigny [23]) does not seem to help much in improving its convergence), together with singular perturbation methods in the limits $R \to -\infty$ and $R \to +\infty$. Our results in the limit $R \to +\infty$ are worth highlighting, being significantly more accurate than those which have been obtained before, the optimal truncation approach enabling us to obtain algebraic accuracy in ϵ despite many of the expansions proceeding in powers of $1/\log(1/\epsilon)$. Such approaches to problems involving logarithmic expansions should prove much more generally applicable and related developments will be reported in more detail elsewhere, our main focus here being on the time-dependent problem discussed in the remaining sections.

4. Time-dependent flows: numerical results

4.1. Numerical scheme for solving the initial-value problem

Our numerical scheme for computing time-dependent flows is broadly based on one of Glenn Ierley (personal communication). Equation (1) (or in practice (3)), together with boundary conditions (2), is solved using the Chebyshev tau method [24]. We represent functions of y as sums of Chebyshev polynomials, truncated at some finite order. We find it convenient then to consider any function as being equivalent to the vector of coefficients in this finite sum. We represent numerically the solution $\mathcal{F}(y, t_n)$ at the nth time step as the vector $\boldsymbol{f}^n = (f_0^n, \dots, f_K^n)^T$, where the superscript T denotes the transpose, with

$$\mathcal{F}(y, t_n) = \sum_{k=0}^{\infty} f_k^n T_k(y), \tag{32}$$

$T_k(y)$ being the k-th Chebyshev polynomial. Derivatives of $\mathcal{F}$ are calculated in spectral space, while products are calculated in physical space.

We discretise (3) in time as

$$\mathcal{F}_y^{n+1} - \mathcal{F}_y^{n-1} = \epsilon\Delta_t(\mathcal{F}_{yyy}^{n+1} + \mathcal{F}_{yyy}^{n-1}) + 2\Delta_t(\mathcal{F}_y^2 - \mathcal{F}\mathcal{F}_{yy} + p(t))^n, \tag{33}$$

where Δ_t is the time step and the superscript $n-1$, n or $n+1$ indicates evaluation at the corresponding time step. This expression is then readily rearranged to give an approximate equation for the Chebyshev coefficients $\boldsymbol{f}^{n+1}$ of the form

$$(D - \epsilon\Delta_t D^3)\boldsymbol{f}^{n+1} = \left[(D + \epsilon\Delta_t D^3)\boldsymbol{f}^{n-1} + 2\Delta_t \boldsymbol{g}^n\right] + (c, 0, 0, \dots, 0)^T, \tag{34}$$

where $c = 2\Delta_t p(t_n)$ and $\boldsymbol{g}^n = (g_0^n, \dots, g_K^n)^T$ is the vector of Chebyshev coefficients such that at the n-th time step

$$\mathcal{F}_y^2 - \mathcal{F}\mathcal{F}_{yy} = \sum_{k=0}^{\infty} g_k^n T_k(y). \tag{35}$$

The matrix D is the differentiation matrix, such that if the vector $\boldsymbol{a}$ represents the function $a(y)$ then $D\boldsymbol{a}$ represents $\mathrm{d}a/\mathrm{d}y$. In the tau method, the final three rows of the vector equation (34) are changed in order to force $\mathcal{F}^{n+1}$ to satisfy appropriate boundary conditions. Since we integrate an equation that is third-order in y, it is appropriate to apply three boundary conditions to (34); we choose to impose the homogeneous conditions from (2). To do this, the last three rows of the matrix $D - \epsilon\Delta_t D^3$ are replaced by

$$\begin{pmatrix} 1 & -1 & 1 & -1 & \cdots & (-1)^{K-1} & (-1)^K \\ 0 & -1 & 4 & -9 & \cdots & (-1)^{K-1}(K-1)^2 & (-1)^K K^2 \\ 0 & 1 & 4 & 9 & \cdots & (K-1)^2 & K^2 \end{pmatrix} \tag{36}$$

to give a modified matrix $\mathcal{D}$, and on the right-hand side of (34) the corresponding rows are replaced by zeros. The matrix $\mathcal{D}$ is then invertible, and so

$$\boldsymbol{f}^{n+1} = \mathcal{D}^{-1}\left[(D + \epsilon\Delta_t D^3)\boldsymbol{f}^{n-1} + 2\Delta_t \boldsymbol{g}^n\right]^{\dagger} + \mathcal{D}^{-1}(c, 0, \dots, 0)^T, \tag{37}$$

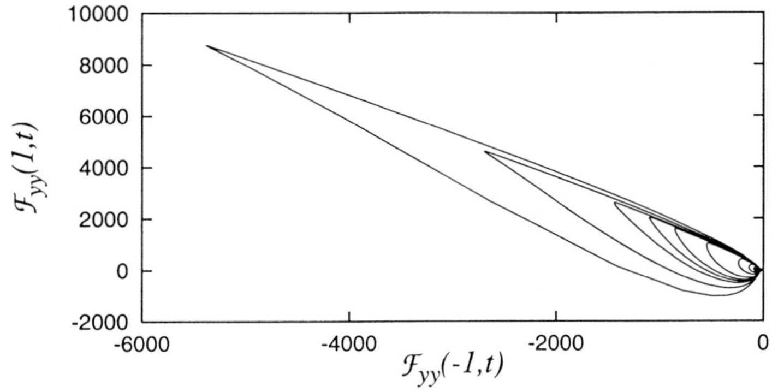

Figure 3. Limit-cycle solutions to (1–2) for $\epsilon = 0{\cdot}04$, 0·035, 0·03, 0·025, 0·02, 0·018, 0·017, 0·016, 0·014 and 0·012 (the cycles grow in size as ϵ decreases). Plotted are the quantities $\mathcal{F}_{yy}(-1,t)$ and $\mathcal{F}_{yy}(1,t)$, which are proportional to the wall stresses. Evolution around the limit cycle is clockwise. Note that the extreme values of $\mathcal{F}_{yy}(\pm 1,t)$ achieved around the cycle increase rapidly as ϵ is reduced.

where it is understood that the quantity $[\cdots]^{\dagger}$ in (37) is equal to the corresponding term in (34), modified by replacing the last three rows with zeros. At this point the quantity c is unknown, and must be determined by applying the remaining boundary condition, $\mathcal{F}(1,t_n) = 1$. Since $T_k(1) = 1$ for all k, we have

$$c = \frac{1 - \sum_{k=0}^{K} \alpha_k}{\sum_{k=0}^{K} \beta_k}, \tag{38}$$

where

$$\alpha = \mathcal{D}^{-1}\left[(D + \epsilon \Delta_t D^3)\boldsymbol{f}^{n-1} + 2\Delta_t \boldsymbol{g}^n\right]^{\dagger}$$

and $\beta = \mathcal{D}^{-1}(1, 0, 0, \cdots, 0)^T$. This completes the evaluation of $\boldsymbol{f}^{n+1}$.

Our implementation of the numerical scheme is rather unsophisticated: for example, we use a constant time step Δ_t in each calculation, although, as we shall see, the nature of the time-dependent solutions suggests that an adaptive time-stepping algorithm would lead to a more efficient numerical solution. For each choice of ϵ, the time step Δ_t is chosen so that calculations using a smaller value of Δ_t give essentially the same results, and the numerical solution is considered to be converged.

4.2. Numerical results for the limit cycle

For $R < R_3$, the system (1–2) has at least one stable steady solution, and the numerical simulation of the initial-value problem appears to converge to a steady state at large time. The bifurcation of the steady state at $R = R_3$ is a Hopf bifurcation, at which two complex conjugate eigenvalues cross the imaginary axis. We find that the time-dependent numerical solution is attracted to the steady state for $R < R_3$ but approaches a small limit cycle for R just beyond R_3, and so we infer that the Hopf bifurcation is supercritical.

Figure 3 shows the limit cycle (by which we mean a spatially non-uniform solution that is periodic in time) for a variety of values of ϵ. It is notable that the greatest wall stresses around the limit cycle grow rapidly with decreasing ϵ. This feature of the solution makes reliable

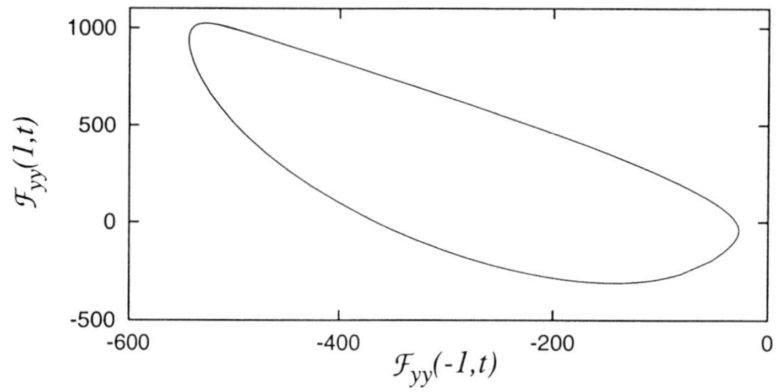

Figure 4. Limit-cycle solution to (1–2) for $\epsilon = 0{\cdot}02$. Plotted are the quantities $\mathcal{F}_{yy}(-1, t)$ and $\mathcal{F}_{yy}(1, t)$, which are proportional to the wall stresses.

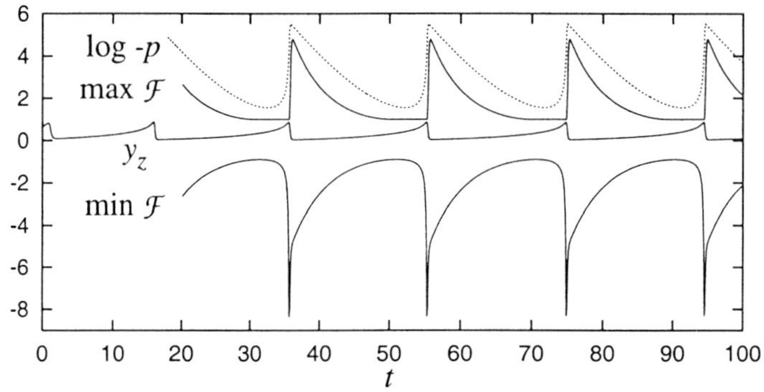

Figure 5. Time evolution of $\max_y \mathcal{F}(y, t)$, $\min_y \mathcal{F}(y, t)$, $y_z(t)$ and $\log -p(t)$ for the limit-cycle solution to (1–2) for $\epsilon = 0{\cdot}02$. Note the separation of the cycle into slow and fast phases.

computation of the solution difficult in the limit $\epsilon \to 0$. Another difficulty in computing these limit cycles for small ϵ is that evolution of $\mathcal{F}$ is extremely rapid around a small part of the limit cycle. Our method of non-adaptive time-stepping therefore becomes very inefficient in this limit.

In Figure 4 we show the limit cycle for $\epsilon = 0{\cdot}02$. More detail of this solution is shown in Figure 5, where we plot the maximum and minimum values of $\mathcal{F}(y, t)$ across the channel as functions of time. Also shown are the evolution of the pressure coefficient $p(t)$ and the interior zero $y_z(t)$ of $\mathcal{F}$, which satisfies $\mathcal{F}(y_z(t), t) = 0$ with $-1 < y_z(t) < 1$. Already for this quite moderate value of ϵ, the separation of the evolution of $\mathcal{F}(y, t)$ into slow and fast phases can clearly be discerned in Figure 5.

For small ϵ, it is a feature of the solution that the interior zero of $\mathcal{F}$ comes very close to the upper wall $y = 1$ near the start of the fast phase of the limit cycle. The zero is then swept rapidly back almost to the midline of the channel, $y = 0$. To characterise this feature, for a given limit cycle we let the minimum value of $y_z(t)$ be y_*; Figure 6 shows how y_* and $\min(1 - y_z(t))$ vary with ϵ.

4.3. PROFILES OF $\mathcal{F}(y, t)$

We now turn to the profile of $\mathcal{F}(y, t)$ during various phases of the limit cycle. To illustrate the different profiles, we show in Figure 7 the limit cycle for $\epsilon = 0{\cdot}011$, which is the smallest value

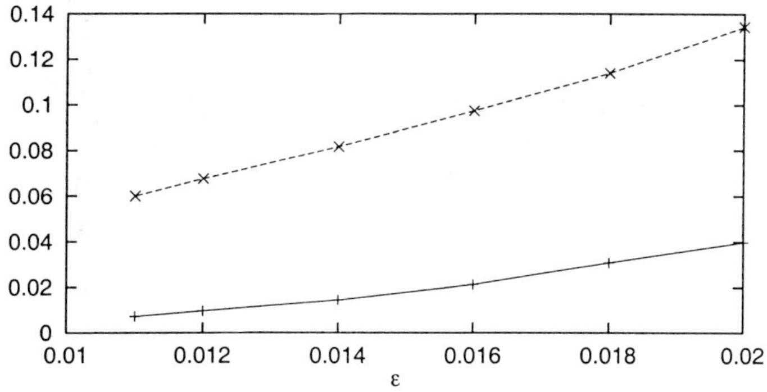

Figure 6. Lower line: plot of y_* against ϵ, where y_* is the minimum value of $y_z(t)$ over one cycle, and $\mathcal{F}(y_z(t), t) = 0$. Upper line: minimum value of $1 - y_z(t)$ over one cycle, plotted against ϵ.

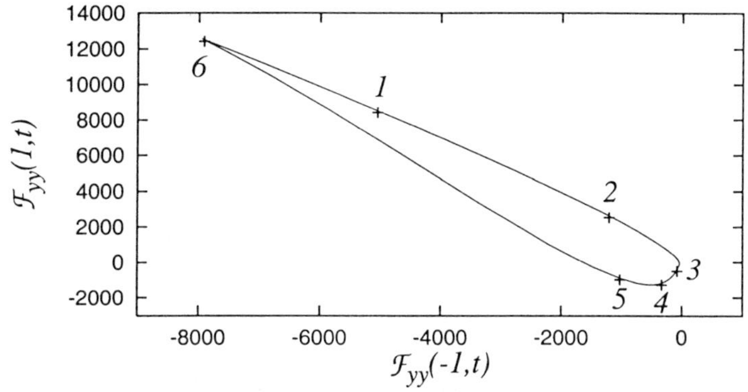

Figure 7. The limit cycle for $\epsilon = 0{\cdot}011$. The evolution of the solution from point 4 to point 6 is very rapid: for example, the time interval from point 1 to point 2 is 9 time units, while that from point 4 to point 6 is 0·15 time units. Evolution around the limit cycle is clockwise.

of ϵ for which we have reliable numerical results, and in subsequent figures the corresponding fixed-t profiles of $\mathcal{F}$.

Figure 8 shows the form of $\mathcal{F}(y, t)$ at time intervals of 0·5 between points 1 and 2 in Figure 7. Here, the solution is approximately sinusoidal away from the walls and decays

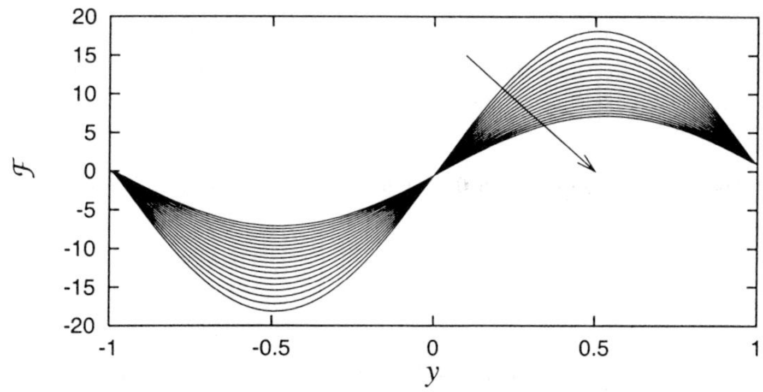

Figure 8. Profiles of $\mathcal{F}$ on the limit cycle for $\epsilon = 0{\cdot}011$. The profiles are shown at equal time intervals between points 1 and 2 in Figure 7. The arrow shows increasing t. The time interval between successive profiles is 0·5. Evolution of $\mathcal{F}$ during this phase of the limit cycle is relatively slow.

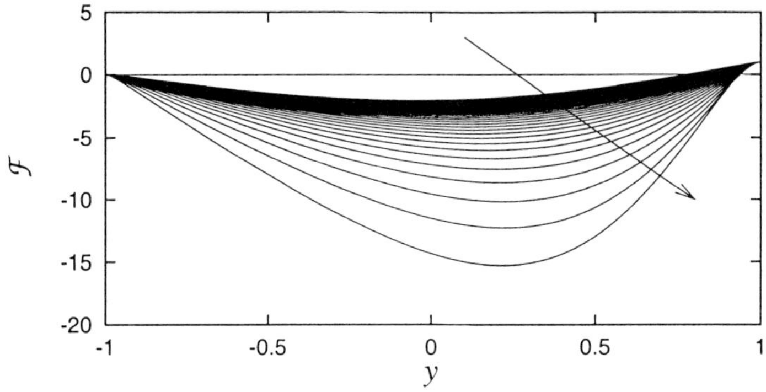

Figure 9. Profiles of $\mathcal{F}$ on the limit cycle for $\epsilon = 0{\cdot}011$. Also plotted is the y axis, for ease of identifying the interior zero of $\mathcal{F}$, $y_z(t)$. The profiles are shown at equal time intervals between points 3 and 5 in Figure 7. The arrow shows increasing t. The time interval between successive profiles is 0·01. Note the rapid evolution of $\mathcal{F}$ as y_z approaches its minimum distance from the upper wall.

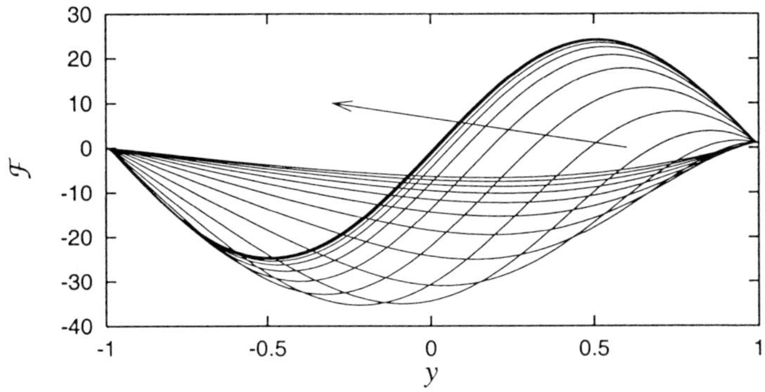

Figure 10. Profiles of $\mathcal{F}$ on the limit cycle for $\epsilon = 0{\cdot}011$. The profiles are shown at equal time intervals between points 4 and 6 in Figure 7. The arrow shows increasing t. The time interval between successive profiles is 0·01. Note the rapid evolution of $\mathcal{F}$ during this phase.

slowly. The approach of the interior zero y_z to the upper wall is illustrated in Figure 9, which shows the evolution of $\mathcal{F}$ at time intervals of 0·01 between points 3 and 5. A fast phase, during which y_z decreases almost to zero, is shown in Figure 10, where the profile of $\mathcal{F}$ is plotted at time intervals of 0·01 between points 4 and 6.

Such numerical results are very helpful in clarifying the spatio-temporal evolution of $\mathcal{F}(y, t)$ and now we turn to a small-ϵ asymptotic description of the limit cycle, restricting the analysis to solutions that are periodic in time.

5. Time-dependent flows: asymptotics of the slow phases

5.1. An exact solution

5.1.1. *Formulation*

We start this section by noting a class of exact solutions to (1); we shall subsequently explain the asymptotic relevance of the results to the boundary-value problem (1–2) in the limit $\epsilon \to 0$. The class of solutions we consider takes the form

$$\mathcal{F} = A(t) + B(t)y + C(t)\sin(\lambda(t)y) + D(t)\cos(\lambda(t)y), \tag{39}$$

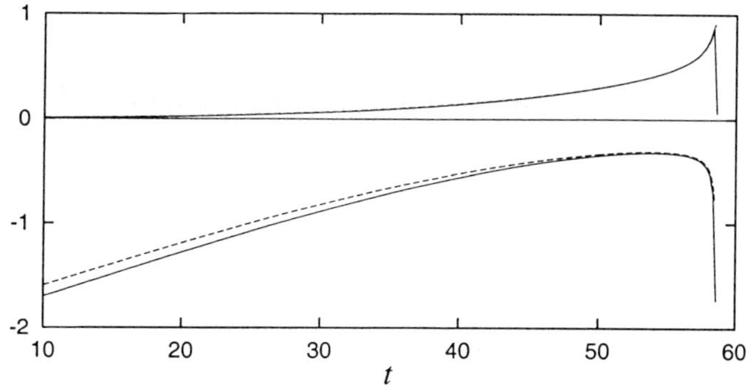

Figure 11. Comparison between asymptotics and numerics for $\epsilon = 0{\cdot}011$. The solid lines represent numerical simulations of (1–2), while the dashed lines give asymptotic results. The left- and right-hand sides of the plot correspond, respectively, to points just after and just before point 6 on Figure 7; we thus show almost a complete cycle. The upper curves show $y_z(t)$. The dashed line is the corresponding quantity calculated asymptotically by solving $\mathcal{F}(y_z(t), t) = 0$, with $\mathcal{F}(y, t)$ given in (39) or (68); the quantities A and B are given in terms of C, D and λ in (44), these last three quantities evolving according to (45). The lower curves show $-\log(-p(t))/5$, where the dashed line shows p as given in (41).

this functional form having originally been identified through its repeated occurrence when pursuing an asymptotic analysis of (1); once identified, it enables us to give a much more concise presentation of that analysis than would otherwise be the case. Equation (1) is of a quadratically nonlinear form, numerous other such systems amenable to low-dimensional reductions akin to (39) having been previously identified (see, for example, [25] and [26]). Unlike most earlier examples, however, (39) yields an underdetermined system when substituted in (1), enabling two boundary conditions to be imposed (see below; most applications of previous examples were restricted to initial-value problems). Another noteworthy feature of the current example is the appearance of a time-dependent wavelength, implying in particular the possibility of a smooth transition from trigonometric to hyperbolic functions with λ becoming imaginary on passing through zero (a scenario we shall need to make use of later).

While the reduction (39) contains five degrees of freedom, substitution in (1) leads (somewhat remarkably) to three constraints only, namely

$$\begin{aligned} \frac{\mathrm{d}\lambda}{\mathrm{d}t} &= -\lambda B, \\ \frac{\mathrm{d}C}{\mathrm{d}t} &= -\epsilon\lambda^2 C + 3BC + \lambda AD, \\ \frac{\mathrm{d}D}{\mathrm{d}t} &= -\epsilon\lambda^2 D + 3BD - \lambda AC. \end{aligned} \tag{40}$$

From (3) we have

$$\frac{\mathrm{d}B}{\mathrm{d}t} = B^2 + \lambda^2(C^2 + D^2) + p, \tag{41}$$

which gives $p(t)$ in terms of the other unknowns. It follows from (40) that

$$\frac{\mathrm{d}}{\mathrm{d}t}\left(\lambda^3(C^2 + D^2)^{1/2}\right) = -\epsilon\lambda^5(C^2 + D^2)^{1/2}, \tag{42}$$

a combination which will prove important in what follows. Because the system (40) represents only three constraints, we are able to impose two boundary conditions, the appropriate choice at this stage (for reasons indicated below) being the 'inviscid' conditions

$$\mathcal{F}(-1,t) = 0, \quad \mathcal{F}(1,t) = 1, \tag{43}$$

giving

$$A = \tfrac{1}{2} - D\cos\lambda, \quad B = \tfrac{1}{2} - C\sin\lambda, \tag{44}$$

and hence

$$\begin{aligned} \frac{d\lambda}{dt} &= -\lambda\left(\tfrac{1}{2} - C\sin\lambda\right), \\ \frac{dC}{dt} &= -\epsilon\lambda^2 C + 3\left(\tfrac{1}{2} - C\sin\lambda\right)C + \lambda\left(\tfrac{1}{2} - D\cos\lambda\right)D, \\ \frac{dD}{dt} &= -\epsilon\lambda^2 D + 3\left(\tfrac{1}{2} - C\sin\lambda\right)D - \lambda\left(\tfrac{1}{2} - D\cos\lambda\right)C. \end{aligned} \tag{45}$$

We shall discuss the solution to (45) subject to (for reasons which will again become clearer later)

$$\text{at } t = 0 \quad \lambda = \pi, \quad C = C_I(\epsilon), \quad D = -\tfrac{1}{2}, \tag{46}$$

where $C_I(\epsilon) \to +\infty$ as $\epsilon \to 0$ in a manner which will be determined in Section 7. A numerical solution to (45–46), together with a corresponding numerical simulation of (1–2), is given in Figure 11. We now describe the limiting behaviour of (45–46) as $\epsilon \to 0$.

5.1.2. *Slowest phases*

The results for the first (large-amplitude) phase, corresponding to solutions roughly between points 6 and 2 in Figure 7, are implicit in [15]. The appropriate scalings are (from (45))

$$t = \epsilon^{-1}\hat{T}, \quad C = C_I(\epsilon)\hat{C}, \quad \lambda \sim \pi, \quad D \sim -\tfrac{1}{2}, \tag{47}$$

it then following immediately from (42) that the leading order solution is simply

$$\hat{C}_0(\hat{T}) = e^{-\pi^2\hat{T}}, \tag{48}$$

with

$$\lambda(\hat{T}) \sim \pi - \frac{1}{2C_I(\epsilon)\hat{C}_0} \qquad \text{as } \epsilon \to 0. \tag{49}$$

The other phase of slowest evolution corresponds to $T = O(1)$, where

$$\hat{T} = \frac{1}{\pi^2}\log(C_I(\epsilon)) + T \tag{50}$$

with

$$\lambda \sim \lambda_0(T), \quad C \sim C_0(T), \quad D \sim D_0(T) \qquad \text{as } \epsilon \to 0,$$

so that (from (42) and (45))

$$C_0\sin\lambda_0 = \tfrac{1}{2}, \qquad D_0\cos\lambda_0 = \tfrac{1}{2}, \tag{51}$$

and

$$\frac{\mathrm{d}}{\mathrm{d}T}\left(\frac{\lambda_0^3}{\sin 2\lambda_0}\right) = -\frac{\lambda_0^5}{\sin 2\lambda_0}, \qquad (52)$$
$$\text{as } T \to -\infty \quad \lambda_0 \sim \pi - \tfrac{1}{2}\mathrm{e}^{\pi^2 T},$$

where we have matched with (49); we note that $\sin 2\lambda_0 < 0$. We have

$$p \sim -\left(\frac{4\lambda_0}{\sin 2\lambda_0}\right)^2.$$

It follows from (52) that λ_0 decreases monotonically until it reaches $\lambda_0 = \lambda_c$, where

$$3\tan 2\lambda_c = 2\lambda_c, \qquad \lambda_c \approx 2{\cdot}039, \qquad (53)$$

at which point the solution suffers finite-time breakdown, with

$$\lambda \sim \lambda_c + \frac{\lambda_c^2}{(2\lambda_c^2+3)^{1/2}}(T_{c0}-T)^{1/2} \qquad \text{as } T \to T_{c0}^- \qquad (54)$$

for some T_{c0}. This breakdown of the solution to (52) leads on to a sequence of increasingly shorter timescales and we now describe the first of these.

5.1.3. *Intermediate phase*
The appropriate scalings here are

$$T = T_c(\epsilon) + \epsilon^{2/3}\tau, \qquad \lambda = \lambda_c + \epsilon^{1/3}\Lambda, \qquad (55)$$

where $T_c(0) = T_{c0}$, and

$$C = \frac{1}{2\sin\lambda} + \epsilon^{2/3}c, \qquad D = \frac{1}{2\cos\lambda} + \epsilon^{2/3}d. \qquad (56)$$

The prediction of much faster evolution when λ reaches the particular value λ_c is noteworthy. To leading order, (45) yields only two independent constraints, but a third is provided at once by (42) (which thus acts as a solvability condition), giving the system

$$\frac{\mathrm{d}\Lambda_0}{\mathrm{d}\tau} = \lambda_c \sin\lambda_c\, c_0, \qquad \frac{\mathrm{d}\Lambda_0}{\mathrm{d}\tau} = \lambda_c\cos\lambda_c\, d_0,$$
$$\frac{2(2\lambda_c^2+3)}{\sin 2\lambda_c}\Lambda_0\frac{\mathrm{d}\Lambda_0}{\mathrm{d}\tau} + \lambda_c^2\left(\cos\lambda_c\frac{\mathrm{d}c_0}{\mathrm{d}\tau} + \sin\lambda_c\frac{\mathrm{d}d_0}{\mathrm{d}\tau}\right) = -\frac{\lambda_c^4}{\sin 2\lambda_c}. \qquad (57)$$

Matching back as $\tau \to -\infty$ we have, by suitable choice of the origin of τ in (55),

$$\sin\lambda_c\, c_0 = \cos\lambda_c\, d_0, \qquad \frac{2\lambda_c^2+3}{2\sin\lambda_c}\Lambda_0^2 + \lambda_c^2 c_0 = \frac{\lambda_c^4}{2\sin\lambda_c}(-\tau) \qquad (58)$$

and

$$\frac{\mathrm{d}\Lambda_0}{\mathrm{d}\tau} = \frac{1}{2\lambda_c}\left(\lambda_c^4(-\tau) - (2\lambda_c^2+3)\Lambda_0^2\right), \qquad (59)$$

so that

$$\Lambda_0 = \frac{2\lambda_c}{(2\lambda_c^2+3)}\frac{\mathrm{d}}{\mathrm{d}\tau}\left(\log\left(\mathrm{Ai}\left(\left(\frac{\lambda_c^2(2\lambda_c^2+3)}{4}\right)^{1/3}(-\tau)\right)\right)\right). \tag{60}$$

It follows from (59) that

$$\Lambda_0 \sim -\frac{2\lambda_c}{(2\lambda_c^2+3)}\frac{1}{(\tau_{c0}-\tau)} \qquad \text{as } \tau \to \tau_{c0}^- \tag{61}$$

for some τ_{c0} which can be expressed using (60) in terms of the first zero of the Airy function. This blow-up in the current formulation requires consideration of the next (and final) phase of this section, during which viscous effects play no role at leading order.

5.1.4. *Inviscid phase*
We write

$$\tau = \tau_c(\epsilon) + \epsilon^{1/3}\hat{t}, \tag{62}$$

where $\tau_c(0) = \tau_{c0}$, and let

$$\lambda \sim \lambda_0(\hat{t}), \quad C \sim C_0(\hat{t}), \quad D \sim D_0(\hat{t}) \qquad \text{as } \epsilon \to 0,$$

these functions being distinct from those which appeared in Section 5.1.2. In (45) it is then only the viscous terms that are negligible, so that

$$\begin{aligned}
\frac{\mathrm{d}\lambda_0}{\mathrm{d}\hat{t}} &= -\lambda_0\left(\tfrac{1}{2} - C_0\sin\lambda_0\right),\\
\frac{\mathrm{d}C_0}{\mathrm{d}\hat{t}} &= 3\left(\tfrac{1}{2} - C_0\sin\lambda_0\right)C_0 + \lambda_0\left(\tfrac{1}{2} - D_0\cos\lambda_0\right)D_0,\\
\frac{\mathrm{d}D_0}{\mathrm{d}\hat{t}} &= 3\left(\tfrac{1}{2} - C_0\sin\lambda_0\right)D_0 - \lambda_0\left(\tfrac{1}{2} - D_0\cos\lambda_0\right)C_0,
\end{aligned} \tag{63}$$

subject to $\lambda_0 \to \lambda_c$, $C_0 \to \frac{1}{2}\sin\lambda_c$, $D_0 \to \frac{1}{2}\cos\lambda_c$ as $\hat{t} \to -\infty$, implying that

$$\lambda_0^3(C_0^2 + D_0^2)^{1/2} = \lambda_c^3/(-\sin 2\lambda_c), \tag{64}$$

a constant (*cf.* (42)). The solution to (63) also suffers finite-time blow-up at $\hat{t} = \hat{t}_{c0}$, say, with

$$C_0 \sim \frac{\lambda_c^3}{\lambda_0^3\sin 2\lambda_c}, \qquad D \sim \frac{D_c}{\lambda_0^2}, \tag{65}$$

for some constant D_c, and

$$\lambda_0 \sim \left(\frac{2\lambda_c^3}{-\sin 2\lambda_c}\right)^{1/2}\left(\hat{t}_{c0} - \hat{t}\right)^{1/2} \qquad \text{as } \hat{t} \to \hat{t}_{c0}^-. \tag{66}$$

It follows from (39), (44) that at $\hat{t} = \hat{t}_c$

$$\mathcal{F} \sim \tfrac{1}{2}(1+y) + \tfrac{1}{2}D_c(1-y^2) - \tfrac{1}{6}C_c y(1-y^2), \tag{67}$$

where $C_c = \lambda_c^3/(-\sin 2\lambda_c)$. In terms of the solution to the inviscid RPJ equation, the apparent singularity in (65–66) is thus illusory; in other words, there is no breakdown in the partial differential equation itself but only in the *ansatz* needed to describe its solution, which ceases to be of the form (39) and becomes

$$\mathcal{F} = A(t) + B(t)y - \Gamma(t)\sinh(\mu(t)y) + D(t)\cosh(\mu(t)y), \tag{68}$$

which corresponds to setting $\lambda = i\mu$, $C = i\Gamma$ in (39), the asymptotic solution shown in Figure 11 having been constructed by solving (45) numerically for $t < t_c$ and the analogous system for (68–69) for $t > t_c$, proceeding through $t = t_c$ via (67); blow-up behaviour of the form (65–66) applies to the full system (45) as well as to the $\epsilon \to 0$ limit. We now have

$$A = \tfrac{1}{2} - D\cosh\mu, \quad B = \tfrac{1}{2} + \Gamma\sinh\mu \tag{69}$$

and, instead of (63),

$$\begin{aligned}
\frac{d\mu_0}{d\hat{t}} &= -\mu_0\left(\tfrac{1}{2} + \Gamma_0\sinh\mu_0\right), \\
\frac{d\Gamma_0}{d\hat{t}} &= 3\left(\tfrac{1}{2} + \Gamma_0\sinh\mu_0\right)\Gamma_0 + \mu_0\left(\tfrac{1}{2} - D_0\cosh\mu_0\right)D_0, \\
\frac{dD_0}{d\hat{t}} &= 3\left(\tfrac{1}{2} + \Gamma_0\sinh\mu_0\right)D_0 + \mu_0\left(\tfrac{1}{2} - D_0\cosh\mu_0\right)\Gamma_0,
\end{aligned} \tag{70}$$

with, in place of (64),

$$\mu_0^3\left(\Gamma_0^2 - D_0^2\right)^{1/2} = \lambda_c^3/(-\sin 2\lambda_c). \tag{71}$$

Initial conditions on (70) are

$$\mu_0 \sim \left(\frac{2\lambda_c^3}{-\sin 2\lambda_c}\right)\left(\hat{t} - \hat{t}_{c0}\right)^{1/2} \qquad \text{as } \hat{t} \to \hat{t}_{c0}^+$$

with

$$\Gamma_0 \sim \frac{\lambda_c^3}{\mu_0^3\sin 2\lambda_c}, \qquad D \sim -\frac{D_c}{\mu_0^2}.$$

The solution to (70) blows up in finite time, at $\hat{t} = \hat{t}_{b0}$ say, and we need to describe this blow-up behaviour in some detail. As $\hat{t} \to \hat{t}_{b0}^-$, we have $\mu_0 \to +\infty$ with $\Gamma_0 \sim -D_0$ so that

$$\frac{d\mu_0}{d\hat{t}} \sim \tfrac{1}{2}\mu_0 D_0 e^{\mu_0}, \qquad \frac{dD_0}{d\hat{t}} \sim \tfrac{1}{2}\mu_0 D_0^2 e^{\mu_0} - \tfrac{3}{2}D_0^2 e^{\mu_0}, \tag{72}$$

from which it follows that

$$D_0 \sim D_b\mu_0^{-3}e^{\mu_0}$$

for some constant D_b and hence that as $\hat{t} \to \hat{t}_{b0}^-$,

$$\begin{aligned}
\mu_0 &\sim -\tfrac{1}{2}\log(\hat{t}_{b0} - \hat{t}) + \log(-\log(\hat{t}_{b0} - \hat{t})) - \tfrac{1}{2}\log D_b - \log 2, \\
D_0 &\sim \frac{4D_b^{1/2}}{(\hat{t}_{b0} - \hat{t})^{1/2}(-\log(\hat{t}_{b0} - \hat{t}))^2}.
\end{aligned} \tag{73}$$

The asymptotic behaviour of

$$\mathcal{F}_0 = \left(\tfrac{1}{2} - D_0 \cosh \mu_0\right) + \left(\tfrac{1}{2} + \Gamma_0 \sinh \mu_0\right) y - \Gamma_0 \sinh(\mu_0 y) + D_0 \cosh(\mu_0 y) \tag{74}$$

therefore subdivides into three regions as $\hat{t} \to \hat{t}_{b0}^-$, as follows.
(I) $y = O(1)$ with $y < 1$.

$$\mathcal{F}_0 \sim -\frac{1}{(\hat{t}_{b0} - \hat{t})(-\log(\hat{t}_{b0} - \hat{t}))}(1 + y). \tag{75}$$

(II) $1 - y = O(1/(-\log(\hat{t}_{b0} - \hat{t})))$.

$$\mathcal{F}_0 \sim -\frac{2}{(\hat{t}_{b0} - \hat{t})(-\log(\hat{t}_{b0} - \hat{t}))}\left(1 - \exp(\log(\hat{t}_{b0} - \hat{t})(1 - y)/2)\right). \tag{76}$$

(III) $1 - y = O(\hat{t}_{b0} - \hat{t})$.

$$\mathcal{F}_0 \sim 1 - \frac{1 - y}{\hat{t}_{b0} - \hat{t}}, \tag{77}$$

so that the interior zero of $\mathcal{F}$ satisfies

$$y_z \sim 1 - (\hat{t}_{b0} - \hat{t}) \qquad \text{as } \hat{t} \to \hat{t}_{b0}^-. \tag{78}$$

It is noteworthy that none of these leading-order expressions depends on D_b.

Unlike (65–66), this form of singularity in the solution is genuine, rather than simply reflecting a straightforward change in the appropriate representation of the solution. In other words, when $\hat{t}$ becomes sufficiently close to $\hat{t}_{b0}$, the *ansatz* (68) ceases to apply and a quite different approach is needed. We discuss the behaviour of this next phase in Section 6. Having been of $O(1)$ for $\hat{t} = O(1)$ (during which it reaches its minimum value), $-p$ blows up as $\hat{t}_{b0}$ is approached (*cf.* Figure 11) according to

$$p \sim -\frac{1}{(\hat{t}_{b0} - \hat{t})^2(-\log(\hat{t}_{b0} - \hat{t}))^2} \qquad \text{as } \hat{t} \to \hat{t}_{b0}^-. \tag{79}$$

5.2. Other regions

5.2.1. *Preliminaries*

The solution (39), (68), subject to (43), provides an asymptotic description of the behaviour away from the boundary layers at $y = \pm 1$. The purpose of the current subsection is to describe these boundary layers and to indicate why the interior layer located about the zero of $\mathcal{F}$ (which played such a crucial role in Section 3.4) does not need separate treatment here. The other issue which needs to be resolved in order to demonstrate the applicability of (39) concerns the initial data; it is necessary for the above reduction to be valid that (39), (44) (together with (46)) apply to leading order at $t = 0$. Since we are seeking solutions which are periodic in time, this is a point to which we shall have to return at the end of the analysis.

It is instructive to note next the balances in (1) which correspond to the various timescales in Section 5.1. On the slow timescales of Section 5.1.2, the dominant balance is inviscid and quasi-steady; from

$$\epsilon \mathcal{F}_{yyT} = \epsilon \mathcal{F}_{yyyy} + \mathcal{F}_y \mathcal{F}_{yy} - \mathcal{F} \mathcal{F}_{yyy} \tag{80}$$

we have at leading order

$$\mathcal{F}_{0y}\mathcal{F}_{0yy} - \mathcal{F}_0\mathcal{F}_{0yyy} = 0, \tag{81}$$

so, imposing (43), we find

$$\mathcal{F}_0 = \frac{\sin \lambda_0(T)(y+1)}{\sin 2\lambda_0(T)}; \tag{82}$$

the quantity $\lambda_0(T)$ is not determined by the inviscid balance (81) but requires consideration of the interior layer (as outlined below); its calculation is automatically included in the analysis of (39) because the *ansatz* encompasses both the outer regions $y = O(1)$, $-1 < y < y_z$ and $y_z < y < 1$, and the viscous interior layer about $y = y_z$.

The timescale of Section 5.1.3 is the one for which the use of the *ansatz* (39) leads to the greatest gain in brevity; basing the discussion on (1) instead would require the consideration of the first two correction terms to the leading-order outer solution

$$\mathcal{F}_0 = \frac{\sin \lambda_c(y+1)}{\sin 2\lambda_c},$$

which is independent of τ, together with their matching into the interior layer. Corresponding to Section 5.1.4, we have the inviscid leading-order balance

$$\mathcal{F}_{0y\hat{t}} = \mathcal{F}_{0y}\mathcal{F}_{0yy} - \mathcal{F}_0\mathcal{F}_{0yyy}, \tag{83}$$

$$\mathcal{F}_0(-1,\hat{t}) = 0, \quad \mathcal{F}_0(1,\hat{t}) = 1. \tag{84}$$

Two boundary conditions are sufficient here, despite (83) being third order in y; as explained in [15] (see also Appendix A.1), (83) has partly hyperbolic nature, with characteristic projections given by

$$\frac{\mathrm{d}y}{\mathrm{d}\hat{t}} = \mathcal{F}_0$$

and, because of (84), these carry information out of the fluid domain (in other words, the conditions (84) and the initial data are sufficient to determine $\mathcal{F}_0$, no new characteristics being generated at the boundaries). A crucial (and rather delicate) issue later will be to find balances whereby information is propagated back into the fluid from a viscous boundary layer (were this not possible the blow-up in (75–77) could not be obviated, because viscous effects could not influence (83), the solution to which suffers blow-up at $\hat{t} = \hat{t}_{b0}$).

We are now in a position to discuss the two boundary layers and the interior layer.

5.2.2. *Boundary layer at $y = -1$*

If we write $\mathcal{F} = \epsilon^{1/2}\bar{\mathcal{G}}$, $y = -1 + \epsilon^{1/2}\bar{Y}$, Equation (80) yields for $T = O(1)$

$$\begin{aligned}
&\bar{\mathcal{G}}_{0\bar{Y}\bar{Y}\bar{Y}\bar{Y}} + \bar{\mathcal{G}}_{0\bar{Y}}\bar{\mathcal{G}}_{0\bar{Y}\bar{Y}} - \bar{\mathcal{G}}_0\bar{\mathcal{G}}_{0\bar{Y}\bar{Y}\bar{Y}} = 0,\\
&\text{at } \bar{Y} = 0 \qquad \bar{\mathcal{G}}_0 = \bar{\mathcal{G}}_{0\bar{Y}} = 0,\\
&\text{as } \bar{Y} \to +\infty \qquad \bar{\mathcal{G}}_0 \sim \mathcal{F}_{0y}(-1,T)\bar{Y},
\end{aligned}$$

where $\mathcal{F}_0$ is given by (82). The scalings

$$\bar{\mathcal{G}}_0 = \left(-\mathcal{F}_{0y}(-1,T)\right)^{1/2}\bar{q}(\bar{\xi}), \quad \bar{Y} = \left(-\mathcal{F}_{0y}(-1,T)\right)^{-1/2}\bar{\xi}$$

reduce this to a standard Falkner–Skan problem (*cf.* Section 3.3 above and Section 3 of [7]).

For $\hat{t} = O(1)$ we obtain the full balance in (1), with

$$\begin{aligned} &\bar{\mathcal{G}}_{0\bar{Y}\bar{Y}\hat{t}} = \bar{\mathcal{G}}_{0\bar{Y}\bar{Y}\bar{Y}\bar{Y}} + \bar{\mathcal{G}}_{0\bar{Y}}\bar{\mathcal{G}}_{0\bar{Y}\bar{Y}} - \bar{\mathcal{G}}_{0}\bar{\mathcal{G}}_{0\bar{Y}\bar{Y}\bar{Y}}, \\ &\text{at } \bar{Y} = 0 \qquad \bar{\mathcal{G}}_0 = \bar{\mathcal{G}}_{0\bar{Y}} = 0, \\ &\text{as } \bar{Y} \to +\infty \qquad \bar{\mathcal{G}}_0 \sim \mathcal{F}_{0y}(-1, \hat{t})\bar{Y}, \end{aligned} \tag{85}$$

where $\mathcal{F}_0$ is the solution to (83–84). Little more can usefully be said about (85); its solution does not in any case influence the behaviour elsewhere. The other timescales of Section 5.1 do not warrant separate discussion.

5.2.3. *Boundary layer at $y = 1$*

From

$$\mathcal{F} = 1 + \epsilon\mathcal{G}, \quad y = 1 + \epsilon Y \tag{86}$$

we find, on any of the timescales of Section 5.2,

$$\begin{aligned} &\mathcal{G}_{0YYYY} - \mathcal{G}_{0YYY} = 0, \\ &\text{at } Y = 0 \qquad \mathcal{G}_0 = \mathcal{G}_{0Y} = 0, \\ &\text{as } Y \to -\infty \qquad \mathcal{G}_0 \sim \mathcal{F}_{0y}\big|_{y=1} Y, \end{aligned}$$

so that

$$\mathcal{G}_0 = \mathcal{F}_{0y}\big|_{y=1}\left(Y + 1 - \mathrm{e}^Y\right). \tag{87}$$

5.2.4. *Interior layer at $y = y_z$*

With the interior zero of $\mathcal{F}$ at $y = y_z(t; \epsilon)$, for $T = O(1)$ the inner scalings are

$$y = y_z(T; \epsilon) + \epsilon^{1/2} z, \quad \mathcal{F} = \epsilon^{1/2} g, \tag{88}$$

giving a balance in which the viscous term appears at leading order, namely

$$\epsilon g_{zzT} - \epsilon^{1/2}\frac{\mathrm{d}y_z}{\mathrm{d}T} g_{zzz} = g_{zzzz} + g_z g_{zz} - g g_{zzz}. \tag{89}$$

From (82) we have

$$y_z(T; 0) = \frac{\pi}{\lambda_0} - 1 \equiv y_{z0}$$

and

$$\mathcal{F}_0 = \mathcal{F}_{0yy} = \mathcal{F}_{0yyyy} = 0 \qquad \text{at } y = y_{z0}(T),$$

so if in (89) we write

$$g \sim g_0 + \epsilon^{1/2} g_1 + \epsilon g_2 + \epsilon^{3/2} g_3 + \epsilon^2 g_4$$

we obtain

$$\begin{aligned} g_0 = \mathcal{F}_{0y}(y_{z0}, T)z, \quad g_1 = \alpha_1(T)z, \quad g_2 = \alpha_2(T)z + \tfrac{1}{6}\mathcal{F}_{0yyy}(y_{z0}, T)z^3, \\ g_3 = \alpha_3(T)z - \tfrac{1}{2}\frac{\mathrm{d}y_{z0}}{\mathrm{d}T}\frac{\mathcal{F}_{0yyy}(y_{z0}, T)}{\mathcal{F}_{0y}(y_{z0}, T)} z^2 + \beta_3(T)z^3, \end{aligned} \tag{90}$$

where the calculation of α_1, α_2, α_3 and β_3 would require the construction of correction terms to the outer solution; they need not concern us here. The crucial equation arises from the z derivative of (89) at $O(\epsilon^2)$, giving

$$G_{4z} - \mathcal{F}_{0y}(y_{z0}, T)zG_4 = g_{2zzzT} - g_{2zz}^2 = \frac{\mathrm{d}}{\mathrm{d}T}\left(\mathcal{F}_{0yyy}(y_{z0}, T)\right) - \mathcal{F}_{0yyy}^2(y_{z0}, T)z^2, \tag{91}$$

where $G_4 = g_{4zzzz}$. Since G_4 is required not to grow exponentially as $|z| \to \infty$, we obtain from (91) that

$$G_4 = \frac{\mathcal{F}_{0yyy}^2(y_{z0}, T)}{\mathcal{F}_{0y}(y_{z0}, T)} z$$

together with the solvability condition

$$\frac{\mathrm{d}}{\mathrm{d}T}\left(\mathcal{F}_{0yyy}(y_{z0}, T)\right) = \frac{\mathcal{F}_{0yyy}^2(y_{z0}, T)}{\mathcal{F}_{0y}(y_{z0}, T)}; \tag{92}$$

since

$$\mathcal{F}_{0y}(y_{z0}, T) = -\frac{\lambda_0}{\sin 2\lambda_0}, \qquad \mathcal{F}_{0yyy}(y_{z0}, T) = \frac{\lambda_0^3}{\sin 2\lambda_0}$$

we thus recover (52), thereby completely specifying the outer solution (82).

There are two important points to make about the above. The first is that at each order the inner solution is simply a polynomial in z, so the upshot of the inner analysis (*i.e.* of the effects of viscosity) is that the outer solution is required at each order to be analytic at $y = y_{z0}$; given this constraint, (92) can instead be derived solely by consideration of the outer expansion. Secondly, the time derivative in (91) is crucial in suppressing exponential growth via the solvability condition (92). As is implicit in Section 3.4, in the steady-state problem exponential growth cannot in general be eliminated in an interior layer about a zero of f at which $f_y > 0$ and this feature is crucial in showing that the interior layer described in Section 3.4.2 must lie close to $y = 1$. By contrast, in the time-dependent problem y_z can be placed anywhere, but must evolve in a way which suppresses the exponential growth that would otherwise be present (namely, according to (92)).

The above solvability condition can be viewed as arising from the viscous effects in the interior layer. The situation for $\hat{t} = O(1)$ is rather different, the interior layer being passive as far as the matching is concerned, viscous effects playing no role in the solution to (83–84). In this case the inner scalings are

$$y = y_z(\hat{t}; \epsilon) + \epsilon\hat{z}, \qquad \mathcal{F} = \epsilon\hat{g}$$

with

$$\epsilon\hat{g}_{\hat{z}\hat{z}\hat{t}} - \frac{\mathrm{d}y_z}{\mathrm{d}\hat{t}}\hat{g}_{\hat{z}\hat{z}\hat{z}} = \hat{g}_{\hat{z}\hat{z}\hat{z}\hat{z}} + \epsilon(\hat{g}_{\hat{z}}\hat{g}_{\hat{z}\hat{z}} - \hat{g}\hat{g}_{\hat{z}\hat{z}\hat{z}}). \tag{93}$$

Hence at each order one has

$$\hat{g}_{n\hat{z}\hat{z}\hat{z}\hat{z}} + \frac{\mathrm{d}y_{z0}}{\mathrm{d}\hat{t}}\hat{g}_{n\hat{z}\hat{z}\hat{z}} = \text{polynomial in } \hat{z},$$

which itself has a polynomial solution and does not lead to a solvability condition. The corresponding equation for (89) reads

$$g_{nzzzz} + \mathcal{F}_{0y}(y_{z0}, T)g_{nzz} - \mathcal{F}_0(y_{z0}, T)zg_{nzzz} = \text{ polynomial in } z,$$

a solvability condition on the right-hand side arising at each order on requiring that g_n be a polynomial (or, equivalently, that exponential growth be absent) because the left-hand side vanishes for $g_n = z^3$.

This completes our description of the slower phases of evolution, which can be described most concisely via (39) and (68). We shall shortly see that a similar expression (see (154)) plays an important role during one of the faster phases, though the first (and most important) of these, discussed next, requires a quite different approach.

6. Time-dependent flows: the phases of closest approach

6.1. Formulation

The results (77–78) and (86–87) each indicate that the preceding analysis breaks down on the timescale $\hat{T} = O(1)$, where

$$\hat{t} = \hat{t}_b(\epsilon) + \epsilon\hat{T}, \tag{94}$$

with $\hat{t}_b(0) = \hat{t}_{b0}$. On this timescale we have at leading order

$$\begin{aligned}
&\mathcal{F}_{0Y\hat{T}} = \mathcal{F}_{0YYY} + \mathcal{F}_{0Y}^2 - \mathcal{F}_0\mathcal{F}_{0YY}, \\
&\text{at } Y = 0 \qquad \mathcal{F}_0 = 1, \quad \mathcal{F}_{0Y} = 0, \\
&\text{as } Y \to -\infty \qquad \mathcal{F}_0 \sim Y/(-\hat{T}), \\
&\text{as } \hat{T} \to -\infty \qquad \begin{cases} \mathcal{F}_0 \sim 1 + (Y + 1 - e^Y)/(-\hat{T}) & \text{for } Y = O(1), \\ \mathcal{F}_0 \sim 1 + Y/(-\hat{T}) & \text{for } Y = O(-\hat{T}), \end{cases}
\end{aligned} \tag{95}$$

where we have matched with (77) and (87). The initial–boundary-value problem (95) is a crucial one, describing the manner in which viscous effects serve to reduce the rate at which the interior zero approaches the right-hand boundary, y_z achieving its closest approach to $y = 1$ (whereby $1 - y_z = O(\epsilon)$) on this timescale.

The formulation (95) is the one of most significance for $\hat{T} = O(1)$ and is discussed is Sections 6.2–6.4 below; the scalings which apply on this timescale in the other regions are as follows.

(A) $Y = \hat{Y}/\epsilon\log(1/\epsilon), \quad \mathcal{F} = \hat{\mathcal{F}}/\epsilon\log(1/\epsilon),$
giving

$$\begin{aligned}
&\hat{\mathcal{F}}_{0\hat{Y}\hat{T}} = \hat{\mathcal{F}}_{0\hat{Y}}^2 - \hat{\mathcal{F}}_0\hat{\mathcal{F}}_{0\hat{Y}\hat{Y}}, \\
&\text{at } \hat{Y} = 0 \qquad \hat{\mathcal{F}}_0 = 0, \\
&\text{as } \hat{Y} \to -\infty \qquad \hat{\mathcal{F}}_{0\hat{Y}} \to 0, \\
&\text{as } \hat{T} \to -\infty \qquad \hat{\mathcal{F}}_0 \sim -2(1 - e^{\hat{Y}/2})/(-\hat{T}),
\end{aligned} \tag{96}$$

where we have matched with (76); the solution to (96) is separable, giving

$$\hat{\mathcal{F}}_0 = -2(1 - e^{\hat{Y}/2})/(-\hat{T}). \tag{97}$$

(B) $y = O(1)$, $\quad \hat{\mathcal{F}} \sim \tilde{\mathcal{F}}_0(y, \hat{T})$,
implying

$$\tilde{\mathcal{F}}_{0yy\hat{T}} = 0, \tag{98}$$

so imposing $\tilde{\mathcal{F}}_0 = 0$ on $y = -1$ and matching with (75) and (97), we find

$$\tilde{\mathcal{F}}_0 = -(1+y)/(-\hat{T}). \tag{99}$$

The pressure coefficient is determined by this region, giving

$$p \sim -\frac{1}{\epsilon^2 \log^2(1/\epsilon)(-\hat{T})^2}.$$

(C) $y = -1 + \epsilon \bar{y}$, $\quad \mathcal{F} = \hat{\mathcal{G}}/\log(1/\epsilon)$,
where, at leading order,

$$\begin{aligned} &\hat{\mathcal{G}}_{0\bar{y}\bar{y}\hat{T}} = \hat{\mathcal{G}}_{0\bar{y}\bar{y}\bar{y}\bar{y}} \\ &\text{at } \bar{y} = 0 \qquad \hat{\mathcal{G}}_0 = \hat{\mathcal{G}}_{0\bar{y}} = 0, \\ &\text{as } \bar{y} \to +\infty \qquad \hat{\mathcal{G}}_0 \sim -\bar{y}/(-\hat{T}); \end{aligned} \tag{100}$$

initial conditions on (100) as $\hat{T} \to -\infty$ must be deduced from the behaviour of (85) as $\hat{t} \to \hat{t}_{b0}^-$, namely

$$\bar{\mathcal{G}}_0 \sim \frac{1}{\left(\hat{t}_{b0} - \hat{t}\right)^{1/2} \left(-\log\left(\hat{t}_{b0} - \hat{t}\right)\right)} \Phi\left(\frac{\bar{Y}}{\left(\hat{t}_{b0} - \hat{t}\right)^{1/2}}\right) \tag{101}$$

with

$$\Phi(\eta) = -\eta + \sqrt{\pi}\left(1 - e^{\eta^2/4}\operatorname{erfc}(\eta/2)\right), \tag{102}$$

the nonlinear terms in (85) being negligible. From this it follows that the required solution to (100) is

$$\hat{\mathcal{G}}_0 = \frac{1}{(-\hat{T})^{1/2}} \Phi\left(\frac{\bar{y}}{(-\hat{T})^{1/2}}\right), \tag{103}$$

with $\Phi(\eta)$ again given by (102).

We now return to (95), the most important feature of which is its behaviour in the limit $\hat{T} \to 0^-$; however, to obtain a sufficiently complete description of this we must first analyse the limits $\hat{T} \to -\infty$ and $Y \to -\infty$, considering an exponentially small correction term to

$$\mathcal{F}_0 \sim Y/(-\hat{T}) + Y_\infty(\hat{T}) \qquad \text{as } Y \to -\infty \tag{104}$$

where Y_∞ is determined as part of the solution to (95).

6.2. The limit $\hat{T} \to -\infty$

In (104), it follows from (95) that

$$Y_\infty(\hat{T}) \sim 1 + 1/(-\hat{T}) \qquad \text{as } \hat{T} \to -\infty. \tag{105}$$

The formulation (95) already contains the behaviour as $\hat{T} \to -\infty$ with $Y = O(1)$; since we shall also require an exponentially small correction term for $Y = O(-\hat{T})$, we write

$$\mathcal{F}_0 \sim Y/(-\hat{T}) + Y_\infty(\hat{T}) + E$$

and linearise in E to give

$$E_{\hat{T}} = E_{YY} + \frac{3}{(-\hat{T})}E - \left(\frac{Y}{(-\hat{T})} + Y_\infty\right)E_Y. \tag{106}$$

Introducing $\zeta = Y/(-\hat{T})$ and using (105), we have

$$E_{\hat{T}} + \frac{1}{(-\hat{T})}(2\zeta + 1)E_\zeta \sim \frac{1}{(-\hat{T})^2}E_{\zeta\zeta} - \frac{1}{(-\hat{T})^2}E_\zeta + \frac{3}{(-\hat{T})}E \qquad \text{as } \hat{T} \to -\infty. \tag{107}$$

In view of the need to match as $\zeta \to 0^-$ into the term $-e^Y/(-\hat{T})$ in (95), the required expansion for (107) in the limit $\hat{T} \to -\infty$ takes the WKBJ form

$$E \sim \frac{1}{(-\hat{T})}a_0(\zeta)e^{-(-\hat{T})\phi(\zeta)}, \tag{108}$$

implying

$$\phi - (2\zeta + 1)\phi_\zeta = \phi_\zeta^2, \tag{109}$$

in which the right-hand side represents the contribution of viscosity, and

$$a_0 + (2\zeta + 1)a_{0\zeta} = -2\phi_\zeta a_{0\zeta} - \phi_{\zeta\zeta}a_0 + \phi_\zeta a_0 + 3a_0, \tag{110}$$

subject to

$$\phi \sim -\zeta, \quad a_0 \to -1 \qquad \text{as } \zeta \to 0^-.$$

Equation (109) can be reduced to autonomous form by writing $\phi = (\zeta + \frac{1}{2})^2\psi(\log|\zeta + \frac{1}{2}|)$, yielding

$$\left(\sqrt{\phi + (\zeta + 1/2)^2} + (\zeta + 1/2)\right)^2\left(2\sqrt{\phi + (\zeta + 1/2)^2} - (\zeta + 1/2)\right) = 1/2, \tag{111}$$

from which it follows that

$$\phi \sim \sqrt{\tfrac{2}{3}}(-\zeta)^{1/2} - \tfrac{1}{4}\sqrt{\tfrac{2}{3}}(-\zeta)^{-1/2} + O(1/(-\zeta)) \qquad \text{as } \zeta \to -\infty. \tag{112}$$

Using (109), we obtain the required solution to (110) as

$$a_0 = -\frac{e^{-\phi_\zeta} - 1}{(-\phi_\zeta)^3(-(2\zeta + 1 + 2\phi_\zeta))^{1/2}}, \tag{113}$$

so that

$$a_0 \sim -\frac{e^{-1}6^{3/2}}{2^{1/2}}(-\zeta) \qquad \text{as } \zeta \to -\infty. \tag{114}$$

6.3. The limit $Y \to -\infty$

The way in which the blow-up behaviour of (95) in the limit $\hat{T} \to 0^-$ is selected is somewhat similar to that in which the minimum-speed travelling wave is selected by appropriate initial-value problems for Fisher's equation (an analogy we expand upon in Appendix A.2 by discussing possible similarity solutions to the inviscid RPJ equation). In particular, a crucial step involves determining an exponentially small term in the far-field behaviour of (95).

The relevant balance is again given by (106), but the viscous term is negligible in the limit $Y \to -\infty$ (as is the case in the terms given in (112)), so that

$$E \sim \frac{1}{(-\hat{T})^3} \hat{E}\left((-\hat{T})Y - \int^{\hat{T}} (-T')Y_\infty(T')\,\mathrm{d}T'\right) \tag{115}$$

for some function $\hat{E}(\hat{\omega})$ which cannot be determined for $\hat{\omega} = O(1)$ without solving the full problem but which, by use of (112–114), satisfies

$$\hat{E}(\hat{\omega}) \sim -\frac{\mathrm{e}^{-1}6^{3/2}}{2^{1/2}}(-\hat{\omega})\mathrm{e}^{-\sqrt{2/3}(-\hat{\omega})^{1/2}} \qquad \text{as } \hat{\omega} \to -\infty. \tag{116}$$

6.4. The limit $\hat{T} \to 0^-$

We are now in a position to address the blow-up behaviour of (95). In the limit $\hat{T} \to 0^-$ we have

$$\mathcal{F}_0 \sim \mathcal{F}_c(Y) \qquad \text{for } Y = O(1) \tag{117}$$

with

$$\mathcal{F}_c(0) = 1, \quad \mathcal{F}_{cY}(0) = 0; \tag{118}$$

$\mathcal{F}_c(Y)$ for $Y > 0$ depends on the evolution over earlier times and cannot be determined without solving the full problem.

For large $-Y$, specifically for $-Y = O(-\log(-\hat{T})/(-\hat{T}))$, the appropriate balance as $\hat{T} \to 0^-$ is of the form

$$\mathcal{F}_0 \sim \frac{-\log(-\hat{T})}{(-\hat{T})^2}\Omega(\xi), \quad \xi = \frac{Y(-\hat{T})}{-\log(-\hat{T})} \tag{119}$$

(see Appendix A.2 for related considerations) with

$$\Omega_\xi - \xi\Omega_{\xi\xi} = \Omega_\xi^2 - \Omega\Omega_{\xi\xi}, \tag{120}$$

the viscous term being negligible, so that

$$\Omega(\xi) = \xi + \xi_0 - \xi_0 \mathrm{e}^{\xi/\xi_0} \tag{121}$$

for some constant ξ_0. Matching with (116) as $\xi \to -\infty$ (a procedure which involves the matching of exponentially small terms) implies that $\xi_0 = 3$ with

$$Y_\infty(\hat{T}) \sim \frac{3(-\log(-\hat{T}))}{(-\hat{T})^2} + \frac{3\log(-\log(-\hat{T}))}{(-\hat{T})^2} + \frac{9\log 3}{2(-\hat{T})^2} \qquad \text{as } \hat{T} \to 0^-, \tag{122}$$

indicating the time-dependence of the next two terms required in the expansion (119). It follows from the behaviour of (121) as $\xi \to 0^-$ that in (117) we have

$$\mathcal{F}_c(Y) \sim -\tfrac{1}{6}Y^2/\log(-Y) \qquad \text{as } Y \to -\infty, \tag{123}$$

so on this timescale the interior zero of $\mathcal{F}$ tends to some fixed location $Y = Y_z(\epsilon) = O(1)$, where $\mathcal{F}_c(Y_{z0}) = 0$, this representing its closest approach to the right-hand boundary $y = 1$.

6.5. TURN-AROUND TIME SCALE: $\hat{T} = O(1/\log(1/\epsilon))$

We conclude this section by discussing an intermediate timescale on which (117) ceases to apply (it does not in fact persist for very long) but (97), (99) and (119) remain applicable. On the timescale $T^\ddagger = O(1)$, where

$$\hat{T} = T^\ddagger/\log(1/\epsilon), \tag{124}$$

we have that the pressure coefficient

$$p \sim -\frac{\log(1/\epsilon)}{\epsilon^2(-T^\ddagger)^2};$$

for $Y = O(1)$ we thus have

$$\mathcal{F}_{0YT^\ddagger} = -\frac{1}{(-T^\ddagger)^2}, \qquad \mathcal{F}_0 = \mathcal{F}_c(Y) - Y/(-T^\ddagger) \tag{125}$$

for $T^\ddagger = O(1)$, showing how the interior zero of $\mathcal{F}$ begins to move inwards (further away from $y = 1$) on this time scale. There is also a subsidiary boundary layer with $Y = O(1/\log^{1/2}(1/\epsilon))$, which is required to satisfy the condition $\mathcal{F}_y = 0$ on $y = 1$.

7. Time-dependent flows: asymptotics of the fastest phases

7.1. $\hat{T} = O(\epsilon\log^2(1/\epsilon))$

7.1.1. *Preamble*

This is the timescale on which it follows from (119) that the boundary layer (95) merges with the intermediate region described by (96–97). We write

$$\hat{T} = \epsilon\log^2(1/\epsilon)\hat{\tau}, \qquad \mathcal{F} = H/\epsilon^2\log^3(1/\epsilon), \tag{126}$$

so (119) implies the matching condition

$$H_0 \sim \frac{1}{(-\hat{\tau})^2}\Omega(\xi), \qquad \xi = \hat{Y}(-\hat{\tau}) \tag{127}$$

as $\hat{\tau} \to -\infty$ with $\hat{Y} = O(1/(-\hat{\tau}))$, where

$$\Omega(\xi) = \xi + 3 - 3e^{\xi/3}. \tag{128}$$

Away from $y = 1$, the analysis of Section 5.1.4 is still pertinent on earlier timescales, so (76) implies the matching condition

$$H_0 \sim -\frac{1}{(-\hat{\tau})}\left(1 - e^{\hat{Y}}\right) \tag{129}$$

as $\hat{\tau} \to -\infty$ with $\hat{Y} = O(1)$.

We have for $\hat{Y} = O(1)$ the inviscid balance that

$$H_{0\hat{Y}\hat{\tau}} = H_{0\hat{Y}}^2 - H_0 H_{0\hat{Y}\hat{Y}}. \tag{130}$$

In view of (125) the boundary conditions on (130) are

$$\left.\begin{array}{ll} \text{as } \hat{Y} \to 0^- & H_0 \sim -\frac{1}{6}\hat{Y}^2, \\ \text{as } \hat{Y} \to -\infty & H_{0\hat{Y}} \to 0, \end{array}\right\} \tag{131}$$

where we have matched as $\hat{Y} \to -\infty$. We now present the solution to (130) subject to (127), (129) and (131).

7.1.2. *Boundary layer solution*

The approach outlined in Appendix A.1 can be used to derive the required solution. Omitting the details of the derivation, we find the solution in terms of a parameter ζ_i (which parametrises the characteristics of (130)) in the form

$$\hat{Y} = -\int_{\zeta_i}^{\infty} \frac{d\zeta}{e^{\zeta/3} - \zeta - \hat{\tau}}, \qquad H_0 = -\int_{\zeta_i}^{\infty} \frac{d\zeta}{(e^{\zeta/3} - \zeta - \hat{\tau})^2}. \tag{132}$$

The crucial characteristics have

$$\hat{Y}(-\hat{\tau}) = -3\log(-\hat{\tau}) + O(1) \qquad \text{as } \hat{\tau} \to -\infty, \tag{133}$$

and thus lie in the overlap regime between (127) and (129); (132) is thus best derived from a composite (uniformly valid) representation such as

$$H_0 \sim \frac{3}{(-\hat{\tau})^2}\left(1 - e^{(-\hat{\tau})\hat{Y}/3}\right) - \frac{1}{(-\hat{\tau})}\left(1 - e^{\hat{Y}}\right) \qquad \text{as } \hat{\tau} \to -\infty. \tag{134}$$

In addition to (132), we have

$$H_{0\hat{Y}} = \frac{1}{e^{\zeta_i/3} - \zeta_i - \hat{\tau}}, \qquad H_{0\hat{Y}\hat{Y}} = -\frac{\frac{1}{3}e^{\zeta_i/3} - 1}{e^{\zeta_i/3} - \zeta_i - \hat{\tau}}, \tag{135}$$

so that

$$H_{0\hat{Y}\hat{Y}} = 0 \qquad \text{on the characteristic } \zeta_i = 3\log 3. \tag{136}$$

For matching purposes we need the following consequences of (132). Firstly,

$$H_0 \sim -\int_{-\infty}^{\infty} \frac{d\zeta}{(e^{\zeta/3} - \zeta - \hat{\tau})^2} \qquad \text{as } \hat{Y} \to -\infty. \tag{137}$$

Secondly, H_0 blows up as $\hat{\tau} \to \hat{\tau}_c^-$, where

$$\hat{\tau}_c = -3(\log 3 - 1). \tag{138}$$

It then follows, if we use (132) along the characteristics with $\zeta_i = 3\log 3 + O((\hat{\tau}_c - \hat{\tau})^{1/2})$ (*cf.* (136)), that

$$H_0 \sim \frac{\hat{Y}}{2(\hat{\tau}_c - \hat{\tau})} - \frac{3^{1/2}}{2^{3/2}(\hat{\tau}_c - \hat{\tau})^{3/2}} \sin\left(\frac{2^{1/2}(\hat{\tau}_c - \hat{\tau})^{1/2}\hat{Y}}{3^{1/2}}\right), \tag{139}$$

for $\hat{Y} = O((\hat{\tau}_c - \hat{\tau})^{-1/2})$ with $-6^{1/2}\pi < \hat{Y}(\hat{\tau}_c - \hat{\tau})^{1/2} < 0$, and

$$H_0 \sim -\frac{3^{1/2}\pi}{2^{1/2}(\hat{\tau}_c - \hat{\tau})^{3/2}} \tag{140}$$

for $\hat{Y} = O((\hat{\tau}_c - \hat{\tau})^{-1/2})$ with $\hat{Y}(\hat{\tau}_c - \hat{\tau})^{1/2} < -6^{1/2}\pi$, consistent with (137). Appendix A.2 clarifies why (139–140) is the self-similar form appropriate for describing the current intermediate asymptotic behaviour.

As $\hat{\tau} \to \hat{\tau}_c^-$ with $\hat{Y} = O(1)$ we have that

$$H_0 \sim H_c(\hat{Y}), \tag{141}$$

where H_c is given by

$$\hat{Y} = -\int_{\zeta_i}^{\infty} \frac{d\zeta}{e^{\zeta/3} - \zeta + 3(\log 3 - 1)}, \qquad H_c = -\int_{\zeta_i}^{\infty} \frac{d\zeta}{(e^{\zeta/3} - \zeta + 3(\log 3 - 1))^2}, \tag{142}$$

so that

$$H_c \sim -\tfrac{1}{6}\hat{Y}^2 \quad \text{as } \hat{Y} \to 0^-, \qquad H_c \sim \tfrac{1}{18}\hat{Y}^3 \quad \text{as } \hat{Y} \to -\infty, \tag{143}$$

consistent with (131) and matching with (139). Finally, there is a similar region with

$$\hat{Y} = \hat{Y}_c(\hat{\tau}) + O(1), \qquad \text{where } \hat{Y}_c \sim -6^{1/2}\pi(\hat{\tau}_c - \hat{\tau})^{-1/2},$$

which provides a smooth transition between (139) and (140); we omit details, however.

7.1.3. *Other regions*

For $y = O(1)$ we have (*cf.* (98–99)), writing $H \sim \hat{H}_0(y, \hat{\tau})$, that

$$\hat{H}_{0yy\hat{\tau}} = 0$$

so that

$$\hat{H}_0 = -\frac{1}{2}\int_{-\infty}^{\infty} \frac{d\zeta}{(e^{\zeta/3} - \zeta - \hat{\tau})^2}(y+1), \tag{144}$$

where we have matched with (137). This implies that

$$\hat{H}_0 \sim -\frac{3^{1/2}\pi}{2^{3/2}(\hat{\tau}_c - \hat{\tau})^{3/2}}(y+1) \qquad \text{as } \hat{\tau} \to \hat{\tau}_c^-. \tag{145}$$

Lastly, the analogous region to that described by (100) has

$$y = -1 + \epsilon^{3/2}\log(1/\epsilon)\tilde{y}, \qquad \mathcal{F} = \tilde{\mathcal{H}}/\epsilon^{1/2}\log^2(1/\epsilon) \tag{146}$$

with, by use of (101),

$$\left.\begin{array}{ll} & \tilde{\mathcal{H}}_{0\tilde{y}\tilde{y}\hat{\tau}} = \tilde{\mathcal{H}}_{0\tilde{y}\tilde{y}\tilde{y}\tilde{y}}, \\ \text{at } \tilde{y} = 0 & \tilde{\mathcal{H}}_0 = \tilde{\mathcal{H}}_{0\tilde{y}} = 0, \\ \text{as } \tilde{y} \to +\infty & \tilde{\mathcal{H}}_0 \sim -\frac{1}{2}\displaystyle\int_{-\infty}^{\infty} \frac{d\zeta}{(e^{\zeta/3} - \zeta - \hat{\tau})^2}\tilde{y}, \\ \text{as } \hat{\tau} \to -\infty & \tilde{\mathcal{H}}_0 \sim \dfrac{1}{2(-\hat{\tau})^{1/2}}\Phi\left(\tilde{y}/(-\hat{\tau})^{1/2}\right), \end{array}\right\} \tag{147}$$

where $\Phi(\eta)$ is given by (102) (we note that, as with (100), the nonlinear terms in (1) are of relative size $O(1/\log(1/\epsilon))$ here, so are only just negligible in (147)). In view of (145) we have blow-up behaviour of the form

$$\tilde{\mathcal{H}}_0 \sim \frac{1}{(\hat{\tau}_c - \hat{\tau})}\Psi\left(\frac{\tilde{y}}{(\hat{\tau}_c - \hat{\tau})^{1/2}}\right) \qquad \text{as } \hat{\tau} \to \hat{\tau}_c^-, \tag{148}$$

where

$$\Psi(\eta) = -\frac{3^{1/2}\pi}{2^{3/2}}\eta(1 - e^{\eta^2/4}\operatorname{erfc}(\tfrac{1}{2}\eta)). \tag{149}$$

7.2. $\hat{\tau} = \hat{\tau}_c + O(1/\log^2(1/\epsilon))$

We now discuss the final, and fastest, phase of evolution, in which the interior zero moves rapidly from the neighbourhood of $y = 1$ to that of $y = 0$. We write

$$\hat{\tau} = \hat{\tau}_c + \bar{\tau}/\log^2(1/\epsilon), \qquad \mathcal{F} = \mathcal{H}/\epsilon^2, \tag{150}$$

this being (in view of (139–140)) the timescale on which the boundary layer of Section 7.1.2 merges with the outer region. At leading order we have in the outer region $y = O(1)$ that

$$\mathcal{H}_{0yy\bar{\tau}} = \mathcal{H}_{0y}\mathcal{H}_{0yy} - \mathcal{H}_0\mathcal{H}_{0yyy}, \tag{151}$$

i.e. an inviscid balance but, in contrast to (130), one in which a non-zero pressure coefficient $\mathcal{P}_0(\bar{\tau})$ is to be determined as part of the solution to (151), with

$$p \sim \mathcal{P}_0(\bar{\tau})/\epsilon^4, \qquad \mathcal{H}_{0y\bar{\tau}} = \mathcal{H}_{0y}^2 - \mathcal{H}_0\mathcal{H}_{0yy} + \mathcal{P}_0(\bar{\tau}). \tag{152}$$

From (145) we have

$$\mathcal{H}_0 \sim -\frac{3^{1/2}\pi}{2^{3/2}(-\bar{\tau})^{3/2}}(y + 1) \qquad \text{as } \bar{\tau} \to -\infty \text{ with } y = O(1) \tag{153}$$

and, since $\mathcal{H}_{0yy} = 0$ holds for all time along characteristics of (151) if it does so initially, we are motivated to seek a solution of the form

$$\left.\begin{aligned} \mathcal{H}_0 &= -\mathcal{A}(\bar{\tau})(1 - y) + \mathcal{B}(\bar{\tau})\sin(\bar{\lambda}(\bar{\tau})(1 - y)) & y > \mathcal{S}(\bar{\tau}), \\ \mathcal{H}_0 &= -\mathcal{E}(\bar{\tau})(y + 1) & y < \mathcal{S}(\bar{\tau}), \end{aligned}\right\} \tag{154}$$

for some $\mathcal{S}(\bar{\tau})$ which is to be determined as part of the solution. The admissibility of such a 'two part' solution stems from the hyperbolic character of (151) and from matching with (139–140), which imply that

$$\left.\begin{aligned} &\mathcal{S}(\bar{\tau}) \sim 1 - 6^{1/2}\pi/(-\bar{\tau})^{1/2} && \text{as } \bar{\tau} \to -\infty \\ &\mathcal{H}_0 \sim -\frac{(1 - y)}{2(-\bar{\tau})} + \frac{3^{1/2}}{2^{3/2}(-\bar{\tau})^{3/2}}\sin\left(\frac{2^{1/2}(-\bar{\tau})^{1/2}}{3^{1/2}}(1 - y)\right) && \text{as } \bar{\tau} \to -\infty,\ y > \mathcal{S}(\bar{\tau}); \end{aligned}\right\} \tag{155}$$

in writing down (154) we have used the fact that $\mathcal{H}_{0yy} = 0$ on $y = 1$ as well as in $y < \mathcal{S}(\bar{\tau})$, the characteristic projections of (151) satisfying

$$\frac{dy}{d\bar{\tau}} = \mathcal{H}_0,$$

so that $y = 1$ is a characteristic, as is $y = \mathcal{S}(\bar{\tau})$. The analyticity of the solution is maintained via an interior layer about $\mathcal{S}(\bar{\tau})$ (*cf.* the discussion of Appendix A.2). The continuity conditions on (154) are

$$\text{at } y = \mathcal{S}(\bar{\tau}) \qquad [\mathcal{H}_0]_-^+ = [\mathcal{H}_{0y}]_-^+ = [\mathcal{H}_{0yy}]_-^+ = 0,$$

and from these and (151) it is straightforward to deduce that, up to translations in $\bar{\tau}$,

$$\left.\begin{aligned} &\tfrac{1}{2}\bar{\lambda}^2 + \pi\bar{\lambda} + \pi^2\log(\bar{\lambda} - \pi) = \tfrac{1}{3}(-\bar{\tau}),\\ &\mathcal{S} = 1 - \frac{2\pi}{\bar{\lambda}}, \quad \mathcal{A} = \frac{\bar{\lambda} - \pi}{3\bar{\lambda}^3}, \quad \mathcal{B} = \frac{1}{3\bar{\lambda}^3}, \quad \mathcal{E} = \frac{\pi}{3\bar{\lambda}^3}. \end{aligned}\right\} \tag{156}$$

It follows that as $\bar{\tau} \to +\infty$

$$\bar{\lambda} \to \pi^+, \qquad \mathcal{S} \to (-1)^+, \tag{157}$$

at an exponentially fast rate, with

$$\mathcal{H}_0 \sim \frac{1}{3\pi^3}\sin(\pi y) \qquad \text{for } y > \mathcal{S}(\bar{\tau}), \tag{158}$$

which brings us full circle back to (46), confirming that the *ansatz* (39) is appropriate and identifying the final unknown as

$$C_I(\epsilon) \sim \frac{1}{3\pi^3\epsilon^2}, \tag{159}$$

and completing the analysis of the limit cycle behaviour. Both $\mathcal{F}$ and $-p$ attain their maximum values as $\bar{\tau} \to +\infty$ (*cf.* Figure 10), with

$$p \sim -\frac{1}{9\pi^4\epsilon^2}.$$

Viscous effects now come back into play, with the solution decaying slowly due to viscous dissipation, as described by (47–48).

The final loose ends concern the behaviour in the boundary layers on this timescale; the relevant scalings are $y = \pm 1 + O(\epsilon^{3/2})$ and $\mathcal{F} = O(\epsilon^{-1/2})$, and the full balance occurs at leading order in both the boundary layers (in view of (157), the range $y < \mathcal{S}(\bar{\tau})$ merges into the left-hand boundary layer for sufficiently large $\bar{\tau}$). In the right-hand boundary layer, there is again an intermediate timescale, analogous to that discussed in Section 6.5, on which (141) ceases to hold. Since

$$\mathcal{P}_0 = -\frac{\pi}{3\bar{\lambda}^5} + \frac{2\pi^2}{9\bar{\lambda}^6}, \tag{160}$$

we have

$$\mathcal{P}_0 \sim \frac{3^{3/2}\pi}{2^{5/2}(-\bar{\tau})^{5/2}} \qquad \text{as } \bar{\tau} \to -\infty$$

and the relevant timescale is

$$\bar{\tau} = \log^{4/3}(1/\epsilon)\tau^{\ddagger}$$

with

$$H_{0\hat{Y}\tau^{\ddagger}} = -\frac{3^{3/2}\pi}{2^{5/2}(-\tau^{\ddagger})^{5/2}}, \qquad H_0 = H_c(\hat{Y}) - \frac{3^{1/2}\pi\hat{Y}}{2^{3/2}(-\tau^{\ddagger})^{3/2}}. \tag{161}$$

A noteworthy reinterpretation of (154) involves consideration of the quantity

$$\Pi(y,t) = \mathcal{F}_{yyy}^2 - \mathcal{F}_{yy}\mathcal{F}_{yyyy}.$$

For $\epsilon = 0$ it follows from (1) that

$$\Pi_t = -\mathcal{F}\Pi_y \tag{162}$$

so that Π is constant along characteristics; moreover, for any ϵ, for which

$$\Pi_t + \mathcal{F}\Pi_y = \epsilon\left(\frac{2}{\mathcal{F}_{yy}}\left(\mathcal{F}_{yyyy}\Pi - \mathcal{F}_{yyy}\Pi_y\right) + \Pi_{yy}\right), \tag{163}$$

the *ansatz* (39) corresponds exactly to Π $(= \lambda^6(C^2 + D^2))$ being independent of y, whereby $\mathcal{F}_{yyyy} = -\lambda^2\mathcal{F}_{yy}$ and (163) reduces to (42). The solution (154) has Π piecewise constant (a feature first established by the intermediate asymptotic similarity solution (139–140)), with its value in $y < \mathcal{S}(\bar{\tau})$ (zero for the leading-order solution $\mathcal{H}_0$) being the remnant after viscous dissipation (*cf.* (42)) of the previous oscillation. This 'old' value is swept out by a 'new' one (the one which holds in $y > \mathcal{S}(\bar{\tau})$) whose value is that required to obtain periodic behaviour by decaying to the 'old' value in the course of the oscillation. With this interpretation, the leading-order value of Π $(\sim 1/9\epsilon^4)$ can be deduced (given (162)) from the second of (143), without any need to solve (151). Indeed, the details of the analysis of Section 7.1 can also be circumvented – the solutions (127) and (129) each correspond to $\Pi = 0$, so the characteristics on which Π is non-zero emanate from the overlap region (133); the blow-up behaviour is dominated by the characteristic on which $H_{0\hat{Y}\hat{Y}} = 0$, so that $H_{0\hat{Y}}$ is maximal, and this satisfies

$$\hat{Y} \sim \frac{3}{(-\hat{\tau})}\log(3(-\hat{\tau})) \qquad \text{as } \hat{\tau} \to -\infty$$

and has, by use of (134), $\Pi \sim 1/9\epsilon^4$. Since the blow-up behaviour (139–140) dictates the subsequent evolution, this determines the 'new' value of Π; the calculation just given emphasises the importance of the composite expression of (134) in describing the overlap between the ranges of validity of (127) and (129), as well as the crucial role played by the timescale of Section 6, since this can be viewed as being responsible for the 'new' value. Plotting Π from the numerical solutions of Section 4 provides (given the relatively large values of ϵ in the simulations) strong support for the scenario described above.

8. Discussion

The asymptotic analysis of the time-dependent problem outlined above leads to a remarkable degree of analytical progress, involving a variety of nonlinear techniques. The key aspects of the evolution are described by the low-dimensional reduction of Section 5 and by the $\alpha = 1$ and $\alpha = 1/2$ similarity solutions of Appendix A.2 (which can be viewed as similarity solutions of the second kind, *cf.* [27], albeit selected in a rather delicate way); in addition the determination of the values of the constants in (121) and (139), corresponding to λ_0 in each of (A12) and (A13), requires respectively the matching of exponentially small terms and the solution of (130) by characteristic methods. Detailed comparison between numerical and asymptotic results is not currently possible, largely because the numerical approach cannot be applied with ϵ small enough for a convincing comparison, particularly given the prevalence of terms in $\log(1/\epsilon)$ in many of the expansions; nevertheless, a comparison based on crude extrapolation to $\epsilon \to 0$ of quantities obtained numerically for a variety of ϵ yields encouraging results. Moreover, the asymptotic results successfully capture all the qualitative features of the numerics (*cf.* Figure 5 in particular), as well as the orders of magnitude of the various quantities. Thus the asymptotic approach is currently the only way to construct the limit cycles for very small ϵ and is very valuable in clarifying how the evolution proceeds and why numerical approaches encounter difficulties when ϵ is small (in particular, the period of the oscillation scales as $t = O(\log(1/\epsilon)/\epsilon)$ (see (50); $(\log(C_I(\epsilon))/\pi^2 + T_{c0})/\epsilon$ provides an estimate of the period), while on the fastest phase (see (150)) t varies by an $O(\epsilon^2)$ amount). While the analysis is sufficiently complicated that the validity of all the algebraic details cannot of course be guaranteed, the approach of Appendix A.2 provides independent verification of the key role played by the $\alpha = 1$ and $\alpha = 1/2$ similarity solutions; the values of λ_0 in (A12) and (A13) cannot be determined solely by such an approach, though, so these two constants are likely to be the quantities most susceptible to error.

The inviscid version of the RPJ equation

$$\mathcal{F}_{0yyt} = \mathcal{F}_{0y}\mathcal{F}_{0yy} - \mathcal{F}_0\mathcal{F}_{0yyy}, \qquad \mathcal{F}_{0yt} = \mathcal{F}_{0y}^2 - \mathcal{F}_0\mathcal{F}_{0yy} + p_0(t)$$

plays a central role in our analysis and has characteristic projections

$$\frac{\mathrm{d}y}{\mathrm{d}t} = \mathcal{F}_0(y, t) \tag{164}$$

on which

$$\frac{\mathrm{d}\mathcal{F}_{0y}}{\mathrm{d}t} = \mathcal{F}_{0y}^2 + p_0(t), \qquad \frac{\mathrm{d}\mathcal{F}_{0yy}}{\mathrm{d}t} = \mathcal{F}_{0y}\mathcal{F}_{0yy} \tag{165}$$

(*cf.* Appendix A.1; the Lagrangian equation for x corresponding to (164) is

$$\frac{\mathrm{d}x}{\mathrm{d}t} = -x\mathcal{F}_{0y}(y, t),$$

it being noteworthy that this furnishes the linearising transformation for the Riccati equation given by the first of (165). Moreover, from (4) we have

$$\frac{\mathrm{d}\mathcal{F}_0}{\mathrm{d}t} = -\hat{p}_y, \qquad \frac{\mathrm{d}}{\mathrm{d}t}(x\mathcal{F}_{0yy}) = 0,$$

the second of which corresponds to vorticity conservation along characteristics). An issue alluded to above concerns the fact that it follows from (164) that the characteristics near $y = 1$ leave the fluid domain (since $\mathcal{F}_0(1, t) = 1$), whereas information needs to be fed back into the fluid for the solution to remain bounded and for the limit cycle to turn around. This is accomplished in two ways. During the later stages of turn-around one has $\mathcal{F} \gg 1$ so that (as in (131)) the leading-order inviscid problem sees zero normal velocity at the wall and there is within the leading-order formulation (*cf.* (127), (129)) an infinite amount of time for the information to propagate in. More noteworthy, however, are the earlier stages of this process; if we write

$$\mathcal{F} \sim \sum_{n=0}^{\mathcal{N}} \mu_n(\epsilon)\mathcal{F}_n(y,t) + \nu_0(\epsilon)\Psi_0(y,t)\mathrm{e}^{-\Phi(y,t)/\epsilon} \qquad \text{as } \epsilon \to 0,$$

where $\mu_0(\epsilon) = 1$ and the summation gives the algebraic expansion for $\mathcal{F}$, with $\mathcal{N}(\epsilon)$ being its optimal truncation point (such a procedure enables the exponentially small term to be legitimately separated off in very general contexts, *cf.* [28]), then (away from Stokes lines, of which we avoid discussion here) we have from (1) that

$$\Phi_t = -\Phi_y^2 - \mathcal{F}_0\Phi_y. \tag{166}$$

Equation (166) has characteristic projections

$$\frac{\mathrm{d}y}{\mathrm{d}t} = \mathcal{F}_0(y,t) + 2\Phi_y(y,t), \tag{167}$$

the final term of which represents the effects of viscosity; since the constraints $\Phi(1,t) = 0$, $\Phi_y(1,t) = -1$ are required in order to match into the viscous boundary layer, (167) implies that information does indeed propagate back into the fluid, a feature which underpins the matching of exponentially small terms in Section 6.

The reduction (39), (68) is noteworthy for a number of reasons. Firstly, it provides a large family of exact solutions to the incompressible Navier–Stokes equations, for which some generalisations are available – for example, via solutions of the form $\psi = \Psi(y,t) - x\mathcal{F}(y,t)$, $p = \hat{p}(y,t) + xP(t) + x^2p(t)/2$ satisfying (3), (4) and

$$\Psi_{yt} = \epsilon\Psi_{yyy} + \Psi_y\mathcal{F}_y - \mathcal{F}\Psi_{yy} - P(t);$$

of course, in the Navier–Stokes (rather than RPJ) context there are important issues concerning the stability or otherwise of the solutions we have described. Secondly, it plays a very valuable (and concise) role in providing a uniformly valid description of the outer behaviour over each of the timescales discussed in Section 5.1; we stress that this reduction is valid in the current context only in the asymptotic limit $\epsilon \to 0$, for which it provides an immense saving in computational time. The concept of a uniformly valid expansion is very well-established (see, for example, [29]), but such examples motivate us to comment on 'uniformly valid equations', whereby the original formulation is replaced by a simpler one which has the same asymptotic behaviour; the resulting simplified formulation may be very convenient for numerical (as well as asymptotic) studies, particularly when the stiffness of the original problem can be removed in this fashion. The concept can be illustrated by giving a reinterpretation of an example discussed in Appendix 1 of [30], namely

$$u_t = \left((u+\epsilon)^m u_x\right)_x \tag{168}$$

with $0 < \epsilon \ll 1$, $m > 0$. For $\epsilon = 0$, Equation (168) has compactly supported solutions and an analysis of the limit $\epsilon \to 0$ clarifies how this property is lost for positive ϵ; a very concise presentation of such a limit (which is guided by the asymptotics) introduces the quantity

$$v = \int_0^u \left(\frac{(u'+\epsilon)^m}{u'} - \frac{\epsilon^m}{u'}\right)\mathrm{d}u' + \epsilon^m \log u, \tag{169}$$

giving the uniformly valid leading-order equation

$$v_{0t} = m v_0 H(v_0) v_{0xx} + v_{0x}^2, \tag{170}$$

where H is the Heaviside step function. Equation (170) encompasses each of the three regions appearing in [30] in a single equation and makes explicit the hyperbolic character of the low concentration tail. Moreover, the simplification to (170) permits a family of similarity solutions, notably

$$v_0 = t^{-m/(m+2)} V(x/t^{1/(m+2)}), \tag{171}$$

not shared by the full equation (168); (171) provides a uniformly valid expression of the asymptotic behaviour for δ-function initial data for u. It is worth noting that, because the nonlinearity in (170) is not analytic in v_0, care is needed when applying classical Lie group techniques to identify such similarity reductions.

We conclude by drawing together the analyses of the steady-state and time-dependent problems by outlining for small ϵ the role played in the transient problem by the unstable equilibrium solution. It is noteworthy that the outer steady-state solution (29) is of the form (39), but the relationship between the two problems is somewhat subtle, with no region completely analogous to the interior layer of Section 3.4.2 featuring in the time-dependent analysis. When the steady-state solution is perturbed, time-dependent effects make their first appearance in the $O(\nu^2)$ term of (23), thereby introducing a time derivative into (28); a number of timescales are needed to describe this, each having $\epsilon t = O(\log^q(1/\epsilon))$ for some q. The conclusion regarding the behaviour subsequent to these timescales is that if the steady state is (roughly speaking) perturbed by moving the interior zero towards $y = 0$, then the solution quickly locks onto the (slow) viscous-dominated phase of evolution (51–52), the instability of the steady state corresponding to the instability from above of the fixed point $\lambda_0 = \pi/2$ of (52), from which λ_0 then increases to λ_c (leading onto the faster phases). Conversely, if the interior zero is perturbed towards $y = 1$, the solution instead enters the (fast) inviscid phase, given by (63) with the $\frac{1}{2}$'s negligible because $D_0 \gg 1$. In view of (29), (31), the relevant scalings are

$$\hat{t} = \epsilon \log(1/\epsilon) t^{\ddagger}, \qquad D_0 = D_0^{\ddagger}/\epsilon \log(1/\epsilon), \qquad C_0 = C_0^{\ddagger}/\epsilon \log(1/\epsilon),$$

with

$$D_0^{\ddagger} \to -\frac{1}{4\pi}, \qquad C_0^{\ddagger} \to 0, \qquad \lambda_0 \to (\pi/2)^- \qquad \text{as } t^{\ddagger} \to -\infty,$$

from which it follows that

$$D_0^{\ddagger} = -\frac{1}{4\pi}\left(\frac{\pi}{2\lambda_0}\right)^3 \sin\lambda_0, \qquad C_0^{\ddagger} = -\frac{1}{4\pi}\left(\frac{\pi}{2\lambda_0}\right)^3 \cos\lambda_0,$$

and

$$\frac{d\lambda_0}{dt^{\ddagger}} = -\frac{1}{8\pi}\left(\frac{\pi}{2}\right)^3 \frac{\sin 2\lambda_0}{\lambda_0^2}, \tag{172}$$

the instability of the steady state corresponding to (172) also having an unstable fixed point at $\lambda_0 = \pi/2$. The solution to (172) then drops to zero (*cf.* (66)), leading onto behaviour comparable to that described above. The unstable steady state can thus be viewed as providing a borderline between initial conditions which evolve onto the slow part of the limit cycle (with

y_z initially decreasing) and those which lead rapidly into the fast phases (with y_z, already close to $y = 1$ for the steady state, initially increasing).

Our ordinary differential equation results of Section 3 are of sufficient accuracy to provide viable representations of the steady-state behaviour for large R, while the partial differential equation results of Sections 4–7 are of practical value in describing the evolution of the RPJ equation in that no straightforward and suitable numerical approach is available for even quite moderately large R; the extent of analytical progress that is possible here makes the asymptotic approach particularly worthwhile. Furthermore, as with many asymptotic studies, valuable insight is provided into the dominant physical balances which occur.

9. Concluding remarks

We hope that the detailed analysis of (1) described above is of interest both from the point of view of asymptotic methods and in view of its relevance as a paradigm problem for the time-dependent Navier–Stokes equations. The variety of asymptotic approaches needed is highlighted at the beginning of Section 8 and the analysis illustrates some of the complexities which arise in studying the asymptotics of time-dependent, spatially heterogeneous solutions to partial differential equations, specifically those exhibiting time dependence in the form of relaxation oscillations.

In the symmetrical version of (1–2), with equal fluid suction velocities at each channel wall, temporally chaotic solutions are found in numerical simulations at moderate Reynolds number [9, 15]. These solutions are analogous to chaotic solutions to the Lorenz system of three ordinary differential equations: the velocity profile may 'flip' about the centre-line $y = 0$ of the channel in an apparently random fashion, just as solutions to the Lorenz system flip between the two wings of the Lorenz attractor. Whether the approaches outlined above would prove equally efficacious in analysing such chaotic behaviour is an intriguing open question. A less speculative extension of our analysis would be in describing the large-amplitude limit-cycle solutions that we have also found in numerical simulations of the symmetrical problem at large Reynolds number, which appear to have a structure close to that of the large-amplitude limit-cycle solutions to the asymmetrical problem given above.

While the asymptotic methods we have used are for the most part well-established, novel features of the current work include the manner in which some of them are applied and the use in combination of such a range; we expect that a similar synthesis of different techniques will prove valuable in many other evolutionary problems. With regard to our steady-state analysis, we would highlight the key role of optimal truncation in making an expansion in inverse powers of the logarithm of the small parameter sufficiently accurate to be useful in practice.

The RPJ equation can play an instructive role as an exact reduction of the Navier–Stokes equations in, for example, studying blow-up phenomena and transitions to chaos. We accordingly hope that some of the apparatus we have developed will also prove useful in the study of the full Navier–Stokes equations or of other similarity reductions thereof. An example of the latter is the flow between a porous rotating disk and an impermeable, fixed plane [31], where a Hopf bifurcation is known to give rise to stable oscillatory solutions, although the time-dependent flow at large Reynolds number remains to be investigated. The natural extension of the RPJ equation to fully three-dimensional flows is also known to exhibit chaotic solutions [32]; these too may be amenable to our approach.

We have given perhaps one of the most detailed asymptotic studies so far undertaken of fully time-dependent nonlinear behaviour in a large Reynolds number flow problem of this type and, as indicated above, we hope that the asymptotic analysis described will be of value in the treatment of a range of related initial–boundary-value problems. Moreover, the current analysis is more broadly of some interest from the point of view of nonlinear parabolic systems, it being rare for scalar parabolic equations to possess temporally periodic ω-limit sets. We hope that the approaches we have outlined will also prove effective in asymptotic studies of a variety of other evolutionary problems arising in a range of contexts.

Acknowledgements

JRK gratefully acknowledges the support of the Leverhulme Trust.

Appendix A. The zero-pressure inviscid Riabouchinsky–Proudman–Johnson equation

A.1. General solution

This appendix is concerned with the partial differential equation

$$\mathcal{F}_{yt} = \mathcal{F}_y^2 - \mathcal{F}\mathcal{F}_{yy}; \tag{A1}$$

this is known to be integrable (*cf.* [15] and references therein) but it is useful in the current context to outline a convenient means by which initial-value problems can be solved for arbitrary initial data. We note that results equivalent to some of those below have also been obtained (in a slightly different context) by Galaktionov and Vazquez [33]; a special case of (39), namely

$$\mathcal{F} = B(t)y + C(t)\sin\lambda(t)y \qquad \text{with } p(t) = 0$$

can be used to exemplify the generic blow-up behaviour they describe. The analysis we give fairly readily generalises to the equation

$$\mathcal{F}_{yt} = \mathcal{F}_y^2 - \mathcal{F}\mathcal{F}_{yy} + p(t)$$

for prescribed $p(t)$ (again see [15]), but in practice the pressure coefficient is typically unknown and we shall not pursue the case $p \not\equiv 0$.

From (A1) we have

$$\mathcal{F}_{yyt} = \mathcal{F}_y\mathcal{F}_{yy} - \mathcal{F}\mathcal{F}_{yyy}, \tag{A2}$$

and so (A1–A2) can be solved by characteristic methods by satisfying

$$\frac{\mathrm{d}y}{\mathrm{d}t} = \mathcal{F}, \tag{A3}$$

$$\frac{\mathrm{d}\mathcal{F}_y}{\mathrm{d}t} = \mathcal{F}_y^2, \qquad \frac{\mathrm{d}\mathcal{F}_{yy}}{\mathrm{d}t} = \mathcal{F}_y\mathcal{F}_{yy} \tag{A4}$$

along the characteristics ((A3) shows that these represent particle paths); given $\mathcal{F}_y \geq 0$, characteristics on which $\mathcal{F}_y$ does not blow up therefore have $\mathcal{F}_y \equiv 0$. For the initial-value problem

$$\text{at } t = 0 \qquad \mathcal{F} = \mathcal{F}_i(y),$$

we impose on each characteristic that

$$\text{at } t = 0 \qquad y = y_i, \qquad \mathcal{F}_y = \mathcal{F}_i'(y_i), \qquad \mathcal{F}_{yy} = \mathcal{F}_i''(y_i),$$

using y_i to parametrise the initial data, and (A4) then gives

$$\mathcal{F}_y = \frac{\mathcal{F}_i'(y_i)}{1 - t\mathcal{F}_i'(y_i)}, \qquad \mathcal{F}_{yy} = \frac{\mathcal{F}_i''(y_i)}{1 - t\mathcal{F}_i'(y_i)}; \tag{A5}$$

moreover, if we write $y = y(t;\, y_i)$, it follows from (A3) that along characteristics

$$\frac{\mathrm{d}}{\mathrm{d}t}\left(\frac{\partial y}{\partial y_i}\right) = \frac{\partial y}{\partial y_i}\mathcal{F}_y$$

so that

$$\frac{\partial y}{\partial y_i} = \frac{1}{1 - t\mathcal{F}_i'(y_i)}$$

and hence

$$y = \int_0^{y_i} \frac{1}{1 - t\mathcal{F}_i'(\xi)}\,\mathrm{d}\xi + y_a(t), \qquad \mathcal{F} = \int_0^{y_i} \frac{\mathcal{F}_i'(\xi)}{(1 - t\mathcal{F}_i'(\xi))^2}\,\mathrm{d}\xi + \dot{y}_a(t), \tag{A6}$$

where y_a is arbitrary except that $y_a(0) = 0$, $\dot{y}_a(0) = -\mathcal{F}_i(0)$. Expressions (A5) and (A6) furnish the general solution to (A1) in terms of the parameter y_i, as is readily confirmed by direct substitution. Imposing the boundary data

$$y = 0 \qquad \mathcal{F} = \mathcal{F}_y = 0,$$

for example (this being relevant to the analysis of Section 7.1), with $\mathcal{F}_i(0) = \mathcal{F}_i'(0) = 0$, implies that $y_a(0) = 0$. In that context, however, we have the difficulty of wishing to impose initial data as $t \to -\infty$ rather than at $t = 0$, with $y \to 0^+$ as $t \to -\infty$ on characteristics; nevertheless, the initial–boundary-value problem in question can similarly be solved to give (132).

A.2. Similarity solutions

Of particular interest in the preceding analysis are the scaling similarity solutions of (A1), namely

$$\mathcal{F} = (-t)^{-(\alpha+1)}\Omega(\eta), \qquad \eta = y(-t)^{\alpha}; \tag{A7}$$

these satisfy

$$\Omega_\eta - \alpha\eta\Omega_{\eta\eta} = \Omega_\eta^2 - \Omega\Omega_{\eta\eta} \tag{A8}$$

so that we require $\Omega_\eta = 0$, $\Omega_\eta = 1$ or

$$\Omega_\eta^{\alpha}(1 - \Omega_\eta)^{1-\alpha} = \lambda_0(\alpha\eta - \Omega) \tag{A9}$$

for some constant of integration λ_0. Writing $\Omega = \alpha\eta - \Phi$ in (A9) yields an autonomous equation whose phase plane is readily analysed; moreover,

$$\int_0^{\Omega_\eta} P^{\alpha-1}(1 - P)^{-\alpha}\,\mathrm{d}P = \lambda_0(\eta + \eta_0) \tag{A10}$$

for some constant η_0. Imposing the conditions

$$\text{at } \eta = 0 \qquad \Omega = \Omega_\eta = 0, \tag{A11}$$

so that $\eta_0 = 0$, we have

$$\Omega \sim \frac{\alpha}{\alpha+1}(\lambda_0\alpha)^{1/\alpha}\eta^{(\alpha+1)/\alpha} \qquad \text{as } \eta \to 0^+,$$

and we thus need $\alpha > 0$, $\lambda_0 > 0$. For our purposes the cases

$$\alpha = 1 \qquad \Omega = -\frac{1}{\lambda_0} + \eta + \frac{1}{\lambda_0}e^{-\lambda_0\eta}, \tag{A12}$$

$$\alpha = \frac{1}{2} \qquad \Omega = \frac{1}{2}\eta - \frac{1}{2\lambda_0}\sin(\lambda_0\eta), \tag{A13}$$

are those of most significance (for reasons which will subsequently become clear). We first consider the condition

$$\text{as } \eta \to +\infty \qquad \Omega_\eta \to 1, \tag{A14}$$

relevant to Section 6.4. This constraint is automatically satisfied for (A9), (A12) when $\alpha \geq 1$, with

$$\Omega_\eta \sim 1 - (\lambda_0(\alpha-1))^{1/\alpha}\eta^{-1/(\alpha-1)} \qquad \text{as } \eta \to +\infty \text{ for } \alpha > 1;$$

the approach to (A14) is therefore most rapid for (A12) and it was shown in Section 6.4 how a 'logarithmically perturbed' version of the $\alpha = 1$ similarity solution is selected by appropriate matching of the exponentially decaying term (in a manner analogous to the way in which a logarithmically perturbed form of the fastest-decaying travelling wave is selected by Fisher's equation; see [34], for example). Equation (A8) also possesses non-analytic solutions which satisfy (A11) and (A14) when $0 < \alpha < 1$; these satisfy (A9) for $\eta < \eta_c$, where $\Omega(\eta_c) = \alpha\eta_c$, $\Omega_\eta(\eta_c) = 1$, and $\Omega = \alpha\eta_c + \eta - \eta_c$ for $\eta > \eta_c$, with

$$\Omega \sim \alpha\eta_c + (\eta - \eta_c) + \frac{1-\alpha}{2-\alpha}(\lambda_0(1-\alpha))^{1/(1-\alpha)}(\eta_c - \eta)^{(2-\alpha)/(1-\alpha)} \qquad \text{as } \eta \to \eta_c^-. \tag{A15}$$

However, for such a solution to be tenable in the current context, it must be realisable as the $\delta \to 0^+$ limit of a solution to the viscous equation

$$\mathcal{F}_{yt} = \delta\mathcal{F}_{yyy} + \mathcal{F}_y^2 - \mathcal{F}\mathcal{F}_{yy}, \tag{A16}$$

with $t \to 0^-$ (where we here use δ to denote the size of the (small) viscous term). Introducing

$$y = s(t) + \delta^{1/2}z, \qquad \mathcal{F} = \dot{s}(t) + \delta^{1/2}\frac{z}{(-t)} + \delta^{(2-\alpha)/2(1-\alpha)}F \tag{A17}$$

to describe the viscous interior layer, where $s(t) \sim \eta_c(-t)^{-\alpha}$, yields to leading order in δ that

$$F_{0zt} = F_{0zzz} + \frac{z}{(-t)}F_{0z} - \frac{z}{(-t)}F_{0zz}. \tag{A18}$$

It is easily seen that the only solutions to (A18) of the form

$$F_0 = (-t)^q\Phi(\zeta), \qquad \zeta = z/(-t)^{1/2} \tag{A19}$$

which do not blow up exponentially in at least one of the limits $\zeta \to \pm\infty$ are polynomials of the form

$$\Phi(\zeta) = k_N \sum_{n=0}^{[N/2]} \gamma_n \zeta^{N-2n}, \qquad q = \tfrac{3}{2}(N-2), \tag{A20}$$

where N is a non-negative integer, k_N is arbitrary, $\gamma_0 = 1$ and the other γ_n can be determined from (A18). It is also readily confirmed that (A19–A20) provide the only relevant solutions to (A18) in the limit $t \to 0^-$; since it is not possible to match any of these both with (A15) as $z \to -\infty$ and with $\Omega = \alpha\eta_c + \eta - \eta_c$ as $z \to +\infty$, the case $\alpha < 1$ can thus be ruled out, further clarifying the reasons why (A12) is the solution that arises when describing the relevant intermediate asymptotic behaviour.

In Section 7.1 a condition corresponding to

$$\text{as } \eta \to +\infty \qquad \Omega \to \Omega_\infty \tag{A21}$$

applies (in addition to (A11)), where the constant Ω_∞ needs to be determined as part of the solution. As already indicated, this condition cannot be satisfied for $\alpha \geq 1$ when (A11) holds; for particular $\alpha < 1$, however, we can construct a suitable solution in two parts, comprising a solution to (A8) with $\Omega_\eta \not\equiv 0, 1$ for $\eta < 2\eta_c$, with $\Omega(2\eta_c) = 2\alpha\eta_c$, $\Omega_\eta(2\eta_c) = 0$, and $\Omega = 2\alpha\eta_c$ for $\eta > 2\eta_c$ (the quantity η_c is given by (A22) below, with $\Omega_\infty = 2\alpha\eta_c$). However, before discussing the acceptability of such a non-analyticity at $\eta = 2\eta_c$, we need to consider the behaviour close to $\eta = \eta_c$, since $\Omega_\eta(\eta_c) = 1$, implying that the analysis of (A18) described above becomes relevant; using (A10) we have

$$\lambda_0\eta_c = \int_0^1 P^{\alpha-1}(1-P)^{-\alpha}\,\mathrm{d}P. \tag{A22}$$

The inner solutions (A20) are now viable, requiring to match with (A15) as $z \to -\infty$ that

$$\alpha = \frac{N-2}{N-1}, \qquad k_N = \frac{1}{N}\left(\frac{\lambda_0}{N-2}\right)^{-(N-2)}. \tag{A23}$$

However, to continue the solution into $\eta > \eta_c$ it is necessary that N be an odd integer; then for $\eta_c < \eta < 2\eta_c$ we can replace λ_0 in (A9) by $-\lambda_0$ and take

$$\Omega(\eta) = 2\alpha\eta_c - \Omega(2\eta_c - \eta) \tag{A24}$$

(such a change of signs in (A9) is already implicit in (A13)). We deduce, therefore, that a similarity solution of the form (A7) satisfying (A10) and (A21) is realisable only when α is given by (A23), wherein $N > 1$ is an odd integer. Moreover, the preceding analysis of (A18) indicates that the generic case is $N = 3$, $\alpha = \frac{1}{2}$, explaining its occurrence in the analysis of Section 7.1. For $\alpha = \frac{1}{2}$, we have from (A13) that $\eta_c = \pi/\lambda_0$.

Finally, we need to explain the reasons for the acceptability of the non-analyticity in the solution at $\eta = 2\eta_c$. (We note that more than one oscillation in, say, (A13) could in principle occur before the solution switches to $\Omega = \Omega_\infty$, but the case $\eta_c = \pi/\lambda_0$ is expected to be generic, rather than $\eta_c = M\pi/\lambda_0$ for some integer $M > 1$.) As $\eta \to 2\eta_c^-$ we have from (A9) that

$$\Omega \sim 2\alpha\eta_c + \frac{\alpha}{\alpha+1}(\alpha\lambda_0)^{1/\alpha}(2\eta_c - \eta)^{(\alpha+1)/\alpha}. \tag{A25}$$

The inner scalings for (A16) thus now read

$$y = s(t) + \delta^{1/2} z, \qquad \mathcal{F} = \dot{s}(t) + \delta^{(\alpha+1)/2\alpha} F \tag{A26}$$

with, in this case, $s(t) \sim 2\eta_c(-t)^{-\alpha}$, leading in place of (A18) to

$$F_{0zt} = F_{0zzz}, \tag{A27}$$

from which it follows that

$$F_0(z,t) \sim \Phi_0(z) \qquad \text{as } t \to 0^-,$$

where viscous effects play no role and $\Phi_0(z)$ is arbitrary (being determined by the evolution over earlier times) except that it satisfies

$$\Phi_0 \sim \frac{\alpha}{\alpha+1}(\alpha\lambda_0)^{1/\alpha}(-z)^{(\alpha+1)/\alpha} \qquad \text{as } z \to -\infty,$$
$$\Phi_0 \to 0 \qquad \text{as } z \to +\infty.$$

Hence the interior layer (A26–A27) places no constraints on the value of α, the only active role played by viscous effects in the discussion of this appendix being confined to the interior layer described by (A17–A18), which plays a crucial role in selecting the value $\alpha = \frac{1}{2}$. Although viscosity plays no explicit role in Section 7.1.2, the inner analysis about $\eta = \eta_c$ described above implies the requirement on the inviscid solution that it be analytic where $\mathcal{F}_y$ is maximal; such a constraint is implicit in the derivation of (132). We note that $\omega = x\mathcal{F}_{yy} = 0$ at both $\eta = \eta_c$ and $\eta = 2\eta_c$ and, in view of (A4), both are therefore characteristics of (A1).

Appendix B. Symmetries of the viscous Riabouchinsky–Proudman–Johnson equation

This appendix is concerned with the symmetries of (1) and briefly indicates some of their implications. This partial differential equation is invariant under translations of both y and t,

$$y^* = y + y_0, \quad t^* = t + t_0, \quad \mathcal{F}^* = \mathcal{F}, \tag{B1}$$

and has a scaling invariant

$$y^* = \sigma y, \quad t^* = \sigma^2 t, \quad \mathcal{F}^* = \mathcal{F}/\sigma \tag{B2}$$

(with the inviscid case $\epsilon = 0$ having another) which implies the existence of the similarity reductions

$$\mathcal{F}(y,t) = (\pm t)^{1/2}\Omega(\eta), \qquad \eta = y/(\pm t)^{1/2} \tag{B3}$$

which fail, however, to feature in our asymptotic analysis (though only just – see (101)). More interesting is the infinite-dimensional Galilean symmetry

$$y^* = y - s(t), \qquad t^* = t, \qquad \mathcal{F}^*(y^*, t^*) = \mathcal{F}(y,t) - \dot{s}(t) \tag{B4}$$

where $s(t)$ is arbitrary. This symmetry has an implicit role in the analysis of a number of the interior layers appearing above and it implies the obvious result that $\mathcal{F} = a(t) + b(t)y$ satisfies (1) for any $a(t)$ and $b(t)$. Moreover, it leads, together with (B1), to the similarity reduction

$$\mathcal{F}(y,t) = \dot{s}(t) + \Omega(\eta), \qquad \eta = y - s(t), \tag{B5}$$

which has steady states and travelling waves as special cases (with the particular implication that steady-state results carry over directly to the more general class of solutions (B5)), and, together with (B2), to

$$\mathcal{F}(y,t) = \dot{s} + (\pm t)^{1/2}\Omega(\eta), \qquad \eta = (y - s(t))/(\pm t)^{1/2}. \tag{B6}$$

Another noteworthy feature of (1) is its quadratically nonlinear form, which we have already exploited to obtain low-dimensional reductions, giving solutions which cannot be obtained by classical similarity methods. The relevant invariant subspaces are preserved under (B1), (B2) and (B3), so (prior to the imposition of boundary conditions) the resulting ordinary differential equations inherit each of these symmetries; it is worth noting that the invariant subspaces include useful special cases of (68) of the form

$$\mathcal{F} = a(t) + b(t)y + c(t)\mathrm{e}^{-\lambda(t)y}, \qquad \mathcal{F} = a(t) + c(t)\mathrm{e}^{-\kappa y}.$$

References

1. D. Riabouchinsky, Quelques considérations sur les mouvements plans rotationnels d'un liquide. *C. R. Hebd. Acad. Sci.* 179 (1924) 1133–1136.
2. I. Proudman and K. Johnson, Boundary-layer growth near a rear stagnation point. *J. Fluid Mech.* 12 (1962) 161–168.
3. K. Hiemenz, Die Grenzschicht an einem in den gleichförmigen Flüssigkeitsstrom eingetauchten geraden Kreiszylinder. *Dinglers Polytech. J.* 326 (1911) 321–324, 344–348, 357–362, 372–376, 391–393, 407–410.
4. A. S. Berman, Laminar flow in channels with porous walls. *J. Appl. Phys.* 24 (1953) 1232–1235.
5. R. M. Terrill, On some exponentially small terms arising in flow through a porous pipe. *Quart. J. Mech. Appl. Math.* 26 (1973) 347–354.
6. R. E. Grundy and H. R. Allen, The asymptotic solution of a family of boundary value problems involving exponentially small terms. *IMA J. Appl. Math.* 53 (1994) 151–168.
7. S. M. Cox and A. C. King, On the asymptotic solution of a high order non-linear ordinary differential equation. *Proc. R. Soc. London* A 453 (1997) 711–728.
8. R. M. Terrill, Laminar flow in a uniformly porous channel. *Aeronaut. Quart.* 15 (1964) 299–310.
9. M. B. Zaturska, P. G. Drazin and W. H. H. Banks, On the flow of a viscous fluid driven along a channel by suction at porous walls. *Fluid Dyn. Res.* 4 (1988) 151–178.
10. I. Proudman, An example of steady laminar flow at large Reynolds number. *J. Fluid Mech.* 9 (1960) 593–602.
11. R. M. Terrill and G. M. Shrestha, Laminar flow through parallel and uniformly porous walls of different permeability. *ZAMP* 16 (1965) 470–482.
12. R. M. Terrill and G. M. Shrestha, Laminar flow through a channel with uniformly porous walls of different permeability. *Appl. Sci. Res.* 15 (1966) 440–468.
13. R. M. Terrill, Flow through a porous annulus. *Appl. Sci. Res.* 17 (1967) 204–222.
14. G. M. Shrestha and R. M. Terrill, Laminar flow with large injection through parallel and uniformly porous walls of different permeability. *Quart. J. Mech. Appl. Math.* 21 (1968) 413–432.
15. S. M. Cox, Two-dimensional flow of a viscous fluid in a channel with porous walls. *J. Fluid Mech.* 227 (1991) 1–33.
16. S. M. Cox, Analysis of steady flow in a channel with one porous wall, or with accelerating walls. *SIAM J. Appl. Math.* 51 (1991) 429–438.
17. R. E. Grundy and R. McLaughlin, Global blow-up of separable solutions of the vorticity equation. *IMA J. Appl. Math.* 59 (1997) 287–307.
18. F. M. Skalak and C.-Y. Wang, On the nonunique solutions of laminar flow through a porous tube or channel. *SIAM J. Appl. Math.* 34 (1978) 535–544.
19. R. M. Terrill, Laminar boundary-layer flow near separation with and without suction. *Phil. Trans. R. Soc. London A* 253 (1960) 55–100.
20. G. N. Mercer and A. J. Roberts, A centre manifold description of contaminant dispersion in channels with varying flow properties. *SIAM J. Appl. Math.* 50 (1990) 1547–1565.

21. C. Domb and M. F. Sykes, On the susceptibility of a ferromagnetic above the Curie point. *Proc. R. Soc. London* A 240 (1957) 214–228.
22. G. K. Batchelor, *An Introduction to Fluid Dynamics.* Cambridge: Cambridge University Press (1967) 615pp.
23. P. G. Drazin and Y. Tourigny, Numerical study of bifurcations by analytic continuation of a function defined by a power series. *SIAM J. Appl. Math.* 56 (1996) 1–18.
24. D. Gottlieb and S. A. Orszag, *Numerical Analysis of Spectral Methods: Theory and Applications.* Philadelphia: SIAM CBMS-MSF Regional Conference Series in Applied Mathematics (1977) 170pp.
25. V. A. Galaktionov and S. A. Posashkov, New exact solutions of parabolic equations with quadratic nonlinearities. *USSR Comput. Maths. Math. Phys.* 21 (1989) 112–119.
26. J. R. King, Mathematical analysis of a model for substitutional diffusion. *Proc. R. Soc. London* A 430 (1990) 377–404.
27. G. I. Barenblatt, *Scaling, Self-Similarity and Intermediate Asymptotics.* Cambridge University Press (1996) 386pp.
28. S. J. Chapman, J. R. King and K. L. Adams, Exponential asymptotics and Stokes lines in nonlinear ordinary differential equations. *Proc. R. Soc. London* A 454 (1998) 2733–2755.
29. M. Van Dyke, *Perturbation Methods in Fluid Mechanics.* Stanford: Parabolic Press (1975) 271pp.
30. J. R. King, 'Instantaneous source' solutions to a singular nonlinear diffusion equation. *J. Eng. Math.* 27 (1993) 31–72.
31. M. A. Goldshtik and N. I. Javorsky, On the flow between a porous rotating disk and a plane. *J. Fluid Mech.* 207 (1989) 1–28.
32. C. L. Taylor, W. H. H. Banks, M. B. Zaturska and P. G. Drazin, Three-dimensional flow in a porous channel. *Quart. J. Mech. Appl. Math.* 44 (1991) 105–133.
33. V. A. Galaktionov and J. L. Vazquez, Blow-up of solutions with free boundaries for the Navier–Stokes equations. Department of Mathematics Preprint 9824, University of Bath, 1998.
34. H. P. McKean, Application of Brownian motion to the equation of Kolmogorov–Petrovskii–Piskunov. *Commun. Pure Appl. Math.* 28 (1975) 323–331. Corrigendum *Commun. Pure Appl. Math.* 29 (1976) 553–554.

Journal of Engineering Mathematics **39:** 131–170, 2001.

Generation of water waves and bores by impulsive bottom flux

M. LANDRINI and P. A. TYVAND[1]
INSEAN, The Italian Ship Model Basin, Via di Vallerano 139, 00128 Roma, Italy
[1] *Department of Agricultural Engineering, Agricultural University of Norway Box 5065, N-1432 Aas, Norway*

Received 28 December 1999; accepted in revised form 6 September 2000

Abstract. The inviscid free-surface flow due to an impulsive bottom flux on constant depth is investigated analytically and numerically. The following classes of two-dimensional flow are considered: an upwelling flow which is uniform over a half-plane, a line source/sink, and a dipole aligned along the bottom. The bottom flux is turned on impulsively and may decay with time. The fully nonlinear problem is solved numerically. A small-time asymptotic expansion to third order is found for the nonlinear problem. An asymptotic large-time solution is found for the linearized problem. A steady source will generate a pair of symmetric bores, and their breaking is investigated. A steady sink generates a depression wave if it is weak, and dip instability if it is strong. Wave breaking will occur for intermediate sink strengths. A decaying source emits solitary waves.

Key words: nonlinear free-surface flows, small-time expansion, boundary-integral equations, impulsive flows, water waves, bores.

1. Introduction

The classical Cauchy-Poisson problem is the linearized free-surface evolution from a prescribed initial elevation and initial velocity at the surface. The basic theory is found in Lamb [1]. For a finite depth we refer to Wehausen and Laitone [2, Section 22], and Whitham [3, Chapter 13]. In the present work we are concerned with the nonlinear Cauchy-Poisson problem due to initial disturbances at a fluid bottom, where an impulsive motion starts suddenly at time zero. Mei [4, Chapter 2] expounded the linear theory of transient free-surface flow due to bottom disturbances, with the application to tsunami modeling. In [5], Hammack discussed tsunami generation and the importance of nonlinearity, which can usually be neglected for tsunamis generated at typical ocean depths.

Tsunamis usually originate from rapid normal deflections at the sea-bottom. The inviscid model is appropriate for tsunami generation because of the small viscous diffusion length for the time scales involved. The flow is assumed incompressible, which means that acoustic phenomena are neglected. Their time scale is of the order of one second or shorter. From Kelvin's theorem it follows that such inviscid incompressible flows started from rest will remain irrotational through their evolution.

Practical tsunami modeling often involves bottom motions that have a horizontal length scale of the order of the fluid depth or larger. This suggests a shallow-water type of theory, although being inadequate during the impulsive start where full potential theory is needed. Vertical accelerations are important for an impulsive flow. If they are neglected, the horizontal length scale for the free-surface flow will be incorrect. However, if the bottom motion varies slowly over a wide horizontal domain, the hydrostatic description of upwelling/downwelling will lead to quantitatively small errors. Hydrostatic theory thus requires small horizontal gradients in the normal velocity at the bottom.

The far-field due to bottom disturbances with short length scale is dominated by singularities. A source contribution is important if the net mass inflow is significant. If the net mass inflow is negligible, the far-field is usually dominated by a horizontal dipole, oriented along the bottom. A strong dipole field will result from a moving mass along the bottom (landslide), and also from a mass-conserving geological fault.

In the present paper we will consider two-dimensional flows forced by a concentrated disturbance at a uniform bottom. The forcing bottom flux, originated by singularities, is impulsively started with a possible subsequent decay.

In our general investigation of surface evolution due to bottom disturbances, we will concentrate on four themes: (I) Impulsive upwelling, which is uniform over a finite or semi-infinite length. This problem introduces itself as a key problem for developing our analytical solution procedure. From a physical point of view, it is interesting to see the nonlinearities evolving at the edge of a uniformly upwelling layer. However, the phenomenon of vortex shedding disturbs the physical realism of the present upwelling model; (II) Bottom sources and dipoles that are turned on impulsively and remain steady afterwards. The source model is not representative for tsunamis, but for a different class of flows: a steady source started from rest defines a basic class of undular bores, where the nonlinear wave front evolves slowly in time as it progresses. Breaking of the bores is also an important issue (see *e.g.* [6, 7]); (III) Bottom sinks that are turned on impulsively and remain steady. This leads to the dip formation phenomenon, which is a gravitational collapse of the free surface into the sink. Previous work has been concerned with dip formation criteria for submerged sinks. In two dimensions this has been investigated analytically and numerically by Tyvand [8] and Kim [9], respectively. The impulsive start of a submerged point sink in three dimensions has been studied analytically and experimentally by Miloh and Tyvand [10] and numerically by Xue and Yue [11]; (IV) The model relevant for tsunami generation, which involves bottom singularities that decay with time. We will concentrate on the nonlinear regime of such flows, and the transient generation of solitons and trains of oscillatory waves.

Our study of potential flows due to impulsively started singularities aims to give an understanding of the basic mechanisms in the fully nonlinear free-surface evolution. The scope of our nonlinear analysis is broader than tsunami modeling, where linear or weakly nonlinear theory is often sufficient. Our work can be regarded as a study of the nonlinear Cauchy-Poisson problem for finite depth in two dimensions. Compared with the classical starting point of initially given surface deflection and surface flow, we take a more causal starting point when the flow is solely due to bottom motion. The issues of energy and momentum transfer to the fluid are outside the scope of the classical Cauchy-Poisson problem, where the initial energy and momentum are given *a priori*.

2. An introductory problem: the free-surface evolution at the edge of an impulsively upwelling fluid layer

We consider an inviscid fluid layer of constant depth $h^\star$ which is at rest at negative times $t^\star < 0$. The gravitational acceleration is $g^\star$. We will consider a two-dimensional flow with surface elevation $\eta^\star(x^\star, t^\star)$ above the undisturbed level $y^\star = 0$. The $x^\star$ axis is horizontal. For positive $x^\star$ an impulsive upwelling flow is started at time zero. It is given mathematically as a forced upwelling velocity $V_+^\star$ through the lower boundary $y^\star = -h^\star$. For negative $x^\star$ there is a forced downwelling velocity $-V_-^\star$. The unit vectors **i**, **j** correspond to the $x^\star$ and $y^\star$ axes. The important dimensionless parameter is the Froude number defined by

$$F = \frac{W^\star}{\sqrt{g^\star h^\star}} , \tag{1}$$

where we have introduced the velocity $W^\star = V_+^\star + V_-^\star$. We will introduce dimensionless (unstarred) quantities by using the initial fluid depth $h^\star$ as the unit of length and $W^\star$ as the unit of velocity. Consistently we introduce units of time $h^\star / W^\star$ and pressure $\rho^\star W^{\star 2}$, $\rho^\star$ being the fluid density.

The inviscid flow is governed by Laplace's equation

$$\nabla^2 \Phi = 0 \tag{2}$$

for the dimensionless velocity potential Φ. The kinematic and dynamic free-surface conditions are:

$$\frac{\partial \eta}{\partial t} + \nabla\Phi \cdot \nabla\eta = \frac{\partial \Phi}{\partial y} , \qquad \frac{\partial \Phi}{\partial t} + \frac{1}{2}|\nabla\Phi|^2 + \frac{1}{F^2}\eta = 0 \qquad \text{on } y = \eta(x, t) , \tag{3}$$

respectively. As initial conditions we take an impulsive start from a situation at rest with an undisturbed free surface:

$$\eta(x, 0) = 0 \qquad \Phi(x, 0, 0) = 0 . \tag{4}$$

The dimensionless upwelling and downwelling velocities for positive and negative x will be denoted by V_+ and $-V_-$, respectively. So the impulsive forced flow is given by:

$$\frac{\partial \Phi}{\partial y} = \begin{cases} V_+ & x > 0 \\ -V_- & x < 0 \end{cases} \qquad \text{on } y = -1, t > 0 \tag{5}$$

and, with our definition of velocity unit, $V_+ + V_- = 1$.

In principle, the discontinuity of the velocity due to the bottom boundary condition implies a shear layer which is not presently modeled. Because of this lack of a fully reliable physical model, the following analysis should be regarded primarily as an example of the small-time expansion procedure. The Fourier expansion here is a necessary basis for the solutions below for a source and a dipole. It should be noted that the forced velocity at y=0 is not purely vertical. There is in fact a singularity in the horizontal velocity at the edge of the upwelling region. However, this bottom singularity does not give any free-surface singularity.

There is one special case which is fully consistent physically: the case of pure downwelling ($V_+ = 0$, $V_- = 1$). But this process has a short life, since the free surface is swallowed into the bottom already at $t = 1$ for $x < 0$. If the Froude number is greater than one, the third-order analysis should be good for most of this lifetime.

2.1. THE SMALL-TIME EXPANSION

We solve our nonlinear initial-boundary-value problem analytically by employing a small-time expansion (*e.g.* [12], [13]). In particular, we postulate:

$$(\Phi, \eta) = H(t)[(\Phi_0, 0) + t(\Phi_1, \eta_1) + t^2(\Phi_2, \eta_2) + ..] \qquad -\infty < t < \infty\,, \tag{6}$$

where the Heaviside unit step function has been introduced:

$$H(t) = 0,\ \ t \le 0 \quad \text{and} \quad H(t) = 1,\ \ t > 0. \tag{7}$$

The governing equation and the conditions implied by the small-time expansion are:

$$\nabla^2 \Phi_n = 0 \qquad -1 < y < 0 \tag{8}$$

$$\frac{\partial \Phi_0}{\partial y} = \begin{cases} V_+ & x > 0 \\ -V_- & x < 0 \end{cases} \qquad \frac{\partial \Phi_n}{\partial y} = 0 \quad n = 1, 2, \ldots \qquad \text{on } y = -1\,. \tag{9}$$

The dynamic and kinematic conditions on the free-surface $y = 0$ are:

$$\Phi_0 = 0, \qquad \Phi_1 = -\frac{1}{2}\eta_1^2, \qquad \Phi_2 = -\eta_1(2\eta_2 + \frac{1}{2F^2}), \tag{10a}$$

$$\eta_1 = \frac{\partial \Phi_0}{\partial y}, \qquad \eta_2 = \frac{1}{2}\frac{\partial \Phi_1}{\partial y}, \qquad \eta_3 = \frac{1}{3}\frac{\partial \Phi_2}{\partial y} + \frac{1}{6}\eta_1^2\frac{d^2\eta_1}{dx^2} + \frac{1}{3}\eta_1\left|\frac{d\eta_1}{dx}\right|^2 \tag{10b}$$

2.2. THE FIRST-ORDER ELEVATION

The nonlinear problem will be solved most conveniently by Fourier series. For this purpose we introduce a dimensionless length L of horizontal periodicity, so that the forced flow alternates between uniform upwelling and uniform downwelling for every length $L/2$ along the bottom. If N is an arbitrary integer, we then consider the periodic boundary condition for the zeroth-order flow:

$$\frac{\partial \Phi_0}{\partial y} = \begin{cases} V_+ & N < x/L < N + 1/2 \\ -V_- & N - 1/2 < x/L < N \end{cases} \qquad \text{on } y = -1\,. \tag{11}$$

This is an artificial periodicity. Our original boundary-value problem is represented as the limit $L \gg 1$. The Fourier-series representation of the boundary condition (11) is then:

$$\frac{\partial \Phi_0}{\partial y} = \frac{V_+ - V_-}{2} + \frac{2}{\pi}\sum_{n\ \text{odd}} \frac{\sin(2\pi n x/L)}{n}$$

where ${}_{n\ \text{odd}}$ is an abbreviation for $n = 1, 3, 5, \ldots$. Therefore the first-order elevation with artificial periodicity reads:

$$\eta_1 = \frac{V_+ - V_-}{2} + \frac{2}{\pi}\sum_{n\ \text{odd}} \operatorname{sech}(2\pi n/L)\frac{\sin(2\pi n x/L)}{n}\,,$$

The corresponding exact solution can be found by integrating up the bottom source Green function for a uniform layer [14]:

$$\eta_1 = \frac{V_+ - V_-}{2} + \frac{2}{\pi}\arctan[\tanh(\pi x/4)] \ . \qquad (12)$$

The exact solution is a constant plus an odd function of x. The constant takes case of the net upwelling mass flux, as the odd function gives zero integral from minus to plus infinity. For later convenience, upon introducing $k = 2\pi/L$ and $k_n = nk$, we rewrite the periodic solution in the form:

$$\eta_1 = \frac{V_+ - V_-}{2} + \frac{2}{\pi}\sum_{n\ \mathrm{odd}} \operatorname{sech} k_n \frac{\sin(k_n x)}{n} \ . \qquad (13)$$

2.3. The second-order elevation

The first-order potential at the free surface $y = 0$ is given by:

$$\Phi_1 = -\eta_1^2/2 = -\frac{(V_+ - V_-)^2}{8} - \frac{V_+ - V_-}{\pi}\sum_{n\ \mathrm{odd}} \operatorname{sech} k_n \frac{\sin(k_n x)}{n} - $$

$$-\frac{1}{\pi^2}\sum_{m\ \mathrm{odd}}\sum_{n\ \mathrm{odd}} \operatorname{sech} k_m \operatorname{sech} k_n \frac{\cos(k_{n-m}x) - \cos(k_{n+m}x)}{nm} \ .$$

We integrate the above expression in the vertical direction, from the free surface up to the bottom, where the condition of zero flux is also applied. Eventually, we obtain the potential $\Phi_1(x, y)$ everywhere in the fluid domain and, by taking the normal gradient at the free surface, the second-order wave profile reads

$$\eta_2(x) = \eta_{2\,\mathrm{odd}} + \eta_{2\,\mathrm{even}} \ ,$$

where

$$\eta_{2\,\mathrm{odd}} = \frac{k}{2\pi}(V_- - V_+)\sum_{n\ \mathrm{odd}} \operatorname{sech} k_n \tanh k_n \sin(k_n x)$$

and

$$\eta_{2\,\mathrm{even}} = -\frac{k}{2\pi^2}\sum_{n\ \mathrm{odd}}\sum_{m\ \mathrm{odd}}\Big[\left(\frac{1}{m} - \frac{1}{n}\right)\tanh k_{n-m}\cos(k_{n-m}x)$$

$$-\left(\frac{1}{m} + \frac{1}{n}\right)\tanh k_{n+m}\cos(k_{n+m}x)\Big] \operatorname{sech} k_n \operatorname{sech} k_m \ .$$

2.4. THE THIRD-ORDER ELEVATION

The second-order potential will be written as the sum $\Phi_2 = \hat{\Phi}_2 + \Phi_2^F$, the terms $\hat{\Phi}_2$ and Φ_2^F being independent and dependent on gravity, respectively. The superscript ^{F} means Froude-number dependent. The surface elevation is decomposed in a similar way: $\eta_3 = \hat{\eta}_3 + \eta_3^F + \eta_{3\,\text{lower}}$. The first two terms are the third-order elevations due to the second-order potential,

$$\hat{\eta}_3 = \frac{1}{3}\frac{\partial \hat{\Phi}_2}{\partial y}\,, \qquad \eta_3^F = \frac{1}{3}\frac{\partial \Phi_2^F}{\partial y}\,, \qquad y = 0\,,$$

independent of and dependent on gravity, respectively. The last term (subscript: $_{\text{lower}}$) is composed from lower-order contributions

$$\eta_{3\,\text{lower}} = \frac{1}{6}\eta_1^2\frac{\mathrm{d}^2\eta_1}{\mathrm{d}x^2} + \frac{1}{3}\eta_1\left(\frac{\mathrm{d}\eta_1}{\mathrm{d}x}\right)^2$$

and can be simplified to:

$$\eta_{3\,\text{lower}} = \frac{1}{6}\eta_1\left(\frac{\mathrm{d}\eta_1}{\mathrm{d}x}\right)^2\left(2 - \pi\eta_1\sinh\left(\frac{\pi}{2}x\right)\right)\,.$$

The second-order potential independent of gravity is given by the first term in third-order boundary condition (10a):

$$\hat{\Phi}_2 = -2\eta_1\eta_2\,.$$

It determines $\hat{\eta}_3$, which is a basic contribution to the third-order elevation, as it takes care of the second level in the hierarchy of nonlinear free-surface interactions.

For a pure upwelling, *i.e.* $V_+ = 1$, $V_- = 0$, the gravity-dependent elevation is found from $3F^2\eta_3^F = 2\eta_{2\,\text{odd}}$ in the form

$$\eta_3^F = -\frac{k}{3\pi F^2}\sum_{n\ \text{odd}} \operatorname{sech} k_n \tanh k_n \sin(k_n x)\,.$$

In the present theory, rather than in the potential, say, ϕ in the x-y plane, we are interested in surface elevation at each order which, in turn, is given through normal derivatives $\partial\phi/\partial y$ on $y = 0$. For this purpose, solving Laplace's equation and applying the non-permeable condition at the bottom and the Dirichlet condition $\phi(x,0) = a_n\sin(k_n x - \alpha)$ at the free surface we obtain the formula $\partial\phi/\partial y = k_n \tanh k_n\ \phi(x,0)$ for the normal derivative on $y = 0$. Therefore, the normal derivative operator on each Fourier component can be replaced by a simple multiplication operator: $\partial/\partial y \quad\longrightarrow\quad k_n \tanh k_n$.

The most difficult term to evaluate in the third-order elevation is given by the second-order potential independent of gravity: $\hat{\Phi}_2 = -2\eta_1\eta_{2\,\text{odd}} - 2\eta_1\eta_{2\,\text{even}}$. These two inhomogeneities are given by the Fourier series:

$$-2\eta_1\eta_{2\,\text{odd}} = \sum_{n\text{ odd}} A_n \sin(k_n x) + \sum_{n\text{ odd}}\sum_{m\text{ odd}} B_{nm}(\cos k_{n-m}x - \cos k_{n+m}x) \;,$$

$$-2\eta_1\eta_{2\,\text{even}} = \sum_{n\text{ odd}}\sum_{m\text{ odd}} C_{nm}\cos k_{n-m}x - D_{nm}\cos k_{n+m}x +$$

$$+ \sum_{q\text{ odd}}\sum_{n\text{ odd}}\sum_{m\text{ odd}} E_{qnm}(\sin k_{n-m+q}x - \sin k_{n-m-q}x) -$$

$$- G_{qnm}(\sin k_{n+m+q}x - \sin k_{n+m-q}x)$$

We have here introduced the Fourier coefficients:

$$A_n = (V_+ - V_-)^2 \frac{k}{2\pi} \tanh k_n \operatorname{sech} k_n \;,$$

$$B_{nm} = 2(V_+ - V_-)\frac{k}{2\pi^2 n} \tanh k_m \operatorname{sech} k_n \operatorname{sech} k_m \;,$$

$$C_{nm} = (V_+ - V_-)\frac{k}{2\pi^2}\left(\frac{1}{m} - \frac{1}{n}\right) \tanh k_{n-m} \operatorname{sech} k_n \operatorname{sech} k_m \;,$$

$$D_{nm} = (V_+ - V_-)\frac{k}{2\pi^2}\left(\frac{1}{m} + \frac{1}{n}\right) \tanh k_{n+m} \operatorname{sech} k_n \operatorname{sech} k_m \;,$$

$$E_{qnm} = \frac{k}{\pi^3 q}\left(\frac{1}{m} - \frac{1}{n}\right) \tanh k_{n-m} \operatorname{sech} k_q \operatorname{sech} k_n \operatorname{sech} k_m \;,$$

$$G_{qnm} = \frac{k}{\pi^3 q}\left(\frac{1}{m} + \frac{1}{n}\right) \tanh k_{n+m} \operatorname{sech} k_q \operatorname{sech} k_n \operatorname{sech} k_m \;.$$

The matrices D and G are very similar to C and E, respectively, the change of sign in the n-terms being the only difference. However, this gives no simplification, because we sum over positive values only, $n = 1, 3, 5, \ldots$

The corresponding third-order elevation is given by:

$$
\begin{aligned}
3\hat{\eta}_3 = {} & k \sum_{n\ \mathrm{odd}} A_n \tanh k_n \ \sin k_n x + \\
& + k \sum_{n\ \mathrm{odd}} \sum_{m\ \mathrm{odd}} (n-m)(B_{nm}+C_{nm}) \tanh k_{n-m} \ \cos k_{n-m} x - \\
& - k \sum_{n\ \mathrm{odd}} \sum_{m\ \mathrm{odd}} (n+m)(B_{nm}+D_{nm}) \tanh k_{n+m} \ \cos k_{n+m} x + \\
& + k \sum_{q\ \mathrm{odd}} \sum_{n\ \mathrm{odd}} \sum_{m\ \mathrm{odd}} E_{qnm} [(n-m+q) \tanh k_{n-m+q} \ \sin k_{n-m+q} x - \\
& \qquad - (n-m-q) \tanh k_{n-m-q} \ \sin k_{n-m-q} x\ ] \\
& - k \sum_{q\ \mathrm{odd}} \sum_{n\ \mathrm{odd}} \sum_{m\ \mathrm{odd}} G_{qnm} [(n+m+q) \tanh k_{n+m+q} \ \sin k_{n+m+q} x - \\
& \qquad - (n+m-q) \tanh k_{n+m-q} \ \sin k_{n+m-q} x\ ] \,.
\end{aligned}
\tag{14}
$$

3. Impulsive waves due to a source at a uniform bottom

We now return to our original problem where a forced impulsive flow is started at time zero. Let us first consider the case of a source with initial volume flux $Q^\star$, given per unit length perpendicular to the flow. It is concentrated at the point $x^\star = 0$, either as an upward flow (bottom source) or downward flow (bottom sink). The Froude number is now defined as

$$
F = \frac{Q^\star}{\sqrt{g^\star h^{\star 3}}} \tag{15}
$$

and an appropriate choice of the unit velocity is $Q^\star/h^\star$. As in the previous sections, the unit of length is the initial fluid depth $h^\star$ and, consistently, $h^{\star 2}/Q^\star$ and $\rho^\star(Q^\star/h^\star)^2$ are the dimensionless time and pressure units. The bottom source is assumed to decay exponentially in time as $Q^\star e^{-\sigma^\star t^\star}$. The dimensionless decay rate is defined as:

$$
\sigma = \sigma^\star h^{\star 2}/Q^\star \,. \tag{16}
$$

So the dimensionless source strength is $e^{-\sigma t}$. The total dimensionless volume (per length of the source) being emitted by the source during its lifespan is given by σ^{-1}. It is of interest to keep this dimensionless area constant and vary the Froude number.

Regardless of whether a bottom source or dipole (*cf.* Appendix A) is considered, the governing equations are the same as in Section 2, except for the bottom boundary condition, which now reads:

$$
\frac{\partial \Phi}{\partial y} = e^{-\sigma t}\, \delta(x) \qquad y = -1,\ t > 0 \,, \tag{17}
$$

where $\delta(x)$ denotes the Dirac delta function.

Following the procedure sketched in the previous section, through the small time expansion (6), we now recast the fully nonlinear problem in the form of a set of problems for the Laplace equation (8) with free-surface boundary conditions (10a), (10b). As the bottom-boundary condition (17) suggests, the potential to each order is written as the sum of a regular potential φ_n and a singular potential ψ_n:

$$\Phi_n = \varphi_n + \psi_n \ . \tag{18}$$

We first consider the singular potential given by the source at the bottom with the zero-potential condition at $y = 0$.

The complex version of the unexpanded singular potential is:

$$\Psi = \frac{1}{2\pi} \mathrm{e}^{-\sigma t} \log \frac{\sinh\left(\frac{\pi}{2}z\right) + \mathrm{i}}{\sinh\left(\frac{\pi}{2}z\right) - \mathrm{i}} \ .$$

Here the complex variable $z = x + \mathrm{i}y$ is introduced. This formula has been derived from the Schwarz-Christoffel confomal mapping for a semi-infinite horizontal layer, which was given in [14]. The normal derivative at the free surface is found to be:

$$\left.\frac{\partial \psi}{\partial y}\right|_{y=0} = -\mathfrak{Im}\left.\left(\frac{\mathrm{d}\Psi}{\mathrm{d}z}\right)\right|_{y=0} = \frac{1}{2}\mathrm{e}^{-\sigma t}\,\mathrm{sech}\left(\frac{\pi}{2}x\right) \ .$$

The same formula for the surface flow can be derived by the image method. But this method has the disadvantage that the series for the potential does not converge.

The bottom conditions for the regular potentials are:

$$\frac{\partial \varphi_n}{\partial y} = 0 \ , \qquad y = -1 \tag{19}$$

for $n = 1, 2, \ldots$

3.1. First-order elevation

The first-order elevation is given by the normal derivative of the singular potential:

$$\eta_1 = \frac{1}{2}\,\mathrm{sech}\left(\frac{\pi}{2}x\right) \ . \tag{20}$$

We note the simple result for the surface elevation just above the source: $\eta_1(0) = 1/2$. In the case of a submerged line source [8] this first-order elevation just above the source is $1/\pi$.

The boundary-value problems will be solved most conveniently by Fourier series. For this purpose we introduce a dimensionless length L of horizontal periodicity. This means that we assume the following boundary condition along the bottom:

$$\frac{\partial \psi}{\partial y} = \mathrm{e}^{-\sigma t} \sum_{n=-\infty}^{+\infty} (-1)^n \,\delta\left(x - n\frac{L}{2}\right) \ , \qquad y = -1, \ t > 0 \ . \tag{21}$$

A Fourier expansion for this boundary condition, which generates the zeroth-order potential by an array of singularities, does not exist. But the potential itself exists, in terms of generalized functions. Its first-order elevation can be found by taking the x-derivative from the upwelling problem investigated by in Section 2. The result is:

$$\eta_1 = \frac{4}{L} \sum_{\text{n odd}} \operatorname{sech}\left(\frac{2\pi n}{L}\right) \cos\left(\frac{2\pi n}{L} x\right) , \tag{22}$$

where n-odd is an abbreviation for $n = 1, 3, 5, \ldots$. In the following, we use the compact notation:

$$k = \frac{2\pi}{L} \qquad k_n = nk \; ; \tag{23}$$

therefore k_n denotes the n-th harmonic wavenumber. The first-order elevation with artificial periodicity is then:

$$\eta_1 = \frac{2}{\pi} k \sum_{\text{n odd}} \operatorname{sech} k_n \cos k_n x \; . \tag{24}$$

As discussed in the previous section, a shortcut to solve the periodic boundary-value problems below is using $\partial/\partial y \to k_n \tanh k_n$, for each Fourier component, to obtain the normal derivative of the potential on $y = 0$ starting from the free-surface potential.

In equivalence with taking the artificial length L to be much greater than one, we have now introduced the lowest-harmonic wave number k as our artificial parameter to be much smaller than one. Its higher harmonics are k_n . We will evaluate the series (24) by truncating it after a number n_{max}. Asymptotically, as $L \gg 1$ (or $k \ll 1$), the approximate solution (24) converges to the exact solution (20).

The relationship between the present problem and the impulsive upwelling problem of Section 2 is intriguing. Although Equation (24) is a Fourier-series solution of our bottom-source problem with a spatial period L, we could find it only indirectly through the upwelling problem. The obtained solution is a Fourier-series solution valid for the free-surface flow generated by an array of bottom sources and sinks, in spite of the fact that it could not be found by Fourier expanding our bottom condition.

These formulas for the first-order elevation represent a bottom source. The same expressions with opposite sign will be valid for a bottom sink.

3.2. Second-order elevation

The regular first-order potential at the free surface $y = 0$ is given by:

$$\varphi_1 = -\frac{1}{2}\eta_1^2 = -\frac{1}{8}\operatorname{sech}^2\left(\frac{\pi}{2}x\right) , \tag{25}$$

which can be recast as a Fourier series with a wave number k:

$$\varphi_1 = -\left(\frac{k}{\pi}\right)^2 \sum_{\text{n odd}} \sum_{\text{m odd}} \operatorname{sech} k_n \operatorname{sech} k_m \left[\cos(k_{n-m}x) + \cos(k_{n+m}x)\right] \tag{26}$$

We integrate the Fourier series and apply the condition of impermeability at the bottom. For the purpose of higher-order analysis we give the second-order elevation by the formula:

$$\begin{aligned} 2\eta_2 = & -\left(\frac{k}{\pi}\right)^2 \sum_{\text{n odd}} \sum_{\text{m odd}} \operatorname{sech} k_n \operatorname{sech} k_m \times \\ & \times \left[k_{n-m} \tanh k_{n-m} \cos(k_{n-m} x) + k_{n+m} \tanh k_{n+m} \cos(k_{n+m} x) \right] \\ & - 2\sigma \frac{k}{\pi} \sum_{\text{n odd}} \operatorname{sech} k_n \cos k_n x \; . \end{aligned} \tag{27}$$

As a last term in Equation (27) we have included the second-order contribution from the decaying source potential. If we introduce its exact instead of periodic version, we have the second-order elevation:

$$\begin{aligned} \eta_2 = & -\frac{1}{2}\left(\frac{k}{\pi}\right)^2 \sum_{\text{n odd}} \sum_{\text{m odd}} \operatorname{sech} k_n \operatorname{sech} k_m \times \\ & \times \left[k_{n-m} \tanh k_{n-m} \cos(k_{n-m} x) + k_{n+m} \tanh k_{n+m} \cos(k_{n+m} x) \right] \\ & - \sigma \frac{1}{4} \operatorname{sech} \left(\frac{\pi}{2} x\right) \; . \end{aligned} \tag{28}$$

3.3. Third-order Elevation

The regular second-order potential at the free surface $y = 0$ is:

$$\begin{aligned} \varphi_2 = & \left(\frac{k}{\pi}\right)^3 \sum_{\text{n odd}} \sum_{\text{m odd}} \sum_{\text{q odd}} \operatorname{sech} k_n \operatorname{sech} k_m \operatorname{sech} k_q \times \\ & \times \{ k_{n-m} \tanh k_{n-m} [\cos(k_{n-m-q} x) + \cos(k_{n-m+q} x)] + \\ & + k_{n+m} \tanh k_{n+m} [\cos(k_{n+m-q} x) + \cos(k_{n+m+q} x)] \; \} \\ & + 2\sigma \left(\frac{k}{\pi}\right)^2 \sum_{\text{n odd}} \sum_{\text{m odd}} \operatorname{sech} k_n \operatorname{sech} k_m [\cos(k_{n-m} x) + \cos(k_{n+m} x)] \\ & - \frac{1}{F^2} \frac{k}{\pi} \sum_{\text{n odd}} \operatorname{sech} k_n \cos k_n x \; . \end{aligned} \tag{29}$$

The total surface elevation is given by:

$$\eta_3 = \underbrace{\frac{1}{3} \frac{\partial \Phi_2}{\partial y}}_{\eta_{3,\infty} + \eta_{3,F}} + \underbrace{\frac{1}{6} \eta_1^2 \frac{d^2 \eta_1}{d x^2} + \frac{1}{3} \eta_1 \left| \frac{d \eta_1}{d x} \right|^2}_{\eta_{3\eta_1}} , \tag{30}$$

where (*cf.* third part of Equation (10a)) the first term is split in a zero-gravity contribution and a truly Froude-number dependent term. The second and last contribution are grouped into $\eta_{3\eta_1}$, the nonlinear contribution coming directly from η_1. Here we will apply the exact formulas for the first-order elevation. We also have to add the contribution from the singular potential. The final result is then:

$$
\begin{aligned}
\eta_3 = {} & \frac{1}{3}\left(\frac{k}{\pi}\right)^3 \sum_{\text{n odd}} \sum_{\text{m odd}} \sum_{\text{q odd}} \operatorname{sech} k_n \operatorname{sech} k_m \operatorname{sech} k_q \times \\
& \times \{k_{n-m} \tanh k_{n-m} [\; k_{n-m-q} \tanh k_{n-m-q} \cos(k_{n-m-q}x) + \\
& \qquad + k_{n-m+q} \tanh k_{n-m+q} \cos(k_{n-m+q}x)] + \\
& + k_{n+m} \tanh k_{n+m} [\; k_{n+m-q} \tanh k_{n+m-q} \cos(k_{n+m-q}x) \\
& \qquad + k_{n+m+q} \tanh k_{n+m+q} \cos(k_{n+m+q}x)] \} \\
& + \frac{2}{3}\sigma \left(\frac{k}{\pi}\right)^2 \sum_{\text{n odd}} \sum_{\text{m odd}} \operatorname{sech} k_n \operatorname{sech} k_m \times \\
& \times [k_{n-m} \tanh k_{n-m} \cos(k_{n-m}x) + k_{n+m} \tanh k_{n+m} \cos(k_{n+m}x)] - \\
& - \frac{1}{3F^2}\frac{k}{\pi} \sum_{\text{n odd}} k_n \operatorname{sech} k_n \tanh k_n \cos k_n x + \frac{\pi^2}{384} \operatorname{sech}^5\left(\frac{\pi}{2}x\right) [\, 3\cosh(\pi x) - 5\,] + \\
& + \frac{\sigma^2}{12} \operatorname{sech}\left(\frac{\pi}{2}x\right) .
\end{aligned}
\tag{31}
$$

Here we have avoided the artificial periodicity wherever an exact formula is available.

4. Numerical solution of the exact problem

The evolution of the flow on a longer time-scale is studied by solving numerically the fully nonlinear problem. Here we briefly sketch the adopted algorithm, more details are provided by Tyvand and Landrini in [15] and in the references cited there and below. We solve an initial value problem for the Euler equations:

$$
\nabla \cdot \boldsymbol{u} = 0 , \qquad \frac{\mathrm{D}\boldsymbol{u}}{\mathrm{D}t} = -\frac{1}{\rho}\nabla p + \boldsymbol{g} , \tag{32}
$$

where $\boldsymbol{u}$ is the fluid velocity, with the no–penetration condition $\boldsymbol{u} \cdot \boldsymbol{\nu} = \mathrm{e}^{-\sigma t}\delta(\boldsymbol{P}_S)$ on the bottom and, respectively, the kinematic and dynamic boundary conditions:

$$
\frac{\partial \boldsymbol{P}(\xi, t)}{\partial t} = \boldsymbol{u} \qquad \frac{\partial u_\xi(\xi, t)}{\partial t} = \frac{\partial}{\partial \xi}\left(\frac{1}{2}u_\xi^2 - gy\right) \tag{33}
$$

on the free surface $\mathcal{F}$. Equations (33) are a Lagrangian description of the free surface $\boldsymbol{P}[x(\xi, t), y(\xi, t)]$, where ξ is a Lagrangian parameter and u_ξ is the covariant component of $\boldsymbol{u}|_{\mathcal{F}}$ along the tangential vector $\partial \boldsymbol{P}/\partial \xi$. This allows to follow large deformations up to possible incipient wave breaking.

The unsteady fully nonlinear problem is numerically solved through a boundary-integral approach coupled with a time marching procedure (a standard fourth-order Runge–Kutta

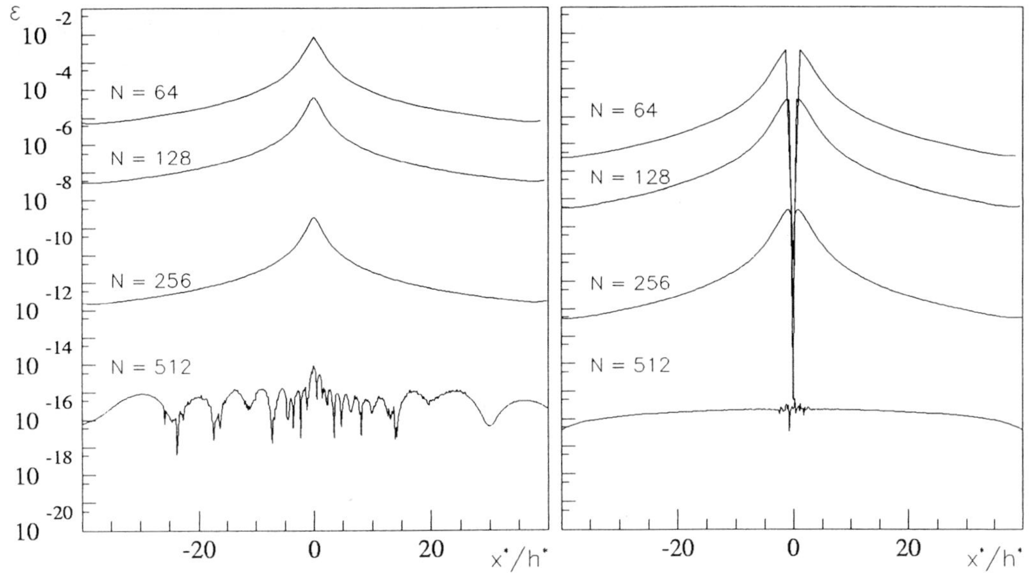

Figure 1. Solution by integral equation. Absolute error ε in determining $\partial\Phi_0/\partial y$ on $y = 0$ for increasing number N of unknowns. Left hand plot: velocity induced by a source (*cf.* Equation 20). Right hand plot: velocity induced by a dipole (*cf.* Equation A4).

scheme). Starting from the Poincaré formula discussed by Bassanini *et al.* in [16], we may express the velocity field by

$$\boldsymbol{u}(\boldsymbol{Q}) = \nabla_Q \int_{\mathcal{F}} \boldsymbol{u}\cdot\boldsymbol{\nu}\,\hat{G}\,\mathrm{d}S_P + \nabla_Q \times \int_{\mathcal{F}} \boldsymbol{u}\times\boldsymbol{\nu}\,\hat{G}\,\mathrm{d}S_P + \boldsymbol{u}_S(\boldsymbol{Q})\,, \tag{34}$$

i.e. 'directly' in terms of the normal $\boldsymbol{u}\cdot\boldsymbol{\nu}$ and tangential $\boldsymbol{u}\times\boldsymbol{\nu}$ velocity components on $\mathcal{F}$. The singular part of the velocity field due to source, dipole or combination of singularities is explicitly given by $\boldsymbol{u}_S(\boldsymbol{Q})$. The presence of the bottom boundary is accounted for through the kernel $\hat{G}$, obtained by modification of the two-dimensional fundamental solution $\log|\boldsymbol{P}-\boldsymbol{Q}|$ for Laplace equation according to the method of images.

At a given instant of time, we assume that u_ξ is known and the unknown velocity component $\boldsymbol{u}\cdot\boldsymbol{\nu}|_{\mathcal{F}}$ is evaluated through the integral equation following from (34) when $\boldsymbol{P}\in\mathcal{F}$ and taking the normal projection. Once the velocity field on $\mathcal{F}$ is evaluated, the solution can be prolonged in time by the free-surface evolution equations (33). Eventually, new geometry and boundary data are obtained and the procedure can be re–initialized.

The contour integrals are discretized by the Euler–McLaurin summation formula which, in combination with trigonometric polynomials, allows for high accuracy (see [17] for a rigorous analysis). The iteration technique developed in [18] is adopted for solving the linear algebraic system. For longer simulations and large number N of unknowns, multipole expansion and fast summation techniques (described in [19]) have been applied to speed up the solution procedure. A typical test of convergence under grid refinement is given in Figure 1, where the error in evaluating $\boldsymbol{u}\cdot\boldsymbol{\nu}\equiv\partial\Phi_0/\partial y$ on the free surface is shown for increasing number N of Lagrangian points on the free surface.

Sawtooth instabilities, first reported by Longuet-Higgins and Cokelet in [20], are removed by a high-order filter [21].

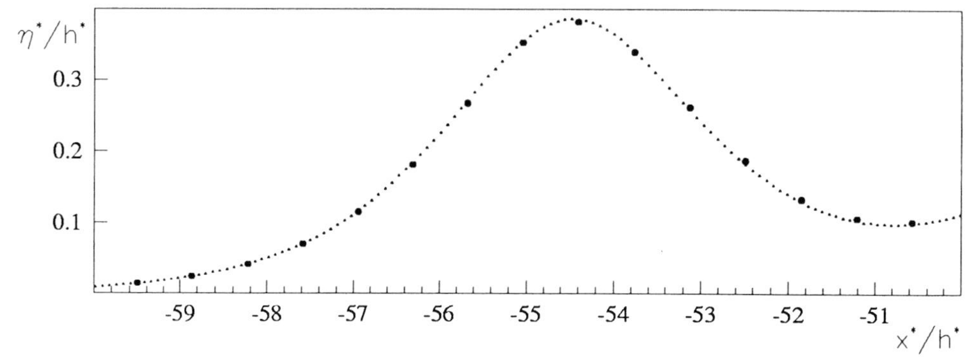

Figure 2. Convergence test under grid and time-step refinement. Comparison of the leading wave of an undular bore ($F = 0{\cdot}5$, $\sigma = 0$ and $t^\star/T_{\mathrm{lw}} = 50$) obtained by the coarsest (large symbols) and the finest (tiny symbols) discretization from Table 4.

During the simulation, for time long enough, the distribution of Lagrangian markers above a submerged source tends to be highly non-uniform. This can cause loss of accuracy and a strongly localized unstable motion of the free surface. We have removed the problem by redistributing periodically the free-surface points around the singularity using a cubic spline based algorithm. This can degrade the overall accuracy up to third-order in mesh-spacing, though the main effect is expected to be around the singularity. To the same purpose, an alternative grid-control procedure has been also developed and tested. In particular, new free-surface particles are inserted as soon as the spacing between the original markers becomes too large. The location and the tangential velocity of the new particles are initialized by cubic fitting. The results obtained by using the two different procuderes are essentially the same, thus confirming each other. The high resolution needed by sink-generated flows has been achieved by the use of tanh-grading for the initial distribution of markers around $x = 0$.

We have checked all the computations shown in the main text by halving both the initial spacing $\delta\ell^\star/h^\star$ of the Lagrangian markers and the time step $\Delta t^\star/T_{\mathrm{lw}}$ until an acceptable convergence is achieved. As a typical example, the Table 4 reports the height $\eta_0^\star/h^\star$ above a constant source, $F = 0{\cdot}5$, for $t^\star/T_{\mathrm{lw}} = 50$ and for increasingly refined discretization. The two

Table 1. Convergence test under refinement of discretization. Effect of main discretization parameters on the wave height $\eta_0^\star/h^\star$ over a constant source, $F = 0{\cdot}5$, for $t^\star/T_{\mathrm{lw}} = 50$.

$\delta\ell^\star/h^\star$ $\Delta t^\star/T_{\mathrm{lw}}$	0·64 1/10	0·32 1/20	0·16 1/40	0·08 1/80
$\eta_0^\star/h^\star$	0·2388911736952251	0·2371882781916181	0·2371157325963261	0·2371139364995115

most refined solutions differ by less than 0·0008 %. Less easy to compare are differences in wave height for those markers largely drifted by the flow. In Figure 2 we simply superimpose the coarsest grid and the most refined (symbols represent the Lagrangian markers) and this gives an idea of the satisfactory convergence achieved. If superimposed, results by the two most refined discretizations would be undistinguishable.

We also recall that, in the numerical simulation, a truncated free surface has to be considered. Therefore, for each case discussed, we ruled out the effect of the finite horizontal extent of the free surface by doubling it, keeping constant Δt and $\delta\ell$, and verifying the invariance of

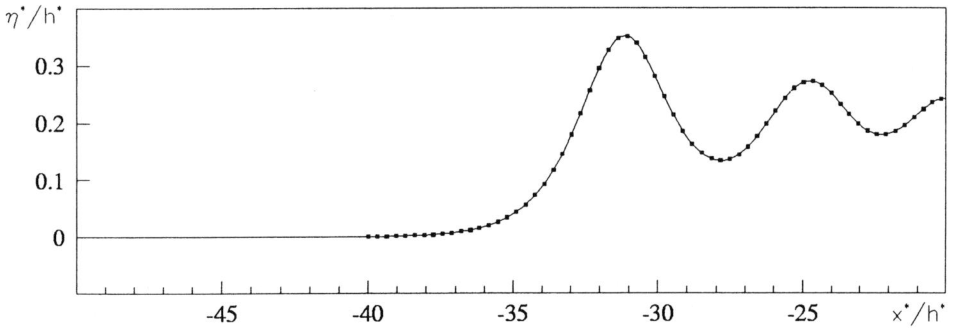

Figure 3. Effect of horizontal truncation of the computational domain. Comparison of the leading wave of an undular bore ($F = 0{\cdot}5$, $\sigma = 0$ and $t^\star/T_{\text{lw}} = 60$) for $L_{\max}/h^\star = -40$ (symbols) and $L_{\max}/h^\star = -80$ (solid line).

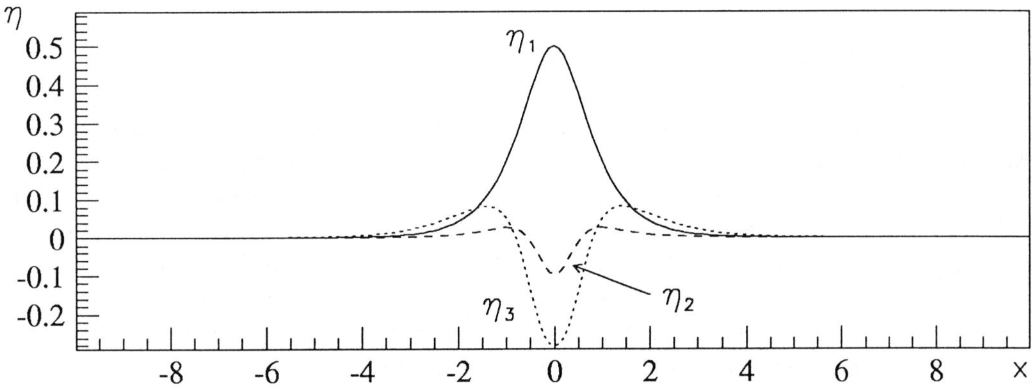

Figure 4. Shapes of η_1, η_2, η_3 for $F = 0{\cdot}5$ and $\sigma = 0$. All variables are non-dimensional. In the series $L = 100$, $n_{\max} = m_{\max} = q_{\max} = 200$. Non-dimensional variables according to the text.

the solution. Actually, Figure 3 shows that, even when the leading wave is close to the edge of the computational domain, the solution are practically the same.

Typically, at the end of the simulations, the error in mass conservation is below 0·001%, while the imbalance between the integrated input of power due to the singularity and the energy in the fluid system is at most of order 1%.

5. Discussion

5.1. Range of Validity of the Small-Time Expansion

Figure 4 shows the height functions η_1, η_2 and η_3 for a constant source, $\sigma = 0$, with strength $F = 0{\cdot}5$. It can be seen that, for this choice of the parameters, higher-order contributions act to reduce the central hump and spreading the disturbance over a larger horizontal extent. This behaviour is not entirely due to the gravity, which actually enters the problem only at third order. Even at this order, a more subtle balance of different tendencies can be observed through Figure 5 where η_3 is split according to Equation (30). Both $\eta_{3\eta_1}$ (pumped by the first-order solution) and $\eta_{3,F}$ (directly related to the gravity) act qualitatively as η_2, with a larger spreading action for the Froude-dependent term. The zero-gravity term gives a larger and narrower positive peak with smaller depression around it.

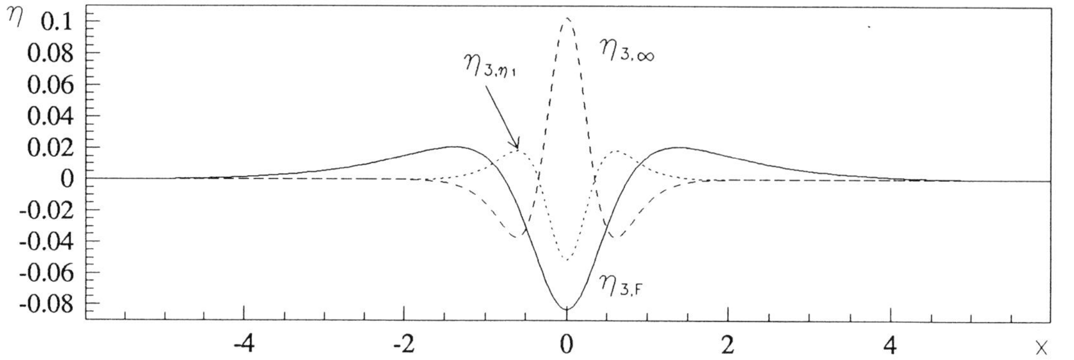

Figure 5. Terms forming η_3 (*cf.* Equation (30)) for $\sigma = 0$. The Froude-dependent term $\eta_{3,F}$ is shown for $F = 1$. Non-dimensional variables according to the text.

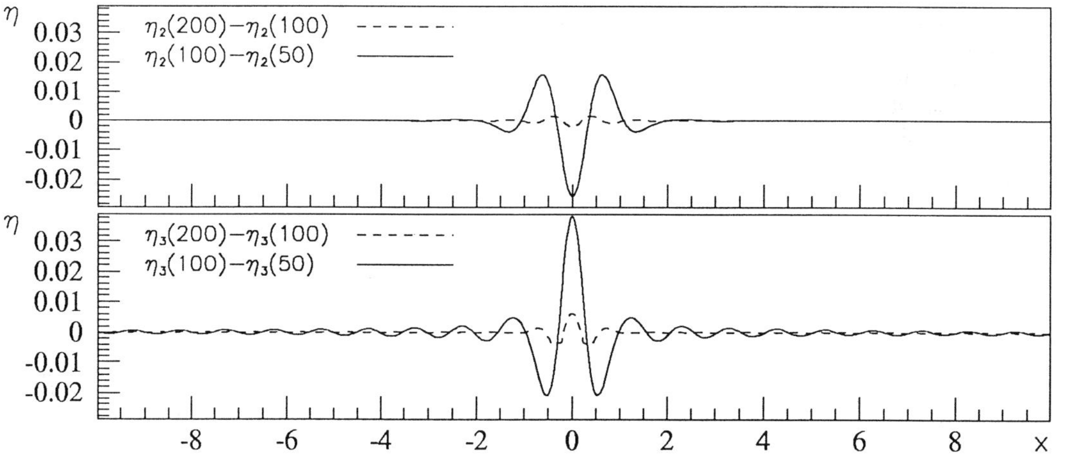

Figure 6. Convergence of series expansion for η_2 (top) and η_3 (bottom). Same of Figure 4. Differences are plotted between the sums obtained by using $n_{\max} = m_{\max} = q_{\max} = 50$, 100 and 200; the same $L = 100$ is adopted. Non-dimensional variables according to the text.

Series entering in Equations (28) and (31) for η_2 and η_3, respectively, cannot be summed up in closed form, and the effect of truncation of the Fourier series is of concern. This is considered in Figure 6 for the same case as above, where the point-wise difference of η_2 and η_3 as obtained by doubling the highest modes $n_{\max} = m_{\max} = q_{\max}$ is plotted against the horizontal coordinate x. Computations have been performed using quadruple precision, without special means to improve the speed of convergence of the series which turned out to be quite slow, with larger variations concentrated around the origin. By further doubling the truncation limits of summations, we detected no changes graphically.

After the heuristic convergence-check of our boundary-integral equation solver by comparing with the analytical solution for Φ_0 (*cf.* Figure 1), we have solved numerically also the velocity fields $\nabla\Phi_n$, $n = 1, 2$, and the same results for η_1, η_2 and η_3 have been obtained with larger efficiency.

The range of validity of our small-time expansion is discussed by comparing the semi-analytical wave profiles with those obtained by solving numerically the fully nonlinear problem. In Figure 7 the two solutions are shown for a given source strength, $F = 1$, and increasing values of the decaying factor σ. For $\sigma \leq 0{\cdot}5$, the small-time expansion (dashed lines) agrees

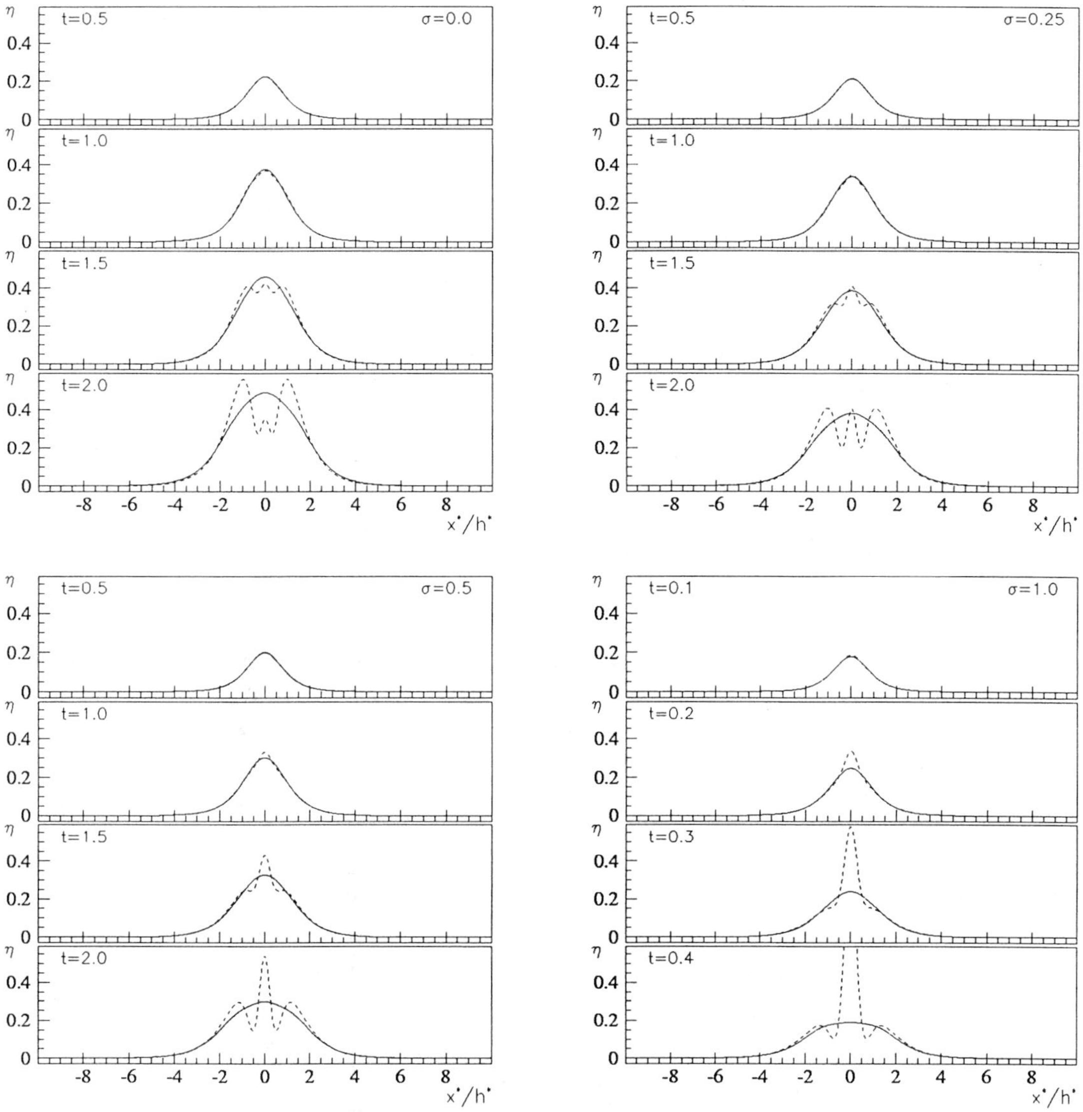

Figure 7. Comparison between the small-time expansion to third order (dashed lines) and the fully nonlinear (numerical) solution for $F = 1{\cdot}0$, $\sigma = 0, 0{\cdot}25, 0{\cdot}5, 1$. Non-dimensional time $t = t^\star Q^\star / h^{\star 2}$.

well with the fully nonlinear solution (solid lines) up to $t \sim \mathcal{O}(1)$. Within this time scale, the behaviour simply consists in the smooth growth of the water level just above the source: a symmetric hump is formed and, after reaching a maximum level, starts to increase its horizontal extent. As can be expected, for non-zero σ the vertical growth is smaller while the horizontal spreading is comparable for all the σ considered.

In all the considered cases, for larger time, the small-time expansion deviates from the nonlinear solution. For $\sigma = 0$, the central hump decreases and two lateral peaks emerge with a separation distance (slightly) increasing in time. For finite σ, the lateral humps still appear, but the central peak remains higher and progressively overestimates the fully nonlinear solution. This trend is magnified in the case $\sigma = 1$ (bottom-right plot), where the exponential behaviour is soon poorly approximated by the Taylor expansion and prevents the semi-analytical solution to agree with the nonlinear one beyond $t \sim \mathcal{O}(0{\cdot}1)$.

5.2. Long-time behaviour for $\sigma = 0$

On the basis of our semi-analytical solution, the initial free-surface evolution can be generally drawn as a process during which the mass injected by the source accumulates in the form of a hump above the initial free surface level. In the following sections we investigate the free-surface evolution on a time-scale beyond the limit of applicability of our small-time expansion analysis. To the purpose, we will mainly rely on the numerical solution of the exact (inviscid) problem, as obtained by the method described in Section 4. Occasionally, we solved also the linearized version of the problem.

Figure 8 provides a first description of the flow for a constant source, $\sigma = 0$, and (relatively) small Froude number, $F = 0{\cdot}25$. The fully nonlinear solution shows how gravity reacts to restore the initial level, resulting in a (weak) decrease of the central peak accompanied by the emission of waves symmetrically propagating outwards. The phenomenon is now dominated by gravity and a more appropriate time-scale is provided by the speed of long waves, $c^\star_{\mathrm{lw}} = \sqrt{g^\star h^\star}$, and a suitable time unit results as $T_{\mathrm{lw}} = h^\star/c^\star = \sqrt{h^\star/g^\star}$. A pair of undular bores emerge, symmetrically propagating outwards with a velocity of order $c^\star_{\mathrm{lw}}$ and leaving beyond an almost flat free surface with a level slightly below $\eta/F \simeq 0{\cdot}5$. In the bottom plot, the six waves profiles are superimposed and the enlarged view of the free surface around the origin witnesses both the presence of a small heap of fluid, above the mean level, and the stationary character quickly reached by the free surface close to the source.

5.2.1. *Asymptotic interpretation of the flow due to a steady source at small Froude number*

Primitive considerations. For a small Froude-number source, outward radiation of gravity waves prevents the accumulation of a high local elevation above the source. The flow is of the impulsive type only for $0 < t < F$, during which a local surface heap of order F is accumulated above the source. Later on, since this amplitude is small, a wave front will develop with a propagation velocity close to $c^\star_{\mathrm{lw}}$, which reads $1/F$ in our non-dimensional variables.

Eventually, the dominating wave will be a bore of almost constant height $\eta^\star_{\mathrm{b}}$, since gravity tends to level out the surface in the bulk of the bore. Mass balance gives the relationship $2\eta^\star_{\mathrm{b}} c^\star_{\mathrm{lw}} = Q^\star$ and the dimensionless bore height then satisfies the equation

$$\eta_{\mathrm{b}} = F/2\ . \tag{35}$$

This almost uniform bore is a shallow-water phenomenon, which establishes after the wave front has traveled several unit lengths horizontally. The bore front is wavy, but the bulk of the bore is increasingly dominated by the average elevation $F/2$. This estimate is better the smaller the Froude number is.

To confirm the simplified analysis, a set of computations for decreasing source strength have been performed and reported in Figure 9 together with the linearized solution (thick-dashed lines). The qualitative behaviour is the same for all of the considered Froude numbers, with the leading crest faster for higher Froude numbers. The enlarged view in the bottom plot clearly shows the tendency to recover the estimate (35) as F decreases, with the bulge of fluid eventually disappearing for smaller and smaller F. Conversely, when the source strength is increased, the central bulge increases its amplitude with respect to the surrounding mean

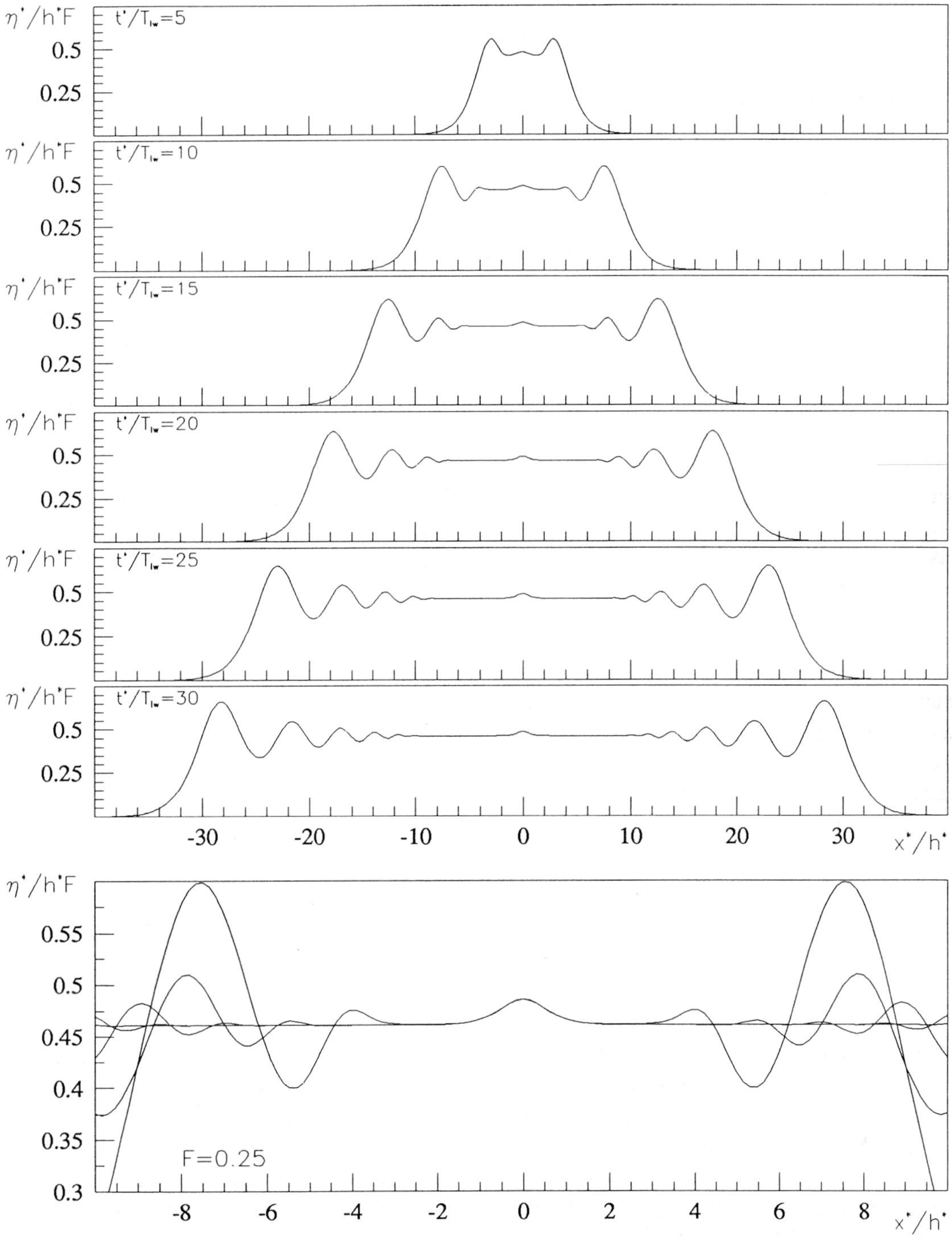

Figure 8. Free-surface motion due to a source of constant strength, $\sigma = 0$ (numerical solution of the fully nonlinear problem). Top: snapshots for $F = 0{\cdot}25$. Bottom: enlarged view around the origin; the same time instants as above are superimposed.

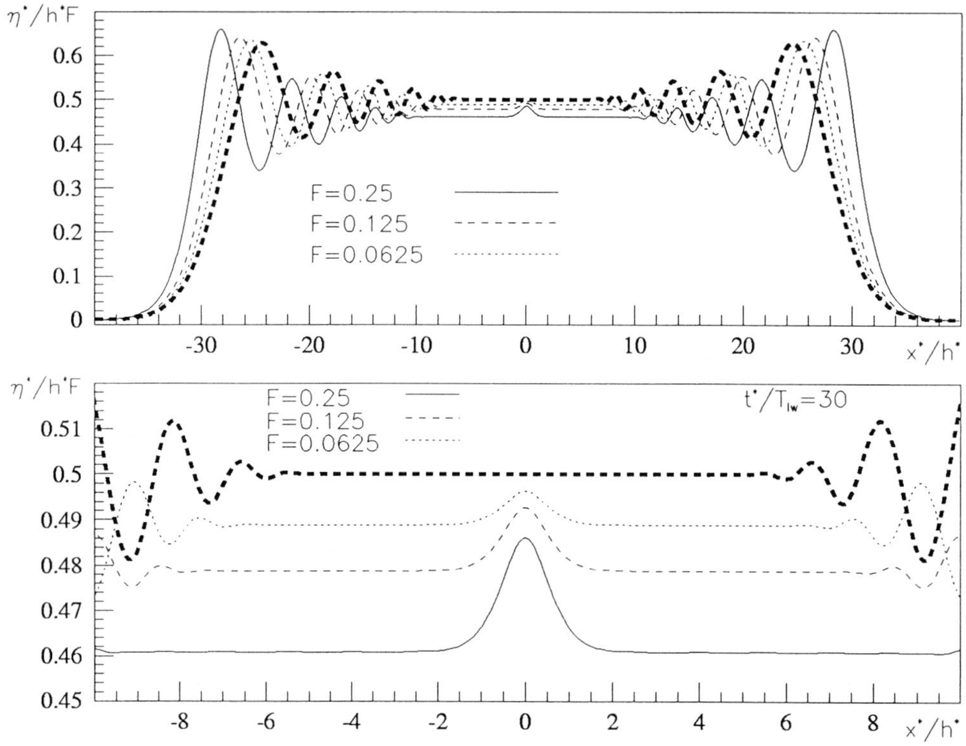

Figure 9. Undular bore due to a constant source, $\sigma = 0$, for decreasing strength F. Thicker dashed lines represent the linear solution. Top: comparison of the free surface at $t = 30$. Bottom: enlarged view around the origin.

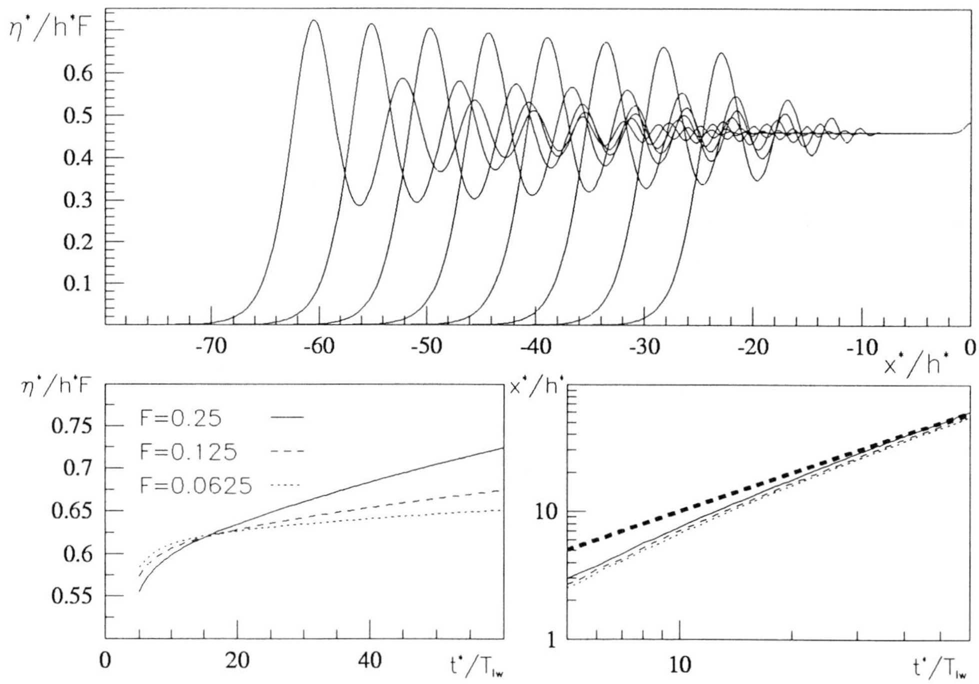

Figure 10. Evolution of the undular bore due to a constant source. Top plot: free surface for $F = 0{\cdot}25$ in the range $t \in [25, 60]$, $\Delta t = 5$. Bottom plots: time evolution of the leading crest of the bore (amplitude and longitudinal position) for decreasing F. The thicker dashed line represents $x = t$.

water level which, in turn, decreases for higher F. It appears, indeed, reasonable that larger bore systems are needed to drain away the greater mass of fluid injected by stronger sources.

The late evolution is discussed in Figure 10. In the top plot, for $F = 0{\cdot}25$, left-hand wave profiles in the range $t^\star/T_{\rm lw} \in [25, 60]$ are superimposed every five time-units, showing that the lengthening of the undular bore is accompanied by the growth in amplitude of the waves forming it. This behaviour is typical also for smaller source strengths. In the same figure, the two lower plots report the history of leading crests for several F. To analyze the numerical simulation, it is assumed that the highest Lagrangian marker represents the actual leading crest well enough. The amplitude increases with a growth rate smaller for smaller F but still positive, even for $F = 0{\cdot}0625$. Finally, the diagram of the horizontal location of maxima gives a rough idea of the propagation velocity of leading crests, which results slightly larger than $c^\star_{\rm lw}$.

Asymptotic analysis for the bore front. Whitham, [3, p. 442], gave an asymptotic analysis for the behaviour of a wave front on constant depth due to an initial hump of fluid released from rest. For long-enough evolution, the flow is governed by the gravitational time scale and, in the following, time and velocity units are changed to $T_{\rm lw}$ and $c^\star_{\rm lw}$, respectively.

The concentrated initial elevation is assumed to have a length scale equal to the depth or smaller. It gives rise to two equal waves, one propagating to the right and one to the left. The problem is governed by Laplace's equation with linearized free-surface conditions. A release of fluid of positive elevation integrated up as unit area (volume per length) at time zero will then, after a time $t \gg 1$, give the surface elevation:

$$\eta = \frac{1}{2}\left(\frac{2}{t}\right)^{1/3} \mathrm{Ai}\left[\left(\frac{2}{t}\right)^{1/3}(|x|-t)\right],$$

where Ai denotes the Airy function (see [22, p. 446]). This formula is asymptotically valid for as long as $t \gg 1$, which means that the wave front must have traveled a horizontal distance much greater than the fluid depth. The positive localized elevation at $t = 0$ which gave rise to this wave, may in fact be due to an impulsive bottom source that put a unit volume into the fluid during the infinitesimal time interval from $t = 0$ to $t = 0^+$.

Now we focus the attention on a steady source impulsively turned on at $t = 0$. The dimensionless volume put into the fluid per time unit is represented by the source Froude number. The idea is to integrate in time all later contributions after the initial one that can have influenced the local motion. This constrains the admissible values of x: we should never go farther than a dimensionless distance $t/2$ away from the wave front located at $|x| = t$, because the asymptotic analysis breaks down. Going further into the bore further reduces the large ratio (propagation length)/(depth) which is vital for the asymptotics. This means that younger and younger wave fronts (not yet fully matured for the asymptotic analysis) are being fed into the integrated elevation. It also means that we are picking contributions from the older waves farther and farther into their oscillatory tails, which will give less accuracy.

The asymptotically integrated elevation for the linearized bore from a source is

$$\eta = \frac{1}{2}F\int_0^t \mathrm{e}^{-\sigma\tau}\left(\frac{2}{t-\tau}\right)^{1/3} \mathrm{Ai}\left[\left(\frac{2}{t-\tau}\right)^{1/3}(|x|+\tau-t)\right]\mathrm{d}\tau\,, \tag{36}$$

where also the effect of the decaying singularity is accounted for. The time interval of integration is the same as the time elapsed since the impulsive start. The singularity in the

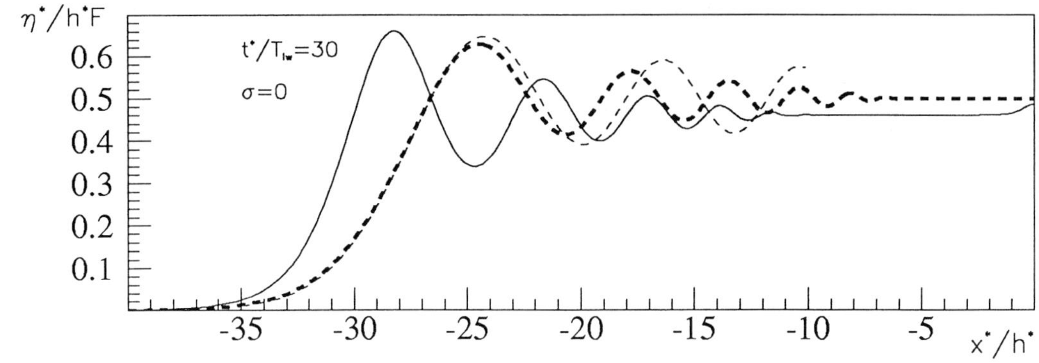

Figure 11. The bore front, for $F = 0{\cdot}25$, as predicted by the asymptotic formula (36) (thin-dashed line) is compared with the linear (thick-dashed line) and the nonlinear (solid line) computations.

integrand occurring as $\tau \longrightarrow t$ does not give any convergence problems. In Figure 11, for $F = 0{\cdot}25$, fully nonlinear and linear (numerical) results are compared with the asymptotic wave form of Equation (36). The asymptotic bore front nicely agrees with the linear solution, and both are slightly retarded with respect to the nonlinear wave profile, which is apparently steeper. Past the leading crest, the Airy function solution exhibits a slower spatial decay and becomes increasingly worse as we move into the bore. As was first shown in [7], inspection of Figures 9 and 11 reveals that for weaker bores the wave system is more stretched as a result of the prevailing dispersive effects. Consistently, the linear solution exhibits, for a given time, a wider distribution of waves.

Improved bore model. We may cope with the nonlinear behaviour of the bore front by accounting for nonlinearities from a kinematic point of view. This is done by including amplitude dispersion as given by the Korteweg-deVries equation [3, p. 463]. In the dimensionless gravitational variables, the average wave velocity in the bulk of a bore of height η_b is:

$$c = 1 + \frac{3}{2}\eta_b \; . \tag{37}$$

From Equation (35) we estimate the wave velocity as a function of the source Froude number F:

$$c = 1 + \frac{3}{4}F + \mathcal{O}(F^2) \; . \tag{38}$$

This is correct to first order in a Froude-number expansion. As an example, we take $F = 0{\cdot}25$ which gives $c = 1{\cdot}19$. This nonlinear kinematic effect of amplitude dispersion explains well the forward displacement of each wave crest in Figure 9, compared with linear theory. However, the dynamical interaction between the different wave crests is not included in this kinematic argument. The numerical solution shows that nonlinear interaction makes the leading wave crest grow at the expense of the following waves, and the bore becomes more flat apart from the increased peak at the front.

A qualitative argument for the dynamical interaction between the wave crests follows from a consideration of group velocity: since the group velocity of the waves in the bulk of the bore is greater than the velocity of the wave front, it is plausible that the wave-energy is being focussed at the leading crest. The finite outward fluid flow inside a nonlinear bore also promotes wave energy transport towards the leading crest.

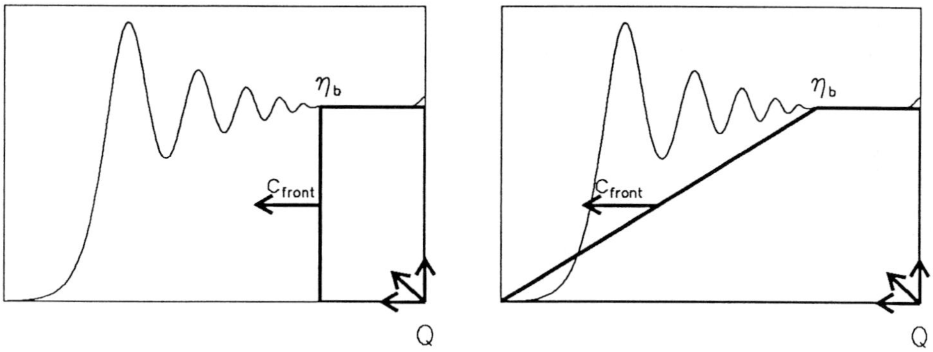

Figure 12. Simplified bore models. Left hand: box approximation. Right hand: trapezoid approximation.

The elementary bore model (35) is the most primitive way to account for the mass balance due to the steady source. The whole bore is imagined as a rectangular box (*cf.* Figure 12) expanding to the left and right with the exact shallow-water velocity. In reality, there is a finite slope at a non-breaking bore front, and the position of the central bore front can be defined by half the bore elevation $\eta_b/2$. The amplitude dispersion vanishes in the beginning of the front and is maximal at the first peak behind the front. The propagation velocity c_{front} of the central bore front is thus given by Equation (37). Because of this, we now imagine the bore as a trapezoid and the mass balance between the source and the expanding bore front gives

$$2\eta_b\, c_{front} = F\ . \tag{39}$$

As front velocity we take the average of the shallow-water propagation velocity (unity) and the wave velocity inside the bore, $c_{front} = (c+1)/2 = 1 + (3/4)\eta_b$, which, when inserted in Equation (39), gives a quadratic equation in η_b with the solution:

$$\eta_b = \frac{2}{3}\left(\sqrt{1 + \frac{3}{2}F} - 1\right)\ . \tag{40}$$

Although it still requires a non-breaking bore, the 'trapezoid model' does not necessarily require $F \ll 1$ as the 'box approximation' above and, in fact, values for η_b/F are in nice agreement with fully nonlinear computations:

	$F = 1{\cdot}0$	$F = 0{\cdot}75$	$F = 0{\cdot}5$	$F = 0{\cdot}25$	$F = 0{\cdot}125$	$F = 0{\cdot}0625$
Equation (40)	0·3874	0·4068	0·4305	0·4603	0·4785	0·4888
Numerical simulation	0·3904	0·4090	0·4318	0·4608	0·4786	0·4888

where in the worst case, $F = 1{\cdot}0$, we have less than 1% error. From this trapezoid estimate of the bore elevation, we find the improved estimate $c = \sqrt{1 + (3/2)F}$ for the wave velocity inside the bore and a Froude number expansion here recovers Equation (38).

5.2.2. *Larger F: the appearance of breaking*

So far, small-time expansion and asymptotic analysis offer a good qualitative understanding of the results obtained by the numerical solutions of the exact problem. We also are able to give fairly precise estimates through relatively simple formulas.

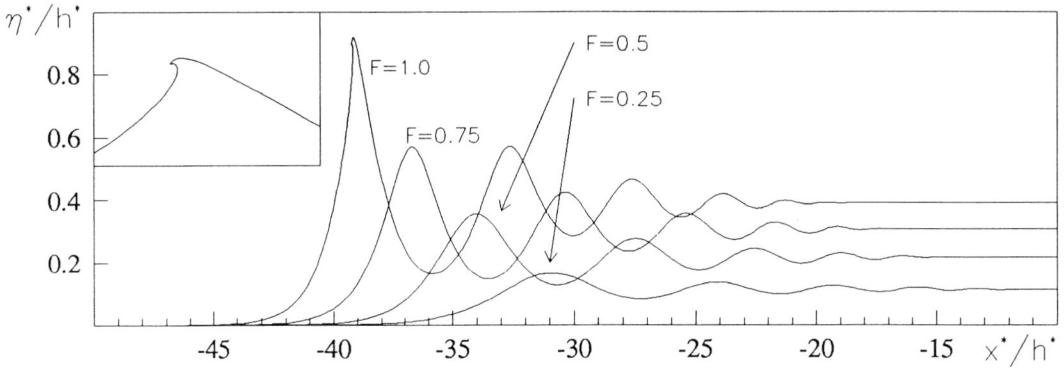

Figure 13. Undular bore generated by constant sources in the range $0{\cdot}25 \leq F \leq 1$. For all cases the free surface refers to $t^\star/T_{\mathrm{lw}} = 32{\cdot}5$ but for $F = 1$ where surface breaking (detailed in the top-left plot) is evidenced for $t^\star/T_{\mathrm{lw}} \simeq 32{\cdot}24$.

Further increasing the Froude number introduces a new feature: the breaking of the leading wave. The problem is numerically studied by Teles da Silva and Peregrine [7] by following the evolution of an already established jump Δ smoothly reconciled through the initial wave profile $\eta = \Delta/2(1 - \tanh(\alpha x))$. In particular, they observed three different regimes: non-breaking undular bores for $\Delta <\sim 0{\cdot}3$, undular breaking bores and breaking bores for $\Delta >\sim 0{\cdot}7$.

For the problem considered here, the first regime has been extensively analyzed in previous paragraphs. In Figure 13 wave profiles in the range $0{\cdot}25 \leq F \leq 1$ are superimposed for $t \sim 32{\cdot}5$ to show the transition from undular to undular breaking bores. As discussed, the mean level above the source increases as $\sim F^{0{\cdot}5}$, Equation (40); the leading crest is steeper the larger F is and propagates faster because of amplitude dispersion. In particular, for the leading peaks shown in the figure, we have:

	$F = 0{\cdot}25$	$F = 0{\cdot}5$	$F = 0{\cdot}75$	$F = 1{\cdot}0$
η	0·1666	0·3563	0·5706	0·9196
x	−30·9368	−33·9954	−36·70476	−39·17567

Actually, for $F = 1$, the simulation has been stopped around $t \sim 32{\cdot}24$ because of the tiny nascent breaker shown in the top-left corner. For $F < 1$, simulations are continued on a longer time-scale. In all cases, no sign of breaking has been detected, though the amplitude of leading crests still (weakly) grows for $t \sim 200$. A similar behaviour can be observed in Teles da Silva and Peregrine results [7, Figure 3], even for the smallest $\Delta = 0{\cdot}1$ and after a quite long evolution, $t = 140$. Moreover, although we expect that the leading crest will be a solitary wave [23], Figure 10 suggests that the time needed to reach this asymptotic behaviour increases nonlinearly as F increases. Finally, it is plausible that on this long time scale viscous effects could play a relevant role.

We have not tried to identify sharply the limiting source strength for the first breaking appearance. In Figure 14, we simply show the time t_{br} requested for the leading-wave front to become vertical as a function of the Froude number. The curve, for $F > 2$, is flattened around $t_{\mathrm{br}} \sim 5$, while, as F decreases, the time for breaking increases sharply, approaching,

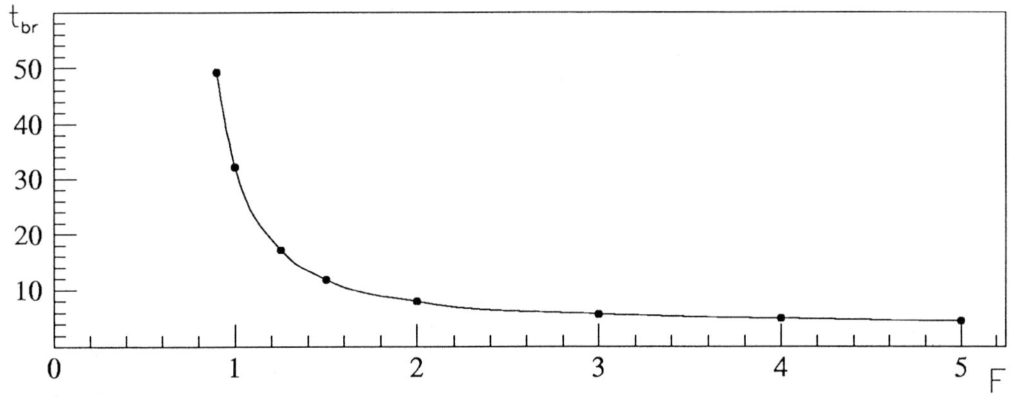

Figure 14. Time t_{br} for the leading-wave front to become vertical as a function of the Froude number. Symbols • denote actual computations.

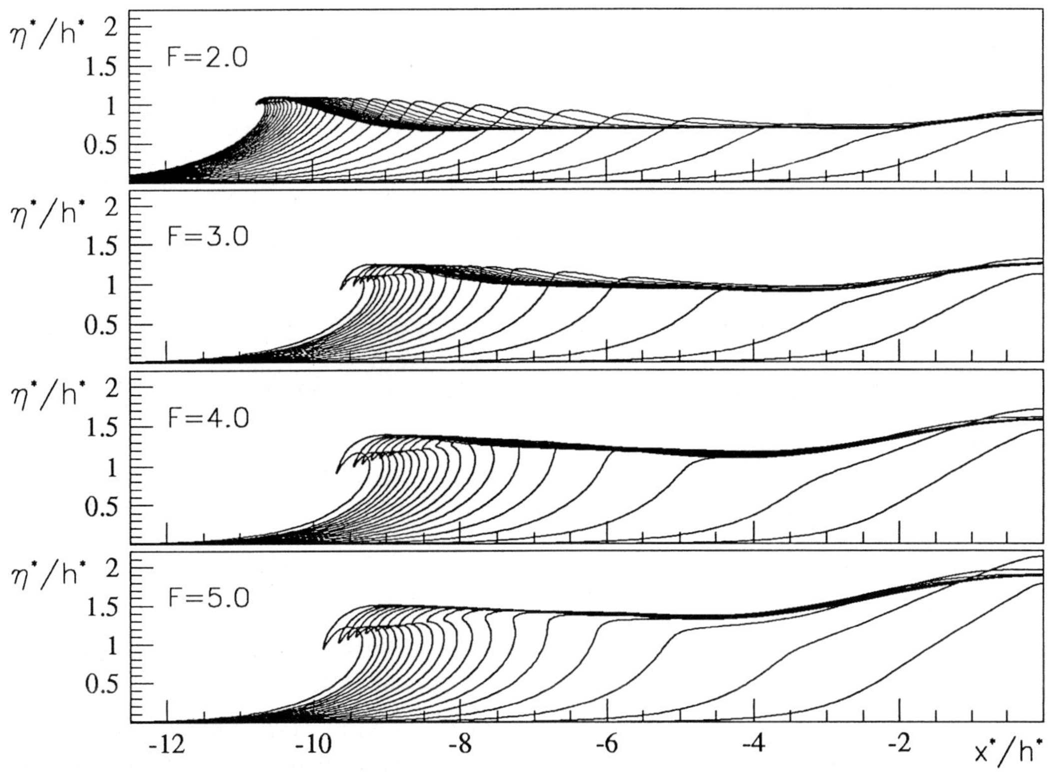

Figure 15. Evolution toward breaking of the bore due to a constant source, $\sigma = 0$, for increasing source strength. Last time reported: $t = 8{\cdot}53$, $7{\cdot}787$, $6{\cdot}239$ and $5{\cdot}94$ from top down, respectively.

according to our simulations, a limiting value in between $F = 0.75$ (non-breaking case) and $F = 0.9$ (breaking case).

For breaking bores in the range $F = 0{\cdot}9, \ldots, 1{\cdot}75$, several undulations are still present behind the bore front but, because wave breaking prevents to proceed further in the simulation, we cannot assess if, in the steady regime, we would observe an undular bore with a weak breaker at the leading front or the undulations would eventually disappear completely.

The plunging breaker is increasingly larger for stronger sources (*cf.* Figure 15), while the time scale needed for its appearance is progressively shorter. Interestingly, the location of the plunger is roughly the same for the cases reported in figure, although the height increases slightly. The undulations behind the bore-front have no time to develop and we enter in the regime of breaking bores with a flat free surface behind. Actually, the hump centered around $x = 0$ witnesses the presence of the submerged source and it is spread over a larger area with respect to cases with smaller F. At this stage the bore has entirely attained the character of breaking bore without residual undulations.

The initial conditions considered in the paper by Teles da Silva and Peregrine are quite different from the case considered here. In our problem, the flow is uniquely determined by F while in the former both Δ and the imbalance between the upstream and downstream currents enter the problem. On the other hand, the source induces local effects which are fairly prominent as F increases, which can hamper the study of the nascent breaker. In spite of this, our present results agree with the findings of Teles da Silva and Peregrine.

5.3. Waves due to a constant sink

Waves generated by a submerged sink are now discussed through the computations reported in Figure 16 for $F = -0{\cdot}25$ and $\sigma = 0$. As the small-time expansion suggests, the motion starts as a growing depression around the sink. On a longer time scale, after reaching a minimum, the induced hollow slightly diminishes as a result of the gravity reaction and tends to attain a constant level. In the meanwhile, the horizontal extent of the depression increases as an obvious consequence of the mass-defect due to the sink. The front of this spreading depression propagates with a velocity of order $c^{\star}_{\mathrm{lw}}$ and appears as a train of waves with decreasing amplitude and wavelength when moving from the front to the location above the sink. Within a linearized theory, the solution can be obtained by simply replacing (η, F) by $(-\eta, -F)$ which clearly justifies the appearance of a depression wave. By allowing F to be negative, the asymptotic formula (36) also covers the case of a weak sink. This is actually done in Figure 17, where for $t = 30$ nonlinear, linear and (linear) asymptotic solutions are compared. As expected, linear solutions agree to the same extent as for $F > 0$ (correct front location and different spatial decaying). The nonlinear wave profile generated by a sink displays different features compared with those for $F > 0$. In particular, the location of leading crest is closer to the singularity compared with the linear solution: the opposite as for the source case. As a consequence, the nonlinear depression wave cannot be classified as a bore because nonlinearity tends to smear out the front instead of steepening it.

The wiggled system inside the depression is unexpected on the ground both of the linear analysis and of the corresponding nonlinear results for a source with $F = 0{\cdot}25$ (*cf.* Figure 8). This feature strikingly appears when we compare nonlinear solutions in Figures 9 and 18. The nonlinear phenomenon of amplitude dispersion offers an explanation: the oscillating wiggles are fed locally by the sink (or the source) and propagate outwards. In a bore, $F > 0$, such short waves travel faster than the bore-front due to amplitude dispersion ($c_{\mathrm{b}} > c_{\mathrm{front}}$) and tend to merge with the leading peak. Eventually they disappear, leaving a waveless region inside the nonlinear bore. For $F < 0$, the front moves faster and the local oscillations in the depression are maintained and fed by the sink.

A closer look at Figures 16–18 reveals that there is a steady positive deflection of the surface inside the depression, just above the sink. It is identical to the positive heap in the surface elevation above a source. The reason that this heap also exists above a sink, is that

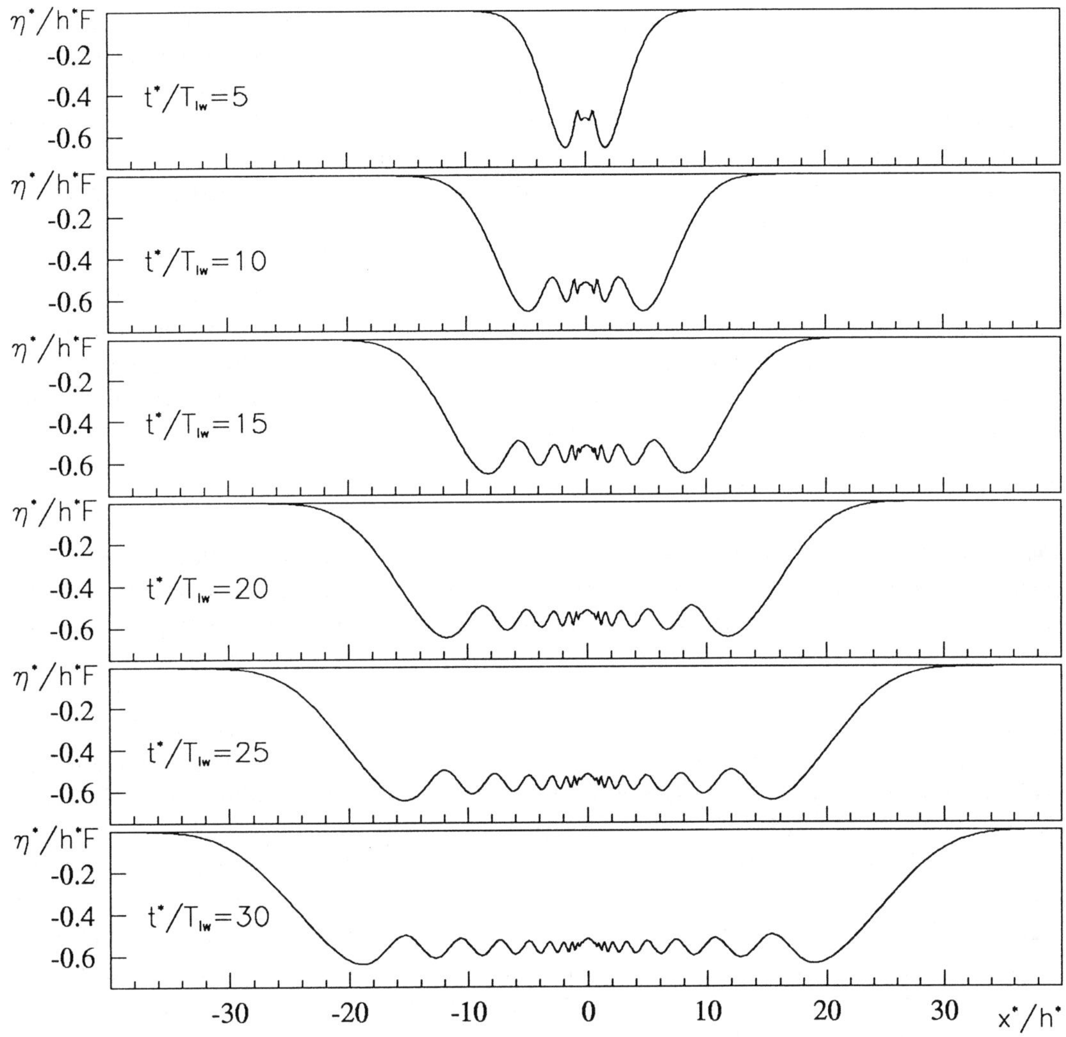

Figure 16. Free-surface motion due to a constant sink ($F = -0{\cdot}25$, $\sigma = 0$).

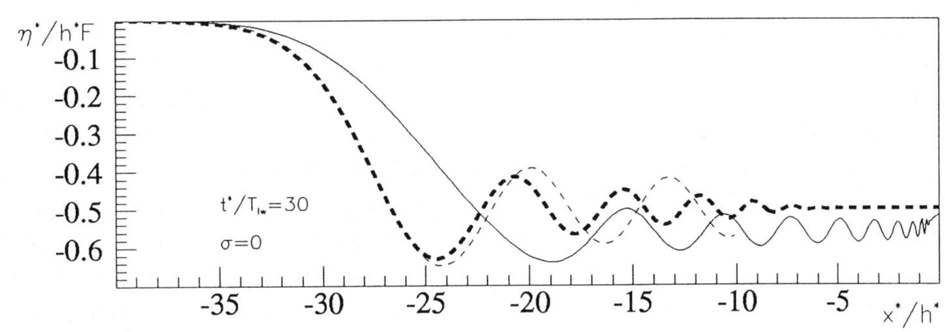

Figure 17. Free-surface motion due to a constant sink ($F = -0{\cdot}25$, $\sigma = 0$). Comparison of asymptotic prediction (thin dashed line), linear (thick dashed line) and nonlinear computations (solid line).

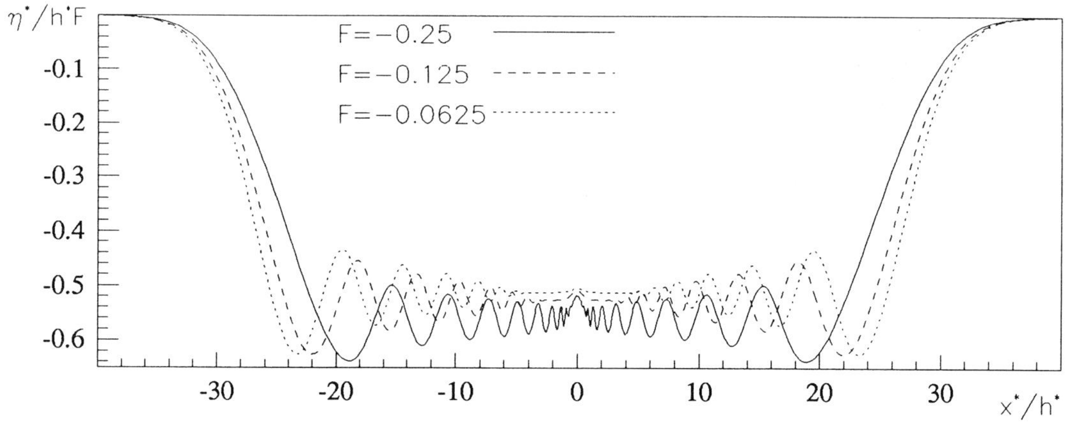

Figure 18. Free-surface motion due to a constant sink ($\sigma = 0$). Comparison of the free surface at $t^\star/T_{\mathrm{lw}} = 30$ for decreasing F.

steady nonlinear free-surface flows are reversible in time (such lack of distinction between steady flows induced by sources or sinks is observed by Koerber and Forbes in [24]). When we compare nonlinear sink flows with corresponding source flows, it is necessary to redefine the Froude number. The effective Froude number for a steady source/sink flow is:

$$F_{\mathrm{eff}} = F\,(1 + \eta_{\mathrm{b}})^{-3/2}\ .$$

This redefinition is needed because there exists no undisturbed fluid level in the steady problem. There is a flat surface at infinity, serving as reference level for steady flow. But it corresponds to the bore level of the transient flow, not the undisturbed level. A good approximation for η_{b} in the definition of F_{eff} is found from Equation (40), valid both for positive F (source) and negative F (sink).

5.3.1. *Moderate sink strength: gravitational restoration and surface breaking*

The weak sinks considered above do not cause dramatic behaviour of the free surface. There is an almost uniform average depression, being modulated by an outgoing wave-train. Nonlinearity promotes these waves and also reduces the slope of the depression front as it propagates into undisturbed fluid. From low to moderate Froude numbers, we observe a relevant gravitational bounce back: the initial downward motion of the surface above the sink is reversed and the surface moves upwards, eventually attaining a stationary value. This bounce-back gravitational restoration is physically very elementary, just like filling in a crater of missing fluid. In Figure 19 we see a very weak bounce back above the sink appearing already at $F = -0{\cdot}2$. The gravitational restoration increases in importance as we increase the sink strength until $F \simeq -0{\cdot}6$. From then on it fades out, and exists no more at $F = -0{\cdot}7$.

When there is no gravitational restoration, one would expect the free surface to collapse toward the sink. This is what happens at infinite depth, [10, 11]. This tendency is apparent in Figure 20 where solutions due to increasingly stronger sinks are superimposed at time instants for which surface breaking appears. The horizontal extent of the displaced portion of the free surface sharply diminishes as the sink strength increases, while the vertical penetration is more than doubled passing from $F = -0{\cdot}4$ to $F = -0{\cdot}8$. The surface starts to break at $-F = 0{\cdot}4$, after a gravitational time about 4·3 (*cf.* Figure 19). Then there is breaking after shorter and shorter time until $F = -0{\cdot}8$. Figure 21 shows the latest evolution of surface breaking for

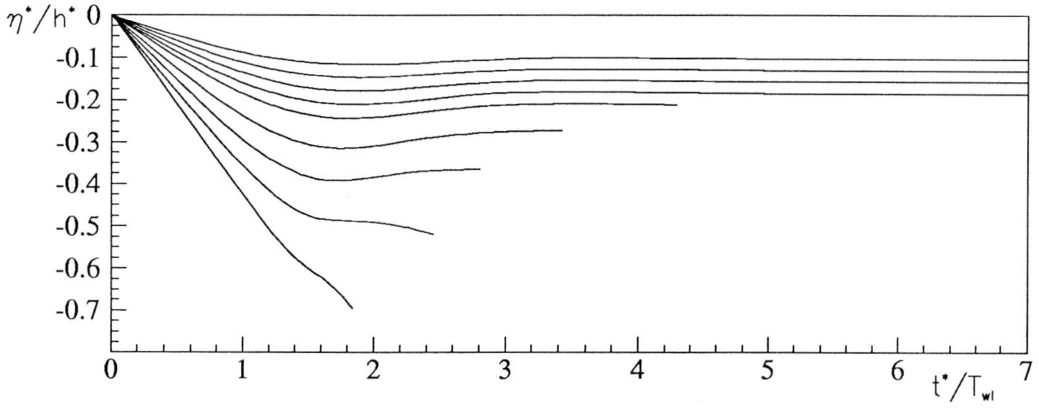

Figure 19. Time evolution of the Lagrangian point above the sink ($x^\star/h^\star = 0$) for $F = -0{\cdot}2, -0{\cdot}25, -0{\cdot}3, -0{\cdot}35, -0{\cdot}4, -0{\cdot}5, -0{\cdot}6, -0{\cdot}7, -0{\cdot}8$. For $|F| \geq 0{\cdot}35$, the numerical simulation is stopped because of breaking.

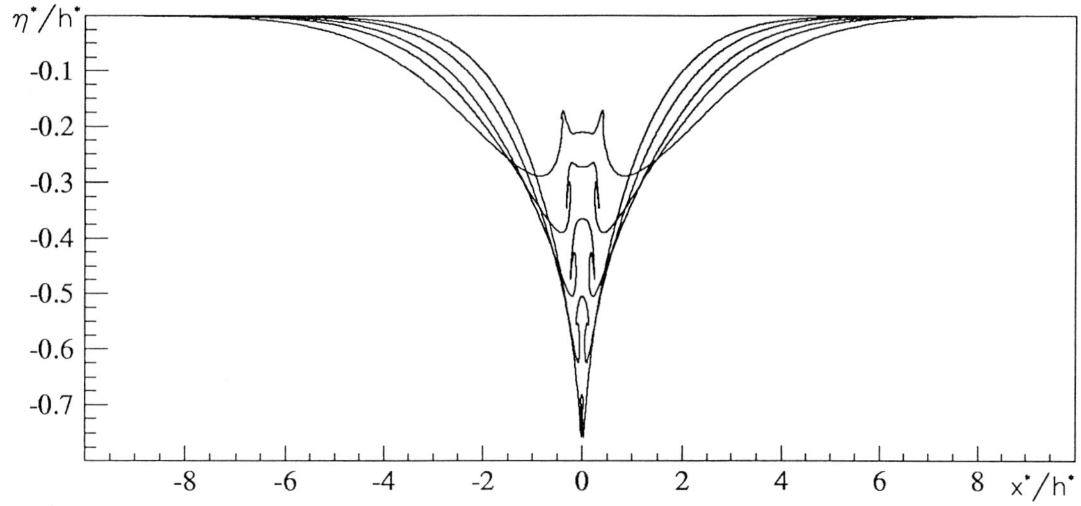

Figure 20. Free-surface profiles for $F = -0{\cdot}4, -0{\cdot}5, -0{\cdot}6, -0{\cdot}7, -0{\cdot}8$ for time $t = 4{\cdot}299, 3{\cdot}423, 2{\cdot}882, 2{\cdot}296$ and $1{\cdot}806$, respectively.

F ranging from $-0{\cdot}4$ to $-0{\cdot}7$. Geometrical details are sensitive with respect to increasing Froude number, the main feature being the reduction of the horizontal separation between the two breakers. A notable feature for $F = -0{\cdot}7$ is the tendency of the positive and negative branches of the free surface to reconcile, almost to isolate the breaking bulge of fluid. The high curvature of the free surface would probably imply a significant role of surface tension effects, presently not modeled. The gravity is still able to produce a tiny jet for $F = -0{\cdot}8$, while for $F = -0{\cdot}9$ computations predict a continuous collapse of the free surface toward the bottom sink.

5.3.2. *Dip instability above a strong sink*

The dip instability is a gravitational collapse of the free surface into the sink. It is a familiar phenomenon in a bath-tub, often accompanied by a vortex. The first scientific investigation of the dip instability in a tank drained from the bottom was done by Lubin and Springer [25]. They performed experiments and presented a simplified analysis for dip instability based on

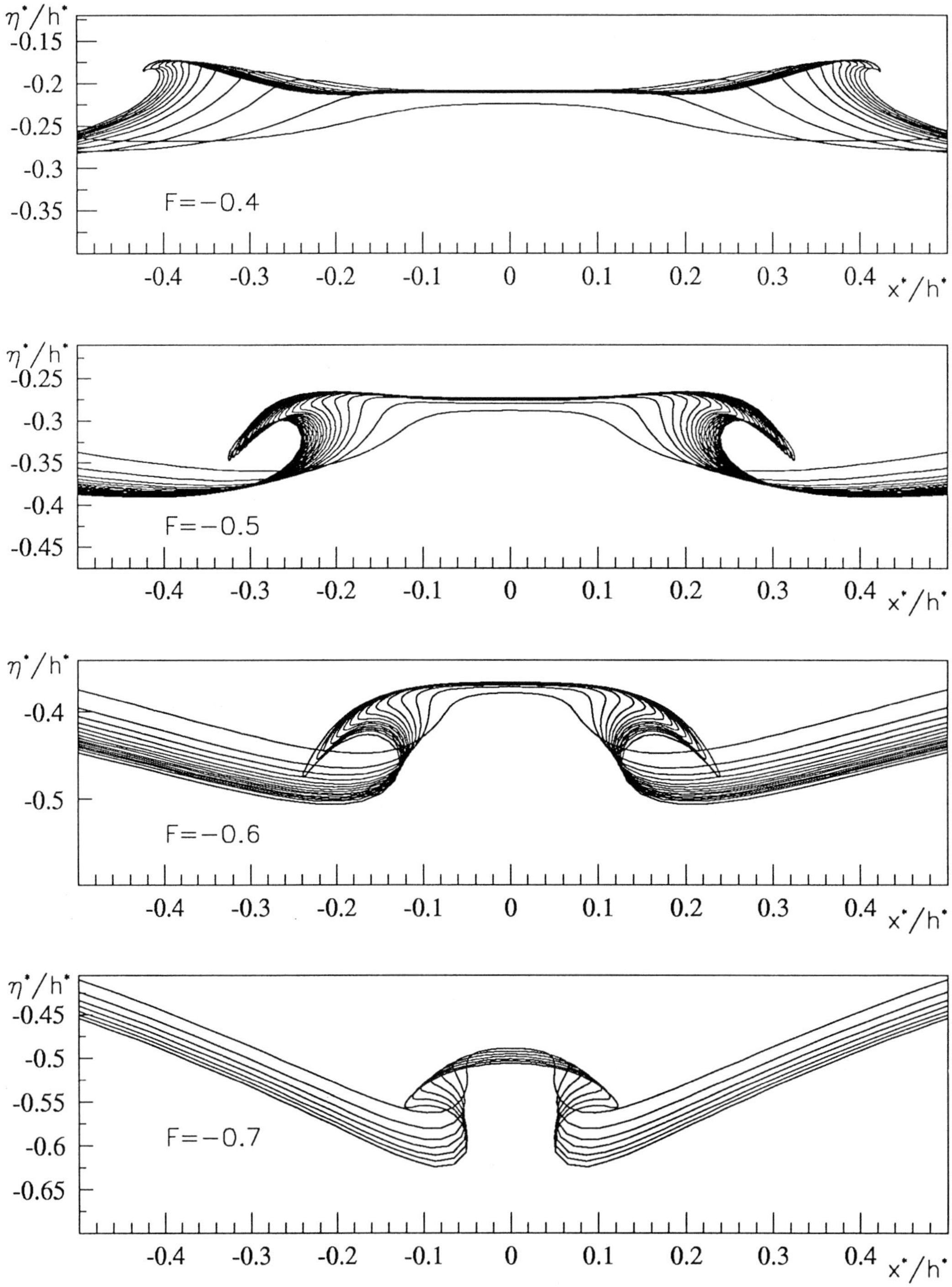

Figure 21. Enlarged view of the free-surface evolution above the sink for $F = -0{\cdot}4, -0{\cdot}5, -0{\cdot}6, -0{\cdot}7$.

Bernoulli's equation. In [26], Zhou and Graebel investigated dip formation numerically for an axisymmetric container, drained impulsively through a hole in the bottom. Tyvand [8] presented a simple analytical model leading to dip formation: the submerged line sink parallel to a free surface. The resulting impulsive flow was calculated to third order in a small-time expansion. A critical Froude number $\hat{F}_c = -1/3$ was found, representing the change of sign of the third-order elevation above the sink. The similar problem for a submerged point sink has been solved by Miloh and Tyvand in [10], where they found a critical Froude number $\hat{F}_c = -0{\cdot}2582$. These critical Froude numbers were taken as criteria for dip formation: there would be an early acceleration of the surface into the sink if the $\hat{F} < \hat{F}_c$. However, Xue and Yue [11] demonstrated numerically that this critical Froude number based on the third-order balance between nonlinearity and gravitational rebound could only be considered as a lower bound. In fact, they solved the fully nonlinear axisymmetric problem and found that dip instability takes place for all $\hat{F} < -0{\cdot}1930$. In [9], Kim performed a numerical study of the two-dimensional submerged sink, in agreement with [8].

The usual definition of the Froude number for these earlier sink papers is:

$$\hat{F} = \frac{Q^\star}{2\pi\sqrt{g^\star h^{\star 3}}} = \frac{1}{2\pi} F \ .$$

This is valid in two dimensions, where the source strength $Q^\star/(2\pi)$ is preferred instead of its flux. In three dimensions the source strength is given by the flux divided by 4π. We will now derive the analytical criterion for dip formation based on the third-order analysis for an impulsive bottom sink in two dimensions.

The third-order elevation above a sink consists of the three terms given in Equation (30), with sign changes everywhere because the source is replaced by a sink. We will evaluate these three terms just above the sink $x = 0$. First we have the term independent of gravity which is given by the second-order potential:

$$\eta_{3,\infty}(0) = -\frac{1}{3}\left(\frac{k}{\pi}\right)^3 \sum_{\text{n odd}} \sum_{\text{m odd}} \sum_{\text{q odd}} \operatorname{sech} k_n \operatorname{sech} k_m \operatorname{sech} k_q \times$$

$$\times \left[k_{n-m} k_{n-m-q} \tanh k_{n-m} \tanh k_{n-m-q} + k_{n-m} k_{n-m+q} \tanh k_{n-m} \tanh k_{n-m+q} + \right.$$

$$\left. + k_{n+m} k_{n+m-q} \tanh k_{n+m} \tanh k_{n+m-q} + k_{n+m} k_{n+m+q} \tanh k_{n+m} \tanh k_{n+m+q} \right]$$

We are interested in the limit $L \longrightarrow \infty$ and $L = 50$ should give a good approximation, especially at $x = 0$ where the influence from the artificial neighboring sinks is the smallest. We have chosen $L = 50$ with each of the triple series truncated after $N = 1001$. The result with 12 digits is: $\eta_{3,\infty}(0) = -0{\cdot}10280837928$ which is equal to: $\eta_{3,\infty}(0) = -\pi^2/96$ within an accuracy of nine nonzero digits.

The contribution from the lower-order potentials is: $\eta_{3\,\eta_1}(0) = \pi^2/192 = 0{\cdot}05140418959$ Numerically, we have now found a very simple relationship: $\eta_{3,\infty}(0)/\eta_{3\,\eta_1}(0) = -2$. The gravity-dependent third-order elevation above the sink is:

$$\eta_{3,F} = \frac{1}{3F^2}\frac{k}{\pi} \sum_{\text{n odd}} k_n \tanh k_n \operatorname{sech} k_n \ . \tag{41}$$

We search for the limit as $L \longrightarrow \infty$. A computation using MATHEMATICA for the case after $L = 50$ gives the result when several thousand terms are included in the series: $F^2 \eta_{3,F}(0) =$

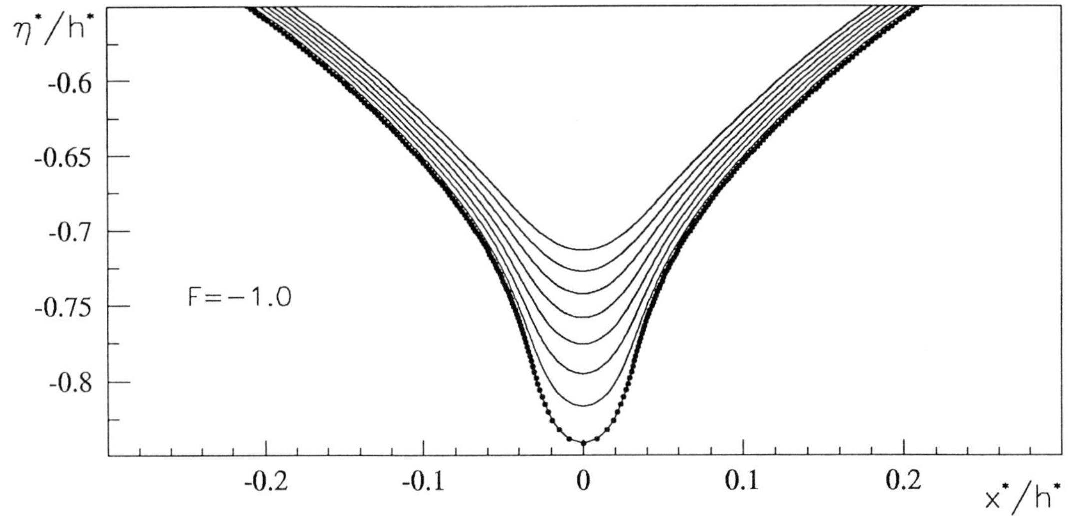

Figure 22. Late evolution of the dip forced by a constant sink. Enlarged view in natural scale for $t^\star/T_{\mathrm{lw}} = 1{\cdot}1, \ldots, 1{\cdot}17\ (0{\cdot}01)$.

1/12. This result remains exactly the same for all L between 26 and 55, with six nonzero digits accuracy. But it starts to diverge for larger L, due to roundoff errors. When $L < 26$, the period length is short enough to influence the result. The total third-order surface elevation above the impulsively-started bottom sink is: $\eta_3(0) = [F^{-2} - \pi^2/16]/12$. A critical Froude number F_c is defined by the third-order elevation equal to zero, $\eta_3(0) = 0$, and this gives the result:

$$F_c = -\frac{4}{\pi} \simeq -1{\cdot}2732395\ . \tag{42}$$

With the same Froude-number definition as [8, 9], we have $\hat{F}_c = -2/\pi^2 \simeq -0{\cdot}20264237$, which is quite far above the value $-1/3$ for a submerged two-dimensional sink. This is reasonable because the bottom sink generates a much more concentrated free-surface flow.

As anticipated, this analytical result has to be considered a lower bound, because dip formation is numerically observed at least for $F < -0{\cdot}9$. Actually, in the range $-0{\cdot}8 < F < -0{\cdot}35$ computations have to be stopped because of surface breaking and it is not possible to determine the evolution on a longer time scale, so physical dip formation could happen even for weaker sinks.

In principle (see [24] for a recent discussion) a steady cusp-flow sink is possible for $F \leq -1$. In the present computations this is not actually observed. As an example, Koerber and Forbes [24, Table 1] give a cusp height of 0·247 for $F = -1$. In our computations the height is 0·158 for $t^\star/T_{\mathrm{lw}} = 1{\cdot}17$ (*cf.* Figure 22). Following a longer evolution would require a dynamic refinement because of the strong stretching of Lagrangian markers around the center of the dip. This has not been attempted, and we can only observe that the two branches of the free surface tend to join each other at a smaller depth. Therefore, it would be possible to have a steady cusped flow consistent with that computed in [24]. Anyway, other issues could be of concern: some of numerical nature (*e.g.* inhability of the present method to develop a singularity in the domain boundary), other related to the stability of such steady solutions (which, though existing, could not be reached through an unsteady evolution).

6. Conclusions

We studied some classes of nonlinear free-surface flows generated by bottom sources and sinks that are started impulsively. Two time scales are involved: first the early impulsive time scale up to $t \sim (h^\star/g^\star)^{1/2}$ since the start, where gravity is of secondary importance. Later on, the flow is governed by the gravitational time scale $(h^\star/g^\star)^{1/2}$.

A constant source may give rise to a progressive bore, where an increasing water level propagates into still water. We show that a seemingly undular bore shape will arise even in linear theory. But it is essentially a transient phenomenon. As time progresses, the undulations will spread and grow farther and farther away from the bore front. The characteristic, almost flat surface in the bulk of the bore is a nonlinear phenomenon. This becomes clear if we compare a steady source flow with a steady sink flow of moderate amplitude.

The sink will develop a propagating depression, which may look like a negative bore if the sink strength is very small. However, a sink of finite strength will develop and maintain a wave pattern within the depression, and its front will not grow steep as for a true bore generated by a source.

A clear explanation for the origin of the undulations in a weak bore seems to be lacking in the literature. We found that the undulations, most pronounced for bores with small amplitude, are gradually suppressed as the bore amplitude increases. We therefore link the bore undulations to those in the classical Cauchy-Poisson problem for a concentrated initial disturbance. Accordingly, we identify the characteristic wavelength of the undulations with the typical wavelength for the linearized Cauchy-Poisson problem. This is asymptotically given by an Airy function, with the length scale of the same order as the undisturbed fluid depth. One may ask if a fully linearized bore exists. It does, to a certain extent, as far as the picking of the typical undulation wavelength and the undulation frequency is concerned. But the steepening of the bore front and the spatial decay of the undulations going into the bulk of the bore are dominated by nonlinear effects. There is no steady solution of a Cauchy-Poisson problem for a concentrated disturbance. Accordingly, there cannot be any steady undular bore either. All bores will be time-dependent, with a time scale slower than the oscillation period for a particle in the waves as they passed by.

We consider a one-parameter family of bores, governed solely by their Froude number, which is the source flux made dimensionless in terms of gravitational units. Each bore has its unique time evolution. Bore breaking is investigated, and the transition between non-breaking and breaking bores takes place for a Froude number slightly smaller than one. The lowest bore height giving breaking is found to be 40% higher than the undisturbed water level. The details of bore breaking are sensitive with respect to the Froude number.

All the considered bores are essentially time-dependent. Undular (non-breaking) bores have been followed up to $t^\star/T_{\mathrm{lw}} \sim 200$. At that time a (small) positive growth rate of the leading crest is still observed. For breaking (undular- or not) bores, simulations were stopped. In both cases, viscous effects could be eventually relevant in determining the steady regime features [7].

A sink with Froude number of order one or greater, will lead to dip formation, which is a gravitational collapse where the free surface is swallowed into the sink. An analytical estimate for dip formation is given which, on the ground of our numerical results, results to be an upper bound for the actual sink strength required to induce the gravitational collapse.

A phenomenon exclusive for finite depth is that sinks somewhat too weak for dip formation will cause surface breaking. This is due to the lack of dispersion of the peak of upward

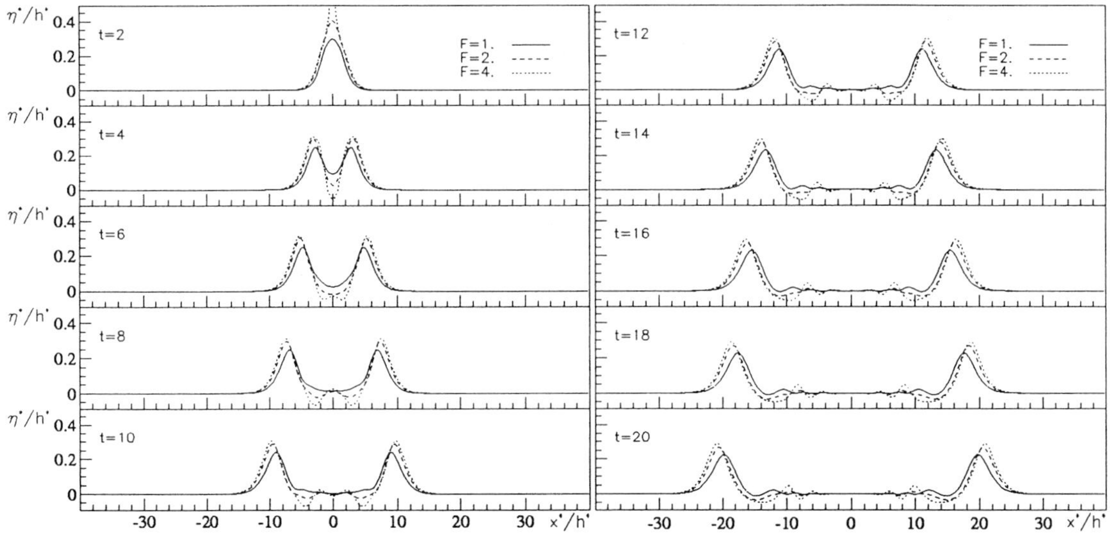

Figure 23. Disintegration of the initial hump due to a decaying source, $\sigma = 0{\cdot}5$. Time evolution of the free surface; non-dimensional $t = t^\star/T_{\mathrm{lw}}$ increases from top down and from left to right.

flow from the gravitational rebound. The horizontal extent of the breaking hump inside the depressed region shrinks gradually, leaving before disappearing a tiny upward jet [27] as the last attempt of the gravity to react to dip formation.

Impulsive sources decaying with time, $\sigma > 0$, have also been considered. A brief example is given in Figure 23, where the raised hump of water disintegrates into a leading solitary wave plus a train of dispersive waves. This behaviour is consistent with that observed for a conventional Cauchy-Poisson problem, where a free surface with finite amplitude is released from rest.

The long-time evolution of the wave system, *e.g.* the number of solitary waves, depends on the total mass of fluid emitted by the source, $1/\sigma$. This is shown in Figure 24, where wave patterns due to stronger sources are compared for $t^\star/T_{\mathrm{lw}} = 200$ ($F/\sigma^\star \sim 1/\sigma$, for $g/h^\star$ fixed). For $F = 0{\cdot}5$, a solitary wave is well separated from its dispersive tail. When F is increased, the leading solitary wave is steeper and faster. In the last case, $F = 1{\cdot}75$, the injected mass is large enough to give rise to a second solitary wave of smaller amplitude. Further increasing σ leads to breaking of the first emitted hump.

Acknowledgements

The research activity of M.L. is supported by the Italian *Ministero dei Trasporti e della Navigazione* through INSEAN Research Program 2000-2002. This work has been partly developed while the first author was visiting researcher at UCSB-OEL. The kind support of ONR, Dr. Ed Rood Program Manager, is gratefully acknowledged.

Appendix A. Formulation for a dipole

The second basic type of singularity is the horizontal dipole located at the bottom. It represents the dominating far-field contribution due to a mass-conserving geological fault. The dipole

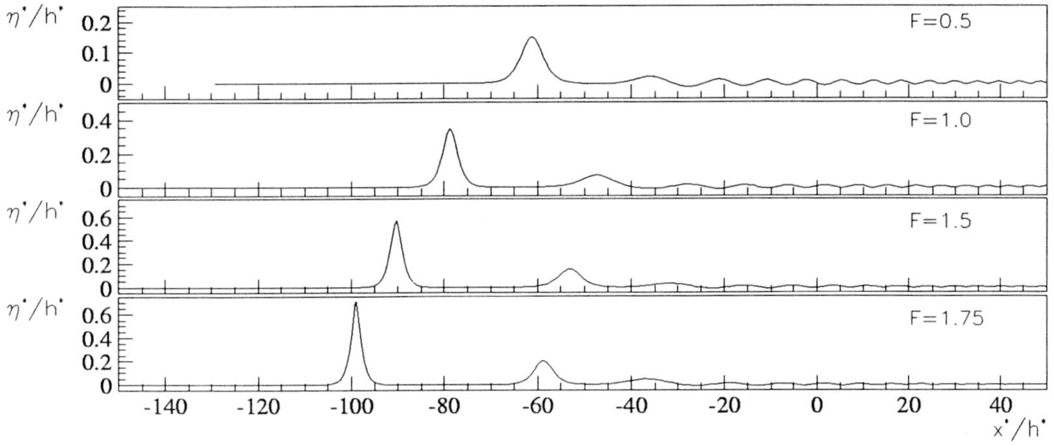

Figure 24. Long time behaviour of fluid packet for $F = 0{\cdot}5, 1, 1{\cdot}5, 1{\cdot}75$ and $\sigma^\star = 0{\cdot}25$. All wave profiles refer to $t^\star/T_{\text{lw}} = 200$. The origin, $x = 0$, is the original location of the source.

strength is $\mu^\star$ and we accordingly redefine the Froude number as:

$$F = \frac{\mu^\star}{\sqrt{g^\star h^{\star 5}}} \tag{A1}$$

and the units of dimensionless velocity, time and pressure as $\mu^\star/h^{\star 2}$, $h^{\star 3}/\mu^\star$ and $\rho^\star\mu^{\star 2}/h^{\star 4}$, respectively. The bottom dipole is assumed to decay exponentially in time as $\mu^\star \mathrm{e}^{-\sigma^\star t^\star}$, with the dimensionless decay rate defined as:

$$\sigma = \frac{\sigma^\star h^{\star 3}}{\mu^\star}\ . \tag{A2}$$

So the dimensionless dipole strength is $\mathrm{e}^{-\sigma t}$. The integrated dipole strength does not have the same clear physical meaning as in the source case. The dimensionless unexpanded singular dipole potential is given by:

$$\psi = \frac{e^{-\sigma t}}{\pi}\sum_{-\infty}^{\infty}(-1)^n \frac{x}{x^2 + (y + 2n - 1)^2}\ . \tag{A3}$$

A.1. First-order elevation

By taking minus the x-derivative of the expression for a bottom source, we find the first-order elevation for a bottom dipole:

$$\eta_1 = \frac{\pi}{4}\ \text{sech}\left(\frac{\pi}{2}x\right)\ \tanh\left(\frac{\pi}{2}x\right)\ . \tag{A4}$$

The first-order elevation with artificial periodicity is now:

$$\eta_1 \ = 2\frac{k}{\pi}\sum_{\text{n odd}} k_n \,\text{sech}\, k_n\ \ \sin(k_n x)\ . \tag{A5}$$

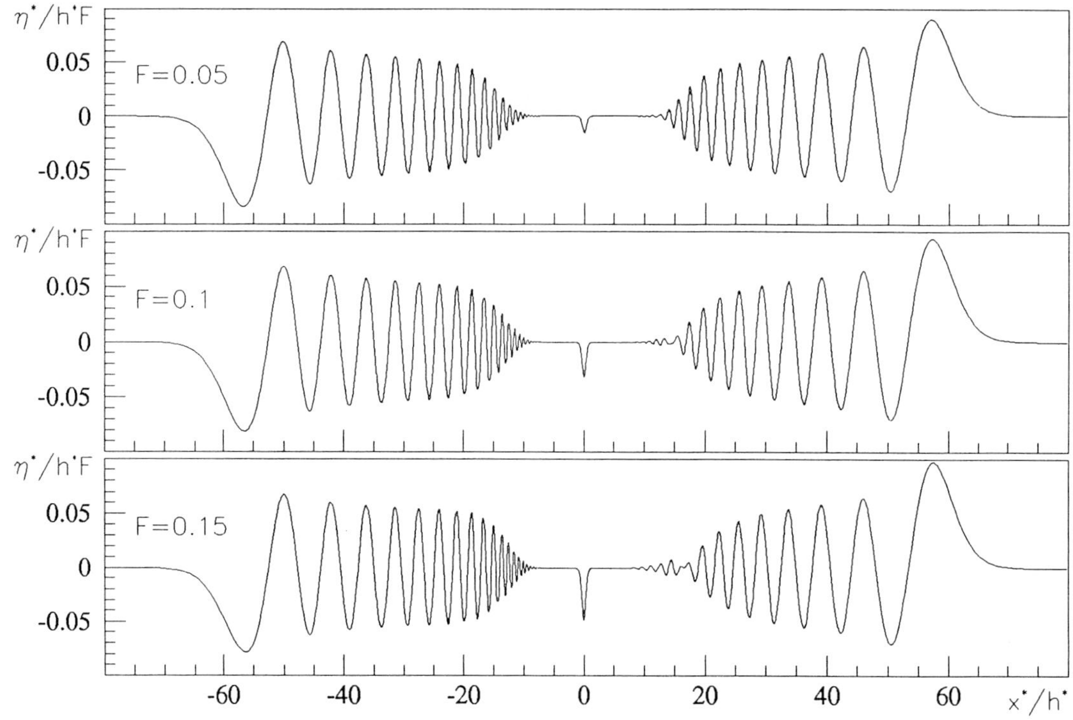

Figure 25. Free surface profiles due to a steady doublet for $t^\star/T_{\mathrm{lw}} = 60$. The strength of the doublet increases from top-down.

A.2. Second-order elevation

In the dipole case, the periodic regular first-order potential at the free surface $y = 0$ is given by:

$$\varphi_1 = -\left(\frac{k}{\pi}\right)^2 \sum_{\text{n odd}} \sum_{\text{m odd}} k_n\, k_m \operatorname{sech} k_n \operatorname{sech} k_m [\cos(k_{n-m}x) - \cos(k_{n+m}x)] \tag{A6}$$

and the resulting second-order elevation by:

$$\begin{aligned}\eta_2 = & -\frac{1}{2}\left(\frac{k}{\pi}\right)^2 \sum_{\text{n odd}} \sum_{\text{m odd}} k_n\, k_m \operatorname{sech} k_n \operatorname{sech} k_m \times \\ & \times [\, k_{n-m} \tanh k_{n-m} \cos(k_{n-m}x) - k_{n+m} \tanh k_{n+m} \cos(k_{n+m}x)\,] - \\ & - \frac{\sigma}{8}\pi \operatorname{sech}\left(\frac{\pi}{2}x\right) \tanh\left(\frac{\pi}{2}x\right)\end{aligned} \tag{A7}$$

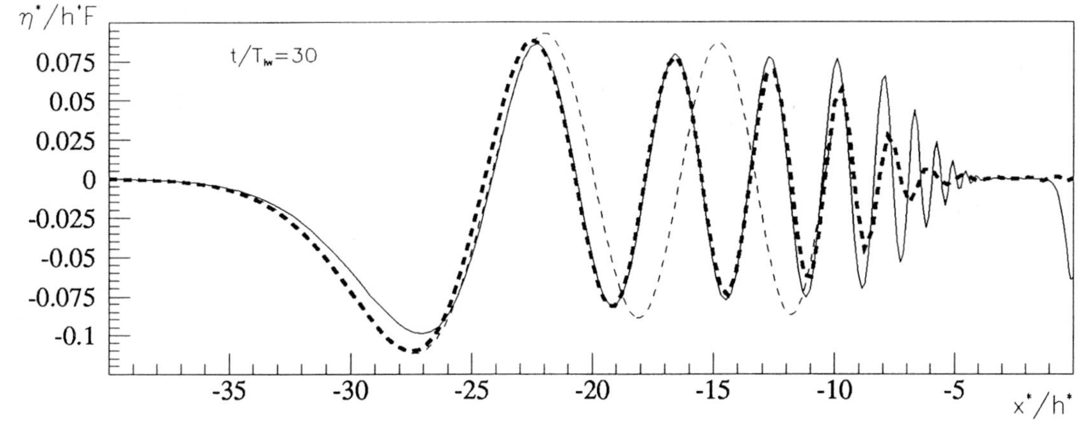

Figure 26. Leading wave system generated by a constant doublet. Comparison between fully nonlinear (solid line, $F = 0{\cdot}2$), linear (thick dashed line) and asymptotic solution (thin dashed line) of Equation (A10).

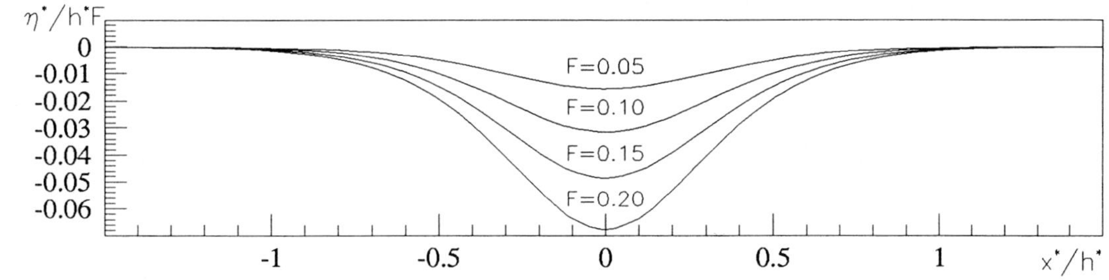

Figure 27. Enlarged view of the free surface about a constant doublet, $t^\star/T_{\mathrm{lw}} = 60$.

A.3. Third-order elevation

The regular second-order potential at the free surface is:

$$
\begin{aligned}
\varphi_2 = &\left(\frac{k}{\pi}\right)^3 \sum_{\text{n odd}} \sum_{\text{m odd}} \sum_{\text{q odd}} k_n k_m k_q \operatorname{sech} k_n \operatorname{sech} k_m \operatorname{sech} k_q \times \\
&\times \left\{k_{n-m} \tanh k_{n-m}[\sin(k_{n-m+q}x) - \sin(k_{n-m-q}x)] + \right. \\
&\left. + k_{n+m} \tanh k_{n+m}[\sin(k_{n+m-q}x) - \sin(k_{n+m+q}x)]\right\} \\
&+ 2\sigma \left(\frac{k}{\pi}\right)^2 \sum_{\text{n odd}} \sum_{\text{m odd}} k_n k_m \operatorname{sech} k_n \operatorname{sech} k_m [\cos(k_{n-m}x) - \cos(k_{n+m}x)] - \\
&- \frac{1}{F^2}\frac{k}{\pi} \sum_{\text{n odd}} k_n \operatorname{sech} k_n \sin(k_n x)\ .
\end{aligned}
\tag{A8}
$$

The third-order elevation is:

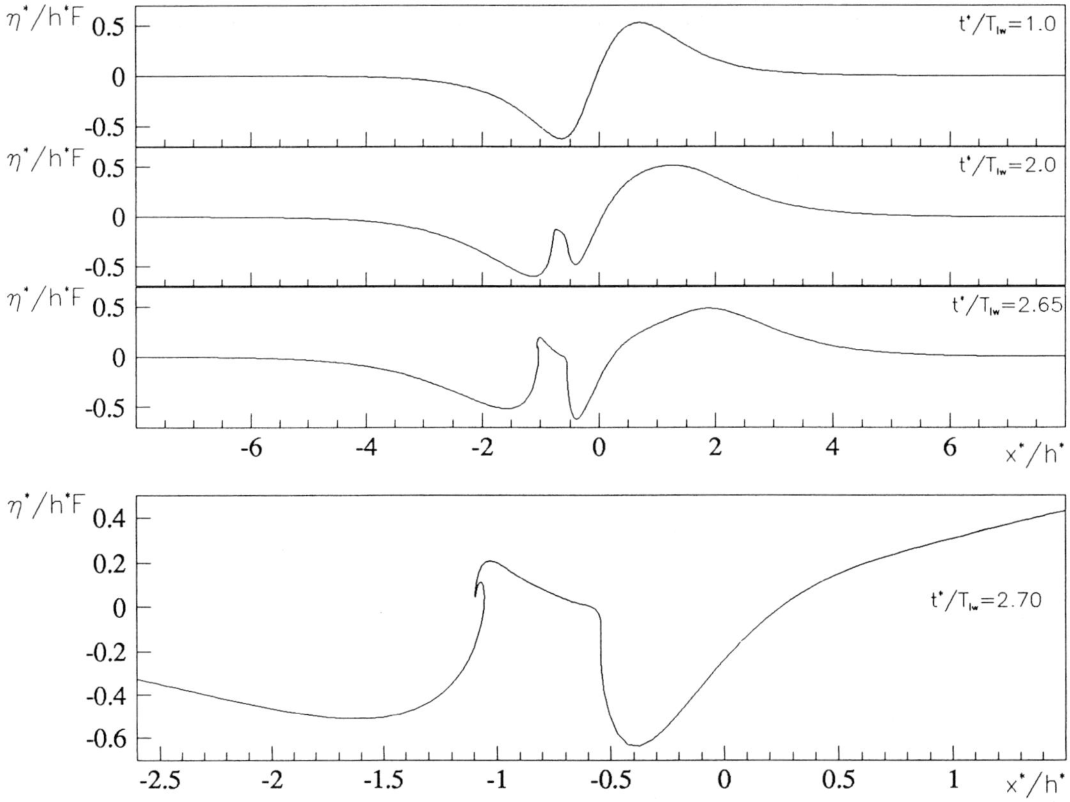

Figure 28. Free surface due to a steady doublet ($F = 0{\cdot}25$). The evolution stops with the clear evidence of surface breaking (see enlarged view in the bottom plot).

$$
\begin{aligned}
\eta_3 = {} & \frac{1}{3}\left(\frac{k}{\pi}\right)^3 \sum_{\text{n odd}} \sum_{\text{m odd}} \sum_{\text{q odd}} k_n k_m k_q \operatorname{sech} k_n \operatorname{sech} k_m \operatorname{sech} k_q \times \\
& \times \{ k_{n-m} \tanh k_{n-m} [k_{n-m+q} \tanh k_{n-m+q} \sin(k_{n-m+q} x) - \\
& \quad - k_{n-m-q} \tanh k_{n-m-q} \sin(k_{n-m-q} x)] + \\
& + k_{n+m} \tanh k_{n+m} [k_{n+m-q} \tanh k_{n+m-q} \sin(k_{n+m-q} x) - \\
& \quad - k_{n+m+q} \tanh k_{n+m+q} \sin(k_{n+m+q} x)] \} \\
& + \frac{2}{3}\sigma \left(\frac{k}{\pi}\right)^2 \sum_{\text{n odd}} \sum_{\text{m odd}} k_n k_m \operatorname{sech} k_n \operatorname{sech} k_m \times \\
& \times [k_{n-m} \tanh k_{n-m} \cos(k_{n-m} x) - k_{n+m} \tanh k_{n+m} \cos(k_{n+m} x)] - \\
& - \frac{1}{3F^2}\frac{k}{\pi} \sum_{\text{n odd}} k_n^2 \operatorname{sech} k_n \tanh k_n \sin(k_n x) + \frac{\sigma^2 \pi}{24} \operatorname{sech}\left(\frac{\pi}{2}x\right) \tanh\left(\frac{\pi}{2}x\right) + \\
& + \frac{\pi^5}{24576}\left(51 \sinh\left(3\frac{\pi}{2}x\right) - 3\sinh\left(5\frac{\pi}{2}x\right) - \right. \\
& \left. - 170 \sinh\left(\frac{\pi}{2}x\right)\right) \operatorname{sech}\left(\frac{\pi}{2}x\right) \sinh^7\left(5\frac{\pi}{2}x\right) .
\end{aligned}
\tag{A9}
$$

Here we have avoided the artificial periodicity wherever an exact formula is available.

A.4. Sample results

In the following, we report a few computations concerning the free surface motion due to a constant doublet. Wave patterns after $t^\star/T_{\mathrm{lw}} = 60$ are shown in Figure 25 for increasing strength of the submerged singularity. A linear analysis would imply a perfect anti-symmetric solution: this qualitatively explains the positive and negative leading waves propagating on the right and on the left, respectively. Following the arguments used for the source, the asymptotic behaviour of the leading wave is given by:

$$\eta = -\frac{1}{2}\mathrm{sign}(x)F\int_0^t \mathrm{e}^{-\sigma t}\left(\frac{2}{t-\tau}\right)^{2/3}\mathrm{Ai}'\left[\left(\frac{2}{t-\tau}\right)^{1/3}(|x|+\tau-t)\right]\mathrm{d}\tau \qquad \text{(A10)}$$

and its comparison with nonlinear, linear computations is offered in Figure 26. The agreement between solutions is reasonably good and consistent with that observed for the source (*cf.* Figures 11, 17).

A closer look at Figure 25 reveals that anti-symmetry is violated, even for the smallest strength. In particular, the right-propagating wave-train is made by fewer and longer waves with respect to the left propagating one. These features recall those observed when wave systems induced by (weak) sources and sinks, respectively, are compared.

The free surface between the two wave systems eventually remains flat and the only signature of the doublet is an unchanging dip which is not predicted on the basis of a linear theory. In particular, for cases considered in Figure 27, this depression results to be symmetric, in spite of the non-symmetric character of the forcing bottom flux considered, and it is deeper the stronger the doublet is.

Last plots, Figure 28, show the initial development of the wave motion forced by a doublet that is slightly stronger. After the early formation of the couple hump-hollow, as soon as the left depression rises, a localized surface breaking is detected and prevents any further computations. Similar behaviour has been observed for stronger singularities.

References

1. H. Lamb, *Hydrodynamics*. New York: Dover (1932) 738pp.
2. J.V. Wehausen and E.V. Laitone, Water waves. In: W. Flugge (ed.) *Handbuck der Physik* (Vol. 9). Berlin: Springer (1960) pp. 446–778.
3. G.B. Whitham, *Linear and Nonlinear Waves*. New York: J. Wiley (1974) 636pp.
4. C.C. Mei, *The Applied Dynamics of Ocean Surface Waves*. Singapore: World Scientific (1983) 740pp.
5. J.L. Hammack, A note on tsunamis: their generation and propagation in an ocean of uniform depth. *J. Fluid Mech.* 60 (1973) 769–799.
6. D.H. Peregrine, Calculations of the development of an undular bore, *J. Fluid Mech.* 25 (1966) 321–330.
7. A.F. Teles da Silva and D.H. Peregrine, Nonsteady computations of undular and breaking bores. In: *Proc. 22nd Int. Conf. Coastal Engng.*, Delft. A.S.C.E. 1 (1990) 1019–1032.
8. P.A. Tyvand, Unsteady free-surface flow due to a line source *Phys. Fluids* A4 (1992) 671–676.
9. M.J. Kim, Numerical study of dip formation using sink array model. *Int. J. Modern Phys.* C8 (1997) 1249–1262.
10. T. Miloh and P.A. Tyvand, Non-linear transient free-surface flow and dip formation due to a point sink. *Phys. Fluids* A5 (1993) 1368–1375.
11. M. Xue and D.K.P. Yue, Nonlinear free-surface flow due to an impulsively-started submerged point sink. *J. Fluid Mech.* 364 (1998) 325–347.

12. D.H. Peregrine, Flow due to a vertical plate moving in a channel. Unpublished note (1972).
13. M. Greenhow and W.-M. Lin, Nonlinear free surface effects: experiments and theory. Report 83-19. MIT, Dept. of Ocean Engineering (1983).
14. P.A. Tyvand and A.R.F. Storhaug, Green functions for impulsive free-surface flows due to bottom deflections in two-dimensional topographies. *Phys. Fluids* 12 (2000), 2819–2833.
15. P.A. Tyvand and M. Landrini, Free-surface flow of a fluid body with an inner circular cylinder in impulsive motion. *J. Eng. Math.* (2000) to appear.
16. P. Bassanini, C.M. Casciola, M.R. Lancia and R. Piva, A boundary integral formulation for kinetic field in aerodynamics. Part I: Mathematical analysis. Part II: Applications to unsteady 2D flows. *Eur. J. Mech. B/Fluids* 10 (1991) 605–627 and 11 (1992) 69–92.
17. A. Sidi and M. Israeli, Quadrature methods for periodic singular and weakly singular Fredholm integral equations. *J. Sci. Comp.* 3 (1988) 201–231.
18. P. Sonneveld, CGS, A fast Lanczos-type solver for nonsymmetric linear systems. *SIAM J. Sci. Stat. Comp.* 10 (1989) 36–52.
19. G. Graziani and M. Landrini, Application of multipoles expansion technique to two-dimensional nonlinear free surface flows. *J. Ship Res.* 43 (1999) 1–13.
20. M.S. Longuet-Higgins and E.D. Cokelet, The deformation of steep surface waves on water. I A numerical method of computation. *Proc. R. Soc. London* A 350 (1976) 1–26.
21. J.W. Dold, An efficient surface integral algorithm applied to unsteady gravity waves. *J. Comp. Phys.* 103 (1992) 90–115.
22. M. Abramowitz and I.A. Stegun, *Handbook of Mathematical Functions*. New York: Dover (1965) 1046pp.
23. T.B. Benjamin and M.J. Lighthill, On cnoidal waves and bores. *Proc. R. Soc. London* A224 (1954) 448–460.
24. A.J. Koerber and L.K. Forbes, Cusp flows due to an extended sink in two dimensions. *J. Eng. Math.* 36 (1999) 349–360.
25. B.T. Lubin and G.S. Springer, The formation of a dip on the surface of a liquid draining from a tank. *J. Fluid Mech.* 29 (1967) 385–390.
26. Q.-N. Zhou and W.P. Graebel, Axisymmetric draining of a cylindrical tank with a free surface. *J. Fluid Mech.* 221 (1990) 511–532.
27. M.S. Longuet-Higgins and H.N. Oguz, Critical jets in surface waves and collapsing cavities. *Phil. Trans. R. Soc. London* A 355 (1997) 625–639.

Journal of Engineering Mathematics **39:** 171–188, 2001.

On the asymptotic analysis of surface-stress-driven thin-layer flow

LEONARD W. SCHWARTZ
Department of Mechanical Engineering The University of Delaware Newark, Delaware 19716, U.S.A.

Received 5 January 2000; accepted in revised form 30 June 2000

Abstract. It has been demonstrated experimentally that thin liquid layers may be applied to a solid surface or substrate if a temperature gradient is applied which results in a surface tension gradient and surface traction. Two related problems are considered here by means of the long-wave or lubrication theory. In the first problem, an improved estimate of the applied liquid coating thickness for a liquid being drawn from a bath is found through asymptotic and numerical matching. Secondly, the theory is extended to consider substrates that are not perfectly wetted but exhibit a finite equilibrium contact angle for the coating liquid. This extension incorporates the substrate energetics using a disjoining pressure functional. Unsteady flows are calculated on a substrate of nonuniform wettability. The finite contact angle value required to stop stress-driven flow is predicted and the resulting steady profiles are compared with experimental results for several values of the applied stress.

Key words: thin-layer flow, asymptotic analysis, Marangoni effect, finite contact angle, numerical simulation.

1. Introduction

For most liquids the surface tension σ is a decreasing function of temperature. We consider a liquid layer that is bounded on one side by a solid surface, the substrate, and on the other side by a surface that is exposed to the atmosphere upon which surface tension acts. If the liquid layer is sufficiently thin, it is possible to assume that the temperature is virtually constant across the thin dimension. Then, if the substrate is nonuniformly heated, the temperature variation will be transferred to the free surface. The dimensionless criterion to be satisfied is that the Biot number, based on the liquid-film thickness, be very small [1]. The resulting temperature gradient on the free surface results in an applied shear stress given simply by

$$\tau = \sigma_x \,, \tag{1.1}$$

where x is measured along the substrate and subscripts are used to indicate differentiation [2].

The temperature-induced surface stress will contribute to the flow within the liquid layer. This is the well-known Marangoni effect. A number of basic theoretical studies have been performed in order to predict the resulting flow for thin liquid layers and isolated droplets on both flat and curved substrates [1, 3–5]. Recently interest has arisen in small devices that use this thermocapillary pumping mechanism to move microscopic quantities of liquid in medical and other applications [6].

Here we will be concerned with a liquid coating layer that is drawn from a bath onto a vertical plate using Marangoni stresses. Several experiments have been performed using a vertical plate that is partially submerged in a liquid bath [7–9]. A temperature gradient is applied to the plate in order to draw liquid upward. Two aspects of this problem will be considered here, as shown schematically in Figure 1. They are, (i) the prediction of the steady-state wetting-layer thickness that is drawn upward by use of a nominally-constant applied shear stress, and (ii) the motion of the front of the rising liquid layer, particularly on a substrate of nonuniform surface energy, *i.e.* for a substrate where the equilibrium contact angle varies with position. In each of these two problems, within the approximation considered here, flow could also be driven by another source of applied surface stress such as a tangential wind stress. As indicated by the broken lines in the figure, the two regions considered are assumed to be separated by a long region over which the coating layer is flat, with thickness h_∞. Thus, for the steady-state problem (i), the layer of thickness h_∞ may be assumed to extend infinitely far upward.

A prediction of the layer thickness h_∞ that is drawn from the bath using the lubrication approximation was made by Fanton *et al.* [10]. The method is similar to the well-known prediction of the coating thickness that is deposited onto a vertical upwardly-moving plate, as treated by Landau and Levich (*viz.* Levich [11]). As in the Landau–Levich analysis, the leading-order prediction neglects the effect of gravity backflow. Compared with the experimental results of Cazabat *et al.* [8, 9] the predicted values of h_∞ in [10] are somewhat high. Here we will find the next term in the asymptotic expansion for layer thickness. The corrected prediction is closer to the experimental measurements. In addition, we solve the problem using a composite differential equation that is correct to the same order in the small parameter $\epsilon = \tau/\sqrt{\rho g \sigma}$. Here ρ is liquid density and g is the acceleration of gravity. The two different estimates are in substantial agreement when ϵ is small. The gravity correction for the stress-driven problem is shown to be of logarithmic order; it is thus a more important correction than the algebraic correction found by Wilson [12] for the Landau–Levich problem. Note that, although the fractional change in surface tension is small, the shear stress or surface tension gradient can be quite large. This is because changes in surface tension occur over quite short distances. Because the fraction change in surface tension is small, assigning a constant average value to σ in the capillary pressure term in each governing equation only introduces a small error.

While not discussed further in this paper, an interesting aspect of the experiments of Cazabat *et al.* [8, 9] is an observed transverse instability of the climbing front. This instability is associated with the appearance of a capillary ridge at the front. Linear stability analyses, performed by Kataoka and Trojan, provide an explanation for the instability [13, 14]. Recently we have modeled this unsteady three-dimensional motion and simulated the onset and nonlinear growth of the 'fingers' resulting from the instability [15]. Predicted finger shapes are in close agreement with the interferometric measurements given in [9].

In an earlier experiment, Ludviksson and Lightfoot [7] used a similar apparatus; however they considered smaller temperature gradients leading to thinner wetting layers. There was no tendency to form fingers; instead they were concerned with the speed and shape of the propagating front. Using observed values of the upstream thickness h_∞, and assuming the front to move at constant speed, Kataoka and Troian employed a lubrication model to get profile shapes similar to those observed experimentally [13, 14]. However, the observed values of h_∞ are significantly larger than those predicted by the bath withdrawal model and these large values remain largely unexplained. It should be noted that the withdrawal model predicts exceedingly small values, virtually of molecular dimensions for the squalane oil that was used. Carles and Cazabat [16] suggest that, since the experiment was performed with a plate that was initially partially submerged, insufficient time may have been taken to allow for complete drainage before the temperature gradient was applied.

In Section 3, we consider a second set of measurements made by Ludviksson and Lightfoot [7]. They applied a barrier coating of nonwetting material to the substrate. The front was found

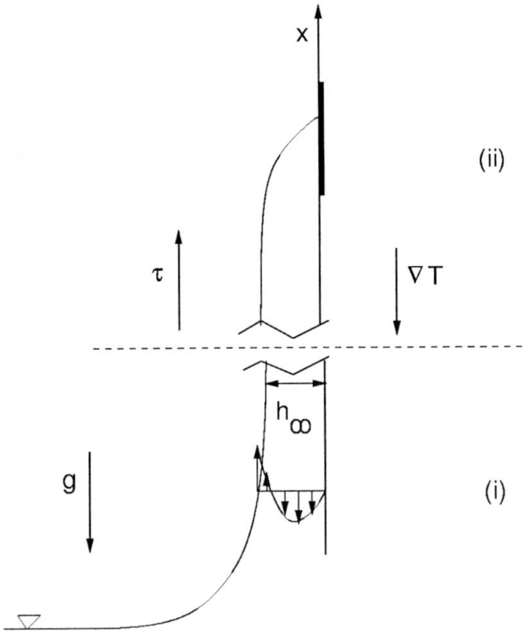

Figure 1. A liquid layer is drawn from a bath by a surface shear stress τ induced by a constant temperature gradient ∇T. The x-axis represents a solid vertical substrate. The darkened portion of the substrate signifies a region of larger equilibrium contact angle. Problem (i) is the prediction of the layer thickness h_∞ in bath withdrawal under steady-state conditions. Problem (ii) concerns the progression of the front tip onto a nonuniform substrate.

to stop at the barrier and the profiles were determined by means of interferometry for several values of the applied stress. We generalize the lubrication model to include the effect of finite equilibrium contact angle θ_e. A universal profile, in terms of dimensionless variables, is found for liquid fronts that come to rest against a barrier. These profiles are in reasonable agreement with the experimental results. We will find an analytic expression for the values of θ_e necessary to stop the frontal motion as a function of the physical parameters in the problem. These angles are remarkably small, much smaller than the contact angle for the strongly nonwetting coating that was applied in the experiment. Because patterns of coating are applied to control capillary-driven motions in real and contemplated microscopic devices, as discussed by Gau *et al.* [17] for example, predictions of barrier strengths are of importance.

Finite contact angle effects are added to the governing unsteady evolution equation through an additional interfacial effect known as disjoining pressure. Derived originally to quantify the magnitude of observed equilibrium contact angles in terms of molecular forces [18, 19], models incorporating disjoining effects can also be used in dynamic simulations that include frontal or contact-line motions on substrates that are not perfectly wetting [20–23]. Also included in Section 3 are dynamic simulations for motion up to and passing a barrier whose strength is insufficient to stop the flow. The general effect of such barrier passage is to increase the thickness of the coating layer. Recently unsteady simulations and analysis of thin-layer stress-driven motions, but without finite contact angle effects, have been reported by Bertozzi and co-workers [24–26].

2. The bath withdrawal problem

We wish to predict the thickness of the coating layer drawn from a bath using a given applied shear stress τ. This is problem (i) in Figure 1. The flow is assumed to be steady. If the x-axis is taken to point downwards, the constant flux Q in the drawn layer is given by

$$Q = \frac{\sigma h^3}{3\mu}\left[\frac{h_{xx}}{(1+h_x^2)^{3/2}}\right]_x - \frac{\tau h^2}{2\mu} + \frac{\rho g h^3}{3\mu} = -\frac{\tau h_\infty^2}{2\mu} + \frac{\rho g h_\infty^3}{3\mu} . \tag{2.1}$$

The idealized problem assumes that the drawn layer extends infinitely far upward, *i.e.* $x \to -\infty$, and the thickness is h_∞ there. The terms represent, respectively, the capillary pressure gradient driven flow, flow due to the upward stress and the downward gravity flow. The capillary pressure is proportional to the surface curvature $\kappa = h_{xx}/(1+h_x^2)^{3/2}$. Far upstream the curvature is zero and $h = h_\infty$. This equation employs the small-slope lubrication approximation and is equivalent to that used by previous workers [9, 13, 14]. We may derive this equation from the exact problem formulation by neglecting inertial effects and performing a systematic expansion in the free-surface slope, a so-called 'long wave' expansion, as demonstrated by Atherton and Homsy [27]. It can be seen from (2.1) that the value of the viscosity μ will not influence the shape of the steady-state profile.

Note that, to consistent order in the surface slope, the curvature gradient in (2.1) can simply be replaced by h_{xxx}. However, by retaining the full curvature, we observe that the equation includes the exact hydrostatic force balance $\sigma\kappa_x + \rho g = \text{const.}$ when $h \to \infty$. This limit corresponds to the static meniscus region where the tin layer meets the bath. The static meniscus represents the 'outer solution' to this problem. Equation (2.1) is thus a composite equation whose solution will give a complete approximation to the profile.

There are three length scales in this problem

$$L_* = \frac{2\sigma}{3\tau}, \quad L_c = \sqrt{\frac{\sigma}{\rho g}}, \quad h_\infty$$

and the scaling is $L_* \gg L_c \gg h_\infty$. Outer dimensionless variables are formed by use of the capillary length L_c,

$$h = L_c H, \quad x = L_c \xi ,$$

and (2.1) may be rewritten as

$$\left[\frac{H_{\xi\xi}}{(1+H_\xi^2)^{3/2}}\right]_\xi = \epsilon \frac{H^2 - H_\infty^2}{H^3} - \left(1 - \frac{H_\infty^3}{H^3}\right) . \tag{2.2}$$

where

$$\epsilon = \frac{L_c}{L_*} = \frac{3}{2}\frac{\tau L_c}{\sigma} = \frac{3}{2}\frac{\tau}{\sqrt{\rho g \sigma}} . \tag{2.3}$$

For given constant values of ϵ, (2.2) may be integrated numerically by a 'shooting method' [28]. The shooting parameter is H_∞ whose values are found by a binary search until $H \to$ "∞" is satisfied. The integration is started on the thin layer. Fourth-order Runge–Kutta integration is used for this and other initial-value problems treated here.

For a slightly-perturbed coating layer, we linearize Equation (2.2) by assuming

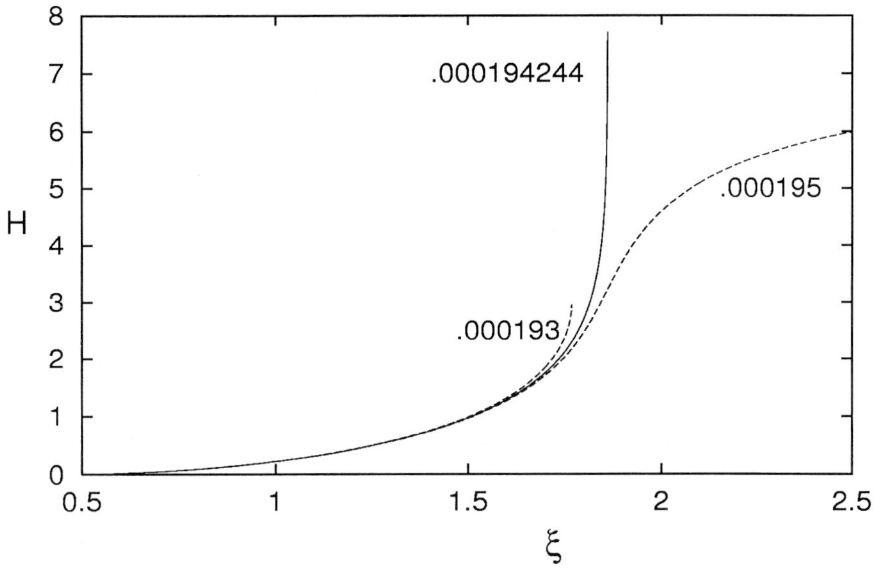

Figure 2. Dimensionless profile $H(\xi)$ determined by a shooting method for $\epsilon = 0{\cdot}01$. Numbers shown are values of the dimensionless wetting layer thickness H_∞. The accurate value is $H_\infty = 1{\cdot}942244\times 10^{-4}$. Incorrect profiles are also shown for H_∞ values that are too small and too large.

$$H = H_\infty + g\ ,$$

where $g/H_\infty << 1$. The perturbation g satisfies

$$g_{\xi\xi\xi} = \left(\frac{2\epsilon}{H_\infty^2} - \frac{3}{H_\infty}\right) g \equiv Zg \tag{2.4}$$

with solution

$$g = \delta \exp(k\xi)\ , \tag{2.5}$$

where k is the positive real root of $k^3 = Z$. Equation (2.5) is used to generate initial conditions for the shooting solution, *i.e.*

$$H(0) = H_\infty + \delta, \quad H_\xi(0) = k\delta, \quad H_{\xi\xi}(0) = k^2\delta, \tag{2.6}$$

where $\delta \ll 1$. In practice δ is taken several orders of magnitude smaller than the shooting parameter H_∞. The criterion for success is simply $H \to \infty$ for some positive value of ξ, corresponding to the meniscus reaching the level of the bath. If H_∞ is too small, the slope H_ξ will become infinite at finite H, while for H_∞ too large, H will attain a maximum value and then decrease with increasing ξ.

A typical result of the shooting method is shown in Figure 2 for $\epsilon = 0{\cdot}01$. The figure shows the dimensionless profile $H(\xi)$ where the vertical plate corresponds to the ξ axis and the bath meniscus is on the right. The accurate value of the wetting layer thickness is $H_\infty = 1{\cdot}94244 \times 10^{-4}$ as indicated. Calculated profiles for larger and smaller H_∞ values are also shown. When H_∞ is too small, the profile achieves a horizontal tangent at a finite H value, while for overly large values, $H(\xi)$ reaches a maximum and then the profile turns back to the substrate. For a range of ϵ values, Table 1 compares the wetting-layer thicknesses found by the numerical shooting method with the results of the matching procedure to be discussed below.

Table 1. Dimensionless coating layer thickness H_∞ as a function of dimensionless shear stress ϵ.

ϵ	Numerical	2-term expansion (2.25)
10^{-5}	$2{\cdot}13518(10^{-10})$	$2{\cdot}13511(10^{-10})$
10^{-4}	$2{\cdot}13210(10^{-8})$	$2{\cdot}13167(10^{-8})$
10^{-3}	$2{\cdot}1078(10^{-6})$	$2{\cdot}1061(10^{-6})$
10^{-2}	$1{\cdot}9424(10^{-4})$	$1{\cdot}9389(10^{-4})$
$0{\cdot}1$	$1{\cdot}2608(10^{-2})$	$1{\cdot}1521(10^{-2})$
$0{\cdot}3$	$0{\cdot}0704$	$0{\cdot}0534$

C.f.: One term result: $H_\infty \approx 2{\cdot}135603\epsilon^2$.

The shooting solution of the composite equation will be compared with analytic approximations by the method of matched asymptotic expansions [29]. An equation valid within the thin layer will be integrated and matched to the capillary-static meniscus at the bath. Define inner variables according to

$$h = h_\infty \eta, \quad x = L\zeta ,$$

where

$$L = h_\infty^{2/3}\left(\frac{2\sigma}{3\tau}\right)^{1/3} = h_\infty^{2/3} L_*^{1/3} . \tag{2.7}$$

Equation (2.1) becomes

$$\left[\frac{\eta_{\zeta\zeta}}{\left(1+\left(\frac{h_\infty}{L_*}\right)^{2/3}\eta_\zeta^2\right)^{3/2}}\right]_\zeta = \frac{\eta^2-1}{\eta^3} - \hat{\epsilon}\left(1-\frac{1}{\eta^2}\right) . \tag{2.8}$$

with

$$\hat{\epsilon} \equiv \frac{2}{3}\frac{\rho g h_\infty}{\tau} = \frac{h_\infty}{L_c}\frac{L_*}{L_c} = \frac{H_\infty}{\epsilon} .$$

$\hat{\epsilon}$ will be shown to be $O(\epsilon)$ while H_∞ is $O(\epsilon^2)$.

The inner expansion is

$$\eta = \eta_0 + \hat{\epsilon}\eta_1 + \cdots , \tag{2.9a}$$

and the first two orders satisfy

$$\eta_{0\zeta\zeta\zeta} = \frac{\eta_0^2-1}{\eta_0^3} , \tag{2.9b}$$

and

$$\eta_{1\zeta\zeta\zeta} = \frac{3-\eta_0^2}{\eta_0^4}\eta_1 - 1 + \frac{1}{\eta_0^3} . \tag{2.9c}$$

Equation (2.9b) is analogous to the Landau–Levich equation for the prediction of the wetting layer on a drawn plate [11, 12] and is integrated in a similar way by the shooting method. Let $\eta = 1 + \phi$, $\phi \ll 1$, to obtain the linearized form

$$\phi_{\zeta\zeta\zeta} = 2\phi + O(\phi^2)\,. \tag{2.10}$$

The exponential growing solution of (2.10) gives the initial conditions

$$\eta_0(0) = 1 + \delta, \quad \eta_{0\zeta}(0) = 2^{1/3}\delta, \quad \eta_{0\zeta\zeta}(0) = 2^{2/3}\delta\,. \tag{2.11}$$

Unlike the Landau–Levich equation [11, 12]

$$\eta_{0\zeta\zeta\zeta} = \frac{\eta_0 - 1}{\eta_0^3}\,, \tag{2.12}$$

for which it is clear that $\eta_{0\zeta\zeta\zeta} \to 0$ as η_0^{-2} for η_0 large, for Equation (2.9b) the limiting values are approached more slowly. Accurate values can still be obtained, however, if we apply a straightforward refinement.

We seek the limiting value P_1 defined by

$$\eta_{0\zeta\zeta} \to P_1 \quad \text{as} \quad \zeta \to \infty\,.$$

Thus, η_0 approaches the parabola

$$\eta_0 \to P_1 \frac{\zeta^2}{2}\,.$$

Using (2.9b), we obtain a more accurate expression:

$$\eta_{0\zeta\zeta} \to P_1 + \int \frac{\mathrm{d}\zeta}{\eta_0} = P_1 + \int \frac{2}{P_1\zeta^2}\,\mathrm{d}\zeta = P_1 - \frac{2}{P_1\zeta}\,.$$

The corrected formula may be reverted to yield

$$P_1 = \left[\eta_{0\zeta\zeta} + \frac{2}{\zeta\,\eta_{0\zeta\zeta}}\right]_{\zeta\to\infty} + O\left(\frac{1}{\zeta^2}\right)\,, \tag{2.13}$$

which gives quadratic convergence, allowing equivalent accuracy to that obtainable for the moving-plate problem that uses (2.12). Marching the numerical solution gives a unique limiting value as $\zeta \to \infty$. It is found to be

$$\eta_{0\zeta\zeta} \to P_1 = 1{\cdot}8211895\,. \tag{2.14}$$

The outer solution is the static meniscus satisfying

$$\left[\frac{H_{\xi\xi}}{(1 + H_\xi^2)^{3/2}}\right]_\xi = -1\,.$$

whose shape is known in closed form. We use the profile $H(\xi)$ that has zero slope at $\xi = 0$. Princen [30] gives expressions for $H(\psi)$ and $\xi(\psi)$ where ψ is the positive downward inclination angle of the static profile with respect to the horizontal. These may be conbined to yield

$$H = H_0 + 0{\cdot}53284 - 2g - \frac{1}{2}\log\frac{1-g}{1+g} \tag{2.15a}$$

where

$$g = g(\xi) = \sqrt{1 - \frac{(\sqrt{2} - \xi)^2}{4}}\,, \tag{2.15b}$$

and $H = H_0$ at $\xi = 0$. H_0 is a matching constant to be determined.

From (2.15) we have $H_{\xi\xi}(0) = \sqrt{2}$ or $h_{xx}(0) = \sqrt{2}/L_c$. In a common set of variables this must be equal to the leading-order curvature found from the inner solution. Thus,

$$\eta_{0\zeta\zeta} = \frac{L^2}{h_\infty} h_{xx} = h_\infty^{1/3} \left(\frac{2\sigma}{3\tau}\right)^{2/3} h_{xx} \to h_\infty^{1/3} \left(\frac{2\sigma}{3\tau}\right)^{2/3} \frac{\sqrt{2}}{L_c}\,.$$

Solving this for h_∞ and using (2.14),

$$h_\infty = P_1^3 \left(\frac{3\tau}{2\sigma}\right)^2 \frac{L_c^3}{2^{3/2}} = \frac{9}{4}\left(\frac{P_1}{\sqrt{2}}\right)^3 \left(\frac{\sigma}{\rho g}\right)^{3/2} \left(\frac{\tau}{\sigma}\right)^2$$

or

$$H_\infty = \frac{h_\infty}{L_c} = \frac{P_1^3}{2^{3/2}}\epsilon^2 \approx 2{\cdot}135603\epsilon^2 = 4{\cdot}80511 \frac{\tau^2}{\rho g \sigma}\,. \tag{2.16}$$

This may be compared with the slightly larger result of Fanton *et al.* [10] which is $H_\infty = 4{\cdot}84$ $\tau^2/(\rho g \sigma)$. The difference between our result and the earlier value is believed to be due to our use of the quadratically-convergent formula (2.13).

The next-order correction to (2.16) will now be found. The limiting equality

$$\eta_{0\zeta\zeta\zeta} \to \frac{2}{P_1 \zeta^2} \tag{2.17a}$$

can be integrated repeatedly to yield

$$\eta_0 \to \frac{P_1}{2}\zeta^2 - \frac{2}{P_1}\zeta \log\zeta + A\zeta + O(\log\zeta)^2\,. \tag{2.17b}$$

The next-order correction for the coating thickness is logarithmically dependent on ϵ and can be found analytically. For large ζ the equality (2.9c) is approximately

$$\eta_{1\zeta\zeta\zeta} = -1 - \frac{\eta_1}{\eta_0^2} + O\left(\frac{1}{\zeta^5}\right)\,. \tag{2.18}$$

Assume a solution of the form

$$\eta_1 = a_0\zeta^3 + a_1\zeta^2 \log\zeta + a_2\zeta^2 + o(\zeta^2);$$

then

$$\eta_{1\zeta\zeta\zeta} = 6a_0 + \frac{2a_1}{\zeta}\,.$$

The right-hand side of (2.18) becomes

$$-1 - \frac{4a_0}{P_1^2\zeta} + O\left(\frac{\log\zeta}{\zeta^2}\right)\,.$$

Thus

$$\eta_1 \to -\frac{\zeta^3}{6} + \frac{1}{3P_1^2}\zeta^2 \log\zeta + O(\zeta^2)\,.$$

and

$$\eta_{1\zeta\zeta} \to -\zeta + \frac{2}{3P_1^2}\log\zeta + \cdots\,.$$

Differentiaton of expansion (2.9a) now yields

$$\eta_{\zeta\zeta} = \eta_{0\zeta\zeta} + \hat{\epsilon}\eta_{1\zeta\zeta} + \cdots = P_1 - \hat{\epsilon}\zeta + \hat{\epsilon}\frac{2}{3P_1^2}\log\zeta + \cdots\,. \tag{2.19}$$

The origin of ζ must be shifted to match the outer solution. With

$$\zeta = \zeta_1 + \bar{\zeta} \tag{2.20}$$

we have

$$\eta_0 = \frac{P_1}{2}\bar{\zeta}^2 + \text{const.} = \frac{P_1}{2}(\zeta - \zeta_1)^2 + \text{const.} = \frac{P_1}{2}\zeta^2 - P_1\zeta_1\zeta + \text{const}\,.$$

Comparing with (2.17b), we find that

$$\zeta_1 = \frac{2}{P_1^2}\log\zeta - \frac{A}{P_1}\,. \tag{2.21}$$

Inserting (2.21) in (2.19), we get

$$\eta_{\bar{\zeta}\bar{\zeta}} = P_1 - \hat{\epsilon}\bar{\zeta} - \hat{\epsilon}\zeta_1 + \hat{\epsilon}\frac{2}{3P_1^2}\log\zeta + \cdots = P_1 - \hat{\epsilon}\bar{\zeta} - \hat{\epsilon}\frac{4}{3P_1^2}\log\zeta + O(\hat{\epsilon})\,. \tag{2.22}$$

We return to outer variables with the transformations

$$\hat{\epsilon} = H_\infty/\epsilon, \quad \eta_{\bar{\zeta}\bar{\zeta}} = \frac{H_\infty^{1/3}}{\epsilon^{2/3}}H_{\xi\xi}, \quad \bar{\zeta} = \frac{\epsilon^{1/3}}{H_\infty^{2/3}}\xi\,.$$

Retaining the $O(\xi)$ term from the outer solution (2.15), we have

$$H_{\xi\xi} \to \sqrt{2} - \xi \quad \text{as} \quad \xi \to 0\,.$$

Thus $\eta_{\bar{\zeta}\bar{\zeta}}$ may be equated to

$$\frac{H_\infty^{1/3}}{\epsilon^{2/3}}(\sqrt{2} - \xi) = P_1 - \frac{H_\infty}{\epsilon}\frac{\epsilon^{1/3}}{H_\infty^{2/3}}\xi - \frac{4}{3P_1^2}\frac{H_\infty}{\epsilon}\log\left(\frac{\epsilon^{1/3}}{H_\infty^{2/3}}\xi\right) + O(\epsilon)\,. \tag{2.23}$$

Let the corrected form of (2.23) be

$$H_\infty = \frac{P_1^3}{2^{3/2}}\epsilon^2(1 + \delta(\epsilon))\,, \tag{2.24}$$

where

$$\delta = O(\epsilon\log\epsilon)\,.$$

Inserting this expression in (2.23) and expanding for small δ, we obtain

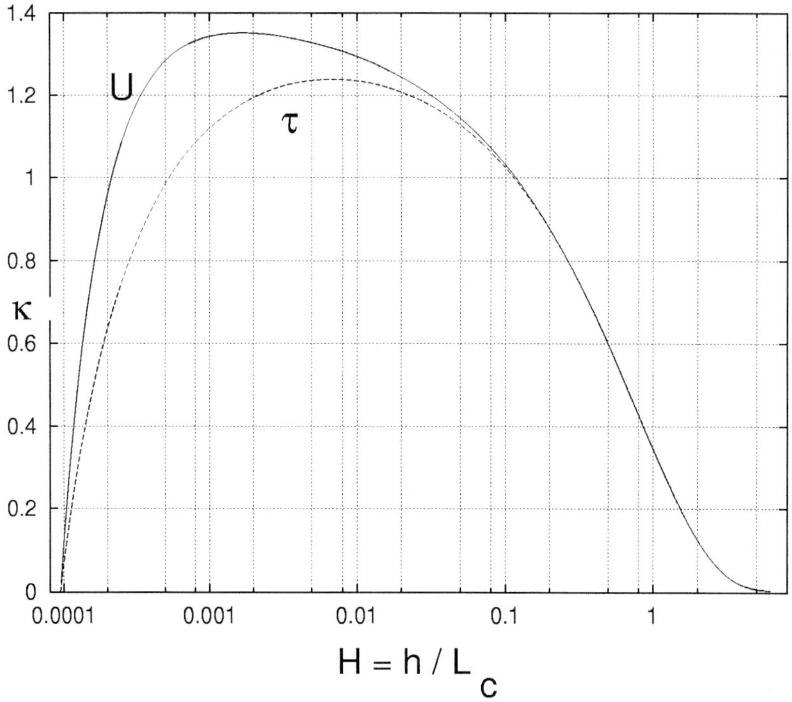

Figure 3. Numerical solutions for the dimensionless curvature κ versus distance from the wall $H = h/L_c$ for the Landau–Levich plate drawing problem, labelled U, and stress-driven problem, labelled τ.

$$P_1(1+\delta)^{1/3} \approx P_1\left(1+\frac{1}{3}\delta\right) = P_1 - \frac{\sqrt{2}}{3}P_1\epsilon\log\left(\frac{2\epsilon^{1/3}}{P_1^2\epsilon^{4/3}}\xi\right) .$$

As $\epsilon \to 0$, this becomes simply

$$\delta = \sqrt{2}\epsilon\log\epsilon + O(\epsilon) .$$

From (2.24) the corrected expression for H_∞ is

$$H_\infty = 2{\cdot}135603\epsilon^2 + 3{\cdot}020199\epsilon^3\log\epsilon + O(\epsilon^3) . \tag{2.25}$$

Values calculated using (2.25) for various ϵ are shown in Table 1 where they may be compared with the shooting-method results.

Compared with the leading-order result, Equation (2.25) provides sowewhat closer agreement with the experimental measurement of Fanton *et al.* [10]. From the experimental values $\tau = 0{\cdot}9$ dynes/cm^2, $\sigma = 21$ dynes/cm, and $\rho = 0{\cdot}965$ gm/cm^3 we have the value $\epsilon = 0{\cdot}00960$. Using this ϵ, we observe that the leading-order theory predicts $h_\infty = 0{\cdot}293$ μm, while both the two-term expansion and the shooting-method give $0{\cdot}266$ μm. The experimentally measured thickness was $0{\cdot}23 \pm 0{\cdot}06$ μm, where the error brackets come from measurement uncertainty. While still too large, the more accurate prediction lies well within the bracketed uncertainty.

It is interesting to compare the profiles $h(x)$ for stress-driven motion as considered here with those obtained for the Landau–Levich plate withdrawal problem [11]. A composite equation for plate withdrawal, analogous to (2.2), is

$$\left[\frac{H_{\xi\xi}}{(1+H_\xi^2)^{3/2}}\right]_\xi = 3\mathrm{Ca}\frac{H-H_\infty}{H^3} - \left(1-\frac{H_\infty^3}{H^3}\right) . \tag{2.26}$$

where $\mathrm{Ca} = \mu U/\sigma$ and U is the constant plate withdrawal speed. Initial conditions for this composite equation are derived from

$$g_{\xi\xi\xi} = \left(\frac{3\mathrm{Ca}}{H_\infty^3} - \frac{3}{H_\infty}\right) g$$

which is analogous to (2.4). Two profiles are compared in Figure 3 where the dimensionless curvature $\kappa = H_{\xi\xi}/(1+H_\xi^2)^{3/2}$ is plotted versus H. A logarithmic scale is invoked to magnify the wall region. The curve labeled U is the plate-withdrawal solution, while τ indicates the stress-driven case. Values of ϵ and Ca are selected so that $h_\infty/L_c \approx 10^{-4}$ for each case. Note that the region in which dynamic effects are important is much larger for the shear-driven problem. In these cases the profile joins the static meniscus at a far greater distance from the wall.

3. Thin-layer motion on nonuniform substrates

In the earlier experiment of Ludviksson and Lightfoot [7] a coating of nonwetting material was applied to the substrate in order to control the motion of the liquid. Here we will generalize the lubrication model to include finite-contact-angle effects. We will be concerned with the unsteady motion of a thin layer whose thickness is constant far upstream; specifically we consider problem (ii) shown in Figure 1. The slope of the free surface relative to the substrate will be assumed to be uniformly small, so that the lubrication approximation can continue to be used.

Problem (ii) involves the motion of a three-phase contact line, *i.e.* a line where the solid substrate, the advancing liquid, and vapor or vacuum meet. It is well known that such a problem requires special treatment. Unless the no-slip boundary condition for the liquid motion on the solid substrate is relaxed in the vicinity of the contact line in some way, one obtains the anomalous result that infinite force is required to move the contact line [31]. A number of modifications that allow motion of the contact line were compared by Moriarty and Schwartz [32]. They found that, apart from the immediate vicinity of the contact line, each choice of relief mechanism yielded the same set of moving liquid profiles. The simplest modification is to 'pre-wet' the substrate ahead of the moving contact line with a very thin layer of the liquid, *i.e.* a slip layer [28]. In addition, for liquids that do not spread spontaneously onto the substrate, a static or equilibrium contact angle θ_e is observed. By appending a disjoining pressure term to the usual capillary pressure, it is possible to prescribe both the thickness of the slip layer and the contact angle θ_e, as explained below.

Returning to the lubrication approximation in dimensional variables, with upward surface traction and downward gravity flow, we observe that the flux is given by

$$Q = -\frac{h^3}{3\mu} p_x + \frac{\tau h^2}{2\mu} - \frac{\rho g h^3}{3\mu} \,. \tag{3.1}$$

The pressure p is composed of the usual capillary pressure associated with free-surface curvature and a disjoining pressure component that can be used to specify an equilibrium contact angle θ_e. θ_e will be taken to vary with position on the substrate. Thus

$$p = -\sigma h_{xx} - \Pi \,, \tag{3.2}$$

where we will use

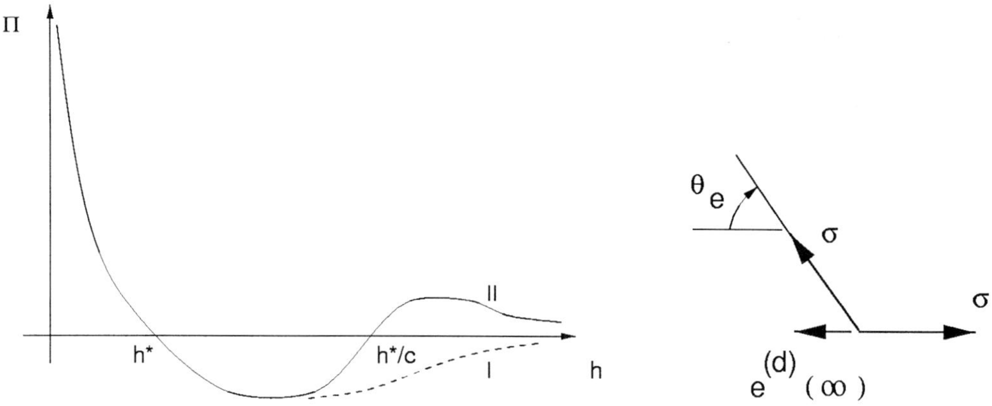

Figure 4. Left: Two different forms of the disjoining pressure function Π plotted versus liquid layer thickness h. *Right*: The contact-line force balance using disjoining pressure. θ_e is the static or equilibrium contact angle.

$$\Pi = K\left(\frac{h^*}{h}\right)^2\left(\frac{h^*}{h}-1\right)\left(\frac{h^*}{h}-c\right) . \tag{3.3}$$

This function represents an assumed interaction between the substrate and the liquid free surface. The constant h^* plays the role of a slip coefficient, while K will be shown to incorporate information about the equilibrium angle θ_e.

The form of Π used here is only one of a number of possible choices, since the full details of the molecular interactions in real systems are not known. Different exponents on h could have been used, for example. The formula (3.3) does include the two basic behaviors that allow for a finite value of θ_e to be specified. Typically schematic plots of $\Pi(h)$ are shown in Figure 4. If the constant c is taken to be zero, $\Pi < 0$ for $h > h^*$. In that case, layers thicker than h_* will be attracted downward. That is the case indicated by (I) in the figure. Alternatively, as indicated by (II), if $c > 0$, sufficiently thick layers will be stable and the disjoining pressure will become repulsive for $h > h^*/c$. Each of these forms is possible and the particular choice depends on the specific liquid-substrate material system, as discussed by Teletzke *et al.* [20] and Churaev and Sobolev [18]. We have performed dynamic simulations in related problems with various exponent choices and find that appreciable differences in profile shape only occur in the immediate neighborhood of the apparent contact line, while the larger scale dynamics is virtually the same [22, 23].

Associated with Π is a disjoining energy density

$$e^{(d)}(h) = -\int_{h^*}^{h} \Pi(h_1)\, dh_1 . \tag{3.4}$$

which represents the work done against disjoining pressure in displacing an element of the liquid-air interface from its equilibrium value $h = h^*$. Consider a wedge of liquid that meets the apparent substrate and whose ultimate thickness is large when measured in units of h^*; *i.e.* we consider $h/h^* \to \infty$. Integration of (3.4) yields

$$e^{(d)}(\infty) = Kh^*\left(\frac{1}{6}-\frac{c}{2}\right) . \tag{3.5}$$

The right side of Figure 4 shows the static force balance at the effective 'contact line,' where a thick layer meets the slip layer $h = h^*$. From the figure, when θ_e is small,

$$e^{(d)}(\infty) = \sigma(1 - \cos\theta_e) \approx \frac{\sigma\theta_e^2}{2} . \tag{3.6}$$

Equating the two expressions for $e^{(d)}(\infty)$, we obtain an explicit expression for K in terms of the contact angle

$$K = \frac{3\sigma\theta_e^2}{h^*(1 - 3c)} ,$$

indicating that the meaningful range for c is $0 \leq c < 1/3$. When Π is inserted in the evolution equation, it provides a method for modeling motion on substrates that exhibit a finite static contact angle. Moreover, it will illustrate the way in which lack of perfect wetting influences motion. Disjoining pressure was introduced in order to explain the molecular origins of static contact angles. For particular material systems, some progress can be made in calculating the exponents and constants that appear in an expression such as (3.3) from first principles [33]. The important features of (3.3), in a dynamic study, are the values of θ_e and h^* and the fact that there is a single stable minimum of the energy functional $e^{(d)}(h; h^*)$.

Ludviksson and Lightfoot [7] include results of a series of experiments where a barrier of non-wetting coating was applied to the substrate. As expected, the wetting front stopped at this barrier. We will consider these results as a special case of unsteady motions on substrates with barriers.

With motion in the direction of x increasing, we may form an unsteady evolution equation using the integral mass conservation condition

$$h_t = -Q_x \tag{3.7}$$

where Q is given by (3.1) and (3.2). A number of parameters can be removed by the scaling

$$t = T^*\tilde{t}, \quad x = L\tilde{x}, \quad h = h_0\tilde{h} , \tag{3.8}$$

where $h_0 = 3\tau/(2\rho g)$, $L = h_0^{1/3} L_c^{2/3}$, $T^* = 3\mu L^4/(\sigma h_0^3)$, and, as before, $L_c = \sqrt{\sigma/(\rho g)}$. The dimensionless evolution equation with tildes omitted is

$$h_t = -\left[h^3 h_{xxx} + h^2 - h^3\right]_x - \left[h^3 \left(D\tilde{\Pi}\right)_x\right]_x . \tag{3.9}$$

Here

$$\tilde{\Pi} = \left(\frac{h^*}{h}\right)\left(\frac{h^*}{h} - 1\right)\left(\frac{h^*}{h} - c\right) . \tag{3.10}$$

Dimensionless variables are used on the right of (3.10) and

$$D = 3\left(\frac{L_c}{h^*}\right)\left(\frac{L_c}{h_0}\right)^{1/3} \frac{\theta_e^2}{1 - 3c} . \tag{3.11}$$

For a steady solution in laboratory-fixed coordinates we have $Q = 0$; then, in dimensionless terms, $h = 1$ is the layer thickness far upstream where $p_x = 0$. Since disjoining effects are only important when $h \sim O(h^*)$ and $h^* << 1$, a universal steady-state differential equation,

$$h_{xxx} = \frac{h - 1}{h} , \tag{3.12}$$

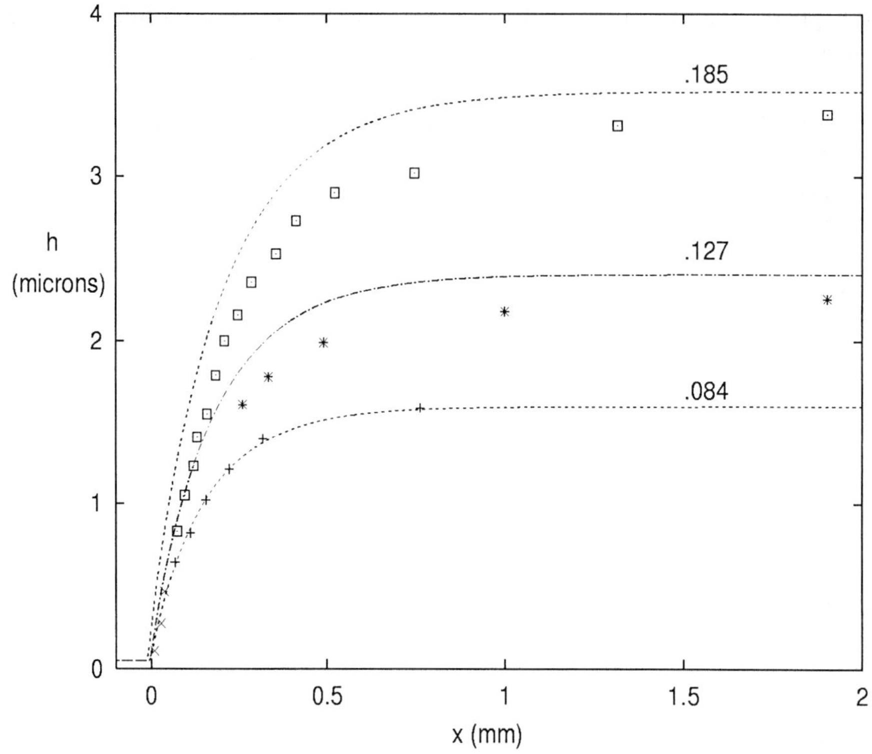

Figure 5. Comparison of calculated steady profiles at a barrier with experimental data of Ludviksson and Lightfoot [7] (symbols). The shear stress values, taken from the experiment results, are $\tau = 0{\cdot}084$, $0{\cdot}127$, and $0{\cdot}185$ dynes/cm^2.

determines the steady profile away from the apparent contact line. There is an essentially unique solution of (3.12) that meets the substrate. The initial conditions are found from the upstream approximate solution

$$h \approx 1 - \delta \exp(kx) \, , \tag{3.13}$$

where δ is a small positive number. The linearized form of (3.12) shows that $k = 1$. Runge-Kutta integration of (3.12) yields the universal static profile. This effective outer solution has finite contact slope K_s at $h = 0$. If the profile is translated so that the contact point corresponds to $x = 0$, a local solution there is

$$h \sim K_s \hat{x} + \frac{\hat{x}^2}{2K_s} \log \hat{x} + O(\hat{x}^2) \, , \tag{3.14}$$

where $\hat{x} = -x$ is used here. From the initial condition (3.13), the universal value

$$K_s = 1{\cdot}34544 \ldots \tag{3.15}$$

is found by integration.

Ultimate profiles at a barrier may be calculated from the evolution equation (3.9) with rather general initial conditions. The equation is written in finite-difference form. An implicit time-marching algorithm has been implemented leading to greatly improved numerical stability and the ability to use much larger time steps. Generally speaking, the method is a higher-order generalization of the Crank–Nicolson method that is commonly used to solve the heat equation. Additional details of the numerical method may be found elsewhere [34, 35]. The contact angle variation $\theta_e(x)$ is prescribed and is taken to rise rapidly from zero to a finite value in a particular region of the substrate. When the moving front reaches the barrier region it stops and, after a period of adjustment to the upstream thickness, the entire profile becomes steady. Calculated profiles are compared with the experimental results of Ludviksson and Lightfoot in Figure 5. Three different values of imposed shear stress are shown; each is derived from the measured temperature gradient and the measured dependence of surface tension on temperature. The agreement is credible but not perfect. Better agreement is observed for the smallest τ value. The three data points marked by (x) are said to be applicable to all profiles. Thus, Ludviksson and Lightfoot [7] maintain that the contact angle is independent of τ; this differs from the result of the present analysis. We can predict the actual contact angle θ_e^* required to arrest the stress-driven motion, using the value for K_s and returning to dimensional quantities via (3.8):

$$\theta_e^* \approx K_s \frac{h_0}{L} = 1{\cdot}762 \left(\frac{\tau}{\sqrt{\sigma \rho g}} \right)^{2/3} . \tag{3.16}$$

For $\tau = 0{\cdot}185$ dynes/cm^2, which is the largest shear stress applied in the barrier experiments of Ludviksson and Lightfoot, the predicted contact angle required to stop the motion is $\theta_e^* \approx 1{\cdot}1$ degrees. Thus, the strongly nonwetting coating that they applied was far stronger than needed to arrest the motion of the contact line.

Figure 5 shows that unsteady solutions each approach the universal steady state if the barrier 'height', meaning the maximum equilibrium contact angle, is sufficiently high. The specific value of maximum contact angle used in the numerical solution of (3.9) was $1{\cdot}2°$. The slip layer thickness h^* was taken to be 50 nanometers, an arbitrarily chosen but realistically small value. In Figure 5 the direction of motion is from right to left. This is the direction used in the original plot of the experimental data given in [7].

Two of the three final profiles from Figure 5 are replotted in Figure 6 in dimensionless variables. The universal solution from (3.12) is also shown as a line. The symbols use the extreme values of shear stress in the experiments, $\tau = 0{\cdot}084$ dynes/cm^2 and $0{\cdot}185$ dynes/cm^2, respectively. The layer thickness h is measured in units of $3\tau/(2\rho g)$ so that the limiting height far away from the barrier is one. In these units, the dimensionless slip layer thickness depends on τ, as shown in the figure. Because the barrier is quite steep, *i.e.* the contact angle changes from zero to $1{\cdot}2°$ over a short distance on the substrate, the unsteady profiles stabilize in almost the same position. The origin of the universal solution is arbitrary. It was shifted so as to illustrate the self similarity of the steady profiles.

Results of another unsteady calculation are shown in Figure 7. Here the barrier 'height' was taken to be smaller than the value required to stop the motion; specifically the contact angle $\theta_{e\,\max}$ is 60 per cent of the critical value of θ_e^*; given by (3.16). The dimensionless slip-layer thickness is $h^* = 0{\cdot}02$. The calculation used 2500 unknown values of h and a point-spacing interval $\Delta x = 0{\cdot}025$.

The direction of motion is from left to right and values of the dimensionless time are shown on each profile. To the left of the barrier $\theta_e = 0$. An initial profile was arbitrarily chosen to be an inverse tangent curve connecting the uniform upstream thickness and the slip layer. The argument of the inverse tangent is scaled so that 70 per cent of the jump in h occurs in $0{\cdot}2$ x units. Similarly, the contact-angle function $\theta_e(x)$ is also an arc tangent whose argument is scaled so that 70 per cent of the jump occurs in $1{\cdot}0$ x units. The profile is seen to quickly

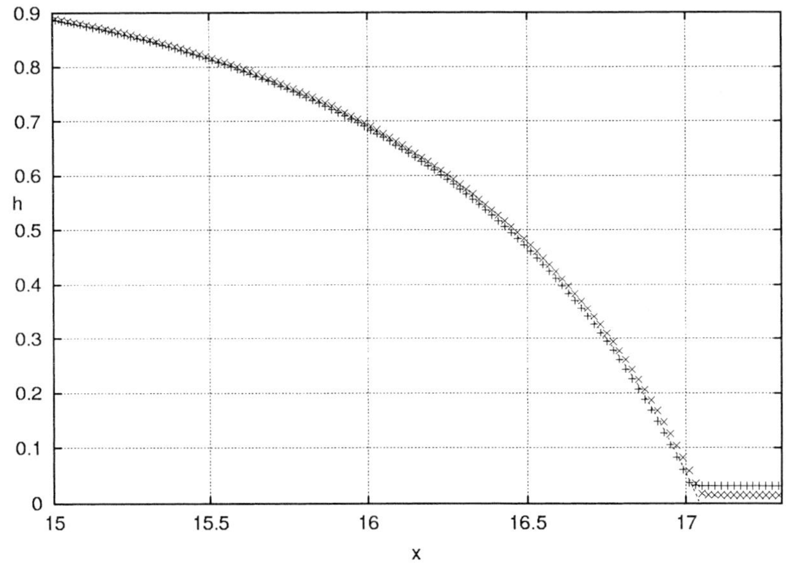

Figure 6. Unsteady calculations are compared with the universal outer solution (line) using dimensionless variables. Each unsteady solution using the same physical slip layer $h^* = 50$ nm. The symbols correspond to the two extreme experimental values of shear stress $\tau = 0{\cdot}084$ dynes/cm^2 and $0{\cdot}185$ dynes/cm^2.

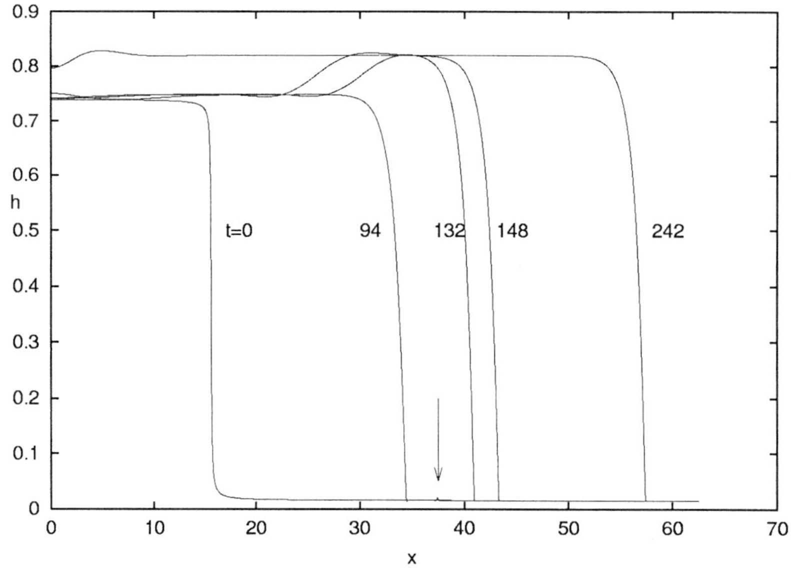

Figure 7. Unsteady calculation of a moving front involving the evolution equation (3.9). The equilibrium contact angle θ_e is a function of substrate position and 'ramps up' from zero, on the left of the arrow to a finite value on the right. The final value of θ_e is too small to stop the motion. Dimensionless time values are shown for each profile.

reform into a unique steadily-propagating shape that is determined by the upstream height and h^*. Upon reaching the location of the barrier, indicated by an arrow in the figure, the profile temporarily slows and forms a hump. Ultimately the upstream height is adjusted upward. This is a simple illustration of the manner in which substrate contact-angle variation may be used to adjust the shape of a flowing stream. The calculation of Figure 7 was repeated using the second form of disjoining pressure with $c = 0{\cdot}1$, as indicated by (II) in Figure 4. Quite similar results were obtained; the major difference was that final height after barrier passage was slightly higher.

Concluding remarks

Two aspects of shear-stress driven flow have been considered. Estimates of coating thicknesses produced under steady-state conditions of bath withdrawal have been refined using two different methods. Each of these account for the gravitational backflow into the bath. Provided that the shear stress is not large, they are in substantial agreement with one another and provide an improved agreement with the experimental result of Fanton *et al.* [10]. The asymptotic expansion has been extended to include a term of order $\epsilon^3 \log \epsilon$. This term is missing from the expansion for the related Landau-Levich problem of plate withdrawal where the correction term is of order ϵ^3, as found by Wilson [12]. The wall region, where viscous effects are important, is thicker for the stress problem. We believe that extension of the solution to order ϵ^3 for the stress case may require a third or intermediate, matching region [29].

An unsteady evolution equation has been introduced to model motion on a substrate of non-zero and variable contact angle. The critical value of contact angle θ_e^* required to stop the motion has been found and is predicted to be proportional to the two-thirds power of applied shear stress. The actual values of θ_e^* are remarkably small, far smaller than that of the 'strongly nonwetting' coating applied by Ludviksson and Lightfoot [7] to arrest the motion. Our results concerning the behavior θ_e^* differ qualitatively from a conjecture given in [7]. Those authors had suggested that there is a universal value of θ_e^*, for a particular liquid-solid system, that is independent of the driving stress τ. For smaller contact angle changes, it has been shown that the shape of the moving liquid stream can be modified and controlled by the pre-set variation in contact angle. Possible applications include thermocapillary pumping devices and other microscopic liquid transport mechanisms.

Acknowledgement

This work is supported by the ICI Business Link Partnership, the NASA Microgravity Program, and the State of Delaware.

References

1. P. Erhard and S. H. Davis, Nonisothermal spreading of liquid drops on horizontal plates. *J. Fluid Mech.* 229 (1991) 365–388.
2. L. D. Landau and E. M. Lifshitz, *Fluid Mechanics*. Oxford: Pergamon Press (1959).
3. S. K. Wilson, The effect of an axial temperature gradient on the steady motion of a large droplet in a tube. *J. Eng. Math.* 29 (1995) 205–217.
4. M. K. Smith, Thermocapillary migration of a two-dimensional liquid droplet on a solid surface. *J. Fluid. Mech.* 294 (1995) 209–230.
5. A. Mazouchi and G. M. Homsy, Thermocapillary migration of long bubbles in cylindrical capillary tubes. *Phys. Fluids* 12 (2000) 542–549.
6. T. S. Sammarco and M. A. Burns, Thermocapillary pumping of discrete drops in microfabricated analysis devices. *AIChE J.* 45 (1999)350–366.
7. V. Ludviksson and E. N. Lightfoot, The dynamics of thin liquid films in the presence of surface-tension gradients. *AIChE J.* 17 (1971 1166–1173.
8. A. M. Cazabat, F. Heslot, S. M. Troian and P. Carles, Fingering instability of thin spreading films driven by temperature gradients. *Nature* 346 (1990) 824–826.

9. A. M. Cazabat, F. Heslot, P. Carles and S. M. Troian, Hydrodynamic fingering instability of driven wetting films. *Adv. Coll. Interf. Sci.* 39 (1992) 61–75.
10. X. Fanton, A. M. Cazabat and D. Quere, Thickness and shape of films driven by Marangoni flow. *Langmuir* 12 (1996) 5875–5880.
11. V. G. Levich, *Physiochemical Hydrodynamics*. Englewood Cliffs: Prentice-Hall (1962) 700 pp.
12. S. D. R. Wilson, The drag-out problem in film coating theory. *J. Eng. Math.* 16 (1982) 209–221.
13. D. E. Kataoka and S. M. Trojan, A theoretical study of instabilities at the advancing front of thermally driven coating films. *J. Colloid Interf. Sci.* 192 (1997) 350–362.
14. D. E. Kataoka and S. M. Troian, Stabilizing the advancing front of thermall driven climbing films. *J. Colloid Interf. Sci.* 203 (1998) 335–344.
15. M. H. Eres, L. W. Schwartz and R. V. Roy, Fingering phenomena for driven coating films. *Phys. Fluids* 12 (2000) 1278–1295.
16. P. Carles and A. M. Cazabat, The thickness of surface-tension-gradient-driven spreading films. *J. Colloid Interf. Sci.* 157 (1993) 196–201.
17. H. Gau, S. Herminghaus, P. Lenz and R. Lipowsky, Liquid morphologies on structured surfaces; from microchannels to microchips. *Science* 283 (1999) 46–48.
18. N. V. Churaev and V. D. Sobolev, Prediction of contact angles on the basis of the Frumkin-Derjaguin approach. *Adv. Colloid Interf. Science* 61 (1995) 1–16.
19. B. V. Derjaguin, The definition and magnitude of disjoining pressure and its role in the statics and dynamics of thin fluid films. *Kolloid Zhurnal* 17 (1955) 205–214.
20. G. F. Teletzke, H. T. Davis and L. E. Scriven, Wetting hydrodynamics. *Revue de Physique Appliquee* 23 (1988) 989–1007.
21. V. S. Mitlin and N. V. Petviashvili, Nonlinear dynamics of dewetting: Kinetically stable structures. *Phys. Lett. A* 192 (1994) 323–326.
22. L. W. Schwartz, Hysteretic effects in droplet motions on heterogeneous substrates: Direct numerical simulation. *Langmuir* 14 (1998) 3440–3453.
23. L. W. Schwartz and R. R Eley, Simulation of droplet motion on low-energy and heterogeneous surfaces. *J. Colloid Interf. Sci.* 202 (1998) 173–188.
24. A. L. Bertozzi, A. Munch, X. Fanton and A. M. Cazabat, Contact line stability and undercompressive shocks in driven thin film flow. *Phys. Rev. Lett.* 81 (1998) 5169–5172.
25. A. L. Bertozzi, A. Munch and M. Shearer, Undercompressive shocks in thin film flows. *Physica D* 134 (1999) 431–464.
26. A. Munch and A. L. Bertozzi, Rarefaction-undercompressive fronts in driven films. *Phys. Fluids* 11 (1999) 2812–2814.
27. R. W. Atherton and G. M. Homsy, On the derivation of evolution equations for interfacial waves. *Chem. Eng. Comm.* 2 (1976) 57–77.
28. E. O. Tuck and L. W. Schwartz, A numerical and asymptotic study of some third- order ordinary differential equations relevant to draining and coating flows. *S.I.A.M. Rev.* 32 (1990) 453–469.
29. M. D. Van Dyke, *Perturbation Methods in Fluid Mechanics*. Annotated Edition. Stanford: Parabolic Press (1975) 271 pp.
30. H. M. Princen, The equilibrium shapes of interfaces, drops and bubbles. *Surf. Colloid Sci.* 11 (1969) 1–84.
31. C. Huh and L. E. Scriven, Hydrodynamic model of steady movement of a solid liquid fluid contact line. *J. Colloid Interf. Sci.* 35 (1971) 85–101.
32. J. A. Moriarty and L. W. Schwartz, Effective slip in numerical calculations of moving-contact-line problems. *J. Eng. Math.* 26 (1992) 81–86.
33. J. N. Israelachvili, Intermolecular and Surface Forces, 2nd Ed. London: Academic Press (1992) 450 pp.
34. J. A. Moriarty and L. W. Schwartz, Dynamic considerations in the closing and opening of holes in thin liquid films. *J. Colloid Interf. Sci.* 161 (1993) 335–342.
35. D. E. Weidner, L. W. Schwartz and R. R. Eley, Role of surface tension gradients in correcting coating defects in corners. *J. Colloid Interf. Sci.* 179 (1996) 66–75.

Journal of Engineering Mathematics **39:** 189–206, 2001.

Matched asymptotic expansions and the numerical treatment of viscous-inviscid interaction

A.E.P. VELDMAN
Institute of Mathematics and Computing Science, University of Groningen, P.O. Box 800, 9700 AV Groningen, The Netherlands (e-mail: veldman@math.rug.nl)

Received 14 January 2000; accepted in revised form 15 August 2000

Abstract. The paper presents a personal view on the history of viscous-inviscid interaction methods, a history closely related to the evolution of the method of matched asymptotic expansions. The main challenge in solving Prandtl's boundary-layer equations has been to overcome the singularity at a point of steady flow separation. Stewartson's triple-deck theory has inspired a solution to this challenge, and thereby it paved the way for industrial use of viscous-inviscid interaction methods.

Key words: boundary-layer separation, viscous-inviscid interaction, matched asymptotic expansions, numerical simulation, transonic airfoils

1. Prandtl's boundary layer

The history of the development of viscous-inviscid interaction methods started in Heidelberg at 11:30 a.m. on August 12, 1904, when Ludwig Prandtl presented the 'boundary layer' before an audience of mathematicians attending the Third International Mathematical Congress [1]. For decades, scientists had been confused by d'Alembert's Paradox ('discovered' in 1752), stating that "there is no drag on a finite body at rest in an infinite, incompressible, inviscid flow otherwise in uniform motion" [2]. Prandtl described how the hardly visible boundary layer near the surface of the body, through the influence of viscosity, can resolve this paradox.

Prandtl's Heidelberg lecture is considered a landmark in the development of a branch of mathematics nowadays called 'matched asymptotic expansions', although various roots of the boundary-layer idea can be found already in the nineteenth century [3]. The method of matched asymptotic expansions treats differential equations where a small parameter multiplies the highest derivative, *i.e.* setting the small parameter at zero implies dropping one (or more) boundary conditions. As a consequence a series development in the small parameter is no longer valid uniformly throughout the domain. Next to the boundary where in the small-parameter limit boundary conditions have to be dropped, a thin layer has to be added where a different series development is required. Prandtl named this thin layer *Grenzschicht* (English translation: boundary layer), a name that has been used ever since for similar thin layers in other applications.

In aerodynamic applications (Figure 1), the boundary layer is driven by the inviscid pressure distribution p_e and its (through Bernoulli's law) related streamwise velocity u_e. In the boundary layer the streamwise velocity component is reduced to zero in order to comply with the no-slip condition at the surface. The lateral coordinate y, together with its corresponding velocity component v, scales with the inverse square root of the Reynolds number Re (defined

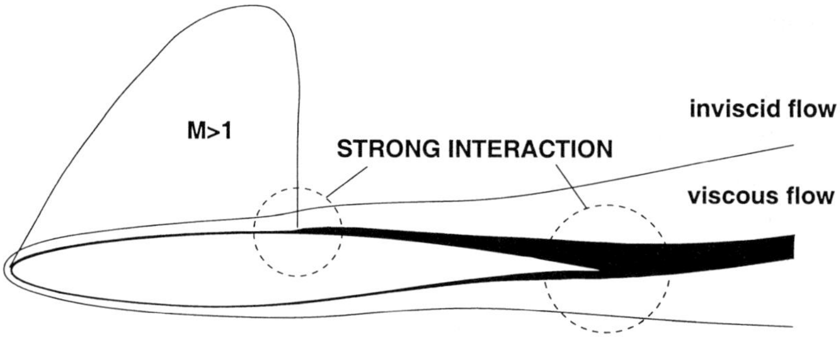

Figure 1. Subdivision of the flow field around an airfoil in an inviscid-flow region and a viscous shear layer (exaggerated in thickness).

in the usual way[1]). The flow equations can be simplified by neglecting the viscous streamwise derivatives, whereas the lateral momentum equation states the pressure to be constant through the boundary layer. Prandtl's *Grenzschichtgleichungen* emerge (in non-dimensional form):

$$\frac{\partial u}{\partial x} + \frac{\partial v}{\partial y} = 0, \quad u\frac{\partial u}{\partial x} + v\frac{\partial u}{\partial y} = u_e\frac{\mathrm{d}u_e}{\mathrm{d}x} + \frac{1}{\mathrm{Re}}\frac{\partial^2 u}{\partial y^2}, \tag{1}$$

with boundary conditions

$$u(x, 0) = v(x, 0) = 0; \quad u(x, y_e) = u_e,$$

where y_e denotes the outer edge of the boundary layer.

Effectively, the boundary layer changes (thickens and smoothes) the shape of the geometry. The resulting effective shape is called the displacement body $y = \delta^*$, which now becomes a streamline for the inviscid flow (*e.g.* Lighthill [4]).

The main advantage of the boundary-layer concept is that the elliptic character of the Navier–Stokes equations is changed into a much easier handled parabolic character. The latter was very relevant in an era when mainly analytical tools were available for solving differential equations. The stable direction of the boundary-layer equations (1) is governed by the sign of u. Hence this direction switches in reversed-flow regions, which has implications for the way they are solved, as we will see below.

2. Flow separation and solution breakdown

For situations with attached flow the boundary layer provides only a small perturbation to the inviscid-flow. However, it is found that as soon as the flow wants to separate from the body surface, the steady boundary-layer calculation breaks down with a solution that tends to become singular (*cf.* the conscientious discussion by Goldstein [5]). A number of possible causes can easily be imagined:

(i) The growth of the solution violates the assumption made in boundary-layer theory that streamwise derivatives should remain small. The remedy would be to include these streamwise derivatives in the equations of motion, but then the elliptic character of Navier–Stokes is retained with its corresponding much higher computational complexity.

[1] All variables have been made dimensionless with a characteristic length scale L and a characteristic velocity scale U. $\mathrm{Re} = UL/\nu$, where ν is the kinematic viscosity.

(ii) The stable parabolic direction of the boundary-layer equations changes locally in reversed-flow regions, with negative streamwise velocity. As a consequence, in these regions the equations should be solved from downstream to upstream, hence one single downstream-marching computational sweep does not suffice any longer.

Now, one has to keep in mind that in the first half of the century there were hardly any appropriate tools to solve the flow equations. At that time it was impossible to check the above two possibilities, and the issue had to remain open.

In 1948 Goldstein presented an in-depth discussion on the breakdown of the boundary-layer equations at separation in which he added some more possible causes [5]. Since then, the singularity at separation bears his name. In particular, on page 50 of his paper, Goldstein formulates the following suggestion:

> "Another possibility is that a singularity will always occur except for certain special pressure variations in the neighbourhood of separation, and that, experimentally, whatever we may do, the pressure variations near separation will always be such that no singularity will occur."

It took twenty more years before algorithms and computers were sufficiently developed to perform some numerical experiments in order to explore the options mentioned above. One of these experiments was described in 1966 by Catherall and Mangler [6], who tried to solve the steady boundary-layer equations with prescribed displacement thickness. Indeed, they succeeded to pass the point of flow separation, but ran into difficulties a bit further downstream. The reason hereof is clear, and in fact was already formulated by the authors [6, p. 178]: 'This is possibly to be expected, since the region of reversal flow should really be integrated in the negative ξ-direction with boundary conditions provided from downstream.' With current computer power, this problem is easily remedied by a downstream discretization of the convective terms and subsequent repeated sweeps through the boundary layer. Nevertheless, as Catherall and Mangler were not convinced of their success, they stopped further research into this subject. In fact, Catherall learned only some twenty years after publication about the large impact their paper had created, as he told me a few years ago.

3. The triple deck

In the late sixties, inspired by ideas put forward by Lighthill in 1953 [7], Stewartson (for steady subsonic and supersonic flow) [8, 9] and, independently, Messiter (subsonic) [10] and Neiland (supersonic) [11], developed asymptotic theories in the neighbourhood of singular points in the flow field, such as a trailing edge or a point of flow separation.

Therefore, let us consider a narrow region around such a singular point S, of extent

$$x - x_S = O(\mathrm{Re}^{-\alpha}),\quad 0 < \alpha < \tfrac{1}{2},\quad \text{with a scaled coordinate } x_\alpha = (x - x_S)\mathrm{Re}^{\alpha},$$

where x-derivatives will be more important than assumed thus far. The restriction $\alpha < \frac{1}{2}$ implies that the width is larger than the boundary-layer thickness; hence x-derivatives remain less important than y-derivatives, which simplifies the analysis. Further it may be anticipated that in vertical direction close to the singular point something happens: say at a y-scale given by $\mathrm{Re}^{-\beta}$ with $\beta > \frac{1}{2}$ (which is smaller than the boundary-layer thickness).

Of course, the oncoming boundary-layer thickness $y = O(\mathrm{Re}^{-1/2})$ will play a role as well. Here the velocity profile immediately before the singular point can be written as ($\tilde{y} = \mathrm{Re}^{1/2} y$)

$$u(x_s, y) = \beta'(\tilde{y}) + O(\mathrm{Re}^{-1/2}),\quad \text{where for } \tilde{y} \downarrow 0 : B(\tilde{y} \sim \tfrac{1}{2} a \tilde{y}^2. \tag{2}$$

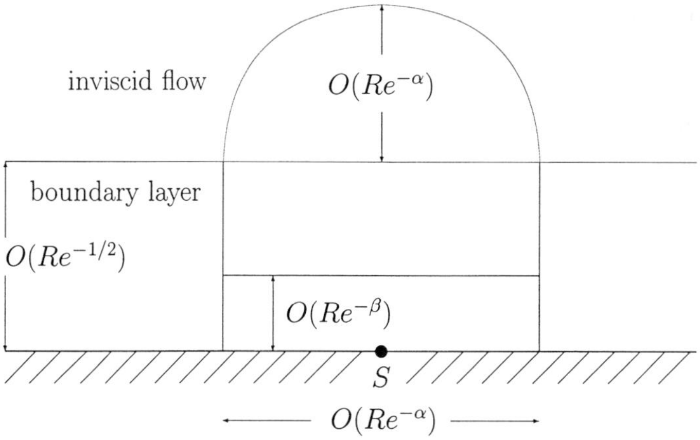

Figure 2. The triple deck describes the asymptotic flow structure near singular points.

The function $B'(\tilde{y})$ is known, *e.g.* it is a Blasius profile where $a = 0{\cdot}332$. Also the y-scale $y = O(\text{Re}^{-\alpha})$ will play a role, since there x-derivatives are as large as y-derivatives.

This three-layered structure, called the triple deck, is sketched in Figure 2. What is left to find are the particular values for α and β, the modelling pertinent to the individual flow domains, and the flow of information between the separate decks. The latter point turns out to be of crucial importance for the design of numerical solution methods.

To obtain the required insight into the triple-deck behaviour it is unavoidable to go into some detail, therefore we will next give a short resume of the derivation of the (incompressible) triple deck.

3.1. The lower deck $y = O(\text{Re}^{-\beta})$

In the lower deck the viscous terms balance with the convective terms. From (2) it follows that the oncoming velocity profile at small values of $\tilde{y}$ is given by $u \sim a\tilde{y}$. Hence for $y = O(\text{Re}^{-\beta})$ the horizontal velocity is of magnitude $u = O(\text{Re}^{1/2-\beta})$. An estimate of the convective and diffusive terms gives for $x = O(\text{Re}^{-\alpha})$:

$$\text{convection: } u\frac{\partial u}{\partial x} = O(\text{Re}^{1-2\beta+\alpha})\,; \qquad \text{diffusion: } \frac{1}{\text{Re}}\frac{\partial^2 u}{\partial y^2} = O(\text{Re}^{\beta-1/2}).$$

Balancing these terms yields a relation between α and β

$$\beta = \frac{\alpha}{3} + \frac{1}{2}. \tag{3}$$

With this value of β the horizontal velocity scales like $u = O(\text{Re}^{-\alpha/3})$, whereas the balancing pressure gradient corresponds with $p = O(\text{Re}^{-2\alpha/3})$.

After substitution of these estimates in the Navier–Stokes equations it follows that the flow in the lower deck is still governed by Prandtl's boundary-layer equations, with a pressure that is again constant in the vertical direction: $p(x, y) = \text{Re}^{-2\alpha/3}P(x_\alpha)$. High in the lower deck, for $y_\beta = \text{Re}^\beta y \to \infty$, we have

$$u(x_\alpha, y_\beta) \sim \text{Re}^{-\alpha/3}\{ay_\beta + aG(x_\alpha) + \cdots\}. \tag{4}$$

The function G is related to the displacement thickness, as will be clear in (6) below.

Next to the solid-wall conditions at $y_\beta = 0$, as a first boundary condition at infinity the coefficient of y_β (*i.e.* a) is given. Additionally, another boundary condition is required. In the classical interpretation this would be the prescription of the pressure, *i.e.* $P(x_\alpha)$, but in the spirit of Catherall and Mangler [6] this could also be a displacement effect, *i.e.* $G(x_\alpha)$.

It is stressed that in this way the lower-deck equations form one relation between P and G. A second relation can be found by matching with the other decks, as will be described next.

3.2. THE MIDDLE DECK $y = O(\mathrm{Re}^{-1/2})$

In order to concentrate on the essential properties of the triple deck, the derivation of the asymptotic expansions in the middle deck is only summarized (refer to the original papers by Stewartson [8] and Messiter [10], or to a later paper by Meyer [12]).

The middle deck is determined through matching with the oncoming flow (2) and the lower deck (3). It 'simply' shows a vertical shift of the oncoming velocity profile, caused by the displacement effect of the lower deck. The expansions in the middle deck are

$$u(x, y) \sim B'(\tilde{y}) + \mathrm{Re}^{-\alpha/3} B''(\tilde{y}) G(x_\alpha) + \cdots, \tag{5}$$

$$v(x, y) \sim -\mathrm{Re}^{2\alpha/3-1/2} B'(\tilde{y}) G'(x_\alpha) + \cdots. \tag{6}$$

Once again, the leading term in the pressure turns out to be constant in the y-direction: $p(x, y) \sim \mathrm{Re}^{-2\alpha/3} P(x_\alpha)$. This information is now passed on to the upper deck.

3.3. THE UPPER DECK $y = O(\mathrm{Re}^{-\alpha})$

In the upper deck the x- and y-dimensions are equal, whereas the viscous effects are not important. It is governed by inviscid flow where Laplace's equation and Bernoulli's law hold.

Since $B'(\tilde{y}) \to 1$ as $\tilde{y} \to \infty$, the vertical velocity (6) induces a vertical velocity $-\mathrm{Re}^{2\alpha/3-1/2}$ $G'(x_\alpha)$ in the upper deck. According to Laplace's, a horizontal velocity perturbation $\tilde{u}$ of the same order is to be expected; in fact it is given by

$$\tilde{u}(x_\alpha, y_\alpha) = -\frac{1}{\pi} \mathrm{Re}^{2\alpha/3-1/2} \int_{-\infty}^{\infty} \frac{G'(\xi)(x-\xi)}{(x_\alpha - \xi)^2 + y_\alpha^2} \, \mathrm{d}\xi. \tag{7}$$

When in Bernoulli's law $p + (u^2 + v^2)/2 = C$ one substitutes $u = 1 + \tilde{u}$ and $v = \tilde{v}$, with $\tilde{u} \ll 1$ and $\tilde{v} \ll 1$, then to first approximation

$$p + \tilde{u} = C - \tfrac{1}{2}. \tag{8}$$

This implies that also the pressure expansion contains a term of order $\mathrm{Re}^{2\alpha/3-1/2}$, which hence is related to displacement effects.

Matching of middle and upper deck now yields two kinds of pressure terms in an expansion that reads

$$p(x, y) = \mathrm{Re}^{-2\alpha/3} p^{(p)}(x_\alpha, y_\alpha) + \mathrm{Re}^{2\alpha/3-1/2} p^{(\delta)}(x_\alpha, y_\alpha) + \cdots. \tag{9}$$

The term $p^{(p)}$ matches the pressure in the middle deck, so that it satisfies $p^{(p)}(x_\alpha, 0) = P(x_\alpha)$; the term $p^{(\delta)}$ arises due to displacement effects and through (8) is related to the horizontal velocity perturbation given by (7).

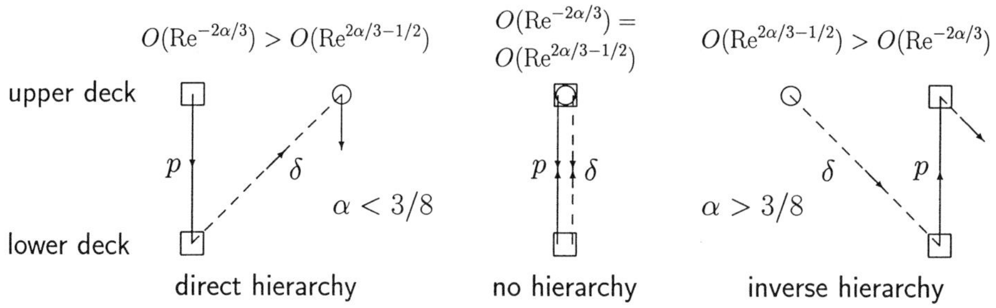

Figure 3. Hierarchy between pressure contributions in lower and upper deck at a streamwise lengthscale $x = O(\mathrm{Re}^{-\alpha})$.

3.4. Loss of hierarchy

Figure 3 shows the relative order of the two terms present in (9), and herewith it reveals the essential character of the triple deck as we will see.

When $\alpha < 3/8$ the term $p^{(p)}$ is the larger one in (9), and it determines, as usual, the pressure P in the boundary layer. The lower-deck equations then provide G after which the second term $p^{(\delta)}$ can be determined, which in turn provides a pressure correction in the boundary layer. The classical hierarchy between inviscid flow and boundary layer is recognized.

This situation changes for $\alpha = 3/8$ when both pressure terms in (9) are equally important, and this is what constitutes the essence of the triple deck. The pressure terms $p^{(p)}$ and $p^{(\delta)}$ have to be identical; thus by equating their values at $y_\alpha = 0$, from (7) and (8) a second relation between P and G is obtained:

$$P(x_\alpha) = \frac{1}{\pi}\int_{-\infty}^{\infty} \frac{G'(\xi)}{x_\alpha - \xi}\, \mathrm{d}\xi. \tag{10}$$

The triple-deck equations now consist of Prandtl's boundary-layer equations, with boundary condition (4) and a second relation between pressure and displacement given by a Cauchy–Hilbert integral (10).

Finally, for $\alpha > 3/8$ the displacement pressure $p^{(\delta)}$ is larger than $p^{(p)}$. Hence the hierarchy inverts, and the pressure is essentially determined in the lower deck.

Parallel to the above subsonic triple-deck, a supersonic version was developed by Stewartson and Williams [9]. Except for the description of the interaction, it is similar, to the inviscid flow. The *global* Cauchy-Hilbert integral (10) is replaced by a *local* Prandtl-Meyer relation

$$P(x_\alpha) = -G'(x_\alpha). \tag{11}$$

The triple deck has inspired a wealth of research on asymptotic descriptions; refer *e.g.* to the review papers by Stewartson [13] and Smith et al. [14–17] and the monograph by Sychev et al. [18].

3.5. Consequences for the numerical treatment

Because of its hyperbolic character the numerical treatment of the supersonic equations is relatively easy: in the absence of reversed flow a single marching sweep through the boundary layer suffices. A solution was presented already, together with their first formulation by

Stewartson and Williams [9], leaving the authors somewhat surprised that, even with reversed flow, their single marching sweep did not tend to become unstable.

The elliptic character of the subsonic triple-deck equations posed a larger challenge, however. It took half a decade before the first full (numerical) solutions were presented [19–21]. With hindsight, today the essence of the numerical obstinacy can 'easily' be understood. In the triple deck the boundary layer is no longer merely providing small corrections to the flow, but instead wants to have an equal say in determining the flow field. In aerodynamical terms, the hierarchy between boundary layer and inviscid flow changes from *weak* interaction into *strong* interaction. Lagerstrom [22, p. 209] in 1975 described the spirit of the triple deck as follows:

> "An important feature is that the pressure is self-induced, that is, the pressure due to displacement thickness is determined simultaneously with the revised boundary-layer solution. [...] this solution exhibits a definite loss of hierarchy."

This lack of hierarchy should also be visible in the numerical information exchange between boundary layer and inviscid flow, thus guiding their appropriate numerical iterative treatment; see Section 5.

Sixteen years earlier, Hayes and Probstein in their monograph on hypersonic flow [23, p. 365] came to a similar conclusion about the boundary-layer interaction near a point of flow separation:

> " ... in general it requires solving simultaneously the integrated momentum and energy equations and the inviscid flow relation describing the pressure along the curve $y = \delta^*(x)$. "

One should realise, of course, that here the inviscid flow is supersonic and the displacement effect of the boundary layer is described by the local Prandtl–Meyer fan (11). Nevertheless, the notion 'simultaneous' was already present!

Remark

The above interactive boundary-layer concept is restricted to mildly separated flows, *i.e.* flows where the thickness of the reversed-flow region is comparable to the boundary-layer thickness. For larger regions of reversed flow (marginal separation or, larger still, massive separation), the asymptotic structure has to be revised [24, 25]. At the same time, the flow in these larger separated-flow regions physically becomes unstable: an unsteady boundary-layer model has to be used, whose validity, in turn, terminates with a Van Dommelen-Shen singularity [26]. In a consistent way, numerical simulation methods based on the thin-boundary-layer concept tend to break down when the thickness of the separated-flow region becomes significant, *e.g.* [27], thus giving a warning that the selected flow modelling should be reconsidered. Readers might wish to consult Chapter 14 of the enlarged edition of Schlichting's 'Boundary-Layer Theory' [28] for a more detailed discussion of these asymptotic issues.

4. Non-asymptotic points of view

In the late seventies the quest for the cause of the singularity has also moved along non-asymptotic lines, which in retrospect can be related to the above. Several investigations into the boundary-layer relation between pressure and displacement thickness have been carried out, which all produced a similar outcome.

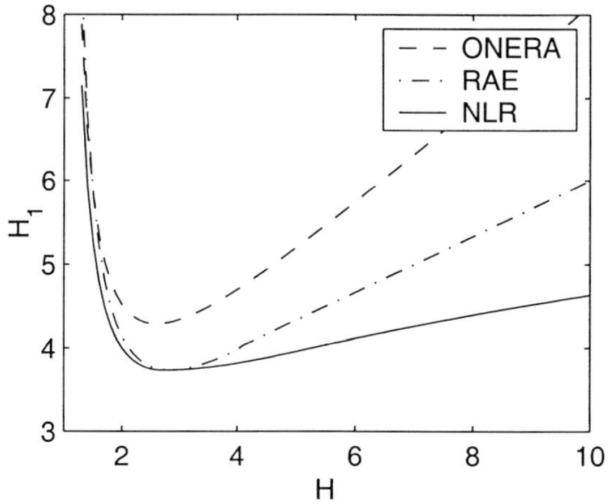

Figure 4. Some H-H_1 relationships as used at ONERA [29], RAE [30] and NLR [31] around 1980. The three relations agree on having a minimum somewhere near $H \approx 2.7$, corresponding with the onset of separation. For larger values of H the curves disagree, but experimental data to support these curves was rare at that time.

First, we will present the reflections of LeBalleur [29] at ONERA in France. He considered an integral formulation of the turbulent boundary-layer equations, consisting of Von Kármán's integral equation and Head's entrainment equation. In case u_e is prescribed, these differential equations are conveniently ordered as

$$\begin{aligned} \text{Von Kármán:} \quad & \frac{\mathrm{d}\theta}{\mathrm{d}x} &&= \tfrac{1}{2}c_f - \frac{\theta}{u_e}(2+H)\frac{\mathrm{d}u_e}{\mathrm{d}x}, \\ \text{Entrainment:} \quad & H_1\frac{\mathrm{d}\theta}{\mathrm{d}x} + \theta\frac{\mathrm{d}H_1}{\mathrm{d}x} &&= \frac{E}{u_e} - \frac{\theta H_1}{u_e}\frac{\mathrm{d}u_e}{\mathrm{d}x}. \end{aligned} \tag{12}$$

Here, θ is the momentum thickness, H the shape factor θ/δ^*, c_f the shear-stress coefficient, E Head's entrainment function and H_1 the entrainment shape factor (which is assumed to be a function of H only). The two differential equations are supplemented with three algebraic relations for determining H, E and c_f.

LeBalleur [29] demonstrated that the numerical problems at separation are caused by the algebraic relation between H and H_1. Since H_1 follows from the two differential equations (12), the algebraic relation should provide H. However, the graph of H_1 as a function of H shows a minimum at (or nearby) a point of flow separation. Figure 4 gives versions of this relation as used at ONERA [29], RAE [30] and NLR [31] around that time; supporting experimental data can be found in the review paper by Lock and Williams [32]. As a consequence, not for every value of H_1 is it possible to find a value for H! LeBalleur further showed that, when also u_e is considered an unknown, no difficulties arise (an extra equation has to be added that describes the coupling between inviscid flow and boundary layer).

As another example, a numerical experiment performed at the National Aerospace Laboratory (NLR) in Amsterdam will be described [33]. In this study the original boundary-layer equations (1), *i.e.* as a field method, were solved with a prescribed displacement thickness chosen such that flow separation occurred. Then at a fixed x-station δ^* was varied, keeping every other station fixed, and the variations of u_e and the shear-stress coefficient c_f were studied. It turned out that in this way u_e as a function of δ^* possessed a minimum that seemed

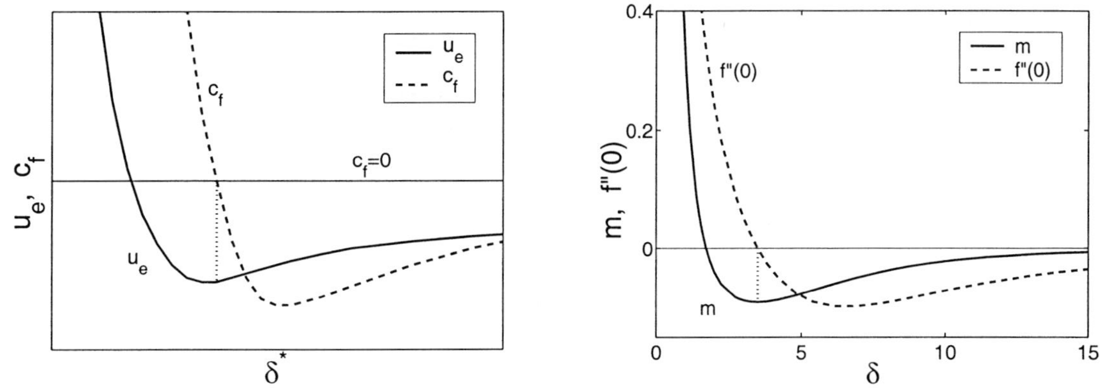

e

Figure 5. Behaviour of u_e and c_f as a function of δ^* at a fixed boundary-layer station (left), and Falkner–Skan relation between pressure parameter m, shear stress $f''(0)$ and displacement parameter δ (right). The left-hand graph is from boundary-layer calculations in 1977; the right-hand graph could have been drawn in 1954.

to correspond (within one or two grid cells) with the point where c_f vanishes, *i.e.* a point of flow separation; Figure 5 (left) gives the idea[2] very resemblant of Figure 4.

All similar studies [34, 35] suggested that the velocity distribution u_e cannot be prescribed arbitrarily near a point of flow separation. There is a certain range in u_e-values outside which no solution seems to exist, a situation correctly predicted by Goldstein some thirty years earlier. In terms of dynamical systems, passing the separation point with u_e prescribed amounts to crossing a saddle point, as explained by Kumar and Yajnik [36]. It would be interesting to study this issue from a theoretical point of view. Only little theory on existence and uniqueness of solutions of the boundary-layer equations exists [37, 38], but with the current numerical evidence it is known what to look for.

In retrospect, it is not difficult to recognize that already earlier similar types of graphs could have been presented, *e.g.* in relation with the family of Falkner–Skan similarity solutions of the boundary-layer equations [39]. This family is governed by the equation

$$f''' + ff'' + \frac{2m}{m+1}(1 - f'^2) = 0, \quad f(0) = f'(0) = 0, \quad f'(\infty) = 1,$$

where m is a parameter related to the pressure gradient through $m = x(\mathrm{d}u_e/\mathrm{d}x)/u_e$. In particular, the main branch of attached flow solutions only exists for $m > -0{\cdot}0904$, whereas for $-0{\cdot}0904 < m < 0$ also a separated flow branch exists; this branch was identified by Stewartson in 1954 [40]. Figure 5 (right) gives an unusual presentation of the Falkner–Skan results: the pressure parameter m and the shear variable $f''(0)$ are shown as a function of the displacement thickness δ (defined through $f(\eta) \sim \eta - \delta$ for $\eta \to \infty$); the resemblance with the much more recent graph in Figure 5 (left) is striking!

[2]The curves have been copied from my research notes of 15 December 1977, drawn in pencil on millimeter paper. The scaling of the axes was not indicated, but it is not relevant: the locations of the minimum in u_e and of the zero of c_f are all that matter.

5. Viscous-inviscid interaction methods

The above, mainly theoretical, considerations brought the insight to tackle engineering boundary-layer problems in an industrial context. The message is twofold: firstly, the boundary-layer approximation is sufficiently accurate to model the flow in mildly separated flow regions; secondly, and most importantly, the hierarchy between boundary layer and inviscid flow is lost. Although for turbulent flow a different asymptotic structure exists [41], the two messages from laminar-flow theory carry over, and in the discussion of Goldstein's singularity the distinction between laminar flow and turbulent flow is irrelevant.

Thus, interactive boundary-layer models were proposed, where Prandtl's boundary-layer equations were coupled with a relation like (10), describing the main interaction with the inviscid flow, or with an accurate inviscid-flow solver. Such a coupled problem can be written as, in principle, two equations with two unknowns:

$$\text{external inviscid flow:} \quad u_e = E[\delta^*], \tag{13}$$

$$\text{boundary-layer flow:} \quad u_e = B[\delta^*]. \tag{14}$$

Here E denotes the external inviscid-flow operator, whereas B is the boundary layer operator for prescribed displacement thickness; note that near flow separation the inverse B^{-1} does not exist.

In the classical, or 'direct', method u_e is computed from the inviscid-flow equation (13), whereas the displacement thickness is determined from the viscous flow (14), with a breakdown of B^{-1} in separation.

Inspired by the theoretical developments described above, in the second half of the seventies a number of ideas have been put forward to circumvent the breakdown singularity. The simplest way is to invert the direction of the iterative process in the classical method. One obtains the so-called 'inverse' method, where, following the idea proposed by Catherall and Mangler [6], the boundary layer is solved with prescribed displacement thickness. An early success was obtained by Carter [42] when he computed the separated flow past an indented plate, by now often used as a benchmark problem [43]. For engineering applications, however, the inverse method converges very slowly and it has not been used on a large scale. To speed up convergence, other methods were developed, of which two have survived [32]: the semi-inverse method of LeBalleur [29, 44] and Carter [45], and the quasi-simultaneous method [46, 47].

5.1. Semi-inverse

The semi-inverse method (Figure 6, left) introduced by LeBalleur in France [29, 44], and independently by Carter in the USA [45], is a mixture of the direct and the inverse method: it solves the boundary-layer equations with prescribed displacement thickness, and the inviscid flow in the traditional way (hence also with prescribed displacement thickness):

$$\begin{gathered} u_e^E = E[\delta^{*(n-1)}] \text{ (direct);} \quad u_e^B = B[\delta^{*(n-1)}] \text{ (inverse);} \\ \delta^{*(n)} = \delta^{*(n-1)} + \omega\left(u_e^B - u_e^E\right). \end{gathered} \tag{15}$$

In order to obtain convergence, some tuning of the relaxation parameter ω is required, and a fair convergence can be obtained.

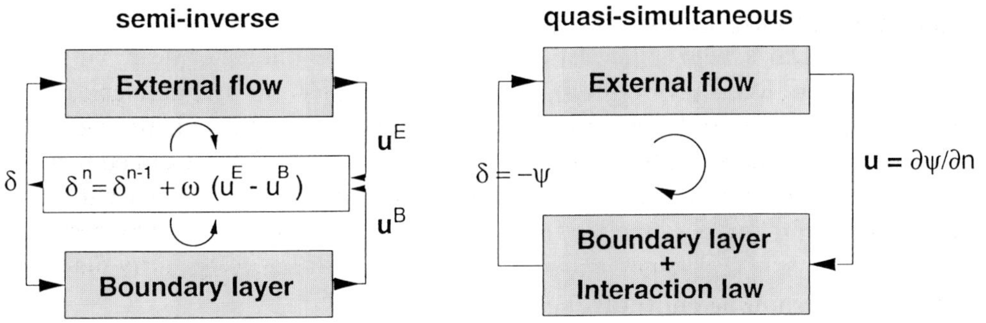

Figure 6. Semi-inverse and quasi-simultaneous VII method.

5.2. Quasi-simultaneous

The quasi-simultaneous method follows the suggestion made by Lagerstrom [22]. It wants to reflect the lack of hierarchy between both subdomains: in principle, it wants to solve both subdomain problems simultaneously. When the boundary layer is modelled by an integral formulation a simultaneous coupling is well feasible, *e.g.* [31, 48]; in principle, such an approach has my preference. However, when in both domains a field formulation is chosen, software complexity may prohibit a practical implementation. Recall that around 1980 mainframe computers possessed a memory of only 1(!) Mbyte. At that time this prevented a fully simultaneous approach, and the idea was born to solve the boundary-layer equations simultaneously with a *simple* but good *approximation* of the inviscid flow, which was termed the interaction law. The difference between this approximation and the 'exact' inviscid flow can then be handled iteratively. In this way, the quasi-simultaneous method (Figure 6, right) can be formulated as

$$\begin{aligned} u_e^{(n)} - I[\delta^{*(n)}] &= E[\delta^{*(n-1)}] - I[\delta^{*(n-1)}], \\ u_e^{(n)} - B[\delta^{*(n)}] &= 0, \end{aligned} \tag{16}$$

where the interaction law reads (compare (10))

$$I[\delta^*] = \frac{1}{\pi} \int_\Gamma \frac{d\delta^*}{d\xi} \frac{d\xi}{x - \xi} \tag{17}$$

(x is still the streamwise boundary-layer coordinate). It is observed in (16) that the interaction law is used in defect formulation, *i.e.* it does not influence the final converged result, but it only enhances the rate of convergence!

Later, when larger computers became available, in engineering applications this simple interaction law has been replaced by more sophisticated ones, for instance interaction laws based on a discrete Laplace description of the inviscid flow (the VILMA method of Arnold and Thiele [49]), or based on an inviscid panel method (*e.g.* Coenen [50]). In this way a better convergence of the iterations in (16) can be obtained. On the other hand, from a scientific point of view, it is interesting to find out how far the above interaction law can be simplified even further. In other words, how close to the traditional 'u_e = prescribed' can the boundary condition for the viscous flow equations be chosen, without being struck by Goldstein's singularity, while at the same time yielding acceptable convergence ot the iterations in (16)?

In three dimensions a similar approach is feasible, with two thin-airfoil expressions like (17) relating the two inviscid surface-velocity components to the shape of the displacement body, as demonstrated, *e.g.*, by Roget *et al.* [51] and Edwards [52]. The latter author was largely inspired by Davis, who at the east coast of the USA has initiated research in interactive boundary layers [53]. Related work in the UK has been carried out by Smith and co-workers [54]. At the USA west coast Cebeci has been an active advocate of quasi-simultaneous VII methods in engineering applications (airfoil analysis and design) [55]. For three-dimensional engineering flow simulations much pioneering work has been done by Cousteix and his colleagues in France [56].

6. Inclusion of streamline curvature

Especially for rear-loaded airfoils the streamlines immediately behind the trailing edge are highly curved. In such a situation the assumption of constant pressure across the boundary layer is no longer justifiable *a priori*. In asymptotic terms it means that higher-order effects, in regions possibly even smaller than the triple-deck, are becoming relevant.

The effect of streamline curvature can be modelled by a jump between the pressure in the boundary layer p_e and the pressure of the inviscid flow $E_p[\delta^*]$ [32]:

$$p_e = E_p[\delta^*] - [p] \quad \text{with} \quad [p] = \kappa \rho_e u_e^2(\delta^* + \theta), \tag{18}$$

where κ represents the streamline curvature. The extended version of (16) thus *could* become

$$\begin{aligned} p_e^{(n)} - I_p[\delta^{*(n)}] &= E_p[\delta^{*(n-1)}] - [p]^{(n-1)} - I_p[\delta^{*(n-1)}], \\ p_e^{(n)} - B_p[\delta^{*(n)}] &= 0. \end{aligned} \tag{19}$$

The change of notation from E to E_p will be clear; the influence of streamline curvature is more easy explained in terms of p_e than in terms of u_e.

The pressure jump (18) is proportional to κ, which can be taken as the curvature of the displacement body [32]. Thus, the pressure jump scales with the 'second derivative' of the streamlines and herewith it introduces a strong-interaction character, as will be explained next.

6.1. Asymptotic view

In asymptotic terms the strong-interaction character can be understood as follows. For a plate under angle of attack, *e.g.*, the inviscid trailing-edge streamline, and hence the displacement body, leaves the edge as $y \propto x^{3/2}$, $x \downarrow 0$; therefore its curvature behaves $\propto x^{-1/2}$. The pressure jump (18) then becomes $O(\text{Re}^{-1/2} x^{-1/2})$, and in the inviscid flow a vertical velocity component of the same magnitude is created. In turn, by integrating this vertical velocity in the x-direction, we will create a disturbance $y = O(\text{Re}^{-1/2} x^{1/2})$ in the inviscid position of the trailing-edge streamline. When $x = O(\text{Re}^0)$, for Re$\rightarrow \infty$ this term is smaller than our original starting point $y \propto x^{3/2}$; however when x is smaller than $O(\text{Re}^{-1})$ the disturbance is larger than the original. As in the above triple-deck derivation, a change of hierarchy occurs, indicating strong interaction and requiring a simultaneous treatment.

6.2. Numerical analysis

This reasoning again has a parallel in numerical terms. When the pressure jump is evaluated from a previous iteration as indicated by (19), through κ, in the discretisation of the second derivative of the displacement body it generates a contribution to the iterative amplification factor proportional to $(\delta x)^{-2}$, where δx is the mesh size. If we allow the mesh size to approach zero, this amplification blows up, as in the above asymptotic view when x tends to zero. This explains the numerical difficulties encountered in the literature when the curvature effect is not treated in a simultaneous way [32, p. 82]. Extensive smoothing and underrelaxation is then required to obtain convergence of the viscous-inviscid iterations, as stated by [57] and references therein.

Once again a simultaneous treatment of the curvature effect is required. The term $[p]^{(n-1)}$ should be shifted to the left-hand side in (19) and evaluated at the new iteration level (n). However, this necessary step to obtain convergence for small grid size turns out not to be sufficient. There is another (delicate) issue, which can be understood through a generalization of the analysis in [58]. Let us assume that the curvature is computed from the displacement body, *i.e.* $\kappa = \mathrm{d}^2\delta^*/\mathrm{d}x^2$. Then the pressure jump (18) becomes

$$[p] = \rho_e u_e^2(\delta^* + \theta)\frac{\mathrm{d}^2\delta^*}{\mathrm{d}x^2} \quad \text{to be abbreviated as} \quad [p] = c\frac{\mathrm{d}^2\delta^*}{\mathrm{d}x^2}.$$

The quasi-simultaneous system (19) now reads

$$\begin{aligned} p_e^{(n)} - I_p[\delta^{*(n)}] + c\frac{\mathrm{d}^2}{\mathrm{d}x^2}\delta^{*(n)} &= E_p[\delta^{*(n-1)}] - I_p[\delta^{*(n-1)}], \\ p_e^{(n)} - B_p[\delta^{*(n)}] &= 0. \end{aligned} \tag{20}$$

Since both I_p and $\mathrm{d}^2/\mathrm{d}x^2$ are negative-(semi)definite operators, and since $c > 0$, the two operators mentioned counteract each other in the left-hand side of (20)! Moreover, since the discrete version of I_p scales with $(\delta x)^{-1}$, whereas the curvature term scales with $(\delta x)^{-2}$, the unfavourable influence of the latter can be made arbitrarily large for vanishing grid size. The envisaged iterations will not converge, and there is only one way out of this problem: discretize the second derivative with a one-sided expression

$$\left.\frac{\mathrm{d}^2\delta^*}{\mathrm{d}x^2}\right|_i = \frac{\delta^*_{i-2} - 2\delta^*_{i-1} + \delta^*_i}{(\delta x)^2} \tag{21}$$

(the downstream-biased version will also work). At least in this way the eigenvalues of the discrete operator in (21) are positive, and hence of the same sign as those of $-I_p$. Numerical experiments with the usual central discretization of the second derivative and with the above upstream-biased one have consistently revealed that only the latter one could be made convergent (at least for not too violent flow cases, which probably is related to the onset of wake instability – more research is required here). An in-depth theoretical discussion of this issue will be presented in a forthcoming publication [59].

7. Application to transonic airfoil flow

The performance of the quasi-simultaneous coupling concept will be demonstrated on a typical calculation of transonic flow past an RAE 2822 airfoil with the NLR Vistrafs code. The boundary layer was modelled by Prandtl's equations with the algebraic Cebeci-Smith turbulence model; effects of streamline curvature were included. The inviscid flow was modelled by transonic full potential theory. As an interaction law, the integral (17) has been used to describe the symmetric displacement effects ('thickness problem'), together with its skew-symmetric

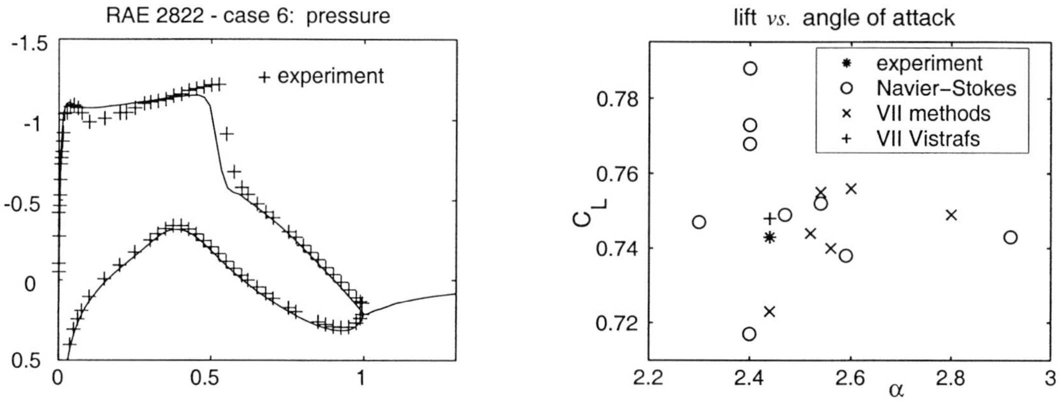

Figure 7. RAE 2822 airfoil at M=0·725, Re=6·5mill., α=2·44, C_L=0·743 (Case 6). Pressure distribution (left) and comparison of lift predictions by various VII and NS methods [61] (right).

counterpart to describe the effects of camber ('lift problem'); for details see Veldman et al. [59, 60].

The flow case presented in Figure 7 is mildly transonic, with a small amount of separated flow near the trailing edge. Pertinent flow parameters are $M = 0{\cdot}725$, Re $= 6{\cdot}5$ million, $\alpha_{\text{exp}} = 2{\cdot}92$ (with a corrected value of $\alpha = 2{\cdot}44$) and fixed transition at 3% chord.

The computational grid consists of 173 × 21 gridpoints (C-type, with 128 points along the airfoil surface) in the boundary layer, and the inviscid-flow grid was 128 × 64 (O-type). The computations require about 10 quasi-simultaneous iterations to converge to 3–4 digits, which on a modern PC takes less than one minute (it was so much different in the mid-eighties when this method was developed ...). The rate of convergence is governed by the difference between the exact inviscid flow E and its approximation I; as a consequence it is independent of grid size. To appreciate the fast convergence even better, one has to realize that the external flow in this example is transonic, with a significant supersonic flow region, whereas the interaction law (17) is based on sub(!)sonic theory.

In 1986 this flow case has been the subject of a workshop [61], where about twenty aerodynamic codes were compared. The lift coefficient predicted by these codes is presented in Figure 7 (right), where a distinction has been made between viscous-inviscid interaction (VII) codes and Navier–Stokes codes. A similar situation was found for the other flow cases investigated in the workshop. The participants were also requested to quantify the computational complexity of their codes: it was found that, depending on the inviscid flow model used, VII codes were one to two orders of magnitude cheaper than Navier–Stokes codes. The latter codes required 10^6–10^7 floating-point operations per grid point; a price tag that is still representative for today's Navier–Stokes codes, as can be inferred from the review data presented by Agarwal [62].

A conclusion of the workshop must be that, in spite of their much smaller complexity, the quality of the VII results is comparable to that of the Navier–Stokes results. In fact, the quality of flow simulations for this type of flow appears to be dominated by the quality of the turbulence model. Any difference between a full Navier–Stokes model and a simplified boundary-layer model just drowns in the uncertainty inherent in turbulence modelling. As stated by Holst [61]: "An engineering turbulence model that can approximately predict the size and extent of separated regions is desperately needed." Since 1986 the situation has not changed significantly. A European CFD validation project in 1992 once again revealed that the uncertainty due to turbulence modelling produces a large spreading of the Navier–Stokes results [63].

Of course, Navier–Stokes modelling is required in situations where the viscous region can no longer be regarded as thin, such as massive flow separation in take-off and landing configurations, or the flow near essentially three-dimensional objects. In this respect it is remarkable that maximum-lift prediction with a VII method appears to be feasible, as demonstrated by Cebeci and co-workers [55, 64].

8. Concluding remarks

Prandtl's 1904 boundary-layer theory formed the starting point for the viscous-inviscid interaction methods that have been developed in the last two decades of the 20th century. They have become very popular, since in comparison with brute-force Navier–Stokes solutions they are about two orders less expensive, whereas for flow conditions with thin shear layers the results are equally useful. Because of their modest computational complexity, they are ideal for use in aerodynamic-airfoil and wing-optimization studies, *e.g.* [65, 66, 67, 68].

The greatest challenge has been to understand and resolve the singularity at separation, which occurs when the boundary-layer equations are solved with prescribed pressure. In 1948, Goldstein already foresaw the possibility that near separation in general no solution does exist, unless the pressure satisfies certain properties. Triple-deck theory provided the insight behind these difficulties, and gave the clue towards their solution. Goldstein turns out to have pointed in the right direction of non-existence; doubts on the validity of the boundary-layer model were found not to be essential here.

Stewartson and his 1969 contemporates have provided the asymptotic framework valid near separation: the triple deck. In 1975 Lagerstrom described his view on the triple deck, and today we know that his paper contains the essential message required to overcome the singularity at a point of flow separation: boundary-layer and inviscid flow have to be solved simultaneously. It is through this type of insight that the use of viscous-inviscid interaction methods in engineering applications can flourish.

Only a small, strongly personally biased, glimpse of the world-wide struggle between Prandtl's boundary-layer concept and the numerical simulation of separated flow could be shown in the paper: we have emphasised steady incompressible flow, and scratched only superficially supersonic inviscid flow and unsteady flow separation. Many instances can be found in the literature which, in retrospect, were close to unraveling the correct view, but lack of computational power preventing further pursuit. It would be interesting to analyse all these 'close encounters', and I hope to find another occasion to dig deeper into this intrigueing 20th-century story.

References

1. L. Prandtl, Ueber Fluessigkeitsbewegung mit kleiner Reibung. In: *Verhandlungen des dritten internationalen Mathematischen Kongresses*. Heidelberg. Leipzig: Teubner Verlag (1905) pp. 484–491.
2. K. Stewartson, D'Alembert's paradox. *SIAM Review* 23 (1981) 308–343.
3. M. Van Dyke, Nineteenth-century roots of the boundary-layer idea. *SIAM Review* 36 (1994) 415–424.
4. M.J. Lighthill, On displacement thickness. *J. Fluid Mech.* 4 (1958) 383–392.
5. S. Goldstein, On laminar boundary layer flow near a point of separation. *Quart. J. Mech. Appl. Math.* 1 (1948) 43–69.

6. D. Catherall and K.W. Mangler, The integration of the two-dimensional laminar boundary-layer equations past the point of vanishing skin friction. *J. Fluid Mech.* 26 (1966) 163–182.
7. M.J. Lighthill, On boundary layers and upstream influence I. A comparison between subsonic and supersonic flows. *Proc. R. Soc. London* A 217 (1953) 478–507.
8. K. Stewartson, On the flow near the trailing edge of a flat plate II. *Mathematika* 16 (1969) 106–121.
9. K. Stewartson and P.G. Williams, Self-induced separation. *Proc. R. Soc. London* A 312 (1969) 181–206.
10. A.F. Messiter, Boundary-layer flow near the trailing edge of a flat plate. *SIAM J. Appl. Math.* 18 (1970) 241–257.
11. V.Ya. Neiland, Towards a theory of separation of the laminar boundary layer in a supersonic stream. *Mekh. Zhid. Gaza,* 4 (1969) 53–57 (English translation in *Fluid Dynamics* 4 (1970) 33-55).
12. R.E. Meyer, A view of the triple deck. *SIAM J. Appl. Math.* 43 (1983) 639–663.
13. K. Stewartson, Multi-structured boundary layers on flat plates and related bodies. *Adv. in Applied Mech.* 14 (1974) 145–239.
14. F.T. Smith, On the high-Reynolds number theory of laminar flows. *IMA J. Appl. Math.* 28 (1982) 207–281.
15. F.T. Smith, Steady and unsteady boundary-layer separation. *Ann. Rev. Fluid Mech.* 18 (1986) 197-220.
16. F.T. Smith and J.H. Merkin, Triple-deck solutions for subsonic flow past humps, steps, concave or convex corners and wedged trailing edges. *Comp. Fluids* 10 (1992) 7–25.
17. A.P. Rothmayer and F.T. Smith, Numerical solution of two-dimensional steady triple-deck problems. In: R.W. Johnson (ed.) *The Handbook of Fluid Dynamics*. Boca Raton: CRC Press (1998) Chapter 25.
18. V.V. Sychev, A.I. Ruban, V.V. Sychev and G.L. Korolev, *Asymptotic Theory of Separated Flows*. Cambridge: Cambridge University Press (1998) 352 pp.
19. C.E. Jobe and O.R. Burggraf, The numerical solution of the asymptotic equations of trailing edge flow. *Proc. R. Soc. London* A 340 (1974) 91–111.
20. R.E. Melnik and R. Chow, *Asymptotic Theory of Two-dimensional Trailing Edge Flows*. Grumman Research Department, Report RJ-510J (1975).
21. A.E.P. Veldman and A.I. van de Vooren, Drag of a finite flat plate. In: R.D. Richtmayer (ed.) Proc. 4th Int. Conf. Num. Meth. Fluid Dyn., *Lecture Notes in Physics* 35. Berlin: Springer Verlag (1976) pp. 422–430.
22. P.A. Lagerstrom, Solutions of the Navier–Stokes equation at large Reynolds number. *SIAM J. Appl. Math.* 28 (1975) 202–214.
23. W.D. Hayes and R.F. Probstein, *Hypersonic Flow Theory.* Academic Press (1959) 464 pp.
24. K. Stewartson, F.T. Smith and K. Kaups, Marginal separation. *Stud. Appl. Math.* 67 (1982) 45–61.
25. J.W. Elliott and F.T. Smith, Dynamic stall due to unsteady marginal separation. *J. Fluid Mech.* 179 (1987) 489–512.
26. L.L. van Dommelen and S.F. Shen, The spontaneous generation of the singularity in a separating laminar boundary layer. *J. Comput. Phys.* 38 (1980) 125–140.
27. R.A.W.M. Henkes and A.E.P. Veldman, On the breakdown of the steady and unsteady interacting boundary-layer description. *J. Fluid Mech.* 179 (1987) 513-529.
28. H. Schlichting and K. Gersten, *Boundary-Layer Theory*, 8th revised and enlarged edition. Berlin: Springer (2000) 799 pp.
29. J.C. LeBalleur, Couplage visqueux-non visqueux: analyse du problème incluant décollements et ondes de choc. *La Recherche Aérospatiale* 1977-6 (1977) 349–358.
30. R.C. Lock, Prediction of the drag of wings at subsonic speeds by viscous-inviscid interaction techniques. In: AGARD-R-723 (1985) paper 10.
31. R. Houwink and A.E.P. Veldman, Steady and unsteady separated flow computation for transonic airfoils. AIAA paper 84–1618 (1984).
32. R.C. Lock and B.R. Williams, Viscous-inviscid interactions in external aerodynamics. *Prog. Aerospace Science* 24 (1987) 51–171.
33. A.E.P. Veldman, Boundary layers with strong interaction: from asymptotic theory to calculation method. In: J.J.H. Miller (ed.), *Proc. BAIL 1 Conf. on Boundary and Interior layers*. Dublin: Boole Press (1980) pp. 149–163.
34. P. Ardonceau, Th. Alziary and D. Aymer, Calcul de l'interaction onde de choc/couche limite aves décollement. In: *Computation of Viscous-Inviscid Interactions*, AGARD-CP-291 (1980) paper 28.
35. H.P. Horton, Numerical investigation of regular laminar boundary layer separation. In: *Flow Separation*, AGARD-CP-168 (1975) paper 7.

36. A. Kumar and K.S. Yajnik, Internal separated flows at large Reynolds number. *J. Fluid Mech.* 97 (1980) 27–51.
37. K. Nickel, Prandtl's boundary-layer theory from the viewpoint of a mathematician. *Ann. Rev. Fluid Mech.* 5 (1973) 405–428.
38. W. Walter, Existence and convergence theorems for the boundary layer equations based on the line method. *Arch. Rat. Mech. Anal.* 39 (1970) 169–188.
39. V.M. Falkner and S.W. Skan, Solutions of the boundary layer equation. *Phil. Mag.* 12 (1931) 865–896.
40. K. Stewartson, Further solutions of the Falkner-Skan equation. *Proc. Cambridge Phil. Soc.* 50 (1954) 454-465.
41. R.E. Melnik, R. Chow and H.R. Mead, Theory of viscous transonic flow over airfoils at high Reynolds number. AIAA paper 77-680 (1977).
42. J.E. Carter and S.F. Wornom, Solutions for incompressible separatedd boundary layers including viscous-inviscid interaction. In: *Aerodynamic Analysis Requiring Advanced Computers.* NASA SP-347 (1975) pp. 125–150.
43. T. Cebeci, H.H. Chen and J.A. Majeski, Indented plate problem revisited. *Int. J. Num. Meth. Fluids* 16 (1993) 391–401.
44. J.C. LeBalleur, Couplage visqueux-non visqueux: méthode numérique et applications aux écoulements bidimensionnels transsoniques et supersoniques. *La Recherche Aérospatiale* 183 (1978) 65–76.
45. J.E. Carter, Viscous-inviscid interaction analysis of turbulent separated flow. AIAA paper 81–1241 (1981).
46. A.E.P. Veldman, A numerical method for the calculation of laminar incompressible bounday layers with strong viscous-inviscid interaction. National Aerospace Laboratory, Report NLR TR 79023 (1979).
47. A.E.P. Veldman, New, quasi-simultaneous method to calculate interacting boundary layers. *AIAA J.* 19 (1981) 79–85.
48. M. Drela and M.B. Giles, Viscous-inviscid analysis of transonic and low Reynolds number airfoils. *AIAA J.* 25 (1987) 1347–1355.
49. F. Arnold and F. Thiele, Laplace interaction law for the computation of viscous airfoil flow in low- and high-speed aerodynamics. *AIAA J.* 31 (1994) 2178–2185.
50. E.G.M. Coenen, Quasi-simultaneous coupling for wing and airfoil flow. In: C.-H. Lai, P.E. Bjorstad, M. Cross and O.B. Widlund (eds.), *Domain Decomposition Methods in Science and Engineering.* Bergen: Domain Decomposition Press (1999) pp. 197–205.
51. C. Roget, J.Ph. Brazier, J. Cousteix and J. Mauss, A contribution to the physical analysis of separated flows past three-dimensional humps. *Eur. J. Mech. B* 17 (1998) 307–329.
52. D.E. Edwards, Analysis of three-dimensional separated flow using interacting boundary-layer theory. In: F.T. Smith and S.N. Brown (eds.), *Boundary-Layer Separation.* Berlin: Springer Verlag (1987) pp. 163–178.
53. M.J. Werle, R. Thomas Davis — His contributions to numerical simulation of viscous flows: Part II — Technical perspective. AIAA paper 88–0602 (1988).
54. F.T. Smith, Steady and unsteady 3-d interactive boundary layers. *Comp. Fluids* 20 (1991) 243–268.
55. K.C. Chang, N. Alemdaroglu, U, Mehta and T. Cebeci, Further comparisons of interactive boundary-layer and thin-layer Navier-Stokes procedures. *J. Aircraft* 25 (1988) 897–903.
56. J. Cousteix, Three-dimensional and unsteady boundary-layer computations. *Ann. Rev. Fluid Mech.* 18 (1986) 173–196.
57. G. Dargel and P. Thiede, Viscous transonic airfoil flow simulation by an efficient viscous-inviscid interaction method. AIAA paper 87–0412 (1987).
58. A.E.P. Veldman, A numerical view on strong viscous-inviscid interaction. In: W.G. Habashi (ed.), *Computational Methods in Viscous Flows.* Southampton: Pineridge Press (1984) pp. 343–363.
59. A.E.P. Veldman and M.A.M. Somers, The inclusion of streamline curvature in a quasi-simultaneous viscous-inviscid interaction method for transonic airfoil flow. Preprint available at `http://www.math.rug.nl/~veldman/preprints.html` (1999).
60. A.E.P. Veldman, J.P.F. Lindhout, E. de Boer and M.A.M. Somers, VISTRAFS: a simulation method for strongly-interacting viscous transonic flow. In: T. Cebeci (ed.), *Numerical and Physical Aspects of Aerodynamic Flow IV.* Berlin: Springer Verlag (1990) pp. 37–51.
61. T.L. Holst, Viscous Transonic Airfoil Workshop - Compendium of Results. AIAA paper 87–1460 (1987).
62. R. Agarwal, Computational fluid dynamics of whole-body aircraft. *Ann. Rev. Fluid Mech.* 31 (1999) 125–169.

63. W. Haase, F. Brandsma, E. Elsholz, M. Leschziner and D. Schwamborn (eds.), *EUROVAL — An European Initiative on Validation of CFD Codes*. Notes on Numerical Fluid Mechanics 42. Braunschweig: Vieweg (1993) 530 pp.
64. T. Cebeci, H. Hefazi, F. Roknaldin and L.W. Carr, Predicting stall and post-stall behaviour of airfoils at low Mach numbers. *AIAA J.* 33 (1995) 595–602.
65. A.J. van der Wees and J. van Muijden, A fast and robust viscous-inviscid interaction solver for transonic flow about wing/body configurations on the basis of full potential theory. AIAA paper 93–3026 (1993).
66. P.D. Smith, A viscous package for attached and separated flows on swept and tapered wings. RAE Technical Report 89027 (1989).
67. M. Drela and M.B. Giles, ISES: a two-dimensional viscous aerodynamic design and analysis code. AIAA paper 87–1118 (1987).
68. A. Verhoff, H.H. Chen, T. Cebeci and T. Michal, An accurate and efficient interactive boundary-layer method for analysis and design of airfoils. AIAA paper 96–0328 (1996).

Journal of Engineering Mathematics **39:** 207–220, 2001.

Stokes flow around an asymmetric channel divider; a computational approach using MATLAB

JOSEPH D. FEHRIBACH and ANTHONY M. J. DAVIS[1]
Mathematical Sciences Department, Worcester Polytechnic Institute, USA and Material Sciences Department, T.U. Delft, Netherlands [1]*Mathematics Department, University of Alabama, U.S.A.*

Received 18 January 2000; accepted in revised form 9 June 2000

Abstract. While computational methods for solving Stokes-flow problems have existed for some time, these have depended on specialized codes developed specifically, for this type of problem. This work shows how to combine traditional applied mathematics and a modern over-the-counter software package MATLAB to solve and study Stokes flow in a channel with a splitter plate. Specifically exact unidirectional flow solutions are used as a basis for choosing boundary conditions for MATLAB to anticipate the boundary conditions of a Stokes flow. A method for selecting zeroth and first-order approximate boundary conditions is presented, along with a suggestion for finding a second-order approximation. It is also shown that small errors made in choosing the approximate boundary conditions do not grow as one moves away from the boundary into the interior of the flow. Finally several computational examples using this approach are presented.

Key words: Stokes flow, biharmonic equation, eigenfunction expansions, numerical computations.

1. Introduction

The goal of this work is not so much to solve a problem that has never been considered before, but rather to show how traditional applied mathematics and modern over-the-counter software packages (in this case, MATLAB) can be combined conveniently to solve common Stokes-flow problems. Special packages for solving Stokes-flow problems have been developed over the years by, among others, Camp and Gipson [1] and Roache [2, Chapter 3]. These are highly tuned, and may well be faster and/or more efficient, but they are not as widely available as the general-purpose packages. It is also possible to use finite-difference methods to solve these sorts of problems (*cf., e.g.* [3, Section 3.10]), but this in general requires users to write their own code and to have access to some sort of graphics package.

When the creeping (Stokes) flow equations are solved in two dimensions or axisymmetric three dimensions by methods that involve in some form the separation of variables, the application of no-slip conditions at rigid boundaries, on which a single coordinate takes distinct values, introduces eigenfunctions that are damped oscillatory in both coordinates and not orthogonal in the usual sense. For example, if the biharmonic stream function $\psi(x, y)$ is such that

$$\psi(x, 0) = \psi_0, \quad \psi(x, 1) = H(x), \quad \frac{\partial\psi}{\partial y}(x, 0) = 0 = \frac{\partial\psi}{\partial y}(x, 1),$$

where $H(x)$ denotes the Heaviside function, then, by Fourier transform methods,

$$\psi(x, y) = \psi_0(1 - 3y^2 + 2y^3) + (3y^2 - 2y^3)H(x) + \\ + \mathrm{sgn}(x)\Re\mathrm{e} \sum_{n=1}^{\infty} \frac{\alpha_n \sin^2 \alpha_n}{\sin 2\alpha_n - 2\alpha_n} \Psi_n(y) \mathrm{e}^{-\alpha_n |x|}. \tag{1}$$

Here the (Papkovitch-Fadle) eigenfunctions in the strip bounded by $y = 0$ and $y = 1$, where $\psi = 0 = \psi'$, are given [4] by

$$\mathrm{e}^{\pm\alpha_n x}\Psi_n(y) = \mathrm{e}^{\pm\alpha_n x}\left[\frac{\sin \alpha_n y - \alpha_n y \cos \alpha_n y}{\sin \alpha_n - \alpha_n \cos \alpha_n} - \frac{y \sin \alpha_n y}{\sin \alpha_n}\right], \quad (n \geq 1) \tag{2}$$

where $\alpha_n^2 = \sin^2 \alpha_n$ and $\Re\mathrm{e}(\alpha_n) > 0$. Odd values of n yield the zeros $\{\lambda_n; n \geq 1\}$ of $\sin \lambda + \lambda$ and even values of n yield the zeros $\{\mu_n; n \geq 1\}$ of $\sin \mu - \mu$, both arranged in order of increasing real part in the first quadrant. Thus $\lambda_1 = 4{\cdot}21 + 2{\cdot}25\mathrm{i}$, $\lambda_2 = 10{\cdot}71 + 3{\cdot}10\mathrm{i}$, etc. and $\mu_1 = 7{\cdot}50 + 2{\cdot}77\mathrm{i}$, $\mu_2 = 13{\cdot}90 + 3{\cdot}35\mathrm{i}$, etc. This solution (1) displays the exact details of the symmetric transition from the source at (0, 1) to the anticipated far field flows. However, if instead the above conditions are restricted to $x > 0$ and end conditions on ψ are applied at $x = 0, 0 < y < 1$, then the structure of ψ is unchanged, *i.e.*,

$$\psi(x, y) = \psi_0(1 - 3y^2 + 2y^3) + (3y^2 - 2y^3) + \Re\mathrm{e} \sum_{n=1}^{\infty} c_n \Psi_n(y) \mathrm{e}^{-\alpha_n |x|},$$

but the determination of the complex-valued coefficients may require more than Fourier-transform methods. The first extensive discussion of such series, in elasticity, was given by Smith [5] and several authors have subsequently presented progressively less restrictive conditions for completeness. Biorthogonality relations are available but are not directly applicable to the common cases in which either the end-velocity components or the end stresses are prescribed. This means that the coefficients are determined by an infinite set of linear equations and Spence [6] demonstrated an 'optimal weighting' scheme that ensures a convergent truncation. A worse scenario occurs if the infinite strip (channel) has a fixed semi-infinite barrier or a sudden change of width. Then the form of the stream function in each semi-infinite region can be written down and uniqueness ensured by requiring bounded velocities. The Spence scheme can be used to match the velocity and stress components but the edge or corner singularities cause the convergence to be slow, as discussed by Phillips [7]. Trogdon and Joseph [8], in their discussion of flow over a slot, avoided this difficulty by including both of the available eigenfunction expansions in their intermediate region. Similarly, Meleshko [9] has found that, for rectangular regions with aspect ratio of order one, a pair of Fourier series suffices for computational purposes.

In the sections below, we attack the barrier problem with a different approach which takes advantage of modern computational software. The next section describes the flow problem at hand in general, giving closed-form solutions for various unidirectional-flow conditions. The final two sections then discuss a computational approach to this problem based on the results from Section 2. The third section describes our computational approach in general, including a proof that errors made in selecting the boundary conditions in our approach do not grow in the interior of the flow. The fourth section presents several specific examples.

2. The flow problem

Consider the unidirectional flow between rigid walls at $y = -1, h(\geq 1)$, where (x, y) are Cartesian coordinates, at which the stream function ψ and velocity $u(= \partial\psi/\partial y)$ have the

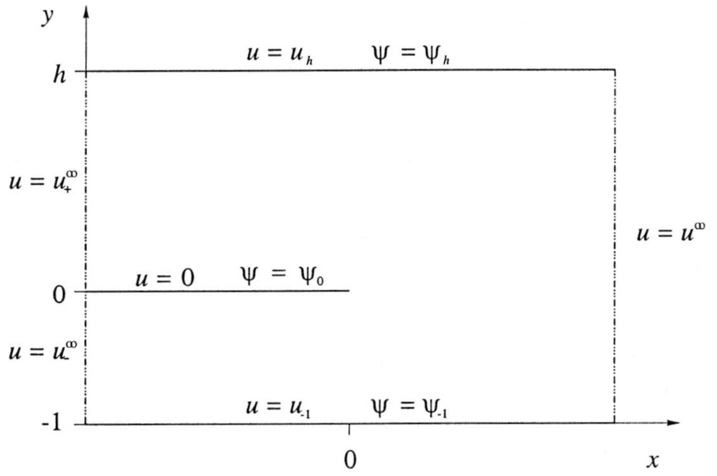

Figure 1. Schematic of the flow-problem domain. The horizontal boundary conditions for ψ and $u = \partial\psi/\partial y$ at $y = -1$ and $y = h \geq 1$ and on the splitter plate at $y = 0$ are shown in the diagram. The dashed vertical boundaries and the associated boundary conditions are taken to be far enough from $x = 0$ so as not to affect the flow. The uniform flow fields u^∞ (upstream) and u_-^∞ and u_+^∞ (downstream) are maintained far from $x = 0$.

prescribed values ψ_{-1}, ψ_h and u_{-1}, u_h, respectively (*cf.* Figure 1). The flow $u^\infty(y)\hat{\mathbf{x}}$, where $\hat{\mathbf{x}}$ is the unit vector in the x direction, is given by

$$\begin{aligned} u^\infty(y) = {} & (u_h - u_{-1})\left(\frac{y+1}{h+1} - \frac{1}{2}\right) + \frac{1}{2}(u_h + u_{-1}) - \\ & - \left[\frac{\psi_{-1} - \psi_h}{h+1} + \frac{1}{2}(u_h + u_{-1})\right]\frac{6(h-y)(y+1)}{(h+1)^2}. \end{aligned} \tag{3}$$

Here the three terms may be, respectively, identified as a shear flow with zero flux due to the different wall speeds, a uniform flow due to the average wall speed and a pressure driven flow due to the prescribed flux being different from the flux generated by the uniform flow. By rewriting (3) in the form

$$\begin{aligned} u^\infty(y) = {} & (u_h - u_{-1})\frac{y}{h+1} + \frac{u_h + hu_{-1}}{h+1} - \\ & - \left[\frac{\psi_{-1}^* - \psi_h^*}{h+1} + \frac{u_h + hu_{-1}}{h+1}\right]\frac{6(h-y)(y+1)}{(h+1)^2}, \end{aligned} \tag{4}$$

where

$$\psi_{-1}^* = \psi_{-1} - \frac{u_h - u_{-1}}{2(h+1)}, \quad \psi_h^* = \psi_h - h^2\frac{u_h - u_{-1}}{2(h+1)},$$

one observes that equal wall velocities suffice for a study of the disturbance flow generated by the introduction of a fixed plate at $y = 0$, $x < 0$. With $u_h = u_{-1} = -U$, this occurs when the flow speed at $y = 0$, namely

$$u^\infty(0) = -U - \left[\frac{\psi_{-1} - \psi_h}{h+1} - U\right]\frac{6h}{(h+1)^2}, \tag{5}$$

is non-zero and or the corresponding stream function $\psi^{\infty}(y)$, given by

$$\psi^{\infty}(y) = -U(y+1) + \psi_{-1} - \left[\frac{\psi_{-1}-\psi_h}{h+1} - U\right]\frac{(3h+1-2y)(y+1)^2}{(h+1)^2}, \tag{6}$$

is such that $\psi^{\infty}(0)$ differs from ψ_0, the stream function value on the plate. So, the second forcing term is, from (6),

$$\psi^{\infty}(0) - \psi_0 = -U + \psi_{-1} - \psi_0 - \left[\frac{\psi_{-1}-\psi_h}{h+1} - U\right]\frac{(3h+1)}{(h+1)^2}, \tag{7}$$

and evidently the sets of values of the flux and wall velocities in (3), for which the presence of the semi-infinite barrier creates a disturbance flow, form a two parameter family described by non-zero values of the vector $[u^{\infty}(0), \psi^{\infty}(0) - \psi_0]$, with only its direction being significant. Thus, any two flows of type (3) that yield parallel values of this vector, determined by Equations (5), (7), may be regarded as equivalent because their suitably weighted difference must be a unidirectional flow with zero velocity at $y = 0$.

For example, the two flows determined by

$$\psi_0 = \psi_{-1}, \quad \psi_{-1} - \psi_h = U(h+1) \tag{8}$$

and

$$\psi_0 - \psi_{-1} = (\psi_{-1} - \psi_h)\frac{3h-1}{(h+1)^3}, \quad U = 0, \tag{9}$$

both yield values of $[u^{\infty}(0), \psi^{\infty}(0) - \psi_0]$ that are parallel to $[1, 1]$, because their suitably weighted difference is the flow $u = Uy(y+1-h)/h$, which is unaffected by the introduction of the splitter plate.

In terms of pressure gradients G, G_-, G_+, the upstream ($x \to \infty$) and downstream ($x \to -\infty$) unidirectional velocity profiles (*cf.* Figure 1) are given by

$$u^{\infty}(y) = -U - \frac{G}{2\mu}(h-y)(y+1), \tag{10}$$

$$u_-^{\infty}(y) = Uy + \frac{G_-}{2\mu}y(y+1), \quad -1 < y < 0, \tag{11}$$

$$u_+^{\infty}(y) = -U\frac{y}{h} - \frac{G_+}{2\mu}y(h-y), \quad 0 < y < h, \tag{12}$$

where

$$G = \left[\frac{\psi_{-1}-\psi_h}{h+1} - U\right]\frac{12\mu}{(h+1)^2}. \tag{13}$$

$$G_- = 12\mu\left[\psi_{-1} - \psi_0 - \tfrac{1}{2}U\right], \tag{14}$$

$$G_+ = \frac{12\mu}{h^2}\left[\frac{\psi_0 - \psi_h}{h} - \tfrac{1}{2}U\right]. \tag{15}$$

Thus,

$$G(h+1)^3 = G_+h^3 + G_- + 6\mu(h+1)U, \tag{16}$$

since the pressure gradients are restricted by the need to achieve flux equality.

In the case of symmetric geometry, $h = 1$ and Equations (5), (7), (11), (12) reduce to

$$u^\infty(0) = \tfrac{1}{2}U + \tfrac{3}{4}(\psi_1 - \psi_{-1}), \tag{17}$$

$$\psi^\infty(0) - \psi_0 = \tfrac{1}{2}(\psi_1 - 2\psi_0 + \psi_{-1}), \tag{18}$$

$$u_-^\infty(y) = Uy + \left[\psi_{-1} - \psi_0 - \tfrac{1}{2}U\right]6y(y+1), \quad -1 < y < 0, \tag{19}$$

$$u_+^\infty(y) = -Uy + \left[\psi_1 - \psi_0 + \tfrac{1}{2}U\right]6y(1-y), \quad 0 < y < 1. \tag{20}$$

Evidently flows with $\psi^\infty(0) = \psi_0$, $u^\infty(0) \neq 0$ are equivalent to the even case:

$$\psi_0 = \tfrac{1}{2}(\psi_1 + \psi_{-1}), \quad u_-^\infty(-y) = u_+^\infty(y), \quad 0 < y < 1, \tag{21}$$

which consists downstream of a shear flow and parallel pressure driven flow, while flows with $\psi^\infty(0) \neq \psi_0$, $u^\infty(0) = 0$ are equivalent to the odd case:

$$U = 0, \quad \psi_1 = \psi_{-1}, \quad u_-^\infty(-y) = -u_+^\infty(y), \quad 0 < y < 1, \tag{22}$$

which is a pressure driven flow out of one channel into the other.

The possible flows that need be considered may be regarded as due solely to the moving walls and various combinations of downstream pressure gradients. The *two cases* are therefore:

(1) $U \neq 0$ and $G_+ = 0 = G_-$.
$[u^\infty(0), \psi^\infty(0) - \psi_0]$ is parallel to $[2(h^2 - h + 1), h(h-1)]$ so its direction depends on h only.

(2) $U = 0$ and various flux ratios.

$$\frac{\psi^\infty(0) - \psi_0}{u^\infty(0)} = \frac{1}{6h}\left[3h + 1 - (h+1)^3\frac{\psi_{-1} - \psi_0}{\psi_{-1} - \psi_h}\right] = \frac{3h + 1 - G_-/G}{6h},$$

which displays a two-parameter (h and the downstream flux ratio) dependence.

In terms of the strip eigenfunctions (2), the upstream ($x > 0$) flow has the structure

$$\psi(x, y) = \psi^\infty(y) + \Re\mathrm{e}\sum_{n=1}^{\infty} a_n \mathrm{e}^{-\alpha_n x/(h+1)}\Psi_n\left(\frac{y+1}{h+1}\right) \tag{23}$$

and the downstream ($x < 0$) flows have the similar structures

$$\psi(x, y) = \psi_+^\infty(y) + \Re\mathrm{e}\sum_{n=1}^{\infty} b_n \mathrm{e}^{\alpha_n x/h}\Psi_n\left(\frac{y}{h}\right) \quad (0 < y < h),$$

$$\psi(x, y) = \psi_-^\infty(y) + \Re\mathrm{e}\sum_{n=1}^{\infty} c_n \mathrm{e}^{\alpha_n x}\Psi_n(-y) \quad (-1 < y < 0), \tag{24}$$

in which the far-field flows are given by (10), (11) and (12).

According to [10], the leading terms in ψ near the barrier edge at the origin are, after requiring finite velocities, such that

$$\psi \sim r^{3/2}\left[A(\cos\tfrac{3}{2}\theta + 3\cos\tfrac{1}{2}\theta) + B(\sin\tfrac{3}{2}\theta + \sin\tfrac{1}{2}\theta)\right]. \tag{25}$$

Hence the vorticity $\omega = -\nabla^2\psi$ and the pressure p are such that

$$\omega \sim -2r^{-1/2}[3A\cos\tfrac{1}{2}\theta + B\sin\tfrac{1}{2}\theta],$$

$$p \sim 2\mu r^{-1/2}[3A\sin\tfrac{1}{2}\theta + B\cos\tfrac{1}{2}\theta]. \tag{26}$$

In particular, on the barrier ($\theta = \pm\pi$),

$$\omega \sim \mp 2Br^{-1/2}, \quad p \sim \pm 6\mu A r^{-1/2}$$

and it is the relative strength of these singularities that determines the contact angle $2\arctan(-A/B)$ because (25) can be rearranged as

$$\psi \sim 4r^{3/2}\cos^2\tfrac{1}{2}\theta(A\cos\tfrac{1}{2}\theta + B\sin\tfrac{1}{2}\theta), \tag{27}$$

which also displays the required double zeros in ψ at $\theta = \pm\pi$.

The flow problem defined above is amenable to the Wiener-Hopf technique [11] but, unless the barrier is centrally placed, a matrix factorization is needed and no general method exists for obtaining this. An accurate approximate method developed by Abrahams [12] has been applied to this flow by Abrahams and Davis [13] who encountered considerable complications due to the need to cater for values of h from 1 to infinity. The next section demonstrates how a simpler problem can produce, computationally, an acceptable approximation to the flow discussed above.

3. Computational problem

The PDE Toolbox of MATLAB is extremely effective in solving a wide variety of two-dimensional boundary-value problems with many types of boundary conditions. To use PDE Toolbox, one must be able to write a problem as either a second-order equation, or a second-order system. As we have seen above, the Stokes-flow problem can easily be written in the latter form. Unfortunately, the types of mixed boundary conditions allowed do not directly include those normally associated with the Stokes flow in Figure 1: providing the value of the stream function and its normal derivative, the flow velocity, on each boundary. It is relatively straightforward, however, to solve the related problem of a second-order system with Dirichlet boundary conditions, *i.e.*, providing the value of the stream function, ψ, and the vorticity, $\omega := \nabla^2\psi$, on each boundary. This, in effect, relaxes the no-slip condition, but it may be possible to anticipate the boundary values of the velocity, $u = \partial\psi/\partial y$, sufficiently well that only acceptably small slip velocities occur.

This modified problem has real eigenfunctions that are quite different because of the second derivative. Thus, (23) and (24) are replaced by

$$\psi(x, y) = \psi^\infty(y) + \sum_{n=1}^{\infty}(A_n x + B_n)e^{-n\pi x/(h+1)}\sin\left(n\pi\frac{y+1}{h+1}\right),$$

$$\psi(x, y) = \psi_+^\infty(y) + \sum_{n=1}^{\infty}(C_n^+ x + D_n^+)e^{n\pi x/h}\sin(n\pi y/h) \quad (0 < y < h),$$

$$\psi(x, y) = \psi_-^\infty(y) + \sum_{n=1}^{\infty}(C_n^- x + D_n^-)e^{n\pi x}\sin n\pi y \quad (-1 < y < 0), \tag{28}$$

in which the coefficients could be determined by matching at $x = 0$. Correspondingly, the matrix to be factorized in the Wiener-Hopf method has a zero off-diagonal element and equal diagonal elements. Hence, a single-function factorization suffices to solve successively for ω and ψ. Double poles in the Fourier transform integral for ψ yield the x terms in the above eigenfunctions.

Despite this contrasting structure, uniqueness ensures that, if the exact boundary values of ω could be prescribed, the resulting solution would be the required Stokes flow. The two principal questions that must be addressed are then (i) how specifically should one set up the computational domain in the PDE Toolbox and give the Dirichlet conditions, particularly the vorticity, to anticipate the no-slip Stokes condition?, and (ii) how does one know that small errors in the prescribed values of the vorticity ω (relative to the exact values for the Stokes flow) do not lead to large errors in ψ in the interior of the flow?

Let us consider the first question: the issues of the computational domain and finding an appropriate approximation for the boundary vorticity (denoted as ω_E). To begin with, because the domain must be simple, the PDE Toolbox does not allow one to directly define a splitter plate of zero thickness at $y = 0$. Instead, one can define a splitter plate of thickness 2δ centered at $y = 0$. The value of δ is then chosen as small as possible within the computational limits of the computer to be used. For the computations presented here, $\delta = 0{\cdot}002$. The next computational issue is the triangulation of the given domain. Fortunately, the PDE Toolbox will do this automatically. For asymmetric domains, however, the triangulation may be far from ideal, presumably due to the presence of the singularity at the origin. This difficulty can be overcome by 'drawing' the downstream regions not as single rectangles, but rather rectangle pairs with symmetric rectangles next to the splitter plate. In the computations below, the upper downstream region consists of two rectangles, one from $y = \delta$ to $y = 0{\cdot}5$, the other from $y = 0{\cdot}5$ to $y = h$, while the lower downstream region consists of a symmetric rectangle from $y = -\delta$ to $y = -0{\cdot}5$, and a final rectangle from $y = -0{\cdot}5$ to $y = -1$. Finally, the length of both the upstream channel and the downstream channels must be sufficient to ensure that the asymptotic behavior is essentially achieved inside the computational domain. The decay rate can be estimated based on the lowest eigenvalues for each of the problems discussed above. For the computations presented below, the downstream and upstream computational limits are, respectively, at $y = -10$ and $y = 10$. The boundary conditions at these ends are given so as not to interfere with the established flow; hence, the normal derivatives of both the stream function and the vorticity are set to zero.

Next, one must determine ω_E, the approximate vorticity along the outer boundaries at $y = -1$ and $y = h$, and along the splitter plate. First observe that the upstream and downstream vorticity values must correspond to the unidirectional velocity profiles given in (10), (11) and (12). Indeed, these far-field values can be thought of (and used) as zeroth-order approximations to the horizontal-boundary and splitter-plate vorticities. Somewhat surprisingly, as we shall see in the examples below, even these zeroth-order approximations give a very reasonable approximation to the stream function ψ. For the record, one should also note that the values of ψ and ω used on the splitter plate are those of the unidirectional velocity profiles at $y = \pm\delta$, not at $y = 0$. This choice implies that the computed solution is extended by the unidirectional flow solutions in the layers $-\delta < y < 0$ and $0 < y < \delta$. To match these thin layers of unidirectional flow, the boundary conditions on the very short vertical 'boundary' near the origin ($x = 0, -\delta < y < \delta$) are the same as on the outer vertical boundaries, namely, the normal derivatives of both the stream function and vorticity are set equal to zero. Our decision to use these boundary conditions on the splitter plate is based on our desire to match the unidirectional-flow problem as closely as possible. One could also consider using no-flow boundary conditions along the splitter plate at $y = \pm\delta$, thereby giving the splitter plate physical thickness in the computational problem. As we should see in the examples below, this is very close to what we used, provided that $\delta \ll 1$.

To obtain a first-order approximation for these vorticities, one can use the eigenvalue calculations discussed above. Even though none of the eigenvalue problems exactly correspond to the region where the channel splits, one can use the lowest eigenvalues as a starting point for numerical experiments to anticipate the no-slip conditions. One can also confirm these results by replacing the Dirichlet conditions along the horizontal boundaries and the splitter plate by Neumann conditions specifying the normal derivatives of the stream function and the vorticity match those of the unidirectional velocity profiles. The problem resulting from using these Neumann conditions will be termed the *associated Neumann problem*. Based on all this, first-order approximations for the wall and splitter-plate vorticities of the following forms were used:

$$\omega_E = \omega_\pm^\infty + \exp(\alpha_- x)(\omega^\infty - \omega_\pm^\infty)/2 \qquad (x < 0)$$

$$\omega_E = \omega^\infty - \exp(-\alpha_+ x)(\omega^\infty - \omega_\pm^\infty)/2 \qquad (x > 0), \tag{29}$$

where ω^∞, $\omega_\pm^\infty$ are the far-field vorticities discussed above, and α_+ and α_- are determined in each case by considering both the eigenvalue problems and the associated Neumann problem. Second-order approximations, which take into account the small oscillations seen both in the eigenvalue problems and the associated Neumann problems, could also be computed but will not be considered here.

Now we shall consider the second question: since the exact value of the Stokes flow vorticity is unattainable using the above procedure, it is useful to show that errors in these boundary values cannot propagate. Suppose that (ψ, ω) is the exact solution of the Stokes system with $\omega(x, -1) = \omega_0(x)$ the (unknown) value of the vorticity on the lower wall (a similar argument can be given for the splitter plate and the upper boundary). Introduce a small error by setting $\omega_E(x, -1) = \omega_0(x) + \epsilon(x)$ and let (ψ_E, ω_E) be the solution of the approximate problem. Suppose also that $\| \epsilon \|_\infty = \epsilon_\infty \ll 1$. Then the error $E(x, y) = \omega_E - \omega$ satisfies

$$\nabla^2 E = 0 \quad (y > -1), \qquad E(x, -1) = \epsilon(x). \tag{30}$$

The splitter plate at $y = 0$ and the wall at $y = 1$ are being ignored here because the smallness of $\epsilon(x)$ enables the problem to be localized near the wall at $y = -1$. Thus, let

$$W = \frac{E}{\epsilon_\infty}, \quad X = \frac{x}{\sqrt{\epsilon_\infty}}, \quad Y = \frac{y+1}{\sqrt{\epsilon_\infty}}, \quad W_0(X) = \frac{\epsilon(x)}{\epsilon_\infty}.$$

Then

$$\partial_{xx} E = \partial_{XX} W, \quad \partial_{yy} E = \partial_{YY} W, \quad \| W \|_\infty = 1$$

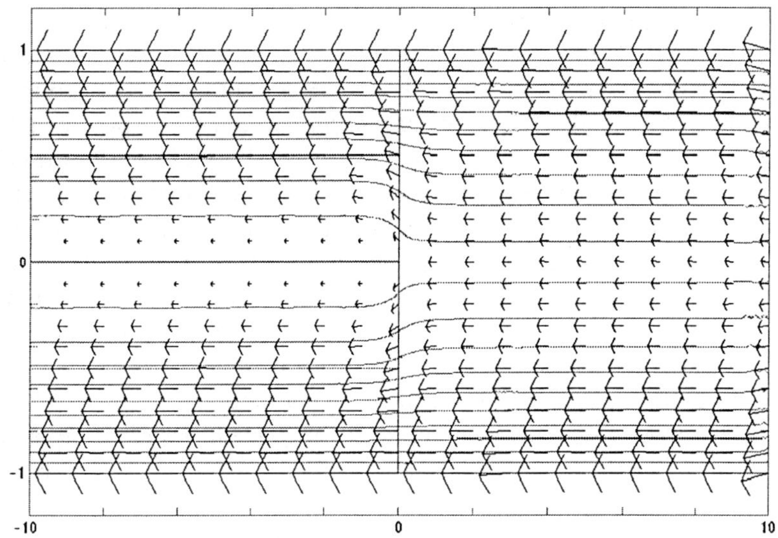

Figure 2. Stream lines and flow field for $h = 1, U = 2$ and $G_+ = 0 = G_-$. Zeroth-order boundary/splitter approximation. Twenty interior stream lines.

and the normalized problem is

$$\nabla^2 W = 0 \quad (Y > 0), \qquad W(X, 0) = W_0(X). \tag{31}$$

Here the effect of the rescaling is that the barrier and the upper wall become more distant as the error diminishes. This allows, when ϵ_∞ is sufficiently small, the solution of (31) for moderate Y to be expressed in terms of the Poisson kernel,

$$P(X, Y) = \frac{1}{\pi} \frac{Y}{X^2 + Y^2},$$

in the form

$$W(X, Y) = (W * P)(X, Y) = \int_{-\infty}^{\infty} W_0(\xi) P(X - \xi, Y)\, \mathrm{d}\xi.$$

An application of Young's inequality (Folland [14, p.14]) (with respect to X) yields

$$\|W(\cdot, Y)\|_2 \le \|W_0\|_1 \|P(\cdot, Y)\|_2 = \frac{\|W_0\|_1}{\pi} \left[\int_{-\infty}^{\infty} \left(\frac{Y}{X^2 + Y^2} \right)^2 \mathrm{d}X \right]^{1/2}$$

$$= \frac{\|W_0\|_1}{\pi} \frac{1}{\sqrt{Y}} \left[\int_{-\infty}^{\infty} \frac{\mathrm{d}\zeta}{(\zeta^2 + 1)^2} \right]^{1/2} = \frac{\sqrt{\epsilon_\infty}}{\sqrt{2\pi}} \frac{\|W_0\|_1}{\sqrt{y + 1}}.$$

So, for a given value of y,

$$\|E(\cdot, Y)\|_2 = \epsilon_\infty \|W(\cdot, Y)\|_2 \le \frac{\epsilon_\infty^{3/2}}{\sqrt{2\pi}} \frac{\|W_0\|_1}{\sqrt{y + 1}},$$

which demonstrates a super-linear decrease in error inside the channel as $\epsilon_\infty \searrow 0$. Thus, small inaccuracies on the boundary cause even smaller errors in the interior. Note that one can increase the power on the ϵ-factor to $\epsilon_\infty^{2-1/p}$ by replacing the two-norm in Young's Inequality with the p-norm.

Since the corresponding stream function ψ_E is determined from ω_E by

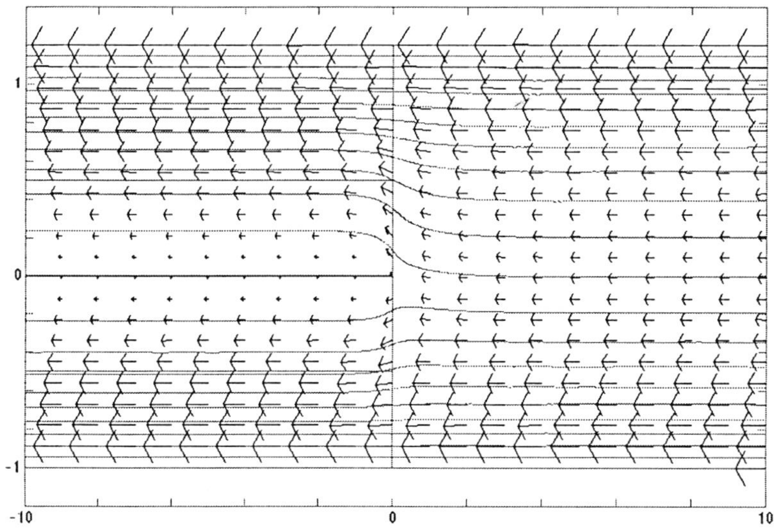

Figure 3. Stream lines and flow field for $h = 1{\cdot}2$, $U = 2$ and $G_+ = 0 = G_-$. Zeroth-order boundary/splitter approximation. Twenty interior stream lines.

$$\nabla^2 \psi_E = -\omega_E \quad (y > -1), \qquad \psi_E(x, -1) = \psi_{-1}(x),$$

and the fundamental solution of Laplace's equation can be used to express ψ_E in terms of ω_E, the bound on the error in ω carries over to a bound on the error in ψ, as required.

4. Computational examples

This section presents several examples of Stokes flow computed using the techniques described in Section 2*. As discussed above, for all computations the splitter plate lies along the negative x-axis, while the upstream and downstream computational limits are at $y = 10$ and $y = -10$, respectively. In each of the plots below, the vertical line at $x = 0$ and the horizontal lines at $y = \pm 0{\cdot}5$, $x < 0$ are part of the skeletal structure used to define the domain in PDE Toolbox; these lines are *not* part of the solution. Finally let $\mu = 1$ throughout.

The first four plots show stream lines and flow fields for the first case described in Section 2*. Here $U = 2$ and $G_+ = 0 = G_-$ while $h = 1$, $1{\cdot}2$ or 3. For each plot, there are twenty-two stream lines (twenty in the interior), evenly distributed from $\psi_{-1} = 1$ on the lower boundary to $\psi_h = -h$ on the upper boundary. Recall that the values of the stream function are also prescribed to be constant on the splitter plate as well as on the upper and lower boundaries, but that the no-slip conditions are not directly prescribed. Instead, the values of ω are prescribed to anticipate the no-slip conditions. So the upper and lower boundaries and the splitter plate are stream lines, and the issue is whether the vectors of the flow field have a constant length of $|U|$ on the outer boundaries near $x = 0$, and are zero on the splitter plate.

For the first two plot ($h = 1$ and $h = 1{\cdot}2$), the zeroth-order approximation for the boundary condition for ω is used, so there is actually a jump in the value of ω on both the lower and upper boundary at $x = 0$. Nevertheless, on the scale of the plots in Figure 2 and Figure 3, it is difficulty or impossible to detect the error in the stream function or flow field caused by this jump. No flow can be seen at the splitter plate, and the flow field appears to have

*Full-colour versions of all plots are available at htt: //www.WPI.edu/~bach/Stokes

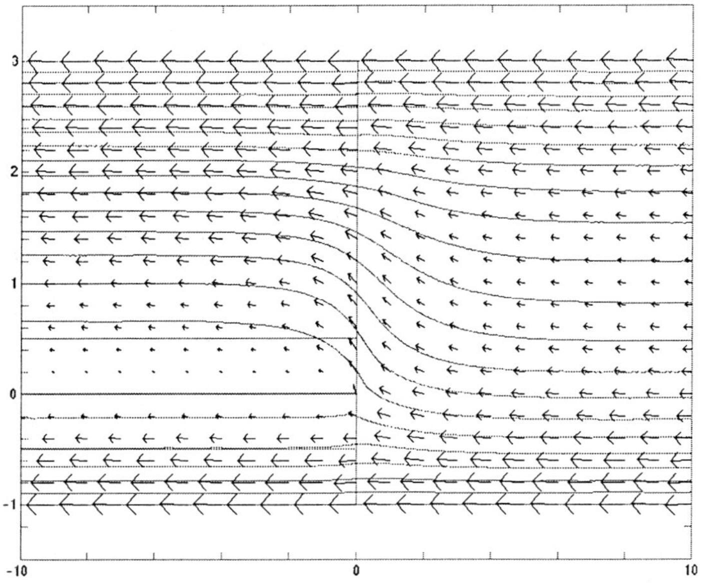

Figure 4. Stream lines and flow field for $h = 3$, $U = 2$ and $G_+ = 0 = G_-$. Zeroth-order boundary/splitter approximation. Twenty interior stream lines.

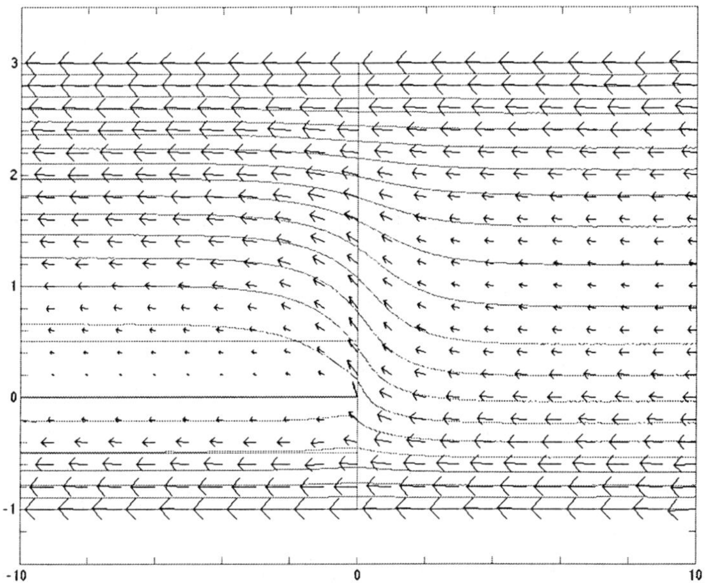

Figure 5. Stream lines and flow field for $h = 3$, $U = 2$ and $G_+ = 0 = G_-$. First-order boundary/splitter approximation. Twenty interior stream lines.

constant length on the outer boundaries. The most obvious errors are at the upstream boundary ($x = 10$) when $h = 1$; here the vector field appears not to satisfy the upstream boundary condition. Some stream lines also are less smooth, tending to have small oscillations. Both of these problems are likely the result of the relatively sparse computational mesh near the up stream boundary. The sparseness of the mesh results both from the computer measuring little error in the solution here relative to the area near the origin, and from way the domain was initially defined. Careful examination of the plot reveals that the up-stream domain ($x > 0$)

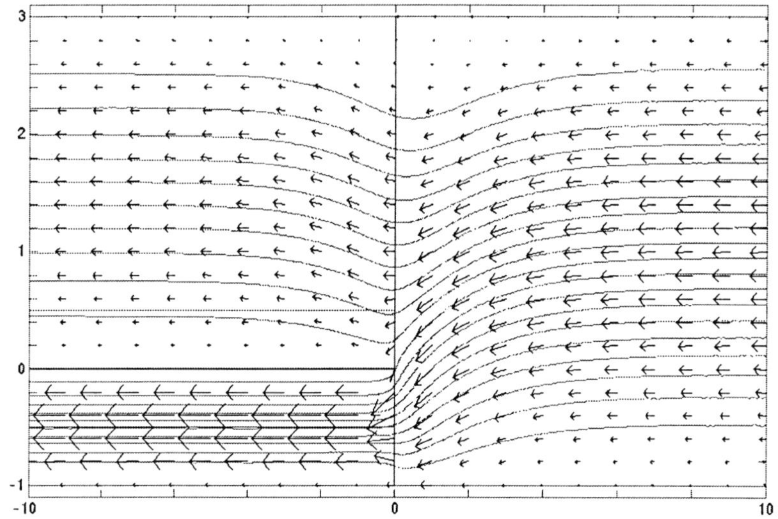

Figure 6. Stream lines and flow field for $h = 3$, $U = 0$, $\psi_0 = 0$, $\psi_{-1} = \frac{1}{2}$ and $\psi_h = -\frac{1}{2}$. Zeroth-order boundary/splitter approximation. Twenty interior stream lines.

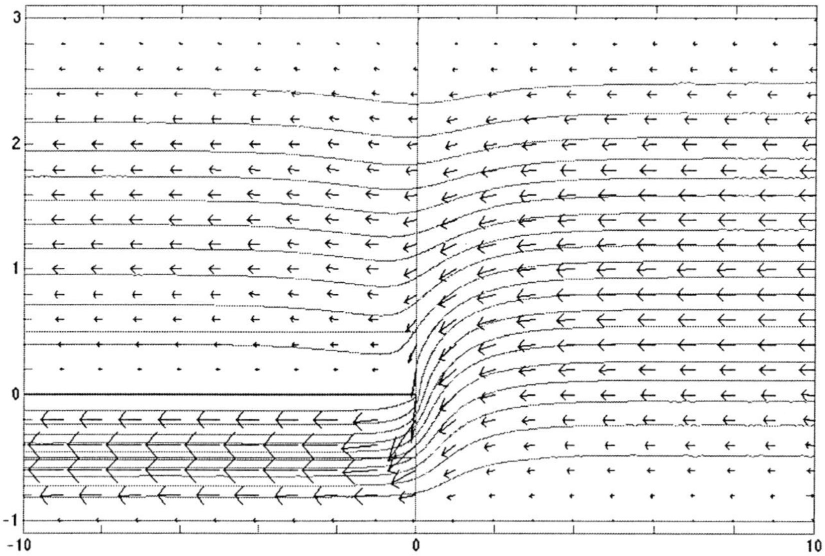

Figure 7. Stream lines and flow field for $h = 3$, $U = 0$, $\psi_0 = 0$, $\psi_{-1} = \frac{1}{2}$ and $\psi_h = -\frac{1}{2}$. First-order boundary/splitter approximation. Twenty interior stream lines.

is composed of a single rectangle while the downstream domain is composed of four rectangles. This domain structure affects the initial triangulation and hence all uniform refinements leading to there always being a finer mesh downstream. Interestingly, the asymmetries in the later plots ($h = 1{\cdot}2$ and $h = 3$) tended to alleviate this problem. For $h = 1{\cdot}2$, the tiny arrows near the splitter plate are in fact just below the splitter plate. They appear here and not in the symmetric case because of the way PDE Toolbox decides to choose base points for the flow field. It is also worth noting that for $h = 1{\cdot}2$, PDE Toolbox plots only one flow vector on the lower boundary; this seems to be simply an error in the plotting software.

For the third plot ($h = 3$), it is no longer quite the case that zeroth-order approximation is sufficient to avoid an observable error on the scale used here. Now there is a small but perceptible error when the zeroth-order approximation is used (*cf.* Figure 4). In particular, note the slight difference in the gap between arrows of the flow field on the lower boundary. To reduce this error, the first-order approximation described in the previous section is used on both the outer boundaries and the splitter plate. When both the associated Neumann problem and the eigenvalues are considered, an appropriate approximation for ω_E is given by (29) with $\alpha_+ = \alpha_- = 1{\cdot}4$. The result is a similar, but slightly more accurate plot shown in Figure 5; in particular, note that the boundary arrows now appear to be of uniform length. Note that the flow in the lower portion of the upstream channel is pulled upward as it approaches the splitter plate and then tends to 'bounce' off as it enters the downstream channel. The slight deviation in the boundary arrows from the horizontal again comes form a plotting or rounding error in PDE Toolbox. Since ψ is defined to be constant on these boundaries, all boundary arrows should be horizontal, even if ω is chosen poorly.

The last two plots show stream lines and flow fields for the second case described in Section 2. Here $U = 0$, $\psi_0 = 0$, $\psi_{-1} = \frac{1}{2}$ and $\psi_h = -\frac{1}{2}$. So there is now unit total flux upstream, with equal portions entering each downstream channel. Then, since $h = 3$ in both plots, the pressure gradients are $G = \frac{3}{16}$, $G_- = 6$ and $G_+ = \frac{2}{9}$.

Figure 6 displays the zeroth-order approximation for this second case; as in the first case, when $h = 3$ there is a small but perceptible error on the outer boundaries. A first-order approximation is again found by considering the eigenvalues and the associated Neumann problem. In this case, the first-order boundary conditions for ω are given by (29) with $\alpha_+ = 2{\cdot}3$ and $\alpha_- = 3{\cdot}3$. Figure 7 shows this first-order solution. Since $U = 0$, there should now be no flow on any boundary or on the splitter plate; the small but uniform flow seen in Figure 7 again indicates the limits of MATLAB and PDE Toolbox. Notice that, as in the first case, a portion of the flow 'bounces' off the splitter plate, this time from above.

5. Concluding remarks

The work presented here demonstrates how MATLAB (and indeed many other over-the-counter software packages) can be used to handle problems for which the package would not at first seem to be appropriate. If one tries to 'force' the Stokes-type boundary conditions directly into MATLAB, the results are very poor. But in this case, and in many others, there is the possibility of posing a slightly different problem that the software can handle and that in fact leads to the solution of the original problem.

This flow problem was specifically chosen for this *Practical Asymptotics* issue of this Journal. Evidently, the known structure of the stream function has played a key role in formulating the approximate boundary conditions for the MATLAB software. The problem nicely demonstrates how such information could be used in further applications. In addition, the convergence proof is asymptotic in nature and demonstrates another use of asymptotics in mathematics and science.

References

1. C.V. Camp and G.S. Gipson, *Boundary Element Analysis of Nonhomogeneous Biharmonic Phenomena.* Berlin: Springer-Verlag (1992) 246 pp.
2. P.J. Roache, *Computational Fluid Dynamics.* Albuquerque: Hermosa Publishers (1985) 446 pp.

3. A.R. Mitchell and D.F. Griffiths, *The Finite Difference Method in Partial Differential Equations*. Chichester: Wiley (1980) 272 pp.
4. V.T. Buchwald, Eigenfunctions of plane elastostatics I. The strip. *Proc. R. Soc. London A* 277 (1964) 385–400.
5. R.C.T. Smith, The bending of a semi-infinite strip. *Aust. J. Sci. Res.* 5 (1952) 227–237.
6. D.A. Spence, A class of biharmonic end-strip problems arising in elasticity and Stokes flow. *IMA J. Appl. Math.* 30 (1983) 107–139.
7. T.N. Phillips, Singular matched eigenfunction expansions for Stokes flow around a corner. *IMA J. Appl. Math.* 42 (1989) 13–26.
8. S.A. Trogdon and D.D. Joseph, Matched eigenfunction expansions for slow flow over a slot. *J. Non-Newtonian Fluid Mech.* 10 (1982) 185–213.
9. V.V. Meleshko, Steady Stokes flow in a rectangular cavity. *Proc. R. Soc. London A* 452 (1996) 1999–2022.
10. D.H. Michael and M.E. O'Neill, The separation of Stokes flows. *J. Fluid Mech.* 80 (1977) 785–794.
11. B. Noble, *Methods Based on the Wiener-Hopf Technique*. 2nd. ed. New York: Chelsea Press (1988) 246 pp.
12. I.D. Abrahams, On the solution of Wiener-Hopf problems involving non-commutative matrix kernel decompositions. *SIAM J. Appl. Math.* 57 (1997) 541–567.
13. I.D. Abrahams and A.M.J. Davis, Asymmetric channel divider and the Stokes entry problem, in preparation.
14. G.B. Folland, *Introduction to Partial Differential Equations*. Princeton: Princeton University. Press (1976) 349 pp.

Journal of Engineering Mathematics **39:** 221–240, 2001.

The frozen-field approximation and the Ginzburg-Landau equations of superconductivity

HANS G. KAPER and HENRIK NORDBORG[1]
Mathematics and Computer Science Division, Argonne National Laboratory, Argonne, IL 60439, U.S.A. (e-mail: kaper@mcs.anl.gov)
[1] *James Frank Institute, The University of Chicago, 5640 South Ellis Avenue, Chicago, IL 60637, U.S.A. (e-mail: Henrik_Nordborg@anl.gov)*

Received 28 January 2000; accepted in revised form 17 July 2000

Abstract. The Ginzburg-Landau (GL) equations of superconductivity provide a computational model for the study of magnetic flux vortices in type-II superconductors. In this article it is shown through numerical examples and rigorous mathematical analysis that the GL model reduces to the frozen-field model when the charge of the Cooper pairs (the superconducting charge carriers) goes to zero while the applied field stays near the upper critical field.

Key words: Ginzburg–Landau equations, superconductivity, frozen-field approximation, asymptotic analysis.

1. Introduction

Superconducting materials hold great promise for technological applications. Especially since the discovery of the so-called high-temperature superconductors in the 1980s, much research has been devoted to understanding the behavior of these new materials. While conventional superconductors require liquid helium (3–4 deg Kelvin) to remain in the superconducting state, high-temperature superconductors: can be cooled with liquid nitrogen (76 deg Kelvin) – a clear economic advantage. Unfortunately, high-temperature superconductors are ceramic materials, which are difficult to manufacture into films and wires, but progress is being made all the time.

High-temperature superconductors belong to the class of type II superconductors. Unlike type-I superconductors, type-II superconductors can sustain a magnetic flux in their interior, but this flux is restricted to quantized amounts–filaments that are encircled by a current. The current shields the magnetic flux from the bulk, which is perfectly superconducting. The configuration resembles that of a vortex in a fluid, and the superconductor is said to be in the *vortex state*.

Figure 1 gives a sketch of the phase diagram of a type-II superconductor in the neighborhood of T_c, the *critical temperature*. The two-dimensional phase space is spanned by the temperature T and the (magnitude of the) magnetic field H and is roughly divided into three subregions. Each subregion corresponds to a particular state: the perfectly superconducting (Meissner) state below the lower critical field H_{c1}, where no magnetic field can penetrate the medium; the normal state above the upper critical field H_{c2}, where the superconductor behaves like a normal metal; and the intermediate vortex state. Above the critical temperature T_c all superconducting properties are lost.

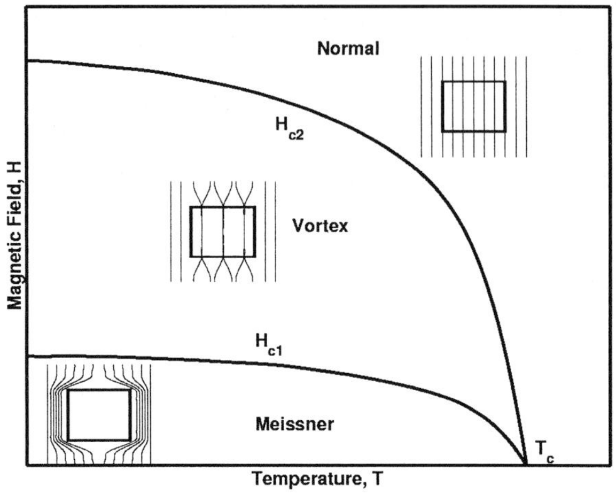

Figure 1. Phase diagram of a type-II superconductor.

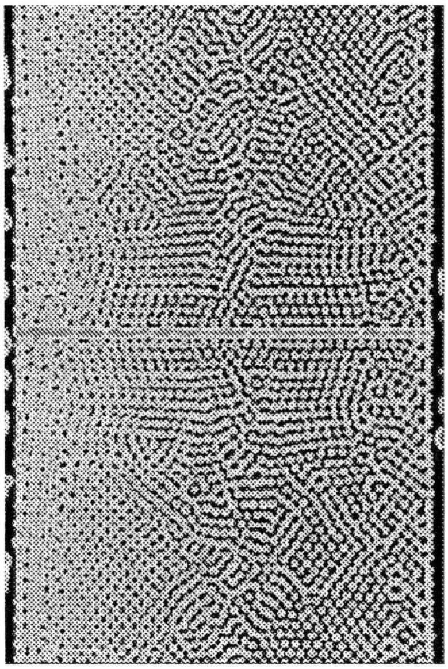

Figure 2. Vortex configuration in two dimensions.

The vortices, and especially their dynamics, determine the current-carrying capabilities of a superconductor. Much effort, both experimental and theoretical, is therefore being spent on the study of vortex dynamics and, especially, mechanisms to inhibit vortex motion when the superconductor is subject to currents and fields. By 'pinning' the vortices, one prevents energy dissipation and, hence, loss of superconductivity.

Vortices can be studied computation ally at various levels of detail using different models. The Ginzburg-Landau (GL) model gives a field (continuum) description that, although phenomenological and not based on any microscopic quantum-mechanical theory, has been used successfully to study both the dynamics and the structure of vortex systems in realistic superconductor configurations [1, 2]. Figures 2 and 3 give two examples of computational results obtained with the GL equations. They illustrate both the effectiveness and the difficulties of such calculations.

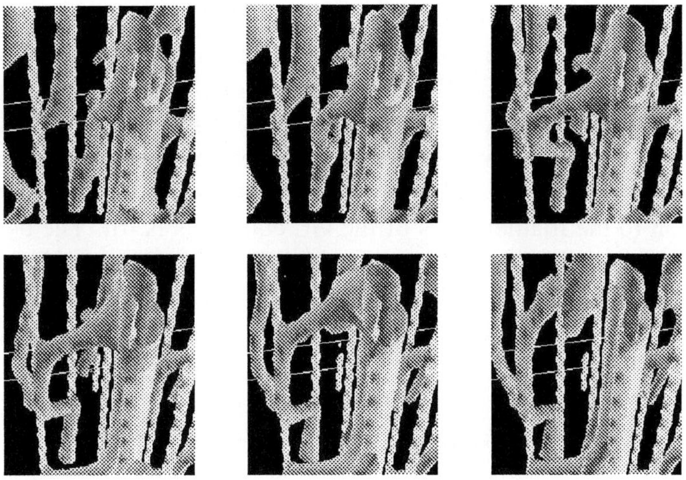

Figure 3. Kinking-induced motion of vortices through splayed columnar defects.

Figure 2 shows a vortex configuration in a two-dimensional cross section of a twinned superconducting crystal, which was computed from a steady-state solution of the GL equations. The twin boundary (an irregularity in the structure of the crystal) is visible in the horizontal line through the center; it acts as a pinning site for the vortices. The field is perpendicular to the plane of the cross section, which measures 128×192 coherence lengths (a characteristic length of the order of microns). Each dot corresponds to a vortex intersecting the plane of the cross section; the entire configuration has approximately 2,700 vortices. The figure shows the level of detail one can achieve with the GL model, given sufficient computing power. At the same time, it illustrates the level of computational complexity one faces if one uses the GL model.

Figure 3 shows a series of snapshots of a vortex configuration in three dimensions, also computed with the GL model. The objective of this computation was to simulate vortex motion through columnar defects and study the potential of the latter as pinning sites. The defects are visible as twisted straight lines. The vortices are the flexible tube-like structures; they move from one defect to another under the influence of external forces. The figure shows the motion of a vortex that is originally pinned on a defect. The vortex develops a loop, the loop peels off, the loop expands in both directions in a traveling-wave-like scenario, and gradually the entire vortex transfers to the next available defect.

Numerical simulations provide the only way to study vortex dynamics at this level of detail. They are an invaluable tool for fundamental research, complementing experiment and theory. Numerical simulations of realistic superconductors based on the GL model, Uke the ones illustrated in Figures 2 and 3 are, however, extremely time consuming, and it is desirable to use simpler models whenever possible. Here, we focus on the 'frozen-field model', which is still a continuum model and the closest approximation to the full GL model. In the frozen-field model, the superconducting phenomena are decoupled from the electromagnetic field, and the latter is prescribed through a vector potential. The frozen-field model is much simpler and has been used successfully for numerical simulations of vortex systems [3].

In this article, we prove that the frozen-field model is obtained as the asymptotic limit of the GL model when the charge of the Cooper pairs (the superconducting charge carriers) goes to zero while the applied magnetic field stays near the upper critical field. (The upper critical field itself depends on the charge of the Cooper pairs and increases as the latter decreases.) Because the temperature is constant in the GL model, this limit corresponds to fixing the temperature T and moving up vertically through the vortex regime to the curve labeled H_{c2} in the phase diagram of Figure 1. The convergence rate is second order in the small parameter.

For more background on the physics of superconductivity we refer the reader to the monograph by Tinkham [4]. The original source for the GL equations of superconductivity is [5]. A good introduction to the mathematics of the GL equations is [6]. The dynamics of the GL equations have been studied by several authors; see [7, 8, 9] and the references cited therein. The present investigation is closely related to the work of Du and Gray [10].

Section 2 introduces the Ginzburg–Landau equations, Section 3 contains the numerical results and Section 4 the analysis.

2. The Ginzburg–Landau equations

In the Ginzburg–Landau theory of superconductivity, the state of a superconducting medium is described by a complex scalar-valued *order parameter* ψ and a real vector-valued *vector potential* $\boldsymbol{A}$. If the state varies with time, a third variable – the *electric potential* ϕ–is necessary to fully describe the electromagnetic field. The evolution of the state variables is governed by the time-dependent Ginzburg–Landau (TDGL) equations,

$$\gamma\hbar\left(\frac{\partial}{\partial t}+\mathrm{i}\frac{q_s}{\hbar}\phi\right)\psi+\frac{1}{2m_s}\left(\frac{\hbar}{\mathrm{i}}\nabla-\frac{q_s}{c}\boldsymbol{A}\right)^2\psi+\alpha\psi+\beta|\psi|^2\psi=0, \tag{2.1}$$

$$\sigma\left(-\frac{1}{c}\frac{\partial \boldsymbol{A}}{\partial t}-\nabla\phi\right)-\frac{c}{4\pi}\nabla\times\nabla\times\boldsymbol{A}+\boldsymbol{J}_s+\frac{c}{4\pi}\nabla\times\boldsymbol{H}=0, \tag{2.2}$$

where the supercurrent density $\boldsymbol{J}_s$ is a nonlinear function of ψ and $\boldsymbol{A}$,

$$\boldsymbol{J}_s=\frac{q_s\hbar}{2\mathrm{i}m_s}(\psi^*\nabla\psi-\psi\nabla\psi^*)-\frac{q_s^2}{m_s c}|\psi|^2\boldsymbol{A}=\frac{q_s}{m_s}\Re\left[\psi^*\left(\frac{\hbar}{\mathrm{i}}\nabla-\frac{q_s}{c}\boldsymbol{A}\right)\psi\right]. \tag{2.3}$$

These equations are supplemented by the boundary conditions,

$$\boldsymbol{n}\cdot\boldsymbol{J}_s=0,\quad \boldsymbol{n}\times(\nabla\times\boldsymbol{A})=\boldsymbol{n}\times\boldsymbol{H}. \tag{2.4}$$

Here, $\boldsymbol{H}$ is the *applied magnetic field*, which we assume to be time independent. The constants m_s and q_s are the mass and charge, respectively, of a Cooper pair (the superconducting charge carriers, also referred to as superelectrons); c is the speed of light; and $\hbar$ is Planck's constant divided by 2π. A Cooper pair is made up of two electrons, each with charge $-e$ (e is the elementary charge); hence, q_s is negative, $q_s=-2e$.

The parameters α and β are material parameters; α changes sign at the critical temperature T_c, $\alpha(T)<0$ for $T<T_c$ (superconducting state) and $\alpha(T)>0$ for $T>T_c$ (normal state); β is only weakly temperature dependent and positive for all T. The remaining parameters are σ, the normal state conductivity, and γ, the mobility coefficient. The latter is dimensionless and related to the diffusion coefficient D, $\gamma=\hbar/2m_sD$.

The boundary conditions (2.4) express the fact that superelectrons cannot leave the superconductor. Also, if no surface currents are present, the tangential components of the magnetic field must be continuous across the boundary.

The parameters α and β are defined phenomenologically, but they can be expressed in terms of measurable quantities, such as the superconducting *coherence length* ξ and the London *penetration depth* λ,

$$\xi = \left(\frac{\hbar^2}{2m_s|\alpha|}\right)^{1/2}, \quad \lambda = \left(\frac{m_s c^2 \beta}{4\pi q_s^2 |\alpha|}\right)^{1/2}. \tag{2.5}$$

The coherence length and the London penetration depth define the respective characteristic length scales for the order parameter and the magnetic induction. Both depend on the temperature T and diverge as T approaches the critical temperature T_c, because of the factor $|\alpha|^{-1/2}$. However, their ratio is, to a good approximation, independent of temperature. This ratio is the *Ginzburg–Landau parameter*,

$$\kappa = \lambda/\xi. \tag{2.6}$$

In high-T_c superconductors, κ is of the order of 50–100.

The electromagnetic variables are the *magnetic induction* $\boldsymbol{B}$, the *current density* $\boldsymbol{J}$, and the *electric field* $\boldsymbol{E}$; they are given in terms of $\boldsymbol{A}$ and ϕ by the expressions

$$\boldsymbol{B} = \nabla \times \boldsymbol{A}, \quad \boldsymbol{J} = \frac{c}{4\pi}\nabla \times \nabla \times \boldsymbol{A}, \quad \boldsymbol{E} = -\frac{1}{c}\frac{\partial \boldsymbol{A}}{\partial t} - \nabla\phi. \tag{2.7}$$

Equation (2.2) is essentially Ampère's law, $\boldsymbol{J} = (c/4\pi)\nabla \times \boldsymbol{B}$, where the current $\boldsymbol{J}$ is the sum of the supercurrent $\boldsymbol{J}_s$, the transport current $\boldsymbol{J}_t = (c/4\pi)\nabla \times \boldsymbol{H}$, and a 'normal' current $\boldsymbol{J}_n = \sigma \boldsymbol{E}$ (Ohm's law). Hence, the GL model uses a quasistatic version of Maxwell's equations, where the time derivative of the electric field is ignored.

The TDGL equations were first given by Schmid [11] in 1966 and subsequently derived from the microscopic theory of superconductivity by Gor'kov and Eliashberg [12]. Our notation is the same as in Gor'kov and Kopnin [13].

The solution of the TDGL equations is not unique. Any solution $(\psi, \boldsymbol{A}, \phi)$ defines a family of solutions $G_\chi(\psi, \boldsymbol{A}, \phi)$ parameterized by a sufficiently smooth function χ of space and time,

$$G_\chi : (\psi, \boldsymbol{A}, \phi) \mapsto \left(\psi \mathrm{e}^{\mathrm{i}(q_s/\hbar c)\chi}, \boldsymbol{A} + \nabla\chi, \phi - \frac{1}{c}\frac{\partial \chi}{\partial t}\right). \tag{2.8}$$

This property is known as *gauge invariance*; the function χ is known as a *gauge function*. Gauge invariance does not affect the physically measurable quantities (the magnetic induction $\boldsymbol{B}$, the magnetization $\boldsymbol{M} = \boldsymbol{B} - \boldsymbol{H}$, and the current density $\boldsymbol{J}$). Uniqueness requires an additional constraint, which is imposed through a gauge choice. The choice of a proper gauge for the TDGL equations has been a subject of considerable debate. The choice is a matter of convenience and may depend on the problem under investigation. In this article we adopt a gauge in which, at any time, the electric potential and the divergence of the vector potential satisfy the identity

$$\sigma\phi + (c/4\pi)\nabla \cdot \boldsymbol{A} = 0 \tag{2.9}$$

everywhere in the domain, while $\boldsymbol{A}$ is tangential at the boundary. This choice is realized by identifying the gauge χ with a solution of the linear parabolic equation

$$\frac{\sigma}{c}\frac{\partial \chi}{\partial t} - \frac{c}{4\pi}\Delta\chi = \sigma\phi + \frac{c}{4\pi}\nabla \cdot \boldsymbol{A}, \tag{2.10}$$

subject to the condition $\boldsymbol{n} \cdot \nabla\chi = -\boldsymbol{n} \cdot \boldsymbol{A}$ on the boundary. In [9], it was shown that the TDGL equations, subject to the constraint (2.9), define a dynamical system under suitable regularity conditions on $\boldsymbol{H}$. (In the more general case, where $\boldsymbol{H}$ varies not only in space but also in time, the TDGL equations define a dynamical process.) This dynamical system has a global attractor, which consists of the stationary points of the dynamical system and the heteroclinic orbits connecting such stationary points. Furthermore, it was shown that every solution on the attractor satisfies the condition $\nabla \cdot \boldsymbol{A} = 0$ (and, therefore, also $\phi = 0$). Thus, in the limit as $t \to \infty$, every solution of the TDGL equations satisfies the GL equations in the London gauge.

2.1. Nondimensional TDGL equations

In this section, we render the TDGL equations dimensionless by choosing units for the independent and dependent variables. Since we are interested in the collective behavior of vortices in the bulk of a superconductor in the limit of weak coupling ($q_s \to 0$), we take care to choose the units in such a way that they remain of order one as $q_s \to 0$. (We recall that q_s is negative, $q_s = -2e$.)

As $q_s \to 0$, the coherence length ξ remains of order one, while the penetration depth λ increases like $|q_s|^{-1}$; see Equation (2.5). This suggests taking the coherence length as the unit of length.

To maintain the diffusion coefficient $D = \hbar/2\gamma m_s, = \xi^2(\gamma\hbar/|\alpha|)^{-1}$ at order one, we measure time in units of $\gamma\hbar/|\alpha|$.

The real and imaginary parts of the order parameter are conveniently measured in units of $\psi_0 = (|\alpha|/\beta)^{1/2}$, which is the value of ψ that minimizes the free energy in the absence of a field.

Next, consider the magnetic field. A fundamental quantity in the theory of type-II superconductors is the flux quantum Φ_0,

$$\Phi_0 = \frac{hc}{|q_s|} = 2\pi\frac{\hbar c}{|q_s|}. \tag{2.11}$$

The flux quantum is the unit of magnetic flux carried by a vortex. Together with the coherence length and penetration depth, it defines three characteristic field strengths: the *lower critical field* H_{c1}, the *thermodynamical critical field* H_c, and the *upper critical field* H_{c2},

$$H_{c1} = \frac{\Phi_0}{4\pi\lambda^2\log\kappa}, \qquad H_c = \frac{\Phi_0}{2\pi\xi\lambda\sqrt{2}}, \qquad H_{c2} = \frac{\Phi_0}{2\pi\xi^2}. \tag{2.12}$$

Below H_{c1}, a superconductor is in the ideal superconducting (Meissner) state, where it does not support magnetic flux in the bulk; above H_{c2}, it is in the normal state, where the magnetic flux is distributed uniformly in the bulk; between H_{c1} and H_{c2}, it is in the vortex state, where magnetic flux is quantized in vortex-like configurations (see Figure 1). The thermodynamical critical field H_c is intermediate between H_{c1} and H_{c2} and is defined by the identity $H_c^2/8\pi = \frac{1}{2}\psi_0^2|\alpha|$; $H^2/8\pi$ is the energy per unit volume associated with a field H, and $\frac{1}{2}\psi_0^2|\alpha|$ is the minimum condensation energy, which is attained when $\psi = \psi_0$, so these two quantities are in balance when $H = H_c$.

As $q_s \to 0$, H_{c1} goes to 0 like $|q_s|$, H_c remains of order one, and H_{c2} grows like $|q_s|^{-1}$. This suggests that we define field strengths in terms of H_c. In fact, it is convenient to absorb a factor $\sqrt{2}$, so we adopt $H_c\sqrt{2}$ or, equivalently, $\hbar c/\xi\lambda|q_s|$ as the unit for the magnetic field strength.

With the coherence length as the unit of length and $H_c\sqrt{2}$ as the unit of field strength, it follows that the vector potential is measured in units of $\xi H_c\sqrt{2}$. Furthermore, energy densities are measured in units of $H_c^2/4\pi$, which is the same as $|\alpha|\psi_0^2$.

Table 1. Nondimensionalization

Independent variables	$\boldsymbol{x} = \xi \boldsymbol{x}'$
	$t = (\gamma\hbar/\lvert\alpha\rvert)t'$
Dependent variables	$\psi = \psi_0\psi'$
	$\boldsymbol{A} = (\xi H_c\sqrt{2})\boldsymbol{A}'$
	$\phi = (1/\gamma\psi_0^2\kappa\lvert q_s\rvert)(H_c^2/4\pi)\phi'$
Electromagnetic variables	$\boldsymbol{B} = (H_c\sqrt{2})\boldsymbol{B}'$
	$\boldsymbol{J} = (cH_c\sqrt{2}/4\pi\xi)\boldsymbol{J}'$
	$\boldsymbol{E} = (1/\gamma\psi_0^2\kappa\lvert q_s\rvert)(H_c^2/4\pi\xi)\boldsymbol{E}'$
Applied field	$\boldsymbol{H} = (H_c\sqrt{2})\boldsymbol{H}'$
Normal conductivity	$\sigma = (\gamma m_s c^2/2\pi\hbar)\sigma'$

Finally, we define the scalar potential ϕ in units of $(1/\gamma\psi_0^2\kappa|q_s|)(H_c^2/4\pi)$. Notice that this unit remains of order one as $q_s \to 0$, because $\kappa|q_s|$ is of order one. On the other hand, the product $q_s\phi$, which represents an energy density, tends to zero as $q_s \to 0$. (It remains finite on the scale of the penetration depth.)

Table 1 summarizes the relations between the original variables and their nondimensional (primed) counterparts. We adopt the latter as the new variables and work until further notice on the nondimensional problem. We omit all primes.

The nondimensional TDGL equations are:

$$\left(\frac{\partial}{\partial t} - \frac{\mathrm{i}}{\kappa}\phi\right)\psi - \left(\nabla + \frac{\mathrm{i}}{\kappa}\boldsymbol{A}\right)^2\psi - (1 - |\psi|^2)\psi = 0, \tag{2.13}$$

$$\sigma\frac{\partial \boldsymbol{A}}{\partial t} - \Delta\boldsymbol{A} - \frac{1}{\kappa}\boldsymbol{J}_s - \nabla\times\boldsymbol{H} = \boldsymbol{0}, \tag{2.14}$$

where

$$\boldsymbol{J}_s = -\frac{1}{2\mathrm{i}}(\psi^*\nabla\psi - \psi\nabla\psi^*) - \frac{1}{\kappa}|\psi|^2\boldsymbol{A} = -\Im\mathrm{m}\left[\psi^*\left(\nabla + \frac{\mathrm{i}}{\kappa}\boldsymbol{A}\right)\psi\right], \tag{2.15}$$

with the corresponding gauge condition,

$$\sigma\phi + \nabla\cdot\boldsymbol{A} = 0. \tag{2.16}$$

In deriving Equation (2.14), we have made use of the gauge condition (2.16) and the vector identity

$$\Delta\boldsymbol{A} = -\nabla\times\nabla\times\boldsymbol{A} + \nabla(\nabla\cdot\boldsymbol{A}). \tag{2.17}$$

If Ω is the domain occupied by the superconducting material (measured in units of ξ), then Equations (2.13)–(2.16) must be satisfied everywhere Ω. At the boundary $\partial\Omega$ of Ω, we have the conditions

$$\boldsymbol{n}\cdot\boldsymbol{J}_s = 0, \quad \boldsymbol{n}\times(\nabla\times\boldsymbol{A}) = \boldsymbol{n}\times\boldsymbol{H}, \quad \boldsymbol{n}\cdot\boldsymbol{A} = 0. \tag{2.18}$$

Here, $\boldsymbol{n}$ is the local unit normal vector.

The electromagnetic variables are given by the expressions

$$\boldsymbol{B} = \nabla \times \boldsymbol{A}, \quad \boldsymbol{J} = \nabla \times \nabla \times \boldsymbol{A}, \quad \boldsymbol{E} = -\partial_t \boldsymbol{A} - \nabla\phi. \tag{2.19}$$

The values of the lower and upper critical fields are

$$H_{c1} = (2\kappa \log \kappa)^{-1}, \quad H_{c2} = \kappa. \tag{2.20}$$

2.2. Link variables

The combination $\nabla + (\mathrm{i}/\kappa)\boldsymbol{A}$ plays a fundamental role; we refer to it as the $\boldsymbol{A}$-gradient and write

$$\nabla_A = \nabla + \frac{\mathrm{i}}{\kappa}\boldsymbol{A}. \tag{2.21}$$

The $\boldsymbol{A}$-gradient defines the $\boldsymbol{A}$-Laplacian (or 'twisted Laplacian'),

$$\nabla_A = \nabla_A \cdot \nabla_A = \left(\nabla + \frac{\mathrm{i}}{\kappa}\boldsymbol{A}\right)^2. \tag{2.22}$$

The relation between the $\boldsymbol{A}$-Laplacian and the ordinary Laplacian is most easily illustrated by means of the *link variables*,

$$\begin{aligned} U_x(x, y, z) &= \exp\left(\frac{\mathrm{i}}{\kappa}\int^x A_x(\xi, y, z)\,\mathrm{d}\xi\right), \\ U_y(x, y, z) &= \exp\left(\frac{\mathrm{i}}{\kappa}\int^y A_y(x, \eta, z)\,\mathrm{d}\eta\right), \\ U_z(x, y, z) &= \exp\left(\frac{\mathrm{i}}{\kappa}\int^z A_z(x, y, \zeta)\,\mathrm{d}\zeta\right). \end{aligned} \tag{2.23}$$

(We omit the argument t.) The integrals are evaluated with respect to an arbitrary reference point. Each $U_\mu (\mu = x, y, z)$ is complex valued and unimodular, $U_\mu^* = U_\mu^{-1}$. The vectors $\boldsymbol{A}$ and $\boldsymbol{U}$ may be used interchangeably. With a slight abuse of notation, we have

$$\boldsymbol{U} = \mathrm{e}^{(\mathrm{i}/\kappa)\int \boldsymbol{A}}, \quad \nabla_A = U^* \nabla U, \quad \Delta_A = U^* \Delta U. \tag{2.24}$$

3. Numerical solution

A parallel code for solving Equations (2.13)–(2.18) has been developed as part of a project for large-scale simulations of vortex dynamics in superconductors. Details on these simulations and on the code will be published elsewhere; here, we give only a brief overview of the numerical methods and the results of numerical simulations showing the behavior of the solution as κ increases.

The algorithm uses finite differences on a staggered grid, making all approximations accurate to second order in the mesh widths, and an implicit method for the time integration, making the algorithm (essentially) unconditionally stable. The code, written in C++, has been designed for a multiprocessing environment; it uses MPI for message passing.

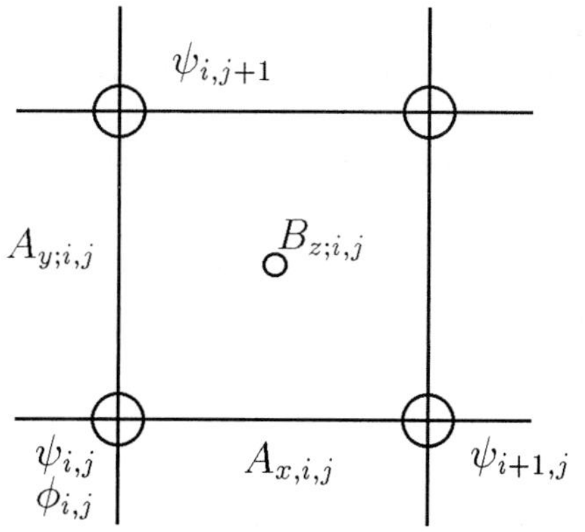

Figure 4. Computational grid cell and definition of the discrete variables.

We restrict the discussion to rectangular two-dimensional configurations that are periodic in one direction and open in the other. The configurations are assumed to be infinite in the third, orthogonal direction, which is also the direction of the applied magnetic field, $\boldsymbol{H} = (0, 0, H_z)$.

3.1. DISCRETIZATION

Computational grid. The computational grid is uniform, with equal mesh sizes in the x and y direction $h_x = h_y = h$. A vertex on the grid is denoted by $\boldsymbol{x}_{i,j} = (x_i, y_j)$ and is the point of reference for the grid cell shown in Figure 4. The indices run through the values $i = 1, \ldots, N_x$ and $j = 1, \ldots, N_y$. We assume periodicity in the x, direction and take the grid so the vertices with $j = 1$ and $j = N_y$ are located on the open boundary of the superconductor. Thus, the size of the domain is $S = N_x(N_y - 1)h^2$.
Variables. The discrete variables are introduced so that all derivatives are given by second-order accurate central-difference approximations. The scalar variables ψ and ϕ are defined on the vertices of the grid,

$$\psi_{i,j} = \psi(\boldsymbol{x}_{i,j}), \quad \phi_{i,j} = \phi(\boldsymbol{x}_{i,j}). \tag{3.1}$$

(We use the same symbol for the original field and its discrete counterpart.) Vectors are defined at the midpoints of the links connecting adjacent vertices,

$$A_{x;i,j} = A_x(\boldsymbol{x}_{i,j} + \tfrac{1}{2}h_x\boldsymbol{e}_x), \quad A_{y;i,j} = A_y(\boldsymbol{x}_{i,j} + \tfrac{1}{2}h_y\boldsymbol{e}_y). \tag{3.2}$$

Here, $\boldsymbol{e}_x$ and $\boldsymbol{e}_y$ denote the unit vectors in the x and y direction, respectively. The definition of the discrete supercurrent $\boldsymbol{J}_s$ is completely analogous. The link variables, defined in Equation (2.24), are obtained from the vector potential,

$$U_{x;i,j} = \mathrm{e}^{(\mathrm{i}/\kappa)A_{x;i,j}h_x}, \quad U_{y;i,j} = \mathrm{e}^{(\mathrm{i}/\kappa)A_{y;i,j}h_y}. \tag{3.3}$$

They are therefore also defined on the links. Finally, the magnetic induction $\boldsymbol{B}$, which is a vector perpendicular to the plane and given by the curl of the vector potential, is defined at the center of a grid cell,

$$B_{z;i,j} = B_z(\boldsymbol{x}_{i,j} + \tfrac{1}{2}h_x\boldsymbol{e}_x + \tfrac{1}{2}h_y\boldsymbol{e}_y). \tag{3.4}$$

The definition of the discrete variables is also illustrated in Figure 4.

Note that, because of the location of the grid relative to the boundaries, all scalar variables, as well as the x components of all vectors (A_x, U_x, $J_{s,x}$, and so forth), are defined on a $N_x \times N_y$ grid, whereas the y components of all vectors and the magnetic induction B_z are defined on a $N_x \times (N_y - 1)$ grid.

Boundary conditions. We assume periodicity in the x direction, so we need to consider the boundary conditions (4.7) only at $y = y_1$ and $y = y_{N_y}$.

The boundary condition for the order parameter, $\boldsymbol{n} \cdot \nabla_{\mathrm{A}} \psi = 0$, becomes

$$U_{y;i,1}\psi_{i,2} - \psi_{i,1} = 0, \quad \psi_{i,N_y} - U^*_{y;i,N_y-1}\psi_{i,N_y-1} = 0, \tag{3.5}$$

for $i = 1, \dots, N_x$. For the vector potential, we require that $\partial_y A_x = H_z$ and A_y is constant ($A_y = 0$) on the boundary.

Operators. The gradient of a scalar is a vector and is therefore defined at the midpoint of a link connecting two adjacent vertices. Thus,

$$(\nabla\phi)_{x;i,j} = (\partial_x\phi)(\boldsymbol{x}_{i,j} + \tfrac{1}{2}h_x\boldsymbol{e}_x) = h_x^{-1}(\phi_{i+1,j} - \phi_{i,j}), \tag{3.6}$$

with an analogous definition for the y component. The gauge-invariant $\boldsymbol{A}$-gradient $\nabla_{\boldsymbol{A}} = \nabla + \mathrm{i}\boldsymbol{A}$ is defined in a similar way, with

$$(\nabla_{\boldsymbol{A}}\psi)_{x;i,j} = h_x^{-1}(\psi_{i+1,j}U_{x;i,j} - \psi_{i,j}). \tag{3.7}$$

Thus, the discrete version of the twisted Laplacian $\Delta_{\boldsymbol{A}}$ is

$$\begin{aligned}(\Delta_{\boldsymbol{A}}\psi)_{i,j} = {} & h_x^{-2}(\psi_{i+1,j}U_{x;i,j} - 2\psi_{i,j} + \psi_{i-1,j}U^*_{x;i-1,j}) \\ & + h_y^{-2}(\psi_{i,j+1}U_{y;i,j} - 2\psi_{i,j} + \psi_{i,j-1}U^*_{x;i,j-1}).\end{aligned} \tag{3.8}$$

The discrete version of the (normal) Laplacian is defined in the usual way,

$$(\Delta\psi)_{i,j} = h_x^{-2}(\psi_{i+1,j} - 2\psi_{i,j} + \psi_{i-1,j}) + h_y^{-2}(\psi_{i,j+1} - 2\psi_{i,j} + \psi_{i,j-1}). \tag{3.9}$$

The magnetic induction, which is the curl of the vector potential, takes the form

$$B_{z;i,j} = h_x^{-1}(A_{y;i+1,j} - A_{y;i,j}) - h_y^{-1}(A_{x;i,j+1} - A_{x;i,j}). \tag{3.10}$$

We also need the divergence of the vector potential, which is given by

$$(\nabla \cdot \boldsymbol{A})_{i,j} = h_x^{-1}(A_{x;i,j} - A_{x;i-1,y}) + h_y^{-1}(A_{y;i,j} - A_{y;i,j-1}). \tag{3.11}$$

Algorithm. For numerical purposes, it is useful to treat the TDGL equations (4.3) and (4.4) as two separate equations, which are coupled only through certain fields and variables. The electromagnetic potentials ϕ and $\boldsymbol{A}$ are treated as static variables in the order parameter equation,

$$(\partial_t - (\mathrm{i}/\kappa)\phi)\psi - \Delta_{\boldsymbol{A}}\psi - (1 - |\psi|^2)\psi = 0. \tag{3.12}$$

The local nonlinear part of this equation,

$$(\partial_t - (\mathrm{i}/\kappa)\phi)\psi - (1 - |\psi|^2)\psi = 0, \tag{3.13}$$

is integrated in the simplest possible manner,

$$\psi_{i,j}(t+\Delta t) = \mathrm{e}^{-(\mathrm{i}/\kappa)\phi_{i,j}\Delta t}\left\{\psi_{i,j}(t) + \Delta t\left(1 - |\psi_{i,j}|^2\right)\psi_{i,j}\right\}. \tag{3.14}$$

The nonlocal part,

$$\partial_t \psi - \Delta_{\boldsymbol{A}} \psi = 0, \tag{3.15}$$

is integrated by using a backward Euler method, where the linear equation system is solved with the conjugate gradient method.

The equation for the vector potential,

$$\sigma\,\partial_t \boldsymbol{A} - \Delta \boldsymbol{A} - (1/\kappa)\boldsymbol{J}_s - \nabla \times \boldsymbol{H} = \boldsymbol{0}, \tag{3.16}$$

is linear and depends only indirectly on the order parameter through the supercurrent. If we treat the supercurrent as a static variable, we can integrate the equation easily, again using the backward Euler method. In the actual implementation, we use the fact that the domain is periodic to do a fast Fourier transform in the x direction, which leaves us with a tridiagonal system to solve in the y direction. This procedure is considerably faster than using an iterative method, such as the conjugate gradient method.

3.2. NUMERICAL RESULTS

We use a rectangular sample, periodic in the x direction and open in the y direction, with $N_x = N_y = 128$. We take $h_x = h_y = \frac{1}{2}\xi$, so the sample measures 64 coherence lengths in the periodic direction and 63·5 coherence lengths across. (The coherence length ξ is defined in (2.5).)

First, we considered this system with $\kappa = 200$ and an applied magnetic field $H_z = 0{\cdot}088\kappa$. With a relatively large value of κ, the surface barrier for vortex entry is low, and the system equilibrates relatively fast [14, 15]. The equilibration required 5×10^4 time steps with $\Delta t = 0{\cdot}4$. The magnetic field produces an almost perfect vortex lattice. Figure 5 gives a contour plot of the density of Cooper pairs $|\psi|^2$ at equilibrium; the zeros correspond to the centers of the vortices.

We then started from the configuration of Figure 5 to find equilibrium configurations for other values of κ, varying κ from $\kappa_{\min} = 40$ to $\kappa_{\max} = 800$. In this range, the ground states are comparable and similar to the one shown in Figure 5. Since the magnetization of a sample is proportional to $1/\kappa^2$, the vortex density decreases with κ; below $\kappa_{\min}$, the equilibrium state has fewer vortices, and a comparison becomes meaningless. Each equilibration required another 3×10^4 time steps.

Figure 6 gives the computed values of the quantities

$$\delta\psi = \|\psi_\kappa - \psi_{\kappa_{\max}}\|_{L^2}, \quad \delta B_z = \frac{\|B_{z,\kappa} - B_{z,\kappa_{\max}}\|_{L^2}}{\|H_z\|_{L^2}}, \tag{3.17}$$

for different values of κ. The data show a behavior like $1/\kappa^2$ down to $\kappa \approx 40$.

Figure 7 shows the average over x of $A_{x,\kappa} - A_{x,\kappa_{\max}}$ as a function of y in the bulk of the sample, for different values of κ.

The numerical results show that the solution of the TDGL equations converges as κ increases; in fact, they show quadratic convergence in the small parameter $1/\kappa$. Given the fact that the Ginzburg–Landau parameter of high-T_c superconducting materials is of the order of 50–100, we conclude that the limiting equation is a practical alternative in many applications.

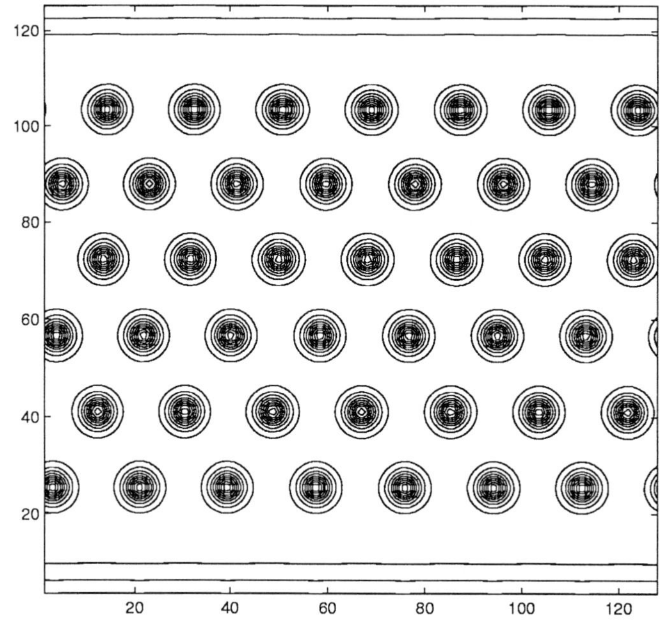

Figure 5. Contours of the density of Cooper pairs, $|\psi|^2$, for a system with $\kappa = 200$.

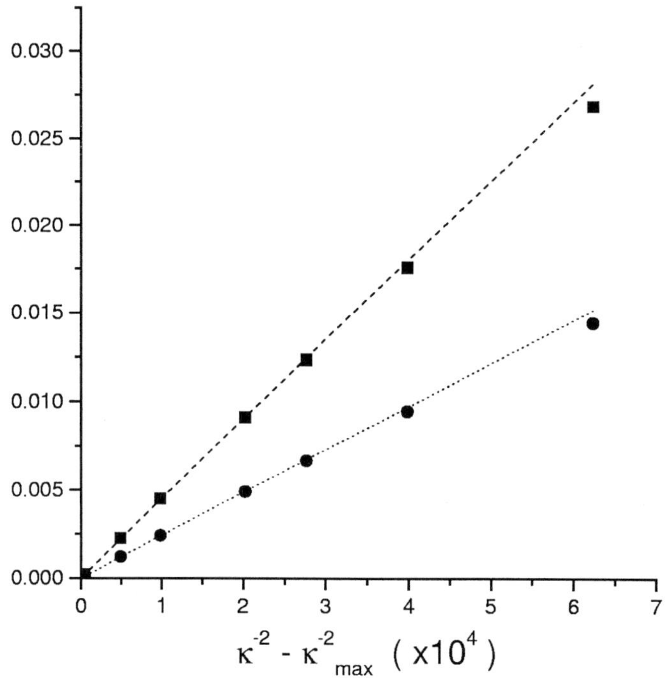

Figure 6. The quantities $\delta\psi$ (solid squares) and δB_z (solid discs) for $\kappa = 40, 50, 60, 70, 100, 140, 400, 800$. The straight lines correspond to $1/\kappa^2$ behavior.

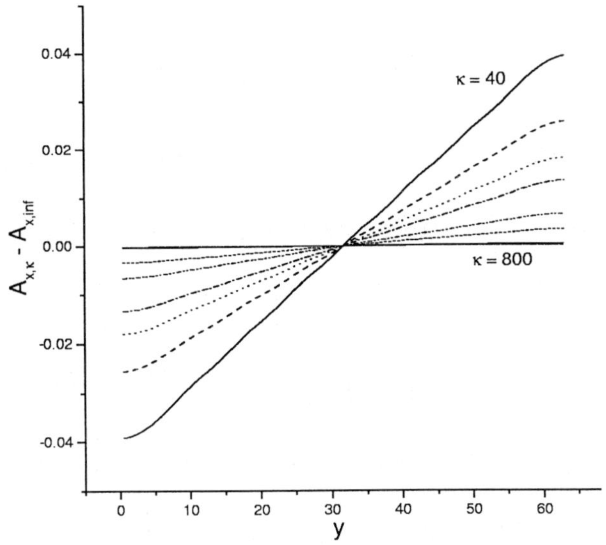

Figure 7. The average $\langle A_{x,\kappa} - A_{x,\kappa_{\max}} \rangle$ vs. y for $\kappa = 40, 50, 60, 70, 100, 140, 400, 800$.

The question thus becomes: What is the limiting equation, and can we confirm the numerical conclusions by rigorous arguments? We address this question in the next section.

4. Asymptotic analysis

We now return to the TDGL equations (2.13)–(2.18) and consider their limit as $\kappa \to \infty$. These are our standing hypotheses:

(H1) Ω is bounded in $\boldsymbol{R}^n (n = 2, 3)$, with a sufficiently smooth boundary $\partial\Omega$, for example, $\partial\Omega$ of class $C^{1,1}$.

(H2) The parameters κ and σ are real and positive.

(H3) $\boldsymbol{H}$ is independent of time; as a function of position, it satisfies the regularity condition $\boldsymbol{H} \in [W^{\alpha,2}(\Omega)]^n$ for some $\alpha \in (\frac{1}{2}, 1)$.

(H4) $\kappa \gg 1$; $\sigma = O(1)$ and $\boldsymbol{H} = O(\kappa)$ as $\kappa \to \infty$.

The assumptions **(H1)-(H3)** suffice to prove that the TDGL equations define a dynamical system in the Hilbert space

$$\mathcal{W}^{1+\alpha,2} = [W^{1+\alpha,2}(\Omega)]^2 \times [W^{1+\alpha,2}(\Omega)]^n; \tag{4.1}$$

see [9]. The space $W^{1+\alpha,2}(\Omega)$ is continuously imbedded in $W^{1,2}(\Omega) \cap L^\infty(\Omega)$, so ψ and $\boldsymbol{A}$ are bounded and differentiable with square-integrable (generalized) derivatives. **(H4)** is the operative hypothesis for the asymptotic analysis.

Table 2. Scaling

Applied Field	$\boldsymbol{H} = \kappa \boldsymbol{H}'$
Dependent variables	$\psi = \psi'$
	$\boldsymbol{A} = \kappa \boldsymbol{A}'$
	$\phi = \kappa^{-1}\phi'$
Electromagnetic variables	$\boldsymbol{B} = \kappa \boldsymbol{B}'$
	$\boldsymbol{J} = \kappa \boldsymbol{J}'$
	$\boldsymbol{E} = \kappa \boldsymbol{E}'$

4.1. Mathematical analysis

Scaling. We begin by scaling the TDGL equations, taking into account the fact that we are interested in the limit as $q_s \to 0$ (*weak coupling*), when the applied field is near the upper critical field. The scaling is done by means of the dimensionless GL parameter κ, which grows like $|q_s|^{-1}$.

Since $\boldsymbol{H} = 0(\kappa)$ as $\kappa \to \infty$, we begin by scaling $\boldsymbol{H}$ by a factor κ, $\boldsymbol{H} = \kappa \boldsymbol{H}'$. By scaling the vector potential by the same factor κ, we achieve that the electromagnetic variables are all of the same order.

The scalar potential is proportional to the charge density of the Cooper pairs, which is $O(|q_s|)$ as $q_s \to 0$. Hence, $\kappa\phi$ remains of order one. This suggests scaling ϕ by a factor κ^{-1}.

Table 2 summarizes the relation between the current (nondimensional) variables and their scaled (primed) counterparts. We adopt the latter as the new variables and work until further notice on the scaled problem. We omit all primes.

After scaling, the relevant parameter is κ^2, rather than κ, so we introduce ε,

$$\varepsilon = \kappa^{-2}. \tag{4.2}$$

The scaled TDGL equations are

$$(\partial_t - \mathrm{i}\varepsilon\phi)\psi - (\nabla + \mathrm{i}\boldsymbol{A})^2\psi - (1 - |\psi|^2)\psi = 0, \tag{4.3}$$

$$\sigma\partial_t \boldsymbol{A} - \Delta\boldsymbol{A} - \varepsilon\boldsymbol{J}_s - \nabla \times \boldsymbol{H} = \boldsymbol{0}, \tag{4.4}$$

where

$$\boldsymbol{J}_s = -\frac{1}{2\mathrm{i}}(\psi^*\nabla\psi - \psi\nabla\psi^*) - |\psi|^2\boldsymbol{A} = -\mathfrak{Im}[\psi^*(\nabla + \mathrm{i}\boldsymbol{A})\psi], \tag{4.5}$$

with the corresponding gauge condition,

$$\varepsilon\sigma\phi + \nabla \cdot \boldsymbol{A} = 0. \tag{4.6}$$

The boundary conditions associated with Equations (4.3) and (4.4) are

$$\boldsymbol{n} \cdot (\nabla + \mathrm{i}\boldsymbol{A})\psi = 0, \quad \boldsymbol{n} \times (\nabla \times \boldsymbol{A}) = \boldsymbol{n} \times \boldsymbol{H}, \quad \boldsymbol{n} \cdot \boldsymbol{A} = 0. \tag{4.7}$$

The electromagnetic variables are given by the expressions

$$\boldsymbol{B} = \nabla \times \boldsymbol{A}, \quad \boldsymbol{J} = \nabla \times \nabla \times \boldsymbol{A}, \quad \boldsymbol{E} = -\partial_t \boldsymbol{A} - \varepsilon \nabla \phi. \tag{4.8}$$

Reduction to homogeneous form. Next, we homogenize the problem. Let $\boldsymbol{A}_0$ be the (unique) minimizer of the convex quadratic form $Q_1 \equiv Q_1[\boldsymbol{A}]$,

$$Q_1[\boldsymbol{A}] = \int_\Omega \left[(\nabla \cdot \boldsymbol{A})^2 + |\nabla \times \boldsymbol{A} - \boldsymbol{H}|^2 \right] \mathrm{d}\boldsymbol{x}, \tag{4.9}$$

on $\mathrm{dom}(Q_1) = \{\boldsymbol{A} \in [W^{1,2}(\Omega)]^n : \boldsymbol{n} \cdot \boldsymbol{A} = 0 \text{ on } \partial\Omega\}$. This minimizer satifies the boundary-value problem

$$\nabla \times \nabla \times \boldsymbol{A} - \nabla \times \boldsymbol{H} = \boldsymbol{0}, \quad \nabla \cdot \boldsymbol{A} = 0 \quad \text{in } \Omega, \tag{4.10}$$

$$\boldsymbol{n} \times (\nabla \times \boldsymbol{A}) = \boldsymbol{n} \times \boldsymbol{H}, \quad \boldsymbol{n} \cdot \boldsymbol{A} = 0 \quad \text{on } \partial\Omega, \tag{4.11}$$

in the dual of $\mathrm{dom}(Q_1)$ with respect to the inner product in $[L^2(\Omega)]^n$. The mapping $\boldsymbol{H} \mapsto \boldsymbol{A}_0$ is linear, time independent, and continuous from $[W^{\alpha,2}(\Omega)]^n$ to $[W^{1+\alpha,2}(\Omega)]^n$ [16]. The contribution of the vector $\boldsymbol{A}_0$ to the magnetic field is

$$\boldsymbol{B}_0 = \nabla \times \boldsymbol{A}_0. \tag{4.12}$$

We substitute variables,

$$\boldsymbol{A} = \boldsymbol{A}_0 + \varepsilon \boldsymbol{A}', \tag{4.13}$$

and rewrite the (scaled) TDGL equations (4.3)–(4.7) in terms of ψ and $\boldsymbol{A}'$ (omitting the primes),

$$\partial_t \psi + \mathrm{i}\sigma^{-1}(\nabla \cdot (\varepsilon \boldsymbol{A}))\psi - (\nabla + \mathrm{i}(\boldsymbol{A}_0 + \varepsilon \boldsymbol{A}))^2 \psi - (1 - |\psi|^2)\psi = 0 \quad \text{in } \Omega, \tag{4.14}$$

$$\sigma \partial_t \boldsymbol{A} - \Delta \boldsymbol{A} - \boldsymbol{J}_s = \boldsymbol{0} \quad \text{in} \quad \Omega, \tag{4.15}$$

where

$$\boldsymbol{J}_s = -\frac{1}{2\mathrm{i}}(\psi^* \nabla \psi - \psi \nabla \psi^*) - |\psi|^2 (\boldsymbol{A}_0 + \varepsilon \boldsymbol{A}), \tag{4.16}$$

and

$$\boldsymbol{n} \cdot \nabla \psi = 0, \quad \boldsymbol{n} \times (\nabla \times \boldsymbol{A}) = \boldsymbol{0}, \quad \boldsymbol{n} \cdot \boldsymbol{A} = 0 \quad \text{on } \partial\Omega. \tag{4.17}$$

Functional formulation. We reformulate the system of Equations (4.14)–(4.17) as an ordinary differential equation for a vector-valued function $u = (\psi, \boldsymbol{A})$ from the time domain $(0, \infty)$ to a space of functions on Ω,

$$u = (\psi, \boldsymbol{A}) : [0, \infty) \to \mathcal{L}^2 = [L^2(\Omega)]^2 \times [L^2(\Omega)]^n. \tag{4.18}$$

The equation is

$$\frac{\mathrm{d}u}{\mathrm{d}t} + Au = f_0(u) + \varepsilon f_1(u), \tag{4.19}$$

where A is the linear operator in $\mathcal{L}^2$ associated with the quadratic form $Q \equiv Q[u]$,

$$Q[u] = \int_{\Omega} \left[|\nabla\psi|^2 + \sigma^{-1} \left((\nabla \cdot \boldsymbol{A})^2 + |\nabla \times \boldsymbol{A}|^2 \right) \right] \, \mathrm{d}\boldsymbol{x}, \tag{4.20}$$

on $\mathrm{dom}(Q) = \{u = (\psi, \boldsymbol{A}) \in \mathcal{L}^2 : \boldsymbol{n} \cdot \boldsymbol{A} = 0 \text{ on } \partial\Omega\}$. The functions f_0 and f_1 are nonlinear,

$$f_i(u) = (\varphi_i(\psi, \boldsymbol{A}), \sigma^{-1}\boldsymbol{F}_i(\psi, \boldsymbol{A})), \quad i = 0, 1, \tag{4.21}$$

where

$$\varphi_0(\psi, \boldsymbol{A}) = 2\mathrm{i}\boldsymbol{A}_0 \cdot (\nabla\psi) - |\boldsymbol{A}_0|^2\psi + (1 - |\psi|^2)\psi, \tag{4.22}$$

$$\varphi_1(\psi, \boldsymbol{A}) = \mathrm{i}(1 - \sigma^{-1})(\nabla \cdot \boldsymbol{A})\psi + 2\mathrm{i}\boldsymbol{A} \cdot (\nabla\psi) - (\boldsymbol{A}_0 \cdot \boldsymbol{A})\psi - |\boldsymbol{A}|^2\psi, \tag{4.23}$$

$$\boldsymbol{F}_0(\psi, \boldsymbol{A}) = 0, \tag{4.24}$$

$$\boldsymbol{F}_1(\psi, \boldsymbol{A}) = -\frac{1}{2\mathrm{i}}(\psi^*\nabla\psi - \psi\nabla\psi^*) - |\psi|^2(\boldsymbol{A}_0 + \varepsilon\boldsymbol{A}). \tag{4.25}$$

Given any $f = (\varphi, \sigma^{-1}\boldsymbol{F}) \in \mathcal{L}^2$, the equation $Au = f$ is equivalent with the system of uncoupled boundary-value problems

$$-\Delta\psi = \varphi \quad \text{in} \quad \Omega, \quad \boldsymbol{n} \cdot \nabla\psi = 0 \quad \text{on} \quad \partial\Omega, \tag{4.26}$$

$$-\Delta\boldsymbol{A} = \boldsymbol{F} \quad \text{in} \quad \Omega, \quad \boldsymbol{n} \times \boldsymbol{A} = \boldsymbol{0}, \quad \boldsymbol{n} \cdot \boldsymbol{A} = 0 \quad \text{on } \partial\Omega. \tag{4.27}$$

in the dual of $\mathrm{dom}(Q)$ with respect to the inner product in $\mathcal{L}^2$. The operator A is selfadjoint and positive definite in $\mathcal{L}^2$; hence, its fractional powers $A^{\theta/2}$ are well defined, they are unbounded if $\theta \geq 0$, and $\mathrm{dom}(A^{\theta/2})$ is a closed linear subspace of $\mathcal{W}^{\theta,2} = [W^{\theta,2}(\Omega)]^2 \times [W^{\theta,2}(\Omega)]^n$; see [17, Section 1.4].

The solution of (4.19) depends on ε; we denote it by u_ε. We compare u_ε with the solution u_0 of the reduced equation

$$\frac{\mathrm{d}u}{\mathrm{d}t} + Au = f_0(u). \tag{4.28}$$

Theorem 4.1 *There exists a positive constant C such that*

$$\|u_\varepsilon(t) - u_0(t)\|_{\mathcal{W}^{1+\alpha,2}} \leq C(\|u_\varepsilon(0) - u_0(0)\|_{\mathcal{W}^{1+\alpha,2}} + \varepsilon), \quad t \in [0, T]. \tag{4.29}$$

Proof. let B_R be the ball of radius R centered at the origin in $\mathcal{W}^{1+\alpha,2}$. Let $u_\varepsilon \in B_R$ and $u_0 \in B_R$ satisfy Equations (4.19) and (4.28), respectively, with initial data $u_\varepsilon(0)$ and $u_0(0)$. The difference $v = u_\varepsilon - u_0$ satisfies the differential equation

$$\frac{\mathrm{d}v}{\mathrm{d}t} + Av = f_0(u_\varepsilon) - f_0(u_0) + \varepsilon f_1(u_\varepsilon) \tag{4.30}$$

or, equivalently, the integral equation

$$v(t) = \mathrm{e}^{-tA}v(0) + \int_0^t \mathrm{e}^{-(t-s)A}[f_0(u_\varepsilon) - f_0(u_0) + \varepsilon f_1(u_\varepsilon)](s) \, \mathrm{d}s. \tag{4.31}$$

From the integral equation we obtain the estimate

$$\begin{aligned}\|v(t)\|_{\mathcal{W}^{1+\alpha,2}} \leq\ & \|\mathrm{e}^{-tA}\|_{\mathcal{W}^{1+\alpha,2}}\|v(0)\|_{\mathcal{W}^{1+\alpha,2}} + \int_0^t \|A^{(1+\alpha)/2}\mathrm{e}^{-(t-s)A}\|_{\mathcal{W}^{1+\alpha,2}} \\ & \times[\|f_0(u_\varepsilon) - f_0(u_0)\|_{L^2} + \varepsilon\|f_1(u_\varepsilon)\|_{L^2}](s)\,\mathrm{d}s.\end{aligned} \tag{4.32}$$

The operator norms satisfy the inequalities

$$\|\mathrm{e}^{-tA}\|_{\mathcal{W}^{1+\alpha,2}} \leq 1, \quad \|A^{(1+\alpha)/2}\mathrm{e}^{-(t-s)A}\|_{\mathcal{W}^{1+\alpha,2}} \leq C(t-s)^{-(1+\alpha)/2}; \tag{4.33}$$

see [17, Theorem 1.4.3]. Furthermore, adding and subtracting terms, we have

$$\begin{aligned}f_0(u_\varepsilon) - f_0(u_0) =\ & \big(2\mathrm{i}\boldsymbol{A}_0\cdot(\nabla(\psi_\varepsilon - \psi_0)) - |\boldsymbol{A}_0|^2(\psi_\varepsilon - \psi_0) \\ & +(1-|\psi_\varepsilon|^2 - |\psi_0|^2)(\psi_\varepsilon - \psi_0) - \psi_\varepsilon\psi_0(\psi_\varepsilon^* - \psi_0^*), 0\big),\end{aligned} \tag{4.33}$$

where

$$\begin{aligned}&\|2\mathrm{i}\boldsymbol{A}_0\cdot(\nabla(\psi_\varepsilon - \psi_0))\|_{L^2} \leq 2\|\boldsymbol{A}_0\|_{L^\infty}\|\psi_\varepsilon - \psi_0\|_{W^{1,2}} \\ &\quad \leq C\|\psi_\varepsilon - \psi_0\|_{W^{1+\alpha,2}} \leq C\|u_\varepsilon - u_0\|_{\mathcal{W}^{1+\alpha,2}}, \\ &\||\boldsymbol{A}_0|^2(\psi_\varepsilon - \psi_0)\|_{L^2} \leq C\|\boldsymbol{A}_0\|_{L^\infty}^2\|\psi_\varepsilon - \psi_0\|_{L^\infty} \\ &\quad \leq C\|\psi_\varepsilon - \psi_0\|_{W^{1+\alpha,2}} \leq C\|u_\varepsilon - u_0\|_{\mathcal{W}^{1+\alpha,2}},\end{aligned}$$

and the other terms are estimated similarly. Here, C is some (generic) positive constant, which may depend on $\boldsymbol{H}$ and Ω but not on u_ε or u_0. (In these inequalities we have used the continuity of the imbedding of $W^{1+\alpha,2}$ into $W^{1,2}\cap L^\infty$.) The result is an inequality of the type

$$\|f_0(u_\varepsilon) - f_0(u_0)\|_{L^2} \leq C\|u_\varepsilon - u_0\|_{\mathcal{W}^{1+\alpha,2}}, \tag{4.35}$$

showing that f_0 is Lipschitz from $\mathcal{W}^{1+\alpha,2}$ to $\mathcal{L}^2$.

Using similar estimates, we show that f_1 is bounded from $\mathcal{W}^{1+\alpha,2}$ to $\mathcal{L}^2$, so there exists a positive constant C such that

$$\|f_1(u_\varepsilon)\|_{L^2} \leq C. \tag{4.36}$$

Combining the estimates (4.33), (4.35), and (4.36) with the inequality (4.32), we conclude that there exist positive constants C_1 and C_2 such that

$$\|v(t)\|_{\mathcal{W}^{1+\alpha,2}} \leq \|v(0)\|_{\mathcal{W}^{1+\alpha,2}} + \varepsilon C_1 t^{(1-\alpha)/2} + C_2\int_0^t (t-s)^{-1+\alpha)/2}\|v(s)\|_{\mathcal{W}^{1+\alpha,2}}\,\mathrm{d}s. \tag{4.37}$$

Applying Gronwall's inequality, we obtain the estimate

$$\|v(t)\|_{\mathcal{W}^{1+\alpha,2}} \leq C(\|v(0)\|_{\mathcal{W}^{1+\alpha,2}} + \varepsilon), \quad t\in[0,T], \tag{4.38}$$

for some positive constant C. ■

It follows from Theorem 4.1 that, if the initial data are such that $\|u_\varepsilon(0) - u_0(0)\|_{\mathcal{W}^{1+\alpha,2}} = o(1)$ as $\varepsilon\downarrow 0$, then

$$\lim_{\varepsilon\to 0} u_\varepsilon = u_0 \tag{4.39}$$

in $C([0,T];\mathcal{W}^{1+\alpha,2})$ for any $T>0$. In particular, if $\|u_\varepsilon(0) - u_0(0)\|_{\mathcal{W}^{1+\alpha,2}} = O(\varepsilon)$, then the convergence in Equation (4.39) is $O(\varepsilon)$.

4.2. Interpretation and Final Remarks

It remains to translate the results hack in terms of the original variables. We denote the solution of the TDGL equations, Equations (2.13)–(2.18), by ψ_κ, $\boldsymbol{A}_\kappa$, ϕ_κ. The variables $\boldsymbol{A}_\kappa$ and ϕ_κ are related by the gauge condition $\sigma\phi_\kappa + \nabla\cdot\boldsymbol{A}_\kappa = 0$ at all times. Let $\boldsymbol{B}_\kappa = \nabla\times\boldsymbol{A}_\kappa$.

Let $\boldsymbol{A}_\infty$ be the solution of the boundary-value problem

$$\nabla\times\nabla\times\boldsymbol{A} - \nabla\times\boldsymbol{H} = \boldsymbol{0}, \quad \nabla\cdot\boldsymbol{A} = 0 \quad \text{in } \Omega, \tag{4.40}$$

$$\boldsymbol{n}\times(\nabla\times\boldsymbol{A}) = \boldsymbol{n}\times\boldsymbol{H}, \quad \boldsymbol{n}\cdot\boldsymbol{A} = 0 \quad \text{on } \partial\Omega, \tag{4.41}$$

and put $\boldsymbol{B}_\infty = \nabla\times\boldsymbol{A}_\infty$. The vector $\boldsymbol{A}_\infty$ and, hence, $\boldsymbol{B}_\infty$ do not vary with time. Let ψ_∞ satisfy the equations

$$\partial_t\psi - \Delta_{A_\infty}\psi - (1-|\psi|^2)\psi = 0 \quad \text{in} \quad \Omega, \qquad \boldsymbol{n}\cdot\nabla_{A_\infty}\psi = 0 \quad \text{on} \quad \partial\Omega. \tag{4.42}$$

Then it follows from Theorem 4.1 that there exists a positive constant C such that

$$\|\psi_\kappa(t) - \psi_\infty(t)\|_{W^{1+\alpha,2}} + \frac{\|\boldsymbol{B}_\kappa(t) - \boldsymbol{B}_\infty\|_{W^{\alpha,2}}}{\|\boldsymbol{H}\|_{W^{\alpha,2}}} \tag{4.43}$$

$$\leq C\left(\|\psi_\kappa(0) - \psi_\infty(0)\|_{W^{1+\alpha,2}} + \frac{\|\boldsymbol{B}_\kappa(0) - \boldsymbol{B}_\infty\|_{W^{\alpha,2}}}{\|\boldsymbol{H}\|_{W^{\alpha,2}}} + \frac{1}{\kappa^2}\right), \tag{4.44}$$

for all $t\in[0,T]$, $T>0$.

The approximation $(\psi_\infty, \boldsymbol{B}_\infty)$ is the 'frozen-field approximation'. Hence, the analysis shows that the solution of the TDGL equations converges to the frozen-field approximation, uniformly on compact time intervals, $[0,T]$ in the topology of $[W^{1+\alpha,2}(\Omega)]^2\times[W^{\alpha,2}(\Omega)]^n$, as soon as the initial data satisfy the asymptotic estimates $\|\psi_\kappa(0) - \psi_\infty\ (0)\|_{W^{1+\alpha,2}} = 0(1)$ and $\|\boldsymbol{B}_\kappa(0) - \boldsymbol{B}_\infty\|_{W^{\alpha,2}} = 0(\kappa)$ as $\kappa\to\infty$. Under slightly sharper conditions we obtain the order of convergence.

Corollary 4.1. *If*

$$\|\psi_\kappa(0) - \psi_\infty(0)\|_{W^{1+\alpha,2}} = O\left(\frac{1}{\kappa^2}\right) \quad \text{and} \quad \frac{\|\boldsymbol{B}_\kappa(0) - \boldsymbol{B}_\infty\|_{W^{\alpha,2}}}{\|\boldsymbol{H}\|_{W^{\alpha,2}}} = O\left(\frac{1}{\kappa^2}\right)$$

as $\kappa\to\infty$, *then*

$$\|\psi_\kappa(t) - \psi_\infty(t)\|_{W^{1+\alpha,2}} + \frac{\|\boldsymbol{B}_\kappa(t) - \boldsymbol{B}_\infty\|_{W^{\alpha,2}}}{\|\boldsymbol{H}\|_{W^{\alpha,2}}} = O\left(\frac{1}{\kappa^2}\right), \tag{4.45}$$

uniformly on compact intervals.

This result explains the numerical results presented in Section 3.

Remark 1. The asymptotic approximation procedure can be continued to higher order, as can be seen from a formal expansion. The equations for the order parameter and the vector potential decouple, and at each order one finds first the vector potential, then the order parameter. The vector potential satisfies a linear heat equation; for example, the first correction beyond $\boldsymbol{A}_\infty$ is $\kappa^{-1}\boldsymbol{A}$ where $\boldsymbol{A}$ satisfies the equation

$$-\sigma\,\partial_t\boldsymbol{A} + \Delta\boldsymbol{A} = \mathfrak{Im}[\psi_\infty^*\nabla_{A_\infty}\psi_\infty]. \tag{4.46}$$

Remark 2. The analysis given here differs at several points from the analysis of Ref. [10]. First, our scaling is slightly different and, we believe, more in tune with the physics; second, our regularity assumptions on the applied field are weaker; third, our proofs are more direct; and fourth, our results hold in a stronger topology.

5. Conclusion

In this article we have been concerned with the Ginzburg-Landau (GL) equations of superconductivity – a set of coupled nonlinear partial differential equations for a complex-valued order parameter and a vector-valued vector potential describing the state of a superconducting medium. The GL equations, although phenomenological, constitute the only model to study the dynamics and structure of magnetic-flux vortices in type-II superconductors. They have been used successfully to study, for example, potential-pinning mechanisms and structures of moving vortex lattices.

Numerical simulations of realistic superconductor configurations are, however, extremely time-consuming, and it is certainly desirable to use simpler models wherever possible. One such model is the 'frozen-field approximation', where one assumes that the vector potential, and thus the electromagnetic field, is given. One is thus left with an equation for the order parameter, whose solution yields the superconducting properties for the material in the given electromagnetic field.

The frozen-field model has been used in the past with some success, but its relation to the GL equations has neverbeen explained satisfactorily. The purpose of this article was to justify the frozen-field model as an asymptotic limit of the GL model. We have shown that the former follows from the latter in the limit when the charge of the Cooper pairs (the superconducting charge carriers) goes to zero while the applied field stays near the upper critical field. The convergence is, in fact, quadratic in the small parameter. The analysis thus confirms the validity of the frozen-field approximation, establishes its domain of applicability, and indicates possible refinements.

Numerical simulations based on the frozen-field approximation are far less demanding of computing resources than the full set of GL equations. Although we have not presented hard numbers here, reductions by an order of magnitude are common. While the solution of the GL equations easily takes of the order of a hundred hous on a massively parallel computer like the IBM SP, and parameters studies with the GL model are still prohibitively expensive, the frozen-field approximation offers an excellent alternative for prototype simulations provided, of course, the conditions for its applicability are met. The asymptotic analysis presented here should be useful in that determination.

Acknowledgements

We thank Professor Todd Dupont (University of Chicago) for stimulating discussions throughout the course of this investigation. We also acknowledge the work of Damien Declat (student from ISTIL, Lyon, France), who assisted in the development of an earlier version of the parallel computer program.

This work was supported by the Mathematical, Information, and Computational Sciences Division subprogram of the Office of Advanced Scientific Computing Research, U.S. Department of Energy, under Contract W-31-109-Eng-38. The second author was partially supported by the University of Chicago/Argonne National Laboratory Collaborative Grant No 96-011.

References

1. D.W. Braun, G.W. Crabtree, H.G. Kaper, A.E. Koshelev, G.K. Leaf, D.M. Levine and V.M. Vinokur, Structure of a moving vortex lattice. *Phys. Rev. Lett.* 76 (1996) 831–834.
2. G.W. Crabtree, D.O. Gunter, H.G. Kaper, A.E. Koshelev, G.K. Leaf and V.M. Vinokur, Numerical solution of driven vortex systems. *Phys. Rev. B* 61 (2000) 1446–1455.
3. I. Aranson and V. Vinokur, Surface instabilities and plastic deformation of vortex lattices. *Phys. Rev. Lett.* 77 (1996) 3208–3211.
4. M. Tinkham, *Introduction to Superconductivity* (2nd ed.). New York: McGraw-Hill (1996) xxi+454 pp.
5. V.L. Ginzburg and L.D. Landau, On the theory of superconductivity. *Zh. Eksp. Teor. Fiz.* (USSR) 20 (1950) 1064–1082; Engl. transl. in: D. ter Haar, *L.D. Landau; Men of Physics* (Vol.1). Oxford: Pergamon Press (1965) pp. 138–167.
6. Q. Du, M.D. Gunzburger and J.S. Peterson, Analysis and approximation of the Ginzburg–Landau model of superconductivity. *SIAM Rev.* 34 (1992) 54–81.
7. Q. Du, Global existence and uniqueness of solutions of the time-dependent Ginzburg–Landau model for superconductivity. *Appl. Anal.* 53 (1994) 1–18.
8. Q. Tang and S. Wang, Time-dependent Ginzburg–Landau equations of superconductivity. *Physica D* 88 (1995) 139–166.
9. J. Fleckinger-Pellé, H.G. Kaper and P. Takáč, Dynamics of the Ginzburg–Landau equations of superconductivity. *Nonlin. Anal.: Theory, Methods & Applic.* 32 (1998) 647–665.
10. Q. Du and P. Gray, High-kappa limits of the time-dependent Ginzburg–Landau model. *SIAM J. Appl. Math.* 56 (1996) 1060–1093.
11. A. Schmid, A time dependent Ginzburg–Landau equation and its application to a problem of resistivity in the mixed state. *Phys. Kondens. Materie* 5 (1966) 302–317.
12. L.P. Gor'kov and G.M. Eliashberg, Generalizations of the Ginzburg–Landau equations for non-stationary problems in the case of alloys with paramagnetic impurities. *Zh. Eksp. Teor. Fiz.* 54 612426 (1968); *Soviet Phys.-JETP* 27 (1968) 328–334.
13. L.P. Gor'kov and N. Kopnin, Vortex motion and resistivity of type-II superconductors in a magnetic field. *Soviet Phys. – Usp.* 18 (1976) 496–516.
14. V.L. Ginzburg, On the destruction and the onset of superconductivity in a magnetic field. *Soviet Phys. – JETP* 34(7) (1958) 78–87.
15. S.J. Chapman, Superheating field of type-II superconductors. *SIAM J. Appl. Math.* 55 (1995) 1233–1258.
16. V. Georgescu, Some boundary value problems for differential forms on compact Riemannian manifolds. *Ann. Mat. Pura Appl.* 122(4) (1979) 159–198.
17. D. Henry, *Geometric Theory of Semilinear Parabolic Equations*. Lecture Notes in Mathematics, Vol. 840. New York: Springer-Verlag (1981) 348 pp.

Journal of Engineering Mathematics **39:** 241–259, 2001.

Analytical approximations to the viscous glass-flow problem in the mould-plunger pressing process, including an investigation of boundary conditions

S.W. RIENSTRA and T.D. CHANDRA[1]
Department of Applied Mathematics and Computing Science, Eindhoven University of Technology, P.O. Box 513, 5600 MB Eindhoven, The Netherlands
[1]*Department of Mathematics of State University of MALANG, Malang, Indonesia*

Received 10 February 2000; accepted in revised form 15 September 2000

Abstract. Industrial glass is produced at temperatures above 600 °C, where glass becomes a highly viscous incompressible fluid, usually considered as Newtonian. In the production two phases may be distinguished, namely the pressing phase and the blowing phase. This study will be concerned with glass flow in the pressing phase, which is called thus because a blob of fluid glass (called a gob) is pressed in a mould by a plunger, such that the glass flows between mould and plunger, in order to obtain the preform of a bottle or jar, called a parison. In the blowing phase (not considered here) the parison is subsequently blown into the final shape of the product. By application of the slender geometry of mould and plunger and a cylindrical symmetry, a form of Reynolds's lubrication flow equations is obtained. These equations are solved by utilizing the incompressibility of the glass, by which the flux at any axial cross section is determined for prescribed plunger velocity, leading to analytical results in closed form for velocity field and pressure gradient. The glass level is implicitly defined by the integral over the varying volume which is to remain constant. The pressure may then be determined by integration. Special attention is given to the required boundary conditions. It is known that, depending on several problem parameters like temperature, pressure, and smoothness of the wall, the glass flow slips, to some extent, along the wall. Therefore, this study includes a general formulation of the boundary condition of partial slip in the form of a linear relation between shear stress and slip velocity, also known as Navier's slip condition. The coefficient of this relation, a positive number, may vary in our solution with axial position, but depends on the problem and is to be obtained from (for example) experiment. Two special cases, which seem to be relevant in practice, are considered as examples: (i) no slip on both plunger and mould; (ii) no slip on the mould and full slip (zero friction) on the plunger. The results are compared with fully numerical (FEM) solutions of a Stokes-flow model, and the agreement is good or excellent. Since in any practical situation it is not the plunger velocity which is prescribed, but (within practical limits) the force applied by the plunger, the problem of a prescribed plunger force has also been investigated.

Key words: glass, viscous flow, slender geometry, slip and no-slip boundary conditions, lubrication theory.

1. Introduction

Glass is a widely used packing material, for example in the form of jars and bottles in the food industry. The production of glass forms like jars proceeds more or less along the following lines [1, pp.612–613]. First, grains and additives, like soda, are heated in a tank. Here gas burners or electric heaters provide the heat necessary to warm the material up to some 1200 °C. At one end the liquid glass comes out and is led to a pressing or blowing machine. To obtain a glass form, a two-stage process is often used. First (Figure 1), a blob of hot glass called a gob falls into a configuration consisting of a *mould* and *plunger*. As soon as the entire glass drop has fallen into this mould, the plunger starts moving to press the glass. This process is

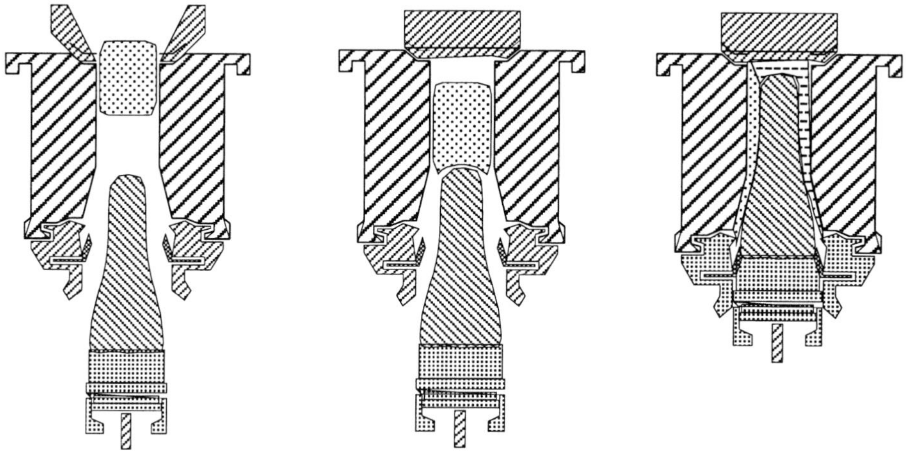

Figure 1. Pressing phase

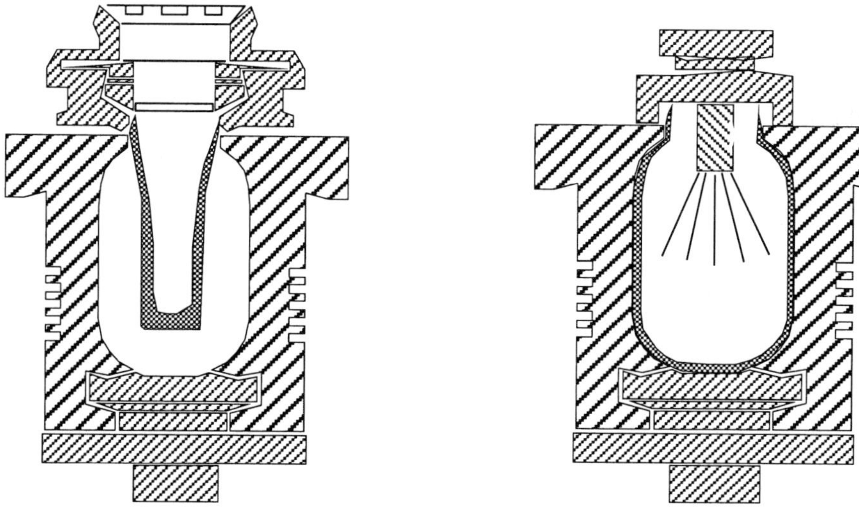

Figure 2. Blowing phase

called *pressing*. At the end, the glass drop is reshaped into a preform of a bottle or jar, called a *parison*. After a short period of time, for cooling purposes (the mould is kept at 500 °C), the parison is blown to its final shape in another mould. This process is called *blowing* (see Figure 2).

The contents of this paper concern glass flow in the pressing phase in manufacturing glass jars or parisons. A typical feature of the shape of a parison is that wall thickness and radius vary very slowly (except for the bottom part), see Figure 3. In the present paper, this slow variation will be utilized to obtain an explicit, analytical description of velocity and pressure gradient of the glass flow.

2. Governing equations

The motion of glass at temperatures above 600°C can be described by the Navier-Stokes equations for incompressible Newtonian fluids [2, p.3], given by

$$\rho\Big(\frac{\partial \boldsymbol{v}}{\partial t} + \boldsymbol{v}\cdot\nabla\boldsymbol{v}\Big) = \nabla\cdot\boldsymbol{\sigma}, \qquad \nabla\cdot\boldsymbol{v} = 0, \tag{1a,b}$$

where $\boldsymbol{v}$ denotes the velocity field, and ρ the density. Further, $\boldsymbol{\sigma}$ is the fluid stress tensor

$$\boldsymbol{\sigma} = -p\boldsymbol{I} + \boldsymbol{\tau}, \quad \text{or} \quad \sigma_{ij} = -p\,\delta_{ij} + \tau_{ij}, \tag{2}$$

where p is the pressure, $\boldsymbol{I} = (\delta_{ij})$ is the unit tensor, defined by $\delta_{ij} = 1$ if $i = j$, and $= 0$ otherwise, and $\boldsymbol{\tau}$ is the deviatoric or viscous stress tensor. In Newtonian fluids, a linear relationship is assumed between $\boldsymbol{\tau}$ and the deformation rate of the fluid element, expressed in the rate-of-strain tensor $\dot{\boldsymbol{\gamma}} = \nabla\boldsymbol{v} + (\nabla\boldsymbol{v})^{\mathrm{T}}$:

$$\boldsymbol{\tau} = \eta\,\dot{\boldsymbol{\gamma}} \quad \text{or} \quad \tau_{ij} = \eta\left(\frac{\partial v_i}{\partial x_j} + \frac{\partial v_j}{\partial x_i}\right), \tag{3}$$

where η is the dynamic viscosity. In general η depends on temperature, and may vary in space and time, but for a uniform temperature as we have here (approximately; see Section 4), η remains constant. Together with $\nabla\cdot\boldsymbol{v} = 0$, $\nabla\cdot\boldsymbol{\sigma}$ reduces to $-\nabla p + \eta\nabla^2\boldsymbol{v}$. As a result, Equations (1*a,b*) assume their common form

$$\rho\Big(\frac{\partial \boldsymbol{v}}{\partial t} + \boldsymbol{v}\cdot\nabla\boldsymbol{v}\Big) = -\nabla p + \eta\nabla^2\boldsymbol{v}, \qquad \nabla\cdot\boldsymbol{v} = 0. \tag{4a,b}$$

In view of the geometry of plunger and mould, we choose cylindrical coordinates (r, θ, z), and v, w, u will denote the r, θ, z components of the velocity $\boldsymbol{v}$. We assume the problem to be axisymmetric, so that both w and all θ-derivatives vanish, and the problem reduces to a two-dimensional one in the (r, z)-plane (see Figure 3).

3. Slender-geometry approximation

We will concentrate our analysis on the flow in the narrow annular duct between plunger and mould ($z > z_p$, see Figure 3). This region is very slender, and therefore amenable to asymptotic analysis [3, p.182], while at the same time the flow in the region between mould bottom and plunger top is practically stagnant, and therefore less important.

We will make Equations (4*a,b*) dimensionless by a suitable scaling. From the geometry of plunger and mould, we have two relevant length scales, the wall thickness of the parison, D, and the length of the plunger, L, while $D \ll L$. Except right near the plunger top, any variation in z scales on L and any variation in r scales on D. Therefore, we scale z with L and r with D, while we introduce the small parameter

$$\varepsilon = \frac{D}{L}. \tag{5}$$

The axial velocity u scales on a typical velocity V, while from the equation of mass conservation it follows that the radial velocity v has to scale on εV. As density ρ and viscosity η are

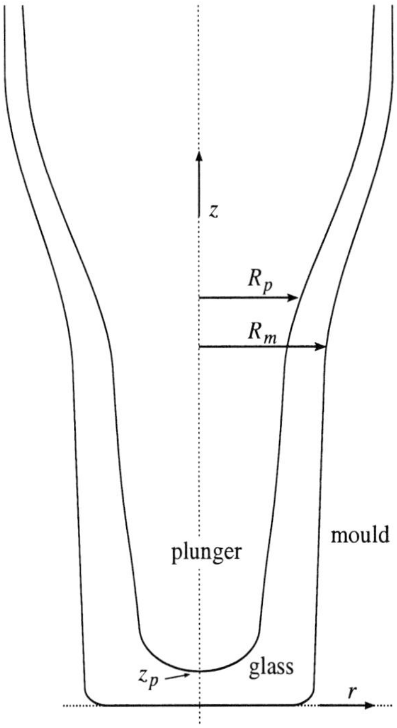

Figure 3. Sketch of configuration. Note that $R_p = R_p(z - z_p)$ is the surface of the plunger, $R_m = R_m(z)$ is the surface of the mould and z_p is position of the top of the plunger.

constant, they are parameters of the problem. Pressure p is to be scaled on $\eta VL/D^2$ (rather than ρV^2), because the glass flow is highly viscous, as we will see below. We have then

$$z = Lz^*, \quad r = Dr^*, \quad u = Vu^*, \quad v = \varepsilon Vv^*, \quad p = \frac{\eta VL}{D^2}p^*, \quad t = \frac{L}{V}t^*.$$

Now we substitute the above scalings in the Navier-Stokes equations and henceforth ignore the asterisks *, to obtain, respectively, the dimensionless z, r-components of the Navier-Stokes equations and the continuity equation, *viz.*

$$\varepsilon\mathrm{Re}\Big(\frac{\partial u}{\partial t} + u\frac{\partial u}{\partial z} + v\frac{\partial u}{\partial r}\Big) = -\frac{\partial p}{\partial z} + \varepsilon^2\frac{\partial^2 u}{\partial z^2} + \frac{1}{r}\frac{\partial}{\partial r}\Big(r\frac{\partial u}{\partial r}\Big), \tag{6a}$$

$$\varepsilon^3\mathrm{Re}\Big(\frac{\partial v}{\partial t} + u\frac{\partial v}{\partial z} + v\frac{\partial v}{\partial r}\Big) = -\frac{\partial p}{\partial r} + \varepsilon^4\frac{\partial^2 v}{\partial z^2} + \varepsilon^2\frac{\partial}{\partial r}\Big(\frac{1}{r}\frac{\partial}{\partial r}(rv)\Big), \tag{6b}$$

$$\frac{\partial u}{\partial z} + \frac{1}{r}\frac{\partial}{\partial r}(rv) = 0, \tag{6c}$$

where $\mathrm{Re} = \rho VD/\eta$ is the Reynolds number.

According to [4, p.7], the velocity of the plunger, which can be used as a characteristic velocity, if the cross section of the annular channel compared to the plunger cross section is not too small[1], is typically $V = 10^{-1}$ m/s. A typical channel width is $D \approx 10^{-2}$m. A typical length of the plunger is 10^{-1} m. The dynamic viscosity of glass varies greatly with temperature, but for a temperature around 800 °C it is of the order of 10^4 kg/ms. The density

[1] as follows from an axial mass flux balance; see Section 6.1

of glass is 2500 kg/m^3 [5, p.4]. Therefore, we obtain typically $\varepsilon = \frac{1}{10}$, $\mathrm{Re} = 2{\cdot}5 \times 10^{-4}$, $\varepsilon\mathrm{Re} = 2{\cdot}5 \times 10^{-5}$, and $\varepsilon^3\mathrm{Re} = 2{\cdot}5 \times 10^{-7}$, and we can ignore the inertia and radial-friction terms to obtain

$$\frac{\partial p}{\partial z} = \frac{1}{r}\frac{\partial}{\partial r}\Big(r\frac{\partial u}{\partial r}\Big), \quad \frac{\partial p}{\partial r} = 0, \quad \frac{\partial u}{\partial z} + \frac{1}{r}\frac{\partial}{\partial r}(rv) = 0. \tag{7a, b, c}$$

with an assumed error of $O(\varepsilon^2)$ because $\mathrm{Re} < \varepsilon$. This set of equations may be recognized as Reynolds's lubrication-flow equations in cylindrical coordinates [6, p.83].

4. The temperature remains constant

Although in viscous flow the energy equation is decoupled from the equation of motion (4*a*), if the flow is incompressible and Newtonian with constant viscosity, the validity of this assumption is not obvious in the present problem. The viscosity is highly temperature-dependent, as is seen from the VFT (Vogel, Fulcher, Tamman)-relation [1, p.936]

$$\log_{10}\eta = A + \frac{B}{T - T_0},$$

where A, B, T_0 are constants depending on the glass composition (for example $A = -2.4$, $B = 4032$, and $T_0 = 170$ for Soda-Silica Glasses [7]), while T is in degrees Celsius. In this case $T = 800$ yields a viscosity of $\eta = 10^4$ kg/ms.

The high viscous forces may generate heat by friction, and the walls may supply or absorb heat by conduction, such that, with the temperature, the viscosity varies along the flow.

To investigate this possibility we will analyse the energy equation and estimate the order of magnitude of the various terms. The energy conservation law for viscous and compressible fluids may be written as [8, p.10]

$$\frac{\partial}{\partial t}(\rho e) + \nabla\cdot(\rho \boldsymbol{v} e) = -p\nabla\cdot\boldsymbol{v} - \nabla\cdot\boldsymbol{q} + \boldsymbol{\tau}:\nabla\boldsymbol{v}, \tag{8}$$

where e denotes the internal energy per unit of mass, $\boldsymbol{q}$ is the heat flux due to heat conduction, and $\boldsymbol{\tau}$ denotes the viscous stress tensor. Since we are only interested in an order-of-magnitude estimate, we assume $e \simeq c_pT$, where c_p is the heat capacity at constant pressure and T is the absolute temperature [9, p.31], and for $\boldsymbol{q}$ Fourier's law $\boldsymbol{q} = -k\nabla T$, where k is the heat conductivity.

In an incompressible flow with constant c_p and k we have

$$\rho c_p\Big(\frac{\partial T}{\partial t} + \boldsymbol{v}\cdot\nabla T\Big) = k\nabla^2 T + \eta(\dot{\boldsymbol{\gamma}}:\nabla\boldsymbol{v}).$$

We non-dimensionalize as before

$$r = Dr^*,\ z = Lz^*,\ v = \varepsilon V v^*,\ u = Vu^*,\ T = T_m + \Delta T\,T^*,\ \eta = \eta_g\eta^*,\ t = \frac{L}{V}t^*,$$

where $\Delta T = T_g - T_m$ and η_g denotes the glass viscosity at the bulk temperature. (Note that in the rest of the paper we use a constant viscosity $\eta = \eta_g$.) We find (ignoring the asterisks $*$)

$$\varepsilon\Big[\frac{\partial T}{\partial t} + \boldsymbol{v}\cdot\nabla T\Big] = \frac{1}{\mathrm{Pe}}\Big[\frac{1}{r}\frac{\partial}{\partial r}\Big(r\frac{\partial T}{\partial r}\Big) + \varepsilon^2\frac{\partial^2 T}{\partial z^2}\Big] + \frac{\mathrm{Ec}}{\mathrm{Re}}\eta\Big[\Big(\frac{\partial u}{\partial r} + \varepsilon^2\frac{\partial v}{\partial z}\Big)^2 + 2\varepsilon^2\Big(\Big(\frac{\partial v}{\partial r}\Big)^2 + \Big(\frac{v}{r}\Big)^2 + \Big(\frac{\partial u}{\partial z}\Big)^2\Big)\Big] \tag{9}$$

with Reynolds number $\mathrm{Re} = \rho V D/\eta_g$, Eckert number $\mathrm{Ec} = V^2/c_p \Delta T$, and Peclet number $\mathrm{Pe} = \rho V D c_p/k$. (Note that Pe and Re are related through the Prandtl number as Pe = RePr, while PeEc/Re = Br is called the Brinkman number.) When we substitute the following values, typical of glass at 800 °C,

D :	typical parison length scale $= 10^{-2}$ m		T_g :	glass temperature	$= 800\,°\mathrm{C}$,
V :	typical plunger velocity	$= 10^{-1}$ m/s	T_m :	mould temperature	$= 500\,°\mathrm{C}$,
ρ :	glass density	$= 2500$ kg/m^3	c_p :	glass heat capacity	$= 1100$ J/kg °C
η_g :	glass dynamic viscosity	$= 10^4$ kg/ms	k :	glass thermal conductivity	$= 1.7$ W/m °C
L :	typical plunger length scale $= 10^{-1}$ m				

we get the following dimensionless numbers:

$$\varepsilon = 10^{-1}, \qquad \frac{1}{\mathrm{Pe}} = 6\cdot 2 \times 10^{-4}, \qquad \frac{\mathrm{Ec}}{\mathrm{Re}} = 1\cdot 2 \times 10^{-4}.$$

Both $1/(\varepsilon\mathrm{Pe})$ and $\mathrm{Ec}/(\varepsilon\mathrm{Re})$ are very small numbers, so in the bulk of the flow we can ignore the heat-conduction and thermal-production terms (the second and third term in Equation 9) against the convection (first) term. Hence, the energy equation simplifies to

$$\frac{\partial T}{\partial t} + \boldsymbol{v}\cdot\nabla T = \frac{\mathrm{d}T}{\mathrm{d}t} = 0,$$

indicating that the temperature is preserved following the flow. So, if we start with a uniform temperature field, it will remain uniform everywhere, and it follows that the viscosity also remains constant.

Note that this is not true in the thin temperature boundary layer along the wall, where the temperature varies from its value at the wall to the bulk temperature, and the conduction term is comparable with the convection term. From this balance,

$$\boldsymbol{v}\cdot\nabla T \sim \frac{1}{\varepsilon\mathrm{Pe}}\left[\frac{\partial^2 T}{\partial r^2} + \frac{1}{r}\frac{\partial T}{\partial r}\right],$$

it follows that the dimensionless boundary-layer thickness is of the order of $O((\varepsilon\mathrm{Pe})^{-1/3})$ (no slip) or $O((\varepsilon\mathrm{Pe})^{-1/2})$ (with slip). For the values considered this corresponds to boundary-layer thicknesses of, respectively, $\sim 2\times 10^{-1}$ and $\sim 0{\cdot}8\times 10^{-1}$, or 20% and 8% of the channel width. These values are, of course, not very small. However, to make progress we will for the moment consider them small enough to be ignored.

A small countereffect that may occur is that very close to the wall, where the glass temperature attains that of the wall, the viscosity increases by several orders of magnitude. For the present example the dimensionless viscosity is equal to $\eta(T_m) = 10^{5\cdot 8}$, which leads to $(\mathrm{Ec}/\varepsilon\mathrm{Re})\eta(T_m) = 760$. So, with little or no slip between glass and wall (otherwise $\dot{\boldsymbol{\gamma}} : \nabla\boldsymbol{v}$ is small) the fluid friction is not negligible very close to the wall and will generate some heat. This may increase the temperature locally, and slightly counteract the lower wall temperature. The resulting decrease of viscosity could be called "self-lubrication".

5. Boundary conditions

Depending on problem parameters such as wall temperature, fluid pressure, surface tension, or the presence of a lubricant like graphite powder [10, 11, 12, 13], the glass flow slips completely, partly or not at all along the wall. This means that the tangential component of the glass velocity $\boldsymbol{v}$ at the wall differs from the wall velocity $\boldsymbol{v}_w$, the difference being called the slip velocity. Since by assumption the glass flows in the (r, z)-plane and the plunger moves in the z-direction, it is sufficient to consider the tangential component in the (r, z)-plane.

Two slip conditions are commonly used. One is based on Coulomb friction [13], which is the assumption of a linear relation between tangential (shear) stress and normal stress (pressure), usually with a threshold value. The other is Navier's slip condition [6, p.87], which assumes a linear relation between the slip velocity and the shear stress. Physically, little is known as to which condition is essentially more correct. Therefore, we will take here Navier's slip condition, as it is mathematically more convenient here, since in our theory the pressure is only available as an integral, while the velocities are explicitly known.

The shear stress (Equations 2,3) applied to a surface with unit normal vector $\boldsymbol{n}$, defined as pointing *out* of the fluid, in the tangential direction $\boldsymbol{t}$, is given by $-(\boldsymbol{\sigma}\cdot\boldsymbol{n})\cdot\boldsymbol{t} = -(\boldsymbol{\tau}\cdot\boldsymbol{n})\cdot\boldsymbol{t}$. The slip velocity $(\boldsymbol{v} - \boldsymbol{v}_w)\cdot\boldsymbol{t}$ will be proportional to this stress in the (r, z)-plane

$$(\boldsymbol{v} - \boldsymbol{v}_w)\cdot\boldsymbol{t} = -s(\boldsymbol{\tau}\cdot\boldsymbol{n})\cdot\boldsymbol{t}\ , \tag{10}$$

where the slip factor s (a positive number) measures the amount of slip. (A better measure is the dimensionless version of s; see below.) There is no slip if $s = 0$, while there is no friction if $s = \infty$. The inverse, s^{-1}, might be called the friction factor. In general, s may be a function of position.

More advanced slip modelling, for example a nonlinear relation with s depending on pressure, is formally included in this way, if we in some way iteratively adapt s and the corresponding solution. We haven't investigated this possibility here, however.

For reference purposes, we note that, in cylindrical coordinates, the components of $\boldsymbol{\sigma}$ are given by

$$\begin{aligned} \sigma_{rr} &= -p + 2\eta\frac{\partial v}{\partial r}, & \sigma_{\theta r} &= \sigma_{r\theta} = 0, & \sigma_{\theta\theta} &= -p + 2\eta\frac{v}{r}, \\ \sigma_{zz} &= -p + 2\eta\frac{\partial u}{\partial z}, & \sigma_{zr} &= \sigma_{rz} = \eta\left(\frac{\partial v}{\partial z} + \frac{\partial u}{\partial r}\right), & \sigma_{\theta z} &= \sigma_{z\theta} = 0. \end{aligned}$$

5.1. Boundary conditions on the plunger

To apply the above conditions to the moving plunger surface, we recall that this is defined as:

$$r = R_p(z - z_p(t)),$$

where $z = z_p(t)$ is the position of the top of the plunger at time t. Subscript "$_p$" denotes the value at the plunger. Unless indicated otherwise, we will write here $R_p = R_p(z - z_p)$.

The unit outward normal $\boldsymbol{n}_p$ and the counterclockwise-directed unit tangent $\boldsymbol{t}_p$ in the (r, z)-plane are given by

$$\boldsymbol{n}_p = \frac{R'_p\boldsymbol{e}_z - \boldsymbol{e}_r}{\sqrt{1 + R'^2_p}}, \qquad \boldsymbol{t}_p = \frac{-\boldsymbol{e}_z - R'_p\boldsymbol{e}_r}{\sqrt{1 + R'^2_p}}. \tag{11}$$

From the definition of stress (2, 3) we have

$$(\boldsymbol{\sigma}\cdot\boldsymbol{n}_p)\cdot\boldsymbol{t}_p = \eta\,\frac{(1-R_p'^2)\Big(\dfrac{\partial u}{\partial r}+\dfrac{\partial v}{\partial z}\Big)-2R_p'\Big(\dfrac{\partial u}{\partial z}-\dfrac{\partial v}{\partial r}\Big)}{1+R_p'^2}. \tag{12}$$

To determine the slip velocity, we note that the plunger is moving down with speed $\frac{\mathrm{d}z_p}{\mathrm{d}t} = u_p$. Thus, the (dimensional) slip velocity at the plunger surface is

$$(\boldsymbol{v}-u_p\boldsymbol{e}_z)\cdot\boldsymbol{t}_p = ((u-u_p)\boldsymbol{e}_z+v\boldsymbol{e}_r)\cdot\frac{-\boldsymbol{e}_z-R_p'\boldsymbol{e}_r}{\sqrt{1+R_p'^2}} = -\frac{(u-u_p)+vR_p'}{\sqrt{1+R_p'^2}}. \tag{13}$$

From Equations (10, 11, 13), we obtain the boundary condition on the plunger

$$\frac{(u-u_p)+vR_p'}{\sqrt{1+R_p'^2}} = s_p\eta\,\frac{(1-R_p'^2)\Big(\dfrac{\partial u}{\partial r}+\dfrac{\partial v}{\partial z}\Big)-2R_p'\Big(\dfrac{\partial u}{\partial z}-\dfrac{\partial v}{\partial r}\Big)}{1+R_p'^2}, \tag{14}$$

where s_p is s at the plunger. The other boundary condition at the plunger follows from the observation that the wall is solid, so $\boldsymbol{v}\cdot\boldsymbol{n}=0$ or

$$v = (u-u_p)R_p'. \tag{15}$$

After introducing the following scaled slip coefficient s^*

$$s = \frac{D}{\eta}s^* \tag{16}$$

and using the same scaling as before, we obtain the dimensionless form of Equation (14) and (15) (with asterisks ignored),

$$\frac{(u-u_p)+\varepsilon^2 vR_p'}{\sqrt{1+\varepsilon^2R_p'^2}} = s_p\,\frac{(1-\varepsilon^2R_p'^2)\Big(\dfrac{\partial u}{\partial r}+\varepsilon^2\dfrac{\partial v}{\partial z}\Big)-2\varepsilon^2R_p'\Big(\dfrac{\partial u}{\partial z}-\dfrac{\partial v}{\partial r}\Big)}{1+\varepsilon^2R_p'^2}, \tag{17}$$

$$v = (u-u_p)R_p'. \tag{18}$$

As the wall temperature or the amount of lubricant may vary along the wall, it is practically important to let s be a function of position. The present solution is perfectly valid for any varying slip factor, as long as axial symmetry is preserved, and s is a smooth function of position z:

$$s = s(z). \tag{19}$$

For small ε we obtain (with an error $O(\varepsilon^2)$) for the boundary condition (17)

$$u-u_p = s_p\frac{\partial u}{\partial r} \qquad \text{at } r=R_p, \tag{20}$$

where s_p, if non-constant, is to be interpreted as a property of the moving surface and therefore to be read as $s_p = s_p(z - z_p)$.

The other boundary condition (18) will be left as it is.

5.2. Boundary conditions on the mould

The surface of the mould, given by

$$r = R_m(z),$$

(subscript "$_m$" denotes the value at the mould) has unit outward normal $\boldsymbol{n}_m$ and counterclockwise-directed unit tangent $\boldsymbol{t}_m$ given by

$$\boldsymbol{n}_m = \frac{-R'_m \boldsymbol{e}_z + \boldsymbol{e}_r}{\sqrt{1 + R'^{\,2}_m}}, \qquad \boldsymbol{t}_m = \frac{\boldsymbol{e}_z + R'_m \boldsymbol{e}_r}{\sqrt{1 + R'^{\,2}_m}}. \tag{21}$$

In a similar way as for the boundary conditions on the plunger, we obtain the boundary condition on the mould as

$$\frac{u + vR'_m}{\sqrt{1 + R'^{\,2}_m}} = -s_m \eta \frac{(1 - R'^{\,2}_m)\left(\dfrac{\partial u}{\partial r} + \dfrac{\partial v}{\partial z}\right) - 2R'_m\left(\dfrac{\partial u}{\partial z} - \dfrac{\partial v}{\partial r}\right)}{1 + R'^{\,2}_m}, \tag{22}$$

and

$$v = uR'_m. \tag{23}$$

In dimensionless form the relations are,

$$\frac{u + \varepsilon^2 vR'_m}{\sqrt{1 + \varepsilon^2 R'^{\,2}_m}} = -s_m \frac{(1 - \varepsilon^2 R'^{\,2}_m)\left(\dfrac{\partial u}{\partial r} + \varepsilon^2 \dfrac{\partial v}{\partial z}\right) - 2\varepsilon^2 R'_m\left(\dfrac{\partial u}{\partial z} - \dfrac{\partial v}{\partial r}\right)}{1 + \varepsilon^2 R'^{\,2}_m}, \tag{24}$$

and

$$v = uR'_m, \tag{25}$$

Similarly, for small ε, boundary condition (24) becomes

$$u = -s_m \frac{\partial u}{\partial r} \qquad \text{at } r = R_m. \tag{26}$$

The other condition (25) will be left as it is.

5.3. The free surface

As the blob of glass does not initially fill the mould completely (see Figure 1), there is a free surface of glass moving into the annular duct between mould and plunger. At the free boundary, the normal stress must be equal to external pressure p_0, which is assumed to be constant

$$(\boldsymbol{\sigma} \cdot \boldsymbol{n}) \cdot \boldsymbol{n} = p_0,$$

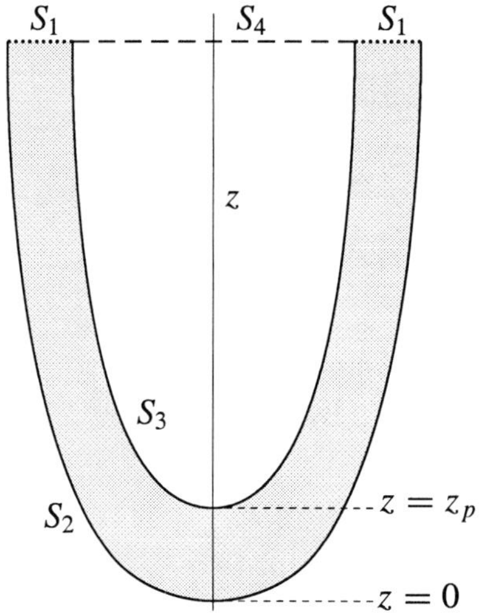

Figure 4. Sketch of the control surfaces to calculate the flux.

and the tangential stress must be equal to zero

$$(\boldsymbol{\sigma} \cdot \boldsymbol{n}) \cdot \boldsymbol{t} = 0.$$

In our slender-geometry approximation the exact shape of the free surface cannot be determined. In its neighbourhood the flow scales both in the r and z directions on the thickness D of the annular channel. In other words, the flow is no longer slowly varying in z, rendering the present approximation invalid.

Within the present approximation (and assumption of circular symmetry), we will deal with the average level b of the free surface as follows (in scaled variables):

$$p = 0 \qquad \text{at } z = b(t), \tag{27}$$

where b is a function of the time-dependent geometry, implicitly defined in such a way that the (incompressible) glass volume between $z = 0$ and $z = b(t)$ is constant for all t.

6. Some auxiliary results

In this section, we derive some expressions (to be used later) for the axial flux and the total force on the plunger.

6.1. THE FLUX

Consider the flux at some level z through cross section S_1 (see Figure 4). The value of this flux depends on z, since the plunger moves downwards and causes the glass to move upwards through a varying cross section. We will use this value later to find the pressure gradient.

Since glass is an incompressible fluid, we have from Gauss's divergence theorem

$$0 = \int_{\Omega} \nabla \cdot \boldsymbol{v} \, \mathrm{d}\boldsymbol{x} = \int_{\partial\Omega} \boldsymbol{v} \cdot \boldsymbol{n} \, \mathrm{d}S = \int_{S_1} \boldsymbol{v} \cdot \boldsymbol{n} \, \mathrm{d}S + \int_{S_2} \boldsymbol{v} \cdot \boldsymbol{n} \, \mathrm{d}S + \int_{S_3} \boldsymbol{v} \cdot \boldsymbol{n} \, \mathrm{d}S \tag{28}$$

where $\boldsymbol{v} = u\boldsymbol{e}_z + \varepsilon v\boldsymbol{e}_r$. Since the mould is stationary and impermeable, we have

$$\int_{S_2} \boldsymbol{v}\cdot\boldsymbol{n}\,\mathrm{d}S = 0,$$

while the flux through S_1 is given by

$$\int_{S_1} \boldsymbol{v}\cdot\boldsymbol{n}\,\mathrm{d}S = 2\pi \int_{R_p}^{R_m} ru(r,z)\,\mathrm{d}r\ .$$

To calculate $\int_{S_3} \boldsymbol{v}\cdot\boldsymbol{n}\,\mathrm{d}S$, we note that it would make no difference for the amount of glass displaced by the plunger if the plunger were filled with glass, because glass is just as incompressible as the solid plunger. Therefore, instead of control surface S_3, we might as well use surface S_4, which yields more easily the result

$$\int_{S_3} \boldsymbol{v}\cdot\boldsymbol{n}\,\mathrm{d}S = \int_{S_4} \boldsymbol{v}\cdot\boldsymbol{n}\,\mathrm{d}S = \pi u_p R_p^2\ .$$

It follows that the flux is given by

$$2\pi \int_{R_p}^{R_m} ru(r,z)\,\mathrm{d}r = -\pi u_p R_p^2\ . \tag{29}$$

6.2. The total force on the plunger

In this section we discuss the total force on the plunger and use the result to find the velocity of the plunger. We return for a moment to a dimensional formulation. Later, we will turn to the dimensionless formulation.

The force (= stress×surface) in direction $\boldsymbol{e}_k$, applied to an infinitesimal surface element $\mathrm{d}S$ with outward normal $\boldsymbol{n}$, is by definition

$$-(\boldsymbol{\sigma}\cdot\boldsymbol{n})\cdot\boldsymbol{e}_k\,\mathrm{d}S = -\sum_j \sigma_{kj}n_j\,\mathrm{d}S.$$

At the plunger surface with unit normal $\boldsymbol{n}_p$, given in (11), the stress in the z-direction is

$$-(\boldsymbol{\sigma}\cdot\boldsymbol{n}_p)\cdot\boldsymbol{e}_z = \left\{\left(p - 2\eta\frac{\partial u}{\partial z}\right)R_p' + \eta\left(\frac{\partial v}{\partial z} + \frac{\partial u}{\partial r}\right)\right\}\frac{1}{\sqrt{1+R_p'^2}},$$

where $R_p = R_p(z - z_p(t))$.

A (circular) surface element is given by

$$\mathrm{d}S = 2\pi R_p\sqrt{1+R_p'^2}\,\mathrm{d}z.$$

Thus, the total force on the plunger between top level $z = z_p(t)$ and glass level $z = b(t)$ is given by

$$f = 2\pi \int_{z_p}^{b} \left[\left(p - 2\eta\frac{\partial u}{\partial z}\right)R_p' + \eta\left(\frac{\partial v}{\partial z} + \frac{\partial u}{\partial r}\right)\right]_{r=R_p} R_p\,\mathrm{d}z. \tag{30}$$

Note that this glass level, as a function of time, is yet to be determined (see next section).

We render Equation (30) dimensionless by using the same scaling as before, with $b = Lb^*$, and scale the force as

$$f = 2\pi\eta V L f^*.$$

This yields (ignoring again the asterisks) :

$$f = \int_{z_p}^{b} \left[\left(p - 2\varepsilon^2 \frac{\partial u}{\partial z} \right) R_p' + \varepsilon^2 \frac{\partial v}{\partial z} + \frac{\partial u}{\partial r} \right]_{r=R_p} R_p \, \mathrm{d}z \simeq \int_{z_p}^{b} \left[p R_p' + \frac{\partial u}{\partial r} \right]_{r=R_p} R_p \, \mathrm{d}z. \tag{31}$$

7. Results

7.1. The velocity and pressure fields

Now we are ready to solve the Equations (**??**a-c) with boundary conditions (20,18) and (26,25). First we note that p is a function of z only. Then Equation (**??**a)

$$\frac{\mathrm{d}p}{\mathrm{d}z} = \frac{\partial^2 u}{\partial r^2} + \frac{1}{r}\frac{\partial u}{\partial r} = \frac{1}{r}\frac{\partial}{\partial r}\left(r \frac{\partial u}{\partial r} \right)$$

has solution

$$u = \frac{1}{4} r^2 \frac{\mathrm{d}p}{\mathrm{d}z} + A(z) \ln(r) + B(z).$$

Using boundary conditions (20) and (26), we obtain:

$$u = \frac{1}{4}\frac{\mathrm{d}p}{\mathrm{d}z}\left[r^2 + \frac{\beta_m(\alpha_p - \log r) - \beta_p(\alpha_m - \log r)}{\alpha_m - \alpha_p} \right] + u_p \frac{\alpha_m - \log r}{\alpha_m - \alpha_p}, \tag{32}$$

where it has been found convenient to introduce auxiliary parameters $\alpha_{p,m}$, $\beta_{p,m}$ and $\gamma_{p,m}$ as follows

$$\sigma_m = s_m / R_m, \ \sigma_p = -s_p / R_p, \ \ \alpha = \log R + \sigma, \ \ \beta = R^2(1 + 2\sigma), \ \ \gamma = R^4(1 + 4\sigma), \tag{33}$$

showing at the same time the curious result that the essential slip parameter in a duct is apparently not s itself, but the product of s and the wall curvature.

Using Equation (32), the expression for u (29), the relation for the flux, and the following integral for ru

$$\int ru(r,z)\,\mathrm{d}r = \tfrac{1}{8} r^2 \frac{\mathrm{d}p}{\mathrm{d}z}\left(\tfrac{1}{2} r^2 + \frac{\beta_m(\alpha_p + \frac{1}{2} - \log r) - \beta_p(\alpha_m + \frac{1}{2} - \log r)}{\alpha_m - \alpha_p} \right] + \dots$$
$$+ \tfrac{1}{2} r^2 u_p \frac{\alpha_m + \frac{1}{2} - \log r}{\alpha_m - \alpha_p}, \tag{34}$$

we can determine the pressure gradient to find

$$\frac{\mathrm{d}p}{\mathrm{d}z} = -4u_p \frac{\beta_m - \beta_p}{(\beta_m - \beta_p)^2 - (\alpha_m - \alpha_p)(\gamma_m - \gamma_p)} \tag{35}$$

leading with Equation (27) to

$$p(z) = -\int_z^b \frac{\mathrm{d}p}{\mathrm{d}z}\,\mathrm{d}z\,. \tag{36}$$

Finally we solve Equation (**??**c) with boundary conditions (18) or (25). (Only one will be needed, as the other has already been used implicitly when we determined the axial flux). We can rewrite Equation (**??**c) as

$$\frac{\partial}{\partial r}(vr) + r\frac{\partial u}{\partial z} = 0,$$

and then use (25), to obtain

$$v = \frac{A(z)}{r} - \frac{1}{r}\int r\frac{\partial u}{\partial z}\,\mathrm{d}r = \frac{1}{r}\Big(R_m R_m' u(R_m, z) + \int_r^{R_m} r\frac{\partial u}{\partial z}\,\mathrm{d}r\Big) = \frac{1}{r}\frac{\mathrm{d}}{\mathrm{d}z}\int_r^{R_m} ru(r,z)\,\mathrm{d}r. \tag{37}$$

Upon substituting (34) in (37) and subsequent differentiation, we obtain (a complicated expression for) v (see appendix). It may be verified that boundary condition (18) indeed is satisfied.

7.2. THE FORCE ON THE PLUNGER

By substituting the above results in Equation (31), we obtain to leading order

$$f = \int_{z_p}^b \Big(p\frac{\mathrm{d}R_p}{\mathrm{d}z} + \Big[\frac{\partial u}{\partial r}\Big]_{r=R_p}\Big) R_p\,\mathrm{d}z = \int_{z_p}^b \Big(-\tfrac{1}{2}R_p^2\frac{\mathrm{d}p}{\mathrm{d}z} + \Big[\frac{\partial u}{\partial r}\Big]_{r=R_p} R_p\Big)\,\mathrm{d}z =$$

$$= u_p \int_{z_p}^b \frac{\gamma_m - \gamma_p}{(\beta_m - \beta_p)^2 - (\alpha_m - \alpha_p)(\gamma_m - \gamma_p)}\,\mathrm{d}z. \tag{38}$$

Although the approximation is strictly speaking no longer valid at $z = z_p$, the integral converges, and the contribution of the plunger-tip area does seem to be asymptotically of lower order.

An interesting conserved quantity of the process is the time integral of the plunger force. (Dimensionally equivalent to a net change of momentum, although inertia plays no rôle in the present model.) It only depends on the end points z_1 and z_2 and not on the way the plunger moves in time, because f depends linearly on u_p.

$$\int_{t_1}^{t_2} f(t)\,\mathrm{d}t = \int_{z_1}^{z_2}\int_{z_p}^b \frac{\gamma_m - \gamma_p}{(\beta_m - \beta_p)^2 - (\alpha_m - \alpha_p)(\gamma_m - \gamma_p)}\,\mathrm{d}z\,\mathrm{d}z_p\,. \tag{39}$$

This is not a unique property of the approximation, but a direct consequence of the fact that in the Stokes equations time is only a parameter, so that velocities may be scaled on u_p, and the plunger force f on $\eta u_p z_p$.

7.3. A PRESCRIBED FORCE OR VELOCITY OF THE PLUNGER

In this section, we consider the relation between time and velocity of the plunger, assuming the velocity of, or the force on the plunger to be a prescribed function of t and the total volume of the glass is constant.

During the pressing phase, the plunger moves downwards as the glass moves upwards. So the plunger velocity u_p or the total force f, the position of the top of the plunger z_p, and the glass level b vary with time. Therefore we have to find a system of three equations for z_p, b, and u_p or f.

First, we observe that for all t the (scaled) volume Ω of the total amount of glass is equal to the constant

$$\Omega = \pi \int_0^b R_m^2 \, \mathrm{d}z - \pi \int_{z_p}^b R_p^2 \, \mathrm{d}z. \tag{40}$$

So, if we could solve this equation explicitly, we would have a functional relationship between b and z_p. This, however, is only possible in the simplest cases, for example when both R_p and R_m are parabolas. In general it has to be solved numerically for each time step. A natural approach is therefore to rewrite this equation as a differential equation in time that can be solved by standard numerical integration routines.

After differentiating Equation (40), using the defining relation between z_p and u_p and the relation (38) between f and u_p, we obtain the following system of equations

$$\frac{\mathrm{d}z_p}{\mathrm{d}t} = u_p. \tag{41a}$$

$$\frac{\mathrm{d}b}{\mathrm{d}t} = -u_p \left. \frac{R_p^2}{R_m^2 - R_p^2} \right|_{z=b}. \tag{41b}$$

$$f = u_p \int_{z_p}^b G(z, z_p) \, \mathrm{d}z, \quad G(z, z_p) = \frac{\gamma_m - \gamma_p}{(\beta_m - \beta_p)^2 - (\alpha_m - \alpha_p)(\gamma_m - \gamma_p)}. \tag{41c}$$

which can be integrated numerically with either $f(t)$ or $u_p(t)$ given. In the simpler case when u_p is given, Equation (41c) only defines f and is decoupled from the others.

8. Examples

In this section we present two examples. In the first we compare the present analytic expressions for u and εv with numerical results obtained from a numerical model for glass flow as developed by the TUE Scientific Computing Group of Professor R.M.M. Mattheij. We use here simple parabolic profiles for the plunger and mould, and the plots are in terms of non-dimensional quantities.

In the second example we consider the velocity of the plunger as it results from a given force. In addition, we use a practically more representative geometry of plunger and mould, as given in Figure 7b. The plots are expressed in dimensional quantities.

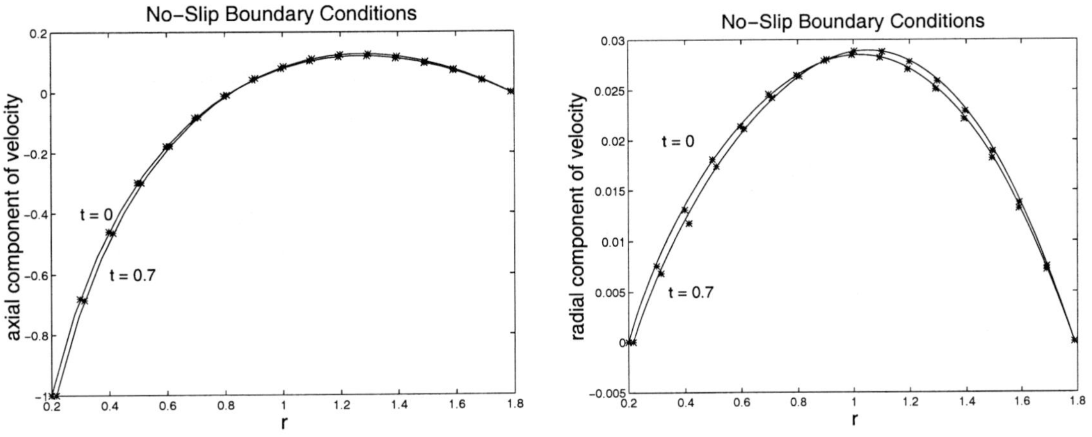

a. Axial velocity u *b.* Radial velocity εv

Figure 5. Axial and radial velocity. Parabolic geometry, no-slip. (Dimensionless.)

8.1. A GIVEN PLUNGER VELOCITY

Define the geometries and motion of plunger and mould by the following (dimensionless) expressions

$$R_p(z) = 0{\cdot}1\sqrt{5}\sqrt{z}, \quad R_m(z) = 0{\cdot}8\sqrt{5}\sqrt{z}, \quad z_p(t) = 1 - t,$$

where $\varepsilon = \frac{1}{5}$ and $u_p = -1$, while we will consider $z = 1$. Two cases are considered: (i) with no-slip boundary conditions ($s_p = s_m = 0$) and (ii) with mixed boundary conditions ($s_p = \infty$ and $s_m = 0$). The results will be presented in dimensionless form, although for a proper comparison the velocities will be scaled both on the same (plunger) velocity. In other words, in the figures the axial velocity u and radial velocity εv are plotted.

In Figure 5 we have case (i). Note that the order of magnitude of the axial velocity u is indeed the same as that of the plunger (dimensionless unity), because of the no-slip condition at both sides. It is seen that there is an exceptionally good agreement between the numerical ($*$) and the analytic solution (solid line).

Next we consider case (ii), with mixed boundary conditions. By taking $s_p \to \infty$ in Equation (20), we obtain $\frac{\partial u}{\partial r} = 0$. Next, we have of course Equation (18). On the stationary mould there is no slip, so here we have simply the boundary condition $u = v = 0$. As there is complete slip at the plunger wall, the velocity of the plunger plays no direct role in the problem. The axial velocity is now determined by the effect of plunger displacement. As a result, the order of magnitude of u is considerably lower than unity. The results of this case are presented in Figure 6. In the axial velocity (Figure 6*a*) the numerical solution ($*$) appears to differ a little from the analytical solution (solid line). This is, however, exactly consistent with the (conjectured) error of the asymptotic approximation, which is $O(\varepsilon^2) \sim 4\%$. Since $s_p = \infty$, Equation (10) becomes $(\boldsymbol{\sigma}\cdot\boldsymbol{n})\cdot\boldsymbol{t} = 0$ at the plunger. For small ε this reduces to the approximate value $\frac{\partial u}{\partial r} = 0 + O(\varepsilon^2)$. This is different from the no-slip case where the condition $u = 0$ is not only satisfied approximately but exactly.

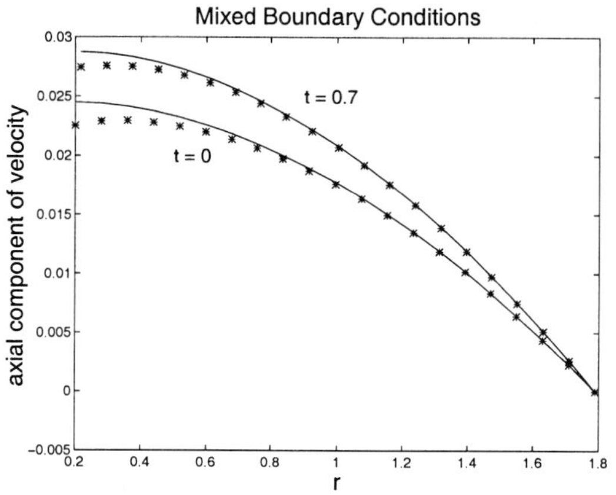

a. Axial velocity u

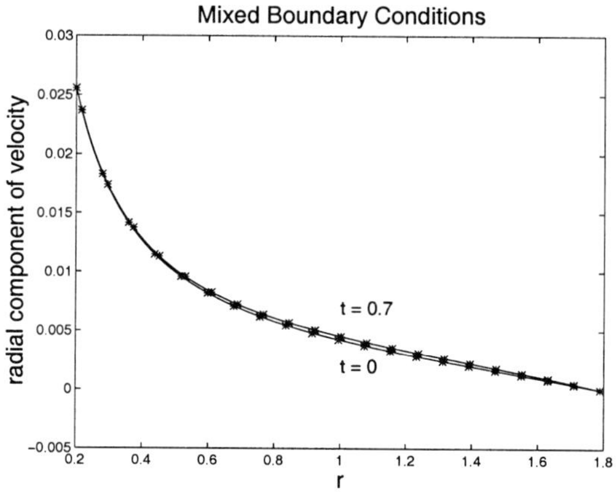

b. Radial velocity εv

Figure 6. Axial and radial velocity. Parabolic geometry, mixed slip. (Dimensionless.)

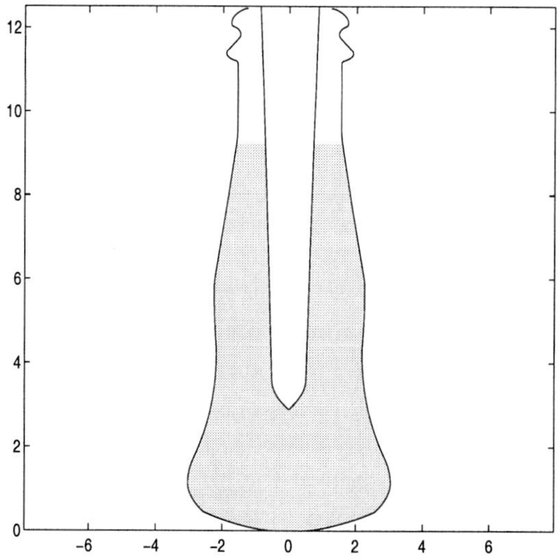

a. Plunger is going down

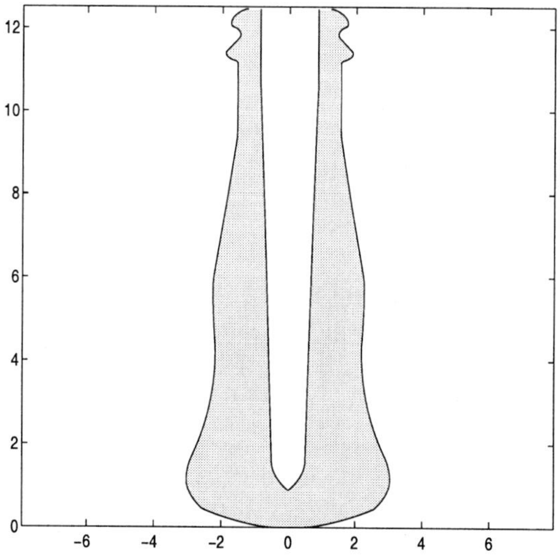

b. Plunger in end position

Figure 7. Realistic geometry of parison (units: cm)

8.2. A GIVEN PLUNGER FORCE

In this example we consider a more realistic situation. We will take a geometry of a real parison (Figure 7*b*), and we will apply a given force from which the plunger velocity results, using the results of Sections 7.2 and 7.3, all in dimensional units. As in example 1, we use two types of boundary condition: no slip ($s_p = s_m = 0$) and mixed ($s_p = \infty$ and $s_m = 0$). The force on the plunger will be assumed constant. Figure 7*b* shows the final position of the plunger in the pressing phase. In order to make the calculation easier, we calculated backwards in time, starting with this final position at $t = 0$. Since the Stokes equations do not include inertial effects, the sign of the time is irrelevant and the results of a forward and a backward calculation are exactly the same. The problem parameters, related to these cases, are the same as given in the table of section 4, namely: $V = 10\,\mathrm{cm/s}$, $D = 1\,\mathrm{cm}$, $\eta = 10^4\,\mathrm{kg/(s\,m)}$, $\varepsilon = 0{\cdot}1$

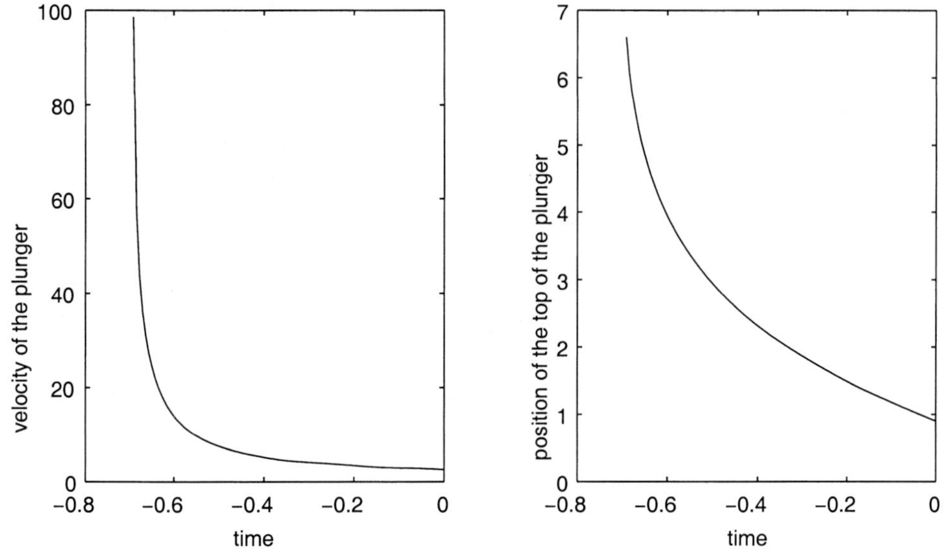

Figure 8. Given force = 7500 N, and $s_p = s_m = 0$ (units: sec and cm/s).

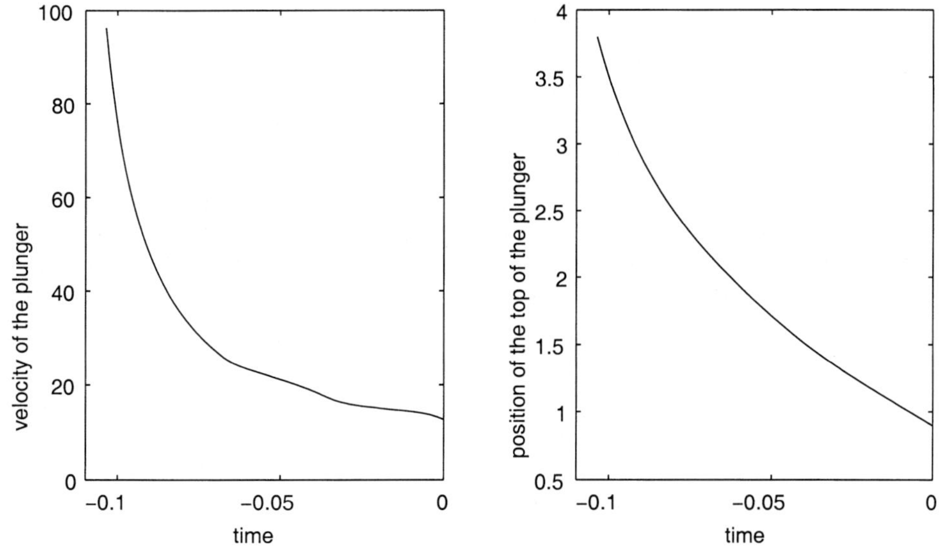

Figure 9. Given force = 5000 N, $s_p = \infty$, and $s_m = 0$ (units: sec and cm/s).

and $z = 1$. This suggests that the results are typically correct with an error of the order of 1%. In the case of Figure 8 (without slip) the applied plunger force is $f = 7500\,\text{N}$, while in the case of Figure 9 (with slip) the force is $f = 5000\,\text{N}$.

Evidently, when we go further back in time until the very early splash of the plunger in the fluid, and assume that the plunger force is still the same, the absence of resistence leads to unlimited high velocities. This, however, is impossible in practice, as the power ($\sim$ force $\times$ velocity) of the equipment is limited, and the inertia of the plunger is nonzero.

8.3. Numerical Method

A detailed description of the numerical method used in example 1 may be found in [14, p.209]. In summary, it is described as follows: the glass flow is modelled by the Stokes equations, *i.e.*, for vanishing Reynolds number. The (self-adaptive) discretization scheme of the domain is based on a triangular mesh, where each triangle has twelve degrees of freedom for the velocity components. On each triangle the pressure is constant and the velocity is piecewise linear. The Finite-Element Method (second order with respect to the size of mesh) is used to solve the Stokes equations, with an indicated accuracy of at least three digits. The time evolution of the glass domain Ω is found by solving the ordinary differential equation

$$\frac{\mathrm{d}\mathbf{x}(t)}{\mathrm{d}t} = \mathbf{v}(\mathbf{x}(t)),\ t \in [t_n, t_{n+1}], \tag{42a}$$

$$\mathbf{x}(t_n) \in \Omega_{t_n}. \tag{42b}$$

9. Conclusions

We have described a model for highly viscous incompressible glass flow during the pressing phase of the production. The model includes a general slip boundary condition. By using a perturbation method based on the slender geometry and low Reynolds number, we obtained explicit expressions for flow velocity and pressure gradient.

Based on these results, we calculated the total force on the plunger for given plunger velocity, and an ordinary differential equation for the resulting plunger velocity when the force is prescribed.

Representative examples showed a very good agreement of the flow velocity between the present solution and numerically obtained "exact" FEM solutions. We conclude that the perturbation method based on the slender geometry is an appropriate approach to this problem.

Appendix : $v(r, z)$

In this appendix, we derive the expression for $v(r, z)$. We already got (Equation 37)

$$v = \frac{1}{r}\frac{\mathrm{d}}{\mathrm{d}z}\int_r^{R_m} ru(r, z)\mathrm{d}r. \tag{A1}$$

Using Equations (34) and (35), we obtain

$$\begin{aligned}\int_r^{R_m} ru(r, z)\,\mathrm{d}r = \frac{u_p}{4}\left(\frac{\beta_m - \beta_p}{\alpha_m - \alpha_p}\right)\Big\{(R_m^4 - r^4)(\alpha_m - \alpha_p)\ldots\\ +2(\beta_m - \beta_p)r^2\log\big(\tfrac{r}{R_p}\big) - 2(\beta_m R_m^2 - \beta_p r^2)\log\big(\tfrac{R_m}{R_p}\big) + \frac{\beta_m\beta_p(R_m^2 - R_p^2)(R_m^2 - r^2)}{R_m^2 R_p^2}\Big\}\times\ldots\\ \left\{(R_m^4 - R_p^4)(\alpha_m - \alpha_p) - 2(\beta_m R_m^2 - \beta_p R_p^2)\log\big(\tfrac{R_m}{R_p}\big) + \beta_m\beta_p\left(\frac{R_m}{R_p} - \frac{R_p}{R_m}\right)^2\right\}^{-1}\ldots\\ -\frac{u_p}{4}\left[\frac{\beta_m(1 - \frac{r^2}{R_m^2}) + 2r^2\log\big(\frac{R_m}{r}\big)}{\alpha_m - \alpha_p}\right].\end{aligned} \tag{A2}$$

From Equation (33), we obtain for the z-derivatives of $\alpha_{m,p}$ and $\beta_{m,p}$

$$\alpha' = \frac{R'}{R}\left(1 - \sigma\right) \pm \frac{s'}{R}, \qquad \beta' = 2RR'(1+\sigma) \pm 2Rs'. \tag{A3}$$

By differentiating Equation (A2) with respect to z and using Equations (A3), we obtain the full expression of v.

Acknowledgements

We are grateful for the cooperation with the TUE Scientific Computing Group of Professor R.M.M. Mattheij, who generously provided the results from the numerical glass-flow simulation. In particular, we want to express our gratitude to Bram van de Broek and L.G.F.C. van Bree. We are grateful for the helpfull discussions with Dr J.K.M. Jansen, Dr Y. Knops, and K. Laevsky. Especially appreciated are the many suggestions and fruitful ideas from Dr Peter Howell. Finally, we are grateful to the referees for their very constructive remarks.

References

1. F.V. Tooley, *The Handbook of Glass Manufacture Vol II*. New York: Books For Industry, Inc, and Glass Industry Magazine (1974) 1147pp.
2. Ph. Simons and R.M.M. Mattheij, *The Cooling of Molten Glass In a Mould*. Rana 95-21, Eindhoven University of Technology (1995) 10pp.
3. J. Kevorkian and J.D. Cole, *Multiple Scale and Singular Perturbation Methods*. New-York: Springer-Verlag (1996) 632pp.
4. W.A. van den Broek, *Glass Morphology in Manufacturing Jars*. Master's Thesis. Eindhoven University of Technology (1996) 77pp.
5. S.L. de Snoo, R.M.M. Mattheij, and G.A.L. van de Vorst, *Modelling of Glass, in Particular Small Scale Surface Changes*. Rana 96-11, Eindhoven University of Technology (1996) 21pp.
6. A.C. Fowler, *Mathematical Models in the Applied Sciences*. Cambridge: Cambridge University Press (1998) 402pp.
7. C.L. Babcock, Viscosity and electrical conductivity of molten glasses. *J. Amer. Ceramic Soc.* 17 (1934) 329–342.
8. L.D. Landau and E.M. Lifshitz, *Fluid Mechanics* (2nd ed.). Oxford: Pergamon (1987) 539pp.
9. Y. Yener and S. Kakac, *Heat Conduction* (3rd ed.). Washington, DC: Taylor & Francis (1993) 363pp.
10. W.C. Dowling, H.V. Fairbanks, and W.A. Koehler. A study of the effect of lubricants on molten glass to heated metals. *J. Amer. Ceramic Soc.* 33 (1950) 269–273.
11. W. Trier, and F. Hassoun, Mechanik des Gleitens heißen, zähflüssigen Glases auf Metalloberflächen. *Glastechnische Berichte* 45 (1972) 271–276.
12. W. Trier, Gleitverhalten von heißem, zähflüssigem Glas auf Metalloberflächen. *Glastechnische Berichte* 51 (1978) 240–243.
13. M. Falipou, F. Sicloroff, and C. Donnet, New method for measuring the friction between hot viscous glass and metals. *Glastechnische Berichte* 72 (1999) pp.59–66.
14. K. Laevsky and R.M.M. Mattheij, Mathematical modeling of some glass problems. In: A. FASANO (ed.), *Complex Flows in Industrial Processes*. Boston: Birkhäuser Boston (2000) pp.191–214.

Journal of Engineering Mathematics **39:** 261–343, 2001.

Asymptotic adaptive methods for multi-scale problems in fluid mechanics

R. KLEIN[1,2], N. BOTTA[2], T. SCHNEIDER[1,3], C. D. MUNZ[4], S. ROLLER[4], A. MEISTER[5], L. HOFFMANN[5] and T. SONAR[5]
[1]*FB Mathematik & Informatik, Freie Universität Berlin, Konrad Zuse Zentrum für Informationstechnik, Takustr. 7, 14195 Berlin, Germany (e-mail: rupert.klein@zib.de)*
[2]*Potsdam Institut für Klimafolgenforschung, Germany*
[3]*Konrad-Zuse-Zentrum für Informationstechnik, Berlin, Germany*
[4]*Institut für Aerodynamik und Gasdynamik, Universität Stuttgart, Germany*
[5]*Institut für Angewandte Mathematik, Universität Hamburg, Germany*

Received 14 February 2000; accepted in revised form: 25 September 2000

Abstract. This paper reports on the results of a three-year research effort aimed at investigating and exploiting the role of physically motivated asymptotic analysis in the design of numerical methods for singular limit problems in fluid mechanics. Such problems naturally arise, among others, in combustion, magneto-hydrodynamics, and geophysical fluid mechanics. Typically, they are characterized by multiple-space and/or -time scales and by the disturbing fact that standard computational techniques fail entirely, are unacceptably expensive, or both. The challenge here is to construct numerical methods which are robust, uniformly accurate, and efficient through different asymptotic regimes and over a wide range of relevant applications. Summaries of multiple-scales asymptotic analyses for low-Mach-number flows, magneto-hydrodynamics at small Mach and Alfvén numbers, and of multiple-scales atmospheric flows are provided. These reveal singular balances between selected terms in the respective governing equations within the considered flow regimes. These singularities give rise to problems of severe stiffness, stability, or to dynamic-range issues in straight-forward numerical discretizations. A formal mathematical framework for the multiple scales asymptotics is then summarized by use of the example of multiple-length-scale single-time-scale asymptotics for low-Mach-number flows. The remainder of the paper focuses on the construction of numerical discretizations for the respective full governing equation systems. These discretizations avoid the pitfalls of singular balances by exploiting the asymptotic results. Importantly, the asymptotics are not used here to derive simplified equation systems, which are then solved numerically. Rather, numerical integration of the full equation sets is aimed at and the asymptotics are used only to construct discretizations that do not deteriorate as a singular limit is approached. One important ingredient of this strategy is the numerical identification of a singular limit regime given a set of discrete numerical state variables. This problem is addressed in an exemplary fashion for multiple-length single-time-scale low-Mach-number flows in one space dimension. The strategy allows a dynamic determination of an instantaneous relevant Mach number, and it can thus be used to drive the appropriate adjustment of the numerical discretizations when the singular limit regime is approached.

Key words: asymptotic analysis, fluid mechanics, singular limit regimes, asymptotics-based numerical methods

1. Introduction

Here we report on the results of a three-year research effort aimed at investigating and exploiting the role of physically motivated asymptotic analysis in the design of numerical methods for singular limit problems in fluid mechanics. Such problems naturally arise, among others, in combustion, magneto-hydrodynamics, and geophysical fluid mechanics. Typically, they are characterized by multiple space and/or time scales and by the disturbing fact that standard computational techniques fail entirely, are unacceptably expensive, or both. The challenge here is to construct numerical methods which are robust, uniformly accurate, and efficient through different asymptotic regimes and over a wide range of relevant applications.

1.1. Key ideas

In order to successfully address the goal stated above one must first accept a shift of paradigm as compared to standard applications of asymptotics: The goal of most of these applications is to derive a simplified asymptotic set of equations which can be solved analytically or with less numerical effort than the original full equation system. The target in this situation is to obtain approximate solutions to a restricted class of problems, which can then be used instead of the unavailable or expensive solutions to the full problem. This utilization of asymptotics could be labeled 'solution oriented'.

We contrast this with a 'structure oriented' exploitation of asymptotics: The ultimate goal here is the numerical solution of the full equation systems. The key questions in this context are: "What are the reasons for the failure of standard numerical schemes in singular limit regimes?" and "How are these reasons related to the asymptotic behavior of the full equation system?". To answer these questions, one must focus on the process of deriving simplified asymptotic limit equations rather than on these equations themselves. In the course of an asymptotic analysis one gains improved insight into the mathematical structure of the singular limit and obtains systematic hints at the origins of the failure of standard numerical methods. It is hoped, and has been verified in a number of examples by now, that this insight can be used to develop improved numerical techniques that avoid the pitfalls of standard schemes and allow one to treat singular and non-singular cases numerically in a unified fashion.

As an example, let us consider the case of low-Mach-number flow, which has been pursued extensively within this project. Asymptotic analysis is most comprehensive, and often considered most elegant, when carried out in terms of a suitably chosen system of coordinates and dependent variables. The probably most efficient choice of dependent variables for low-Mach-number flow are the so-called 'primitive variables', namely density, velocity and pressure. Suppose, however, that we intend to extend a standard compressible flow solver to the low- and zero- Mach-number regime. One quickly finds that modern compressible flow solvers are in conservation form for mass, momentum and energy and that the key quantities of these numerical methods are grid-cell averages of the mass, momentum, and energy densities and their respective flux densities.

The challenge then is to derive and incorporate the singular limit behavior of these fluxes in an extended numerical scheme. Obviously, the most appropriate choice of variables for a related asymptotic analysis are now the conserved quantities, regardless of the efficiency or elegance of the procedure. The results - to be described in more detail below - are somewhat surprising in that the familiar divergence constraint for the velocity field in this formulation has nothing to do with mass continuity, but is rather a consequence of energy conservation. This in turn implies that a numerical scheme in conservation form should automatically incorporate suitable elliptic constraints as part of solving the energy-conservation equation, rather than by imposing velocity-divergence conditions externally.

Technically, the following program needs to be pursued in order to realize the desired merging of numerical and asymptotic methods:

1. Identify an equation system and one of its singular limit regimes in which standard numerical methods deteriorate,

2. Reformulate existing asymptotic analyses to match the numerical framework or, if necessary, invent an appropriate new asymptotic approach such that conclusions can be drawn regarding the failure of the standard numerical method,

3. Develop modifications of the numerical technique to overcome its limitations in the singular regime on the basis of the asymptotic analysis.

The implementation of this program requires the close collaboration of scientists from a wide range of backgrounds: The first step should be driven by applications from natural sciences and engineering. It is hardly possible to cover all conceivable singular limit regimes for any given equation system, so that an educated judgement regarding the importance of any particular regime in applications should guide the selection of the target problem. The second step requires the input of both a skilled 'asymptotician' and a numerical analyst, familiar with the mathematical structure of the considered numerical methods. The third step should be pursued by the same combination of experts, this time with the major work load on the specialist in numerical methods. Ultimately, the applied scientist or engineer is required again in defining suitable tests and applications for the new method that would serve the validation of the method and be sufficiently close to applications to be meaningful in the first place.

An important additional branch of work is concerned with a mathematically precise formulation of both the asymptotic analysis and its translation into extended numerical techniques. The incorporation in a numerical method of asymptotic results, which are generally derived under specific simplifying assumptions regarding the considered physical problems, requires discrete numerical operations that properly match the asymptotic limit processes. Therefore, particular emphasis is given to the derivation of rigorous guidelines and rules which determine under which conditions it is possible to use the asymptotics and under which it is not.

The joint project summarized in this paper has brought together scientists with partly overlapping and partly complementary competences in order to cover the required range of scientific expertise. The authors acknowledge the generous support by the Deutsche Forschungsgemeinschaft under grants KL 611/6-i, MU 1319/2-i and SO 363/2-i ($i \in \{1, 2\}$).

1.2. Overview

There are three main sections to this paper. Section 2 summarizes asymptotic analyses for a number of fluid-mechanical problems. These analyses have been developed as guidelines in the construction of improved numerical methods. Thus, even though some of the considered regimes have been discussed in the literature before, both the presentation and the conclusions drawn here should be new. Specifically we address the following issues:

- low-Mach-number weakly compressible flows in a single-time multiple-space-scale regime,
- Magneto-hydrodynamics for small Mach and Alfvén numbers, and
- Atmospheric flows under a particular distinguished limit for the Mach and Froude numbers, namely $\mathrm{Fr} = \mathrm{M}$ as $\mathrm{M} \to 0$.

Section 3 provides a general mathematical framework for the asymptotics and prepares for its later 'translation' into numerical procedures for the multiple-scales analysis given in Section 2.1. We state in unambiguous terms what assumptions and conditions have to be satisfied by solutions of the systems considered in order to allow an asymptotic (multiple-scales) approximation. Note that the analyses do not prove that such solutions actually do satisfy those constraints. Such proofs are formidable tasks for each application. They are hardly generalizable and beyond the scope of the present paper. Even though the discussion is motivated by the low-Mach-number limit, its results are more generally applicable. These developments are important in the context of transferring asymptotic results into numerical algorithms: Asymptotic multiple-scales procedures typically involve averaging procedures and the extraction of small-scale fluctuations, which are easily defined formally and in the limit of vanishing scale separation parameters. However, in a numerical application one is typically faced with small, *but finite* values of these parameters and the exact definitions of averaging operators and their discrete analogues become crucial. This section as well as Section 4.1 should be considered as exemplary in the sense that such a rigorous, detailed analysis should ideally be pursued for any asymptotics/ numerics merger.

Section 4 describes the announced translation of asymptotic results into improved numerical procedures. Subsection 4.1 addresses an issue that is a cornerstone of any attempt at using asymptotics in the construction of numerical methods. Asymptotic analysis may be described as a systematic procedure for the construction of particular approximate solutions to a given equation system. Such solutions are proper approximations only in the particular limit regime they are designed for. As an example consider low-Mach-number flows: the disparity between flow velocities and the speed of sound can, under suitable initial and boundary conditions, induce a time-scale separation. In this case one finds high-frequency acoustics superimposed over a much slower, quasi-incompressible flow field (see, *e.g.*, [1, 2, 3, 4, 5]). If, on the other hand, acoustic perturbations are of sufficiently long wave length, the material flow and the sound waves evolve on the same time, but on disparate spatial scales (see [6]). These physically very different regimes amount to equally different asymptotic solutions. Thus, if one is interested in exploring asymptotic results to improve numerical methods, it is necessary to employ testing procedures that determine which particular asymptotic regime is valid for the application at hand. The identification of multiple-space-scales asymptotic limit regimes by means of discrete filtering operators is the topic of Subsection 4.1.

Subsection 4.2 describes the necessary steps needed to allow the computation of weakly compressible flows using the principal machinery of standard incompressible flow solvers. The asymptotic analysis developed earlier in Section 2 shows that in weakly compressible flows the pressure plays a multiple role as a thermodynamic variable, as an acoustic mode amplitude, and as a an agent (sometimes called a 'Lagrangian multiplier') guaranteeing that, in the limit of zero Mach number, the velocity field satisfies an elliptic divergence constraint. These three roles are attributed to the leading, first, and second-order solutions in a power-series expansion of the pressure in terms of the Mach number. Only the second-order pressure survives in the limit of zero Mach number, and this is the only pressure variable known to a standard incompressible flow solver. An extension of such a method to weakly compressible flows requires the introduction of additional pressure variables, which account for global thermodynamic pressure changes and acoustic waves. Thus, this subsection describes the development of a Multiple Pressure Variable (MPV) scheme for low-Mach-number flows from the basis of a standard incompressible flow solver. Applications include examples showing baroclinic vorticity generation through long-wave acoustic pulses and a thermally driven cavity flow.

Subsection 4.3 describes the extension of conservative finite-volume methods for compressible flows to zero Mach number from [7]. The key ingredients of this extension are

(i) again the introduction of multiple pressure variables and (ii) a semi-implicit determination of numerical fluxes of mass, momentum, and energy. This procedure involves explicit estimates for the convective flux contributions based on the original upwind machinery of the underlying compressible flow solver and elliptic corrections that are derived directly from low-Mach-number asymptotics. Applications include 'falling droplets' (motion of high-density blobs of fluid under gravitational forces) and a model for thermo-acoustic devices.

Subsection 4.4 reconsiders the extension of a standard numerical method designed for the compressible Euler equations. Emphasis is now on *small, but finite* Mach numbers. In particular, the techniques described in Section 4.1 for the identification of an asymptotic regime through discrete filters is used here to automatically adjust the (small) numerical reference parameter that represents the low-Mach-number effects. A scheme is developed that allows a smooth transition from very weakly compressible to fully compressible flows with shock waves. This transition is associated with a change of the Mach-number reference parameter from very small values, where numerical flux corrections similar to those described in Section 4.3 are applied, to unity, in which case the scheme automatically reduces to the original underlying compressible flow solver. Examples include the passage of a long-wave acoustic pulse over a gas layer with high frequency, large-amplitude density variations and the nonlinear steepening of an acoustic wave into a weak shock.

The aim in Subsection 4.5 is to use results of Section 2.2 in distinguishing between terms related to pure convection and terms related to the fast wave speeds in magneto-hydrodynamics. As in the case of the Euler equations the idea behind the approach is to obtain a splitting of the system that enables one to treat the convection terms, which remain hyperbolic in the limit, and the fast-wave-speed terms, which become elliptic, with different numerical methods optimized for the respective task. In MHD there are additional hyperbolic waves besides the acoustic ones. As a consequence, there are two characteristic numbers, the Mach and Alfvén numbers, which characterize the ratio of a typical flow velocity *vs.* the respective hyperbolic wave speeds. New split systems are proposed that are used in our current work for the construction of efficient and accurate numerical methods.

In Section 5 we present conclusions and an outlook on current and future work.

2. Asymptotic analyses

2.1. Low-Mach-number asymptotics for compressible fluid flows

In this section we discuss low-Mach-number asymptotics in a sufficiently general fashion to include low-speed reacting flows. Much of the discussion will in fact be motivated by applications from combustion, although, by applying suitable simplifications, all the analytical steps carry over to non-reacting flows as well. We begin by introducing the governing equations and some general considerations regarding scaling and non-dimensionalization. This subsection will also define the non-dimensional scalar parameters that will allow us to identify the relevant asymptotic limit regimes.

2.1.1. *Governing equations*

Unscaled formulation. The full governing equations of gas-phase combustion with as little approximation as probably doable can be found in comprehensive textbooks on combustion theory, such as [8]. Here we shall consider a simplified system only, so that the essential lines of thought can be worked out straightforwardly.

The simplified system to be discussed here consists of the conservation equations for mass, momentum and energy

$$
\begin{aligned}
&(\rho)_t + \nabla \cdot (\rho \boldsymbol{v}) = 0, \\
&(\rho \boldsymbol{v})_t + \nabla \cdot (\rho \boldsymbol{v} \circ \boldsymbol{v} + 1 p) + \nabla \cdot \tau = 0, \\
&(\rho e)_t + \nabla \cdot (\boldsymbol{v}[\rho e + p]) + \nabla \cdot \left(\boldsymbol{j}_T + \tau \cdot \boldsymbol{v} + \sum_{i=1}^{n_{\text{spec}}} (\Delta H)_i \boldsymbol{j}_i \right) = 0.
\end{aligned}
\tag{2.1.1}
$$

Here ρ, $\boldsymbol{v}$, p, e are the mass density, fluid flow velocity, pressure, and total energy per unit mass, respectively, and $\boldsymbol{\tau}$, $\boldsymbol{j}_T$, $\boldsymbol{j}_i$ denote the molecular transport of momentum, heat, and of the mass of the ith species. These transport terms and the pressure are related to the mass, momentum, energy and species densities ρ, $\rho\boldsymbol{v}$, ρe, ρY_i through the caloric equation of state

$$
\rho e = \frac{p}{\gamma - 1} + \frac{1}{2} \rho \boldsymbol{v}^2 + \sum_{i=1}^{n_{\text{spec}}} (\Delta H)_i \rho Y_i \tag{2.1.2}
$$

and the transport models

$$
\begin{aligned}
\boldsymbol{\tau} &= -\mu \left(\nabla \boldsymbol{v} + (\nabla \boldsymbol{v})^T \right) - \eta (\nabla \cdot \boldsymbol{v}) 1, \\
\boldsymbol{j}_T &= -\kappa \, \nabla T, \\
\boldsymbol{j}_i &= -D_i \nabla Y_i.
\end{aligned}
\tag{2.1.3}
$$

The temperature T is related to pressure and density via the thermal equation of state

$$
T = \frac{p}{\rho R}. \tag{2.1.4}
$$

The quantities γ, R, μ, η, κ, D_i, $(\Delta H)_i$ are the isentropic exponent, the ideal gas constant, the shear and bulk viscosities, the heat conductivity, the species diffusivities and the species' formation enthalpies, respectively. All of these are assumed constant throughout this text.

The species mass fractions Y_i satisfy the inhomogeneous balance laws

$$
(\rho Y_i)_t + \nabla \cdot (\rho Y_i \boldsymbol{v}) + \nabla \cdot \boldsymbol{j}_i = \rho \omega_i, \quad (i = 1 \ldots n_{\text{spec}}), \tag{2.1.5}
$$

where $\omega_i = \omega_i(p, \rho, \{Y_j\}_{j=1}^{M_{spec}})$ is the net production rate of species i per unit mass of the gas mixture. When n_{spec} actually denotes the total number of chemical species in the system, then the sum of all equations in (2.1.5) leads back to the mass conservation equation in $(2.1.1)_1$ and yields a constraint for the rate expressions

$$
\sum_{i=1}^{n_{\text{spec}}} \rho \omega_i = 0. \tag{2.1.6}
$$

In this case, the mass conservation equation or one of the species balances are redundant.

In order to specify a single solution to these equations uniquely one must provide suitable initial and boundary conditions determining the solution behavior at some start-up time and close to the physical boundaries of the system. Since these conditions define the distinctions between all systems that follow the same set of governing equations, we cannot exhaustively discuss them here.

Non-dimensionalization, similarity and scaling. Within the above governing equations one can identify four fundamental physical dimensions $\{\chi_i\}_{i=1}^4$: Length $\mathcal{L}$, Time $\mathcal{T}$, Mass $\mathcal{M}$ and Temperature Θ. Each physical quantity ϕ_j that appears in the governing equations has a physical dimension that is a product of these fundamental ones, so that

$$\mathrm{Dim}(\phi_j) = \prod_{i=1}^{4} (\chi_i)^{b_j^i} . \tag{2.1.7}$$

To name a few examples,

Quantity ϕ	*Physical dimension* $\mathrm{Dim}(\phi)$
ρ	$\mathcal{M}/\mathcal{L}^3$
$\boldsymbol{v}$	$\mathcal{L}/\mathcal{T}$
e	$\mathcal{L}^2/\mathcal{T}^2$
p	$\mathcal{M}/(\mathcal{L}\mathcal{T}^2)$
R	$\mathcal{L}^2/(\Theta\mathcal{T}^2)$

. (2.1.8)

Once a system of units is chosen based on which these fundamental dimensions shall be measured, each of the physical quantities and coefficients in the governing equations can be quantified by a sole number. The familiar SI-system is one example, where $(\mathcal{L}, \mathcal{T}, \mathcal{M}, \Theta)$ are measured in terms of (Meter, m; Second, s; Kilogram, kg; Kelvin, K). Knowing a quantity's physical dimension and the underlying system of units, one can always transform these non-dimensional numbers back into measurable physical values. Obviously, there is a one-to-one map between any two different systems of units, so that the exact solutions of the governing equations will not depend on which system is chosen.

As they stand, the governing equations given above do not reveal anything besides what was built into them from the start: conservation of mass, momentum and energy (2.1.1), conversion of one set of chemical species into another (2.1.5), transformations between various forms of energy (2.1.2) and some thermodynamic relations between the state quantities (2.1.4). To obtain a somewhat improved intuition about possible solutions one may study classes of solutions distinguished by some particular global mathematical characterization.

For any given solution of the equations one can identify 'characteristic values' $\left[\phi_{j,\mathrm{ref}}\right]_{j=1}^{N}$ of the total of N physical quantities in the system which roughly describe their orders of magnitude throughout the solution or at least during a certain time interval and within a selected region in space. These dimensional characteristic quantities can be combined into non-dimensional characteristic numbers

$$\Pi_k = \prod_{j=1}^{N} \left(\phi_{j,\mathrm{ref}}\right)^{a_k^j} , \tag{2.1.9}$$

with the exponents a_k^j chosen so as to guarantee that the Π_k do not have a physical dimension as will be explained shortly.

These numbers are extremely useful as they provide a comparison between various quantities that may have the same physical dimension but very different physical origin. An example is the Mach number

$$\mathrm{M} := \frac{|v_{\mathrm{ref}}|}{c_{\mathrm{ref}}}, \tag{2.1.10}$$

which compares a typical flow velocity with a characteristic speed of sound or, equivalently, the momentum flux due to convection with the momentum flux due to pressure forces (note that the speed of sound is $c^2 = \gamma p/\rho$. Here we simply take $c_{\mathrm{ref}}^2 := p_{\mathrm{ref}}/\rho_{\mathrm{ref}}$, *i.e.*, $\mathrm{M}^2 = \rho_{\mathrm{ref}} u_{\mathrm{ref}}^2 / p_{\mathrm{ref}}$).

For the non-dimensional Π's to be actually non-dimensional, all the physical dimensions have to cancel exactly in the product. Using (2.1.7), we may rephrase this statement as

$$\mathrm{Dim}(\Pi_k) = \prod_{j=1}^{N} \left[\prod_{i=1}^{4} (\chi_i)^{b^i_j} \right]^{a^j_k} = \prod_{i=1}^{4} \left[\prod_{j=1}^{N} (\chi_i)^{b^i_j a^j_k} \right] = \prod_{i=1}^{4} (\chi_i) \left[\prod_{i=1}^{N} b^i_j \, a^j_k \right] \equiv 1 \,. \tag{2.1.11}$$

For this equation to hold, the respective powers of each of the fundamental dimensions χ_i must vanish independently, so that

$$\sum_{j=1}^{N} b^i_j \, a^j_k \equiv 0 \quad (i = 1 \ldots 4, \quad k \text{ arbitrary}) \,. \tag{2.1.12}$$

These are four linear constraints on the N-tuples $\mathbf{a}_k = (a^1_k, \ldots, a^N_k)$, which therefore span a total space of dimension $N - 4$. This, in turn, is equivalent to the existence of a set of $N-4$ independent characteristic numbers $\{\Pi_k\}_{k=1}^{N-4}$, and we arrive at the famous *Buckingham's π-theorem.*

The existence of these non-dimensional numbers has very deep and important consequences for the set of all possible solutions of the governing equations. To be more specific we introduce the reference quantities $(\rho_{\mathrm{ref}}, p_{\mathrm{ref}}, v_{\mathrm{ref}})$ for density, pressure and velocity, $(t_{\mathrm{ref}}, \ell_{\mathrm{ref}})$ for the time and space coordinates, (ω_{ref}) for chemical reaction rates and $(\mu_{\mathrm{ref}}, \kappa_{\mathrm{ref}}, D_{\mathrm{ref}}, R_{\mathrm{ref}}, (\Delta H)_{\mathrm{ref}})$ for the various parameters in the constitutive equations. Next we define new dependent and independent variables,

$$\rho' = \frac{\rho}{\rho_{\mathrm{ref}}}, \quad p' = \frac{p}{p_{\mathrm{ref}}}, \quad \boldsymbol{v}' = \frac{\boldsymbol{v}}{v_{\mathrm{ref}}}, \quad T' = \frac{T}{p_{\mathrm{ref}}/(\rho_{\mathrm{ref}} R_{\mathrm{ref}})}, \quad e' = \frac{e}{p_{\mathrm{ref}}/\rho_{\mathrm{ref}}} \tag{2.1.13}$$

and

$$\boldsymbol{x}' = \frac{\boldsymbol{x}}{\ell_{\mathrm{ref}}}, \quad t' = \frac{t}{t_{\mathrm{ref}}}. \tag{2.1.14}$$

The governing equations are now transformed into their scaled analogues (with the prime superscript dropped for convenience of notation):

Conservation Laws:

$$\begin{aligned}
&\mathrm{Sr}(\rho)_t \quad + \nabla \cdot (\rho \boldsymbol{v}) = 0, \\
&\mathrm{Sr}(\rho \boldsymbol{v})_t \; + \nabla \cdot (\rho \boldsymbol{v} \circ \boldsymbol{v} + \frac{1}{\mathrm{M}^2} \nabla p) + \frac{1}{\mathrm{Re}} \nabla \cdot \boldsymbol{\tau} = 0, \\
&\mathrm{Sr}(\rho e)_t \; + \nabla \cdot (\boldsymbol{v}[\rho e + p]) + \nabla \cdot \left(\frac{1}{\mathrm{Pe}} \boldsymbol{j}_T + \frac{\mathrm{M}^2}{\mathrm{Re}} \boldsymbol{\tau} \cdot \boldsymbol{v} + \frac{\mathrm{Q}}{\mathrm{Sc}} \sum_{i=1}^{n_{\mathrm{spec}}} \delta h_i \boldsymbol{j}_i \right) = 0;
\end{aligned} \tag{2.1.15}$$

Species Balances:

$$\mathrm{Sr}(\rho Y_i)_t + \nabla \cdot (\rho Y_i \boldsymbol{v}) = -\frac{1}{\mathrm{Sc}} \nabla \cdot \boldsymbol{j}_i + \mathrm{Da}\rho\omega_i \qquad (i = 1 \ldots n_{\mathrm{spec}}); \tag{2.1.16}$$

Caloric Equation of State:

$$\rho e = \frac{p}{\gamma - 1} + \mathrm{M}^2 \frac{1}{2} \rho \boldsymbol{v}^2 + \mathrm{Q} \sum_{i=1}^{n_{\mathrm{spec}}} \delta h_i \, \rho Y_i; \tag{2.1.17}$$

Thermal Equation of State:

$$T = \frac{p}{\rho}; \tag{2.1.18}$$

Transport Models:

$$\begin{aligned} \boldsymbol{\tau} &= -\mu' \left(\nabla \boldsymbol{v} + (\nabla \boldsymbol{v})^T \right) - \eta' (\nabla \cdot \boldsymbol{v}) \mathbf{1}, \\ \boldsymbol{j}_T &= -\kappa' \nabla T, \\ \boldsymbol{j}_i &= -D_i' \nabla Y_i. \end{aligned} \tag{2.1.19}$$

Notice that in (2.1.15), (2.1.17) we have introduced the scaled reaction enthalpies

$$\delta h_i = \frac{(\Delta H)_i}{(\Delta H)_{\mathrm{ref}}}, \tag{2.1.20}$$

and that $\mu' = \mu/\mu_{\mathrm{ref}}$, $\kappa' = \kappa/\kappa_{\mathrm{ref}}$ etc. in (2.1.19) could all be set to unity in case of constant molecular transport coefficients.

The procedure of scaling the equations has led to a set of seven characteristic numbers:

Abbreviation	Definition	Name
Sr	$\ell_{\mathrm{ref}}/(t_{\mathrm{ref}} v_{\mathrm{ref}})$	Strouhal number
M	$v_{\mathrm{ref}}/\sqrt{p_{\mathrm{ref}}/\rho_{\mathrm{ref}}}$	Mach number
Re	$\rho_{\mathrm{ref}} v_{\mathrm{ref}} \ell_{\mathrm{ref}}/\mu_{\mathrm{ref}}$	Reynolds number
Pe	$\rho_{\mathrm{ref}} v_{\mathrm{ref}} \ell_{\mathrm{ref}}/(\kappa_{\mathrm{ref}}/R_{\mathrm{ref}})$	Peclet number
Sc	$\rho_{\mathrm{ref}} v_{\mathrm{ref}} \ell_{\mathrm{ref}}/D_{\mathrm{ref}}$	Schmidt number
Da	$\omega_{\mathrm{ref}} \ell_{\mathrm{ref}}/v_{\mathrm{ref}}$	Damköhler number
Q	$(\Delta H)_{\mathrm{ref}}/(p_{\mathrm{ref}}/\rho_{\mathrm{ref}})$	Heat Release Parameter

. (2.1.21)

Notice that this number was to be expected following the earlier discussion in conjunction with (2.1.11), (2.1.12). We have introduced $N = 11$ reference quantities to characterize the magnitude of all terms and physical variables in the unscaled equations and ended up with $N - 4 = 11 - 4 = 7$ non-dimensional numbers.

Here are the key observations regarding the structure of solutions of the full governing equations to be derived from the exercise of non-dimensionalization and scaling:

1. Given:

- Initial and boundary conditions in terms of the scaled variables,
- A set of non-dimensional parameters (Sr, M, Re, Pe, Sc, Da, Q),
- The solution to the scaled governing equations (2.1.15) – (2.1.19),

Then:

- For any set of reference quantities $(\rho_{\text{ref}}, p_{\text{ref}}, v_{\text{ref}}, t_{\text{ref}}, \ell_{\text{ref}})$ and fluid specific parameters $(\mu_{\text{ref}}, \kappa_{\text{ref}}, D_{\text{ref}}, R_{\text{ref}}, (\Delta H)_{\text{ref}}, \omega_{\text{ref}})$ that are compatible with the given values for (Sr,M,Re,Pe,Sc,Da,Q), the fields obtained by reversing the scaling from (2.1.13), (2.1.14) represent a valid solution to the original unscaled equations!

Thus:

- Each set (Sr,M,Re,Pe,Sc,Da,Q) defines a class of equivalent solutions, which differ from each other merely by a scale transformation.

2. Close inspection of the scaled governing equations reveals that singularities arise when one or more of the non-dimensional characteristic numbers approach zero or infinity. Numerical flow simulation is a particular challenge in these regimes, as will be elucidated by a number of examples throughout this text.

Remark: The set of non-dimensional characteristic numbers chosen above is not unique. In the combustion and heat-transfer literature the Peclet and Schmidt numbers Pe and Sc are often replaced with the Prandtl and Lewis numbers Pr = Pe/Re and Le = Sc/Pe, respectively.

Remark: The Strouhal number Sr is often set to unity when a set of scaled governing equations is formulated. Considering its definition in (2.1.21), we observe that this choice implies the characteristic time scale of flow field evolution to be the convection time scale $\ell_{\text{ref}}/v_{\text{ref}}$. Even though we will also adopt this choice below, this is neither necessary, nor is it meaningful in all cases. Taking, *e.g.*, the limit Sr $\to$ 0 allows one to formulate the governing equations for stationary flows.

2.1.2. *Single-length, single-time-scale asymptotics*

In most real-life applications, such as atmosphere-ocean flows, fluid flows in engineering devices, etc., velocities are small compared with the speed of sound. This fact has profound consequences for both the mathematical behavior of solutions to the governing equations from Section 2.1.1 and their numerical approximate solutions. Physically, in the limit of arbitrarily slow flow (or infinitely fast sound propagation) the elasticity of the gas with respect to bulk compression becomes negligible and sound-wave propagation becomes unnoticeable. Mathematically, as the Mach number M from (2.1.10) tends to zero, the pressure gradient contribution in the momentum equations $(2.1.15)_2$ becomes singular. In order to explore the consequences of this singularity, we consider a formal asymptotic analysis, closely following [9, 6].

A systematic derivation of the governing equations for zero-Mach-number combustion has been given by Majda and Sethian [9]. The formulation adopted below, which explicitly focuses on the conservation equations for mass, momentum and energy, has been introduced in [6] in conjunction with a multiple-length-scale, single-time-scale analysis. In [6, 7] the authors exploit the asymptotics to derive fully conservative numerical methods for low- and zero-Mach-number flows. In Section 3 below, one of the authors describes a rigorous framework for the formal asymptotics presented here with an emphasis on the physical background (see also [10]).

In recounting these earlier results we restrict ourselves to the case of an ideal gas mixture with a simple one-step reaction $F \to P$, where the fuel F is turned into the product species P. The chemical-energy conversion rate then is $\mathrm{Q}\, \rho\omega_F$, where Q quantifies the specific reaction enthalpy of the fuel species and $\rho\omega_F$ its production density. Under these conditions we need to describe the time evolution of only the fuel mass fraction Y_F, using a single transport equation of the type described in (2.1.16).

The asymptotic solution *Ansatz*

$$\begin{aligned} p &= p^{(0)}(\boldsymbol{x},t) + \mathrm{M}p^{(1)}(\boldsymbol{x},t) + \mathrm{M}^2 p^{(2)}(\boldsymbol{x},t) + o(\mathrm{M}^2), \\ \boldsymbol{v} &= \boldsymbol{v}^{(0)}(\boldsymbol{x},t) + \mathrm{M}\boldsymbol{v}^{(1)}(\boldsymbol{x},t) + o(\mathrm{M}), \\ \rho &= \rho^{(0)}(\boldsymbol{x},t) + \mathrm{M}\rho^{(1)}(\boldsymbol{x},t) + o(\mathrm{M}), \\ Y_F &= Y_F^{(0)}(\boldsymbol{x},t) + \mathrm{M}Y_F^{(1)}(\boldsymbol{x},t) + o(\mathrm{M}) \end{aligned} \tag{2.1.22}$$

is introduced into the dimensionless governing equations (2.1.15)–(2.1.16). Following standard procedures of asymptotic analysis one obtains a hierarchy of equations for the various expansion functions $p^{(i)}, \boldsymbol{v}^{(i)}, \rho^{(i)}, Y_F^{(i)}$ by collecting all terms multiplied by equal powers of the Mach number M and separately equating these to zero. The momentum equations to orders M^{-2} and M^{-1} become

$$\nabla p^{(0)}(\boldsymbol{x},t) = 0, \quad \nabla p^{(1)}(\boldsymbol{x},t) = 0. \tag{2.1.23}$$

One concludes that $p^{(0)}$ and $p^{(1)}$ depend on time only in this regime of length and time scales, so that

$$p^{(0)} \equiv P^{(0)}(t) \qquad \text{and} \qquad p^{(1)} \equiv P^{(1)}(t). \tag{2.1.24}$$

The continuity and energy equations at leading order are then

$$\left(\rho^{(0)}\right)_t + \nabla\cdot\left(\rho^{(0)}\boldsymbol{v}^{(0)}\right) = 0, \tag{2.1.25}$$

$$\frac{1}{\gamma-1}\frac{\mathrm{d}P^{(0)}}{\mathrm{d}t} + \nabla\cdot\left(H^{(0)}\boldsymbol{v}^{(0)}\right) = \left(\frac{1}{\mathrm{Pe}}\nabla\cdot\left(\lambda\nabla T^{(0)}\right) + \mathrm{Da}\,\mathrm{Q}(\rho\omega_F)^{(0)}\right). \tag{2.1.26}$$

where

$$H^{(0)}(t) = \frac{\gamma}{\gamma-1}P^{(0)}(t). \tag{2.1.27}$$

To arrive at (2.1.26), insert the expansion (2.1.22) into the energy conservation law $(2.1.15)_3$ and take into account that, according to (2.1.17), the kinetic energy is smaller than the thermal energy for $\mathrm{M} \ll 1$ by a factor of M^2. The contribution of the viscous forces to the energy budget, represented by the term $\nabla\cdot(\frac{\mathrm{M}^2}{\mathrm{Re}}\boldsymbol{\tau}\cdot\boldsymbol{v})$, will appear for the first time in the energy equation at order $O(\mathrm{M}^2)$.

The momentum equation at order M^0 reads

$$\left(\rho^{(0)}\boldsymbol{v}^{(0)}\right)_t + \nabla\cdot\left(\rho^{(0)}\boldsymbol{v}^{(0)}\circ\boldsymbol{v}^{(0)}\right) + \nabla p^{(2)} = -\frac{1}{\mathrm{Re}}\nabla\cdot\boldsymbol{\tau}^{(0)}. \tag{2.1.28}$$

Notice the change in structure of these equations: the pressure-evolution equation does *not* determine the pressure variable $p^{(2)}$ appearing in the momentum equation! The appropriate interpretation, corresponding directly to the theory of incompressible flows, is that the equation for $P^{(0)}$ from (2.1.26) is a divergence constraint for the leading-order energy flux, *i.e.*,

$$\nabla\cdot\left(H^{(0)}(t)\,\boldsymbol{v}^{(0)}\right) = -\left[\frac{1}{\gamma-1}\frac{\mathrm{d}P^{(0)}}{\mathrm{d}t} - \left(\frac{1}{\mathrm{Pe}}\nabla\cdot\left(\lambda\nabla T^{(0)}\right) + \mathrm{DaQ}(\rho\omega_F)^{(0)}\right)\right] \tag{2.1.29}$$

and that the second-order pressure $p^{(2)}$ is responsible for guaranteeing that constraint to be fulfilled. A useful and more familiar interpretation of this equation results from using explicitly that $H^{(0)}(t) = \gamma P^{(0)}(t)/(\gamma-1)$ is a function of time only for the present equation of state and deriving a

velocity divergence constraint

$$\nabla\cdot\boldsymbol{v}^{(0)} = -\frac{1}{\gamma P^{(0)}}\left[\frac{\mathrm{d}P^{(0)}}{\mathrm{d}t} - (\gamma-1)\left(\frac{1}{\mathrm{Pe}}\nabla\cdot\left(\lambda\nabla T^{(0)}\right) + \mathrm{DaQ}\rho\omega_F\right)\right]. \tag{2.1.30}$$

We observe that the velocity divergence is driven by chemical-energy conversion and energy-transport effects: Chemical-heat release, heat conduction and global pressure changes conspire to induce a divergence field for the velocity. As a direct consequence we derive from the mass-continuity equation (2.1.25) an equation that describes the temporal evolution of the density along particle paths

$$\frac{\mathrm{D}\rho}{\mathrm{D}t} := \frac{\partial\rho}{\partial t} + \boldsymbol{v}\cdot\nabla\rho = -\rho\nabla\cdot\boldsymbol{v}. \tag{2.1.31}$$

To summarize, the energy conversion and transport processes drive the divergence of the energy flux, which is related to the velocity divergence. The latter, in turn, leads to compression or expansion of individual mass elements and thus to density variations of individual particles. The original interpretation of (2.1.29) as an energy-flux divergence constraint proves to be useful in the construction of energy conserving finite-volume methods, (see [7, 11]).

Equations (2.1.25)–(2.1.28) form a closed system, provided the temporal evolution of the leading-order pressure $P^{(0)}$ is known and the state dependence of the reaction rate $\rho\omega_F$ is given. For combustion under atmospheric conditions $P^{(0)}$ equals the atmospheric ambient pressure and is constant in time. For combustion in a closed chamber we explore the fact that $P^{(0)}$ is homogeneous in space, integrate (2.1.30) over the total flow domain, use Gauß' theorem to replace the divergence integrals with boundary integrals and obtain a global pressure evolution equation:

$$\frac{\mathrm{d}P^{(0)}}{\mathrm{d}t} = \frac{1}{\Omega}\left[-\oint_{\partial V}\left(\gamma P^{(0)}\boldsymbol{v} - \frac{\gamma-1}{\mathrm{Pe}}\lambda\nabla T^{(0)}\right)\cdot\boldsymbol{n}\mathrm{d}\sigma + \mathrm{Da}\int_V(\gamma-1)\mathrm{Q}(\rho\omega_F)^{(0)}\mathrm{d}V\right], \tag{2.1.32}$$

where $\boldsymbol{n}$ is the outward-pointing unit normal at the boundary and $\Omega = \int_V \mathrm{d}V$ is the total volume of the domain of integration V. Given appropriate velocity and thermal boundary conditions, all changes of the background pressure are thus related to the overall chemical-energy conversion within the domain.

The structure of the above equations is similar to that of incompressible, non-reactive flow, in that there is convection, diffusion, and an explicit velocity-divergence constraint. Thus, appropriate extensions of incompressible flow solvers should, in principle, be able to handle zero-Mach-number reactive flows as well. See [12, 13, 14, 15] for reviews of typical developments based on this *ansatz*.

2.1.3. *Multiple-length, single-time-scale analysis*

A number of interesting engineering applications are characterized by the presence of low-frequency flow-acoustic interactions. A prominent example is the 'rumbling noise' generated in the exhaust systems of large burners. These oscillations can become quite severe when eigenfrequencies of the exhaust system are excited. The origin of these resonances are interactions of a small scale, quasi-incompressible (turbulent) inflow with low-frequency sound waves that have the same time scale as the influx, but much longer wave lengths. A formal asymptotic analysis of this situation using multiple-scales techniques has been presented in [6] and shall be reviewed briefly in this section. For simplicity of exposition we restrict ourselves here to the non-reactive Euler equations for an ideal gas with constant specific-heat capacities.

We consider acoustic oscillations having the same time scale as the underlying small scale, quasi-incompressible flow. Due to the fast propagation of sound waves, their characteristic wavelength must then be larger by a factor of order 1/M, which is the ratio of a typical sound speed over a characteristic flow velocity. A suitable asymptotic solution *ansatz* reads

$$\begin{aligned} p &= p^{(0)}(\boldsymbol{x}, \mathrm{M}\boldsymbol{x}, t) + \mathrm{M}p^{(1)}(\boldsymbol{x}, \mathrm{M}\boldsymbol{x}, t) + \mathrm{M}^2 p^{(2)}(\boldsymbol{x}, \mathrm{M}\boldsymbol{x}, t) + o(\mathrm{M}^2), \\ \boldsymbol{v} &= \boldsymbol{v}^{(0)}(\boldsymbol{x}, \mathrm{M}\boldsymbol{x}, t) + \mathrm{M}\boldsymbol{v}^{(1)}(\boldsymbol{x}, \mathrm{M}\boldsymbol{x}, t) + o(\mathrm{M}), \\ \rho &= \rho^{(0)}(\boldsymbol{x}, \mathrm{M}\boldsymbol{x}, t) + \mathrm{M}\rho^{(1)}(\boldsymbol{x}, \mathrm{M}\boldsymbol{x}, t) + o(\mathrm{M}). \end{aligned} \tag{2.1.33}$$

The *ansatz* is introduced into the dimensionless governing equations (2.1.15)–(2.1.16). In doing so, the following generalized differentiation rule must be used in order to account for the multiple-scale expansion in the spatial coordinate

$$\nabla = \nabla_x + \mathrm{M}\nabla_\xi, \qquad \boldsymbol{\xi} := \mathrm{M}\boldsymbol{x}. \tag{2.1.34}$$

Here ∇_x, ∇_ξ denote gradients including the partial derivatives w.r.t. $\boldsymbol{x}, \boldsymbol{\xi}$ at fixed $(t, \boldsymbol{\xi})$ and $(t, \boldsymbol{x})$, respectively.

Following standard procedures of multiple-scales asymptotics, we require that the sums of all terms in the expanded governing equations that multiply the same powers of the Mach number add up to zero. (Notice that this is not a trivial step, as these terms themselves depend implicitly on the Mach number through their argument $\boldsymbol{\xi} = \mathrm{M}\boldsymbol{x}$. Consult Section 3 for a detailed discussion and mathematical justification of this subtle but important point.) The result for the momentum balances at orders M^{-2} and M^{-1} is

$$\nabla_x p^{(0)}(\boldsymbol{x}, \boldsymbol{\xi}, t) = 0, \quad \nabla_x p^{(1)}(\boldsymbol{x}, \boldsymbol{\xi}, t) = -\nabla_\xi p^{(0)}(\boldsymbol{x}, \boldsymbol{\xi}, t). \tag{2.1.35}$$

One first concludes that $p^{(0)} \equiv \widetilde{P}^{(0)}(\boldsymbol{\xi}, t)$. Next one integrates the second equation in $\boldsymbol{x}$ and, following standard sub-linear growth arguments (see Section 3), concludes that

$$p^{(0)} \equiv P^{(0)}(t) \qquad \text{and} \qquad p^{(1)} \equiv P^{(1)}(\boldsymbol{\xi}, t). \tag{2.1.36}$$

The continuity and energy equations at leading order are unchanged from (2.1.25) and (2.1.26), except that the ∇-operator is to be replaced with ∇_x and that all terms in (2.1.26) due to chemistry and heat conduction may be dropped here for simplicity.

The first major change appears in the momentum equation at order M^0, which now reads

$$\left(\rho^{(0)}\boldsymbol{v}^{(0)}\right)_t + \nabla_x \cdot \left(\rho^{(0)}\boldsymbol{v}^{(0)} \circ \boldsymbol{v}^{(0)}\right) + \nabla_x p^{(2)} = -\nabla_\xi P^{(1)}. \tag{2.1.37}$$

The second-order pressure $p^{(2)}$ still is responsible for guaranteeing compliance with an $\boldsymbol{x}$-scale divergence constraint from the leading-order energy equation (2.1.26), (2.1.30), but there is now a large scale driving force due to $P^{(1)}$. We will see below that this term represents the effects on the momentum balance of long wavelength acoustics.

As before we explore next the fact that $P^{(0)}$ is homogeneous in order to derive an equation for its time dependence. First we integrate (2.1.30) over a finite sub-domain in $\boldsymbol{x}$ at fixed time and fixed $\boldsymbol{\xi}$, use Gauß' theorem to replace the divergence integrals with boundary integrals and obtain

$$\frac{\mathrm{d}P^{(0)}}{\mathrm{d}t} = -\frac{1}{\Omega}\oint_{\partial V} \gamma P^{(0)}\boldsymbol{v} \cdot \boldsymbol{n}\, \mathrm{d}\sigma, \tag{2.1.38}$$

where, as before, $\boldsymbol{n}$ is the outward-pointing unit normal at the boundary and $\Omega = \int_V \mathrm{d}V$ is the total volume of the domain of integration V. The argument now differs from the previous one for the single scale case as follows: Since the total flow domain is large enough to cover long-wavelength acoustics represented by order $O(1)$ values of $\boldsymbol{\xi}$, for every fixed $\boldsymbol{\xi}$ the relevant domain of definition in $\boldsymbol{x}$ is infinite (*i.e.*, of order $O(1/\mathrm{M})$ as $\mathrm{M} \to 0$). Thus, we may let the boundary ∂V of V tend to infinity. Requiring that the flow velocities on the boundary remain bounded or, at least, grow sub-linearly with $|\boldsymbol{x}|$ as $|\boldsymbol{x}| \to \infty$, we conclude that

$$\lim_{\Omega\to\infty}\left[\frac{1}{\Omega}\oint_{\partial V} \gamma P^{(0)}\boldsymbol{v} \cdot \boldsymbol{n}\mathrm{d}\sigma\right] = 0, \tag{2.1.39}$$

due to the vanishing surface-to-volume ratio of the domain V. (A precise formulation would have to specify in more detail how the domain boundary is spread to infinity, so that a vanishing surface-to-volume ratio is actually guaranteed. We skip these details here as they are part of the standard procedures of multiple-scales asymptotics. See, however, also Section 3.) Obviously, (2.1.39) implies that

$$P^{(0)} = P_\infty \equiv \text{constant} \tag{2.1.40}$$

on the time scales considered here. To close the obtained equations, it is necessary to derive additional equations that determine the time evolution of the first-order long-wavelength pressure contribution $P^{(1)}(\boldsymbol{\xi}, t)$. To this end we integrate the momentum equation from (2.1.37) in $\boldsymbol{x}$ over a domain V as done before for the energy equation and again use the sub-linear growth condition. Similarly, we proceed for the first-order energy equation, [6], and find

$$\begin{aligned} \left(\overline{\rho^{(0)}\boldsymbol{v}^{(0)}}\right)_t + \quad & \nabla_\xi P^{(1)} &= 0, \\ \left(P^{(1)}\right)_t + \quad & \nabla_\xi \cdot \left(\gamma P_\infty \overline{\boldsymbol{v}^{(0)}}\right) &= 0, \end{aligned} \tag{2.1.41}$$

where $\overline{\rho^{(0)}\boldsymbol{v}^{(0)}}(\boldsymbol{\xi}, t)$, $\overline{\boldsymbol{v}^{(0)}}(\boldsymbol{\xi}, t)$ denote the small-scale averaged momentum and velocity fields, respectively.

The most striking conclusion from this analysis is that there is a mutual and non-trivial feed-back between the large and small scale flow features in the presence of small-scale, large amplitude density fluctuations, *i.e.*, in the general case of $\rho^{(0)} = \rho^{(0)}(\boldsymbol{x}, \boldsymbol{\xi}, t)$, $\nabla_x \rho^{(0)} \neq 0$. In this case, the change of momentum induced by $\nabla_\xi P^{(1)}$ in (2.1.37) will lead to large velocity changes for low density mass elements and small velocity changes for heavy ones. The result is the so-called baroclinic vorticity generation. This is a manifestation of the generation of small-scale structures by long-wave acoustic pressure gradients in the presence of small-scale density fluctuations.

We consider, on the other hand, (2.1.41) and notice that the momentum equation involves the time derivative of the small-scale average *momentum*, while the energy-flux term in the second equation contains the $\boldsymbol{\xi}$-divergence of the small-scale averaged *velocity* field. As a consequence, in presence of non-trivial small-scale fluctuations of the density $\rho^{(0)}$, these equations are not closed, as they involve the unknowns $P^{(1)}$, $\overline{\rho^{(0)}\boldsymbol{v}^{(0)}}$ and $\overline{\boldsymbol{v}^{(0)}}$. We can reveal the consequences by decomposing the velocity and density fields into mean and fluctuation according to

$$\boldsymbol{v}^{(0)} = \overline{\boldsymbol{v}^{(0)}} + \widetilde{\boldsymbol{v}^{(0)}}, \qquad \rho^{(0)} = \overline{\rho^{(0)}} + \widetilde{\rho^{(0)}} \tag{2.1.42}$$

and rewriting the averaged momentum as

$$\overline{\rho^{(0)}\boldsymbol{v}^{(0)}} = \overline{\rho^{(0)}}\,\overline{\boldsymbol{v}^{(0)}} + \overline{\widetilde{\rho^{(0)}}\widetilde{\boldsymbol{v}^{(0)}}}. \tag{2.1.43}$$

This leads to the revised formulation of the long-wave dynamics

$$\begin{aligned} \left(\overline{\boldsymbol{v}^{(0)}}\right)_t + \frac{1}{\overline{\rho^{(0)}}}\nabla_\xi P^{(1)} &= -\frac{1}{\overline{\rho^{(0)}}}\left(\overline{\widetilde{\rho^{(0)}}\widetilde{\boldsymbol{v}^{(0)}}}\right)_t, \\ \left(P^{(1)}\right)_t + \gamma P_\infty \nabla_\xi \cdot \overline{\boldsymbol{v}^{(0)}} &= 0. \end{aligned} \tag{2.1.44}$$

In the derivation we have anticipated that $(\overline{\rho^{(0)}})_t = 0$, which is an immediate consequence of the small-scale averaged leading-order continuity equation (2.1.25). We observe that the left-hand side of this system represents long-wavelength linear acoustics with variable sound speed, the latter being due to the $\boldsymbol{\xi}$-dependence of $\overline{\rho^{(0)}}(\boldsymbol{\xi})$. The right-hand side of the first equation in (2.1.44) is present only if there are long-wave correlations of the small-scale fluctuations of velocity and density. In this case there is a non-negligible effect of the small scales onto the long-wave acoustics. Together with the previous observations regarding baroclinic vorticity generation this is part of a mutual interaction of small and large scales.

For a more detailed description of the mathematical framework of the multiple-scales techniques used here see Section 3. For developments of new numerical techniques on the basis of this analysis, see Sections 4.2, 4.3.

2.2. The magneto-hydrodynamic-equations (MHD-equations)

The equations of ideal MHD are

$$\frac{\partial}{\partial t}\begin{pmatrix} \rho \\ \rho \boldsymbol{v} \\ \boldsymbol{B} \\ \rho e \end{pmatrix} + \nabla \cdot \begin{pmatrix} \rho \boldsymbol{v} \\ \rho \boldsymbol{v} \circ \boldsymbol{v} + (p + \frac{1}{2}\boldsymbol{B}\cdot\boldsymbol{B})I - \boldsymbol{B}\circ\boldsymbol{B} \\ \boldsymbol{v}\circ\boldsymbol{B} - \boldsymbol{B}\circ\boldsymbol{v} \\ (\rho e + p + \frac{1}{2}\boldsymbol{B}\cdot\boldsymbol{B})\boldsymbol{v}\cdot\boldsymbol{B})\boldsymbol{B} \end{pmatrix} = 0 \tag{2.2.1}$$

with the additional constraint $\nabla \cdot \boldsymbol{B} = 0$. As in the equations governing compressible gas dynamics, ρ, $\boldsymbol{v}$, e and p are the density, the velocity, the total energy and the pressure, respectively. $\boldsymbol{B}$ is the magnetic field. If the plasma behaves like an ideal gas, pressure and total energy are related via a state equation

$$\rho e = \frac{p}{\gamma - 1} + \frac{1}{2}\rho \boldsymbol{v} \cdot \boldsymbol{v} + \frac{1}{2}\boldsymbol{B} \cdot \boldsymbol{B}.$$

Like the equations of gas dynamics, the MHD equations can be written as evolution equations for the primitive variables $\rho, \boldsymbol{v}, \boldsymbol{B}, p$. Let us consider, for simplicity, one-dimensional solutions in a three-dimensional space. In a Cartesian frame of reference with $\boldsymbol{x} = (x, y, z)$, $\boldsymbol{u} = (u, v, w)$, $\boldsymbol{B} = (B_1, B_2, B_3)$ and $\partial/\partial y = \partial/\partial z = 0$ the primitive equations read

$$\begin{aligned}
&\rho_t + u\rho_x + \rho u_x &&= 0\,, \\
&u_t + uu_x + \frac{1}{\mathrm{M}^2}\frac{p_x}{\rho} + \frac{1}{2\mathrm{Av}^2\rho}(\boldsymbol{B} \cdot \boldsymbol{B})_x &&= 0\,, \\
&v_t + uv_x - \frac{B_1}{\mathrm{Av}^2\rho}B_{2x} &&= 0\,, \\
&w_t + uw_x - \frac{B_1}{\mathrm{Av}^2\rho}B_{3x} &&= 0\,, \\
&B_{2t} + uB_{2x} + B_2u_x - B_1v_x &&= 0\,, \\
&B_{3t} + uB_{3x} + B_3u_x - B_1w_x &&= 0\,, \\
&p_t + up_x + \gamma pu_x &&= 0\,.
\end{aligned} \tag{2.2.2}$$

Here we have introduced the scaling described in Section 2.1 and an additional reference value B_{ref} for the magnetic field. Since there is an additional equation for the magnetic field, a new characteristic number appears. This is the Alfvén number Av, the ratio between flow velocity and speed of the magneto-sonic waves, $c_{A,\mathrm{ref}}$:

$$\mathrm{Av} := \frac{|v_{\mathrm{ref}}|}{c_{A,\mathrm{ref}}} = \frac{|v_{\mathrm{ref}}|}{\sqrt{\frac{B_{\mathrm{ref}}^2}{\rho_{\mathrm{ref}}}}}.$$

2.2.1. *The case of small Mach number*

For the case of small Mach number but finite Av, the asymptotic expansion described in Section 2.1 can be extended to the MHD equations. The asymptotic multi-scale *ansatz* (2.1.33), completed by

$$\boldsymbol{B}(x, t; \mathrm{M}) = \boldsymbol{B}^{(0)}(x, \mathrm{M}x, t) + \mathrm{M}\boldsymbol{B}^{(1)}(x, \mathrm{M}x, t) + o(\mathrm{M}), \tag{2.2.3}$$

leads to the following zero-Mach-number limit equations:

$$
\begin{aligned}
&\rho_t + \rho u_x + \rho_x u && = 0\,,\\
&u_t + u u_x + \frac{1}{\rho} p_x^{(2)} + \frac{1}{2\mathrm{Av}^2 \rho} (\boldsymbol{B}\cdot\boldsymbol{B})_x && = -\frac{1}{\rho} p_\xi^{(1)}\,,\\
&v_t + u v_x - \frac{1}{\mathrm{Av}^2 \rho} B_1 B_{2x} && = 0\,,\\
&w_t + u w_x - \frac{1}{\mathrm{Av}^2 \rho} B_1 B_{3x} && = 0\,,\\
&B_{2t} + B_2 u_x - B_1 v_x + u B_{2x} && = 0\,,\\
&B_{3t} + B_3 u_x - B_1 w_x + u B_{3x} && = 0\,,\\
&p_t^{(0)} + \gamma p^{(0)} u_x && = 0\,.
\end{aligned}
\tag{2.2.4}
$$

As in Section 2.1, it turns out that the first three terms of the pressure expansion

$$
p = p^{(0)} + \mathrm{M} p^{(1)} + \mathrm{M}^2 p^{(2)} + o(\mathrm{M}^2) \tag{2.2.5}
$$

carry different physical meanings; $p^{(0)}$ and $p^{(1)}$ represent the thermodynamic pressure and the amplitude of acoustic perturbations, respectively. Their evolution equations are analogous to (2.1.36) and (2.1.38)–(2.1.41). The second-order pressure $p^{(2)}$ is implicitly defined by (2.2.4). The main difference with respect to the case of pure gas dynamics is that, here, magnetic field and velocity field are coupled. Therefore, the magnetic field appears in the elliptic equation for $p^{(2)}$. The one-dimensional analysis can be extended to the multidimensional case in a straightforward way.

The main result of the analysis is that the presence of the magnetic field does not break the structure of the zero-Mach-number limit equations. Therefore, numerical methods for low-Mach-number finite-Alfvén-number inviscid MHD can be developed on the same basis as numerical methods for the low- Mach-numbers Euler equations. One such method is discussed in Section 4.5.

2.2.2. *The case of small Alfvén number*

We consider now the distinguished limit in which the magneto-sonic waves are much faster than the flow, $\mathrm{Av} \to 0$, but the Mach number is finite. The asymptotic *ansatz* now reads

$$
U(x,t;\mathrm{Av}) = U^{(0)}(x,\mathrm{Av}x,t) + \mathrm{Av}U^{(1)}(x,\mathrm{Av}x,t) + \mathrm{Av}^2 U^{(2)}(x,\mathrm{Av}x,t) + o(\mathrm{Av}^2) \tag{2.2.6}
$$

with $U = \{\rho, \boldsymbol{v}, p, \boldsymbol{B}\}$. Introducing this *ansatz* into the one-dimensional MHD equations (2.2.2) and collecting the terms which multiply the same powers of the Alfvén number leads to the basic equations of MHD for zero Alfvén number and to a set of perturbation equations. Using the same techniques described in the analysis of the hydrodynamic equations, small-scales averaging and standard sub-linear growth constraints, one obtains the following results:

Due to the singular terms in the Alfvén number appearing in the MHD equations, the leading-order magnetic field becomes spatially homogeneous. The first-order magnetic field is homogeneous in x but supports non-trivial structures on the large scale ξ:

$$
B_2^{(0)} = B_2^{(0)}(t), \qquad B_2^{(1)} = B_2^{(1)}(\xi, t)\,, \tag{2.2.7}
$$

$$B_3^{(0)} = B_3^{(0)}(t), \qquad B_3^{(1)} = B_3^{(1)}(\xi, t)\,. \tag{2.2.8}$$

The MHD equations for zero-Alfvén-number limit read

$$\begin{aligned}
&\rho_t + \rho u_x + \rho_x u &&= 0\,,\\
&u_t + uu_x + \frac{1}{\mathrm{M}^2}\frac{1}{\rho}p_x + \frac{1}{\rho}(B_2^{(0)}B_{2,x}^{(2)} + B_3^{(0)}B_{3,x}^{(2)}) &&= -\frac{1}{\rho}(B_2^{(0)}B_{2,\xi}^{(1)} - B_3^{(0)}B_{3,\xi}^{(1)})\,,\\
&v_t + uv_x - \frac{1}{\rho}B_1 B_{2,x}^{(2)} &&= \frac{1}{\rho}B_1 B_{2,\xi}^{(1)}\,,\\
&w_t + uw_x - \frac{1}{\rho}B_1 B_{3,x}^{(2)} &&= \frac{1}{\rho}B_1 B_{3,\xi}^{(1)}\,,\\
&p_t + \gamma p u_x + p_x u &&= 0\,.
\end{aligned}$$

and one finds the following additional coupling constraints between velocity and magnetic field

$$\begin{aligned}
B_{2,t}^{(0)} + B_2^{(0)}u_x - B_1 v_x &= 0\,,\\
B_{3,t}^{(0)} + B_3^{(0)}u_x - B_1 w_x &= 0\,,
\end{aligned} \tag{2.2.9}$$

We remark that, according to (2.2.7) and (2.2.8), $B_2^{(i)}$, $B_3^{(i)}$, $i = 0, 1$, are constant in x. Hence (2.2.9) may be interpreted as constraints for the velocity components. The structure here is much more complicated than in the zero-Mach-number limit. It becomes even more complicated in the multidimensional case. Evolution equations for long-wavelength magneto-acoustics are obtained by averaging the velocity and magnetic-field perturbation equations. A further interesting regime would be described by the limit $\mathrm{M} = c\ \mathrm{Av}$ with $c = O(1)$ as $\mathrm{M} \to 0$. As we will see in the next section, such a distinguished limit leads to new interactions between the fast waves that are not captured by any sequential limit or by any two-parameter expansion!

2.3. Slow Atmospheric Motion

Understanding and computing motions in the atmosphere is particularly challenging because of the multitude of physical processes and of the different space and time scales involved. In spite of its physical complexity and variety, however, atmospheric motion is confined to a very special regime. The Mach number M varies from zero at rest to about 0·3 in fast jet streams, hurricanes and tornadoes.

As seen in Section 2.1, in the $\mathrm{M} \to 0$ limit the governing equations for compressible (air) flow become singular. Due to this singularity pressure differences as small as M^2 can generate $O(1)$ accelerations of the horizontal winds, while the pressure forces are in an almost exact balance (see Equation (2.1.28)). In fact, the problem of extracting the 'correct' net effect that modifies an almost perfect balance is a very common one in weather forecasting, see, *e.g.*, [16 p. 187], and both theoretical and numerical investigations of atmospheric motion make use of related approximate *models*. These models achieve a simplification of the governing equations by assuming some kind of balance, *e.g.*, *hydrostatic*, *geostrophic*, *Boussinesq*, *anelastic*, *pseudo-incompressible*. For comprehensive accounts of the matter the reader may want to consult the classical textbooks by Gill [17] and Pedlosky [18], or the monographs by

Zeytounian [19, 20]. Note that Zeytounian uses techniques of asymptotic analysis that are very close to those employed in the present work, albeit with the aim of *asymptotic modelling* (deriving simplified asymptotic equations), instead of guiding the construction of new numerical techniques.

In spite of the outstanding role of approximate models, modern computational approaches in numerical weather forecasting turn back to the full governing equations, see [21]. The reasons are:

1. The full equations are believed to allow a better description of real motions than approximate models do.

2. In contrast to approximate models, the full equations are valid on a wide range of scales. Therefore, numerical methods based on the full equations can be naturally coupled with local-mesh-refinement techniques for, *e.g.*, regional weather forecasting.

3. Both numerical weather forecasting and climate research need fast algorithms. If one knows how to avoid time-step restrictions due to fast modes, the computational cost of integrating the full equations can be significantly lower than the cost of solving the equation of, *e.g.*, an anelastic model.

When attempting to compute numerical solutions of the full compressible equations at different resolutions one has to face two major problems. 1) Standard methods exhibit a breakdown of both accuracy and efficiency in the low-Mach-number regime, see [22], [23], [6], [24], [25]. 2) As the grid size changes, one has to adjust the *parameterizations* of the sub-cell processes. Such parameterizations have been designed (and are considered to be valid) for a given scaling or a well defined grid size. Tautologically, there is very little knowledge on how parameterizations depend on the grid size. It has been observed in climate research and weather forecasting that, for a given model, a simple grid refinement by a factor of two may have drastic impacts on the computed solution and may thus require a complete new adjustment of parameters in the sub-grid scale models.

We propose multiple-scale low-Mach-number low-Froude-number asymptotic analysis as a general framework for understanding the motion in the atmosphere on space scales ranging from a few meters to thousands of kilometers. The theory, detailed in [26], provides a consistent picture of slow atmospheric flows and turns out to be the natural framework in which approximations, traditionally obtained on the basis of simplifying assumptions or *ad hoc* scaling arguments, can be derived. The analysis yields interesting implications for discrete methods aiming at the numerical computation of atmospheric motions as, *e.g.*, in numerical weather prediction or climate modeling. There are two kinds of implications. On one hand one finds side constraints on parameterizations of unresolved physical processes such as turbulent heat transport or velocity boundary conditions. On the other hand, the asymptotic analysis provides guidelines for consistently 'filtering' the equations and suggests that 'singular dynamic-range problems', such as the multiple-pressure problem for low-Mach-number flows from Section 2.1, can be overcome by the introduction of suitable multiple variables that mimic the asymptotic decomposition of the field quantities.

The analyses presented below are designed to support the construction of numerical integration techniques. Thus, we use asymptotics in order to reveal singular behaviour *on the numerically resolved length and time scales*. Importantly, these scales do not necessarily match the length and time scales that a meteorologist or oceanographer would ultimately be interested in. For example, in the near future one may expect horizontal numerical grid resolutions of 5 km or less and numerical time steps on the order of minutes or less. In contrast, the synoptic scales of weather dynamics in the mid-latitudes involve hundreds to thousands of kilometres and time scales up to several hours or days. Our primary interest here is in the much shorter, numerically resolved scales, and this is why some of our results may appear not to be in line with classical results of theoretical meteorology [17, 18] or asymptotic modelling [19, 20] at a first glance. We emphasize, however, that those results can be recovered by systematically employing multiple-scales asymptotics. With the present multiple-*space*-scale analyses we do already recover some of the classical results, while we hope to provide a completely consistent picture based on both multiple-*time* and multiple-*space* scale analyses elsewhere in the near future.

2.3.1. *Scaling, equations of motion*

An atmosphere is said to be in hydrostatic balance when the vertical pressure gradient balances the force of gravity: $\partial p/\partial z = -\rho g$. Let p_{ref} be a reference pressure (*e.g.* some mean sea-level pressure). The scale height $h_{\mathrm{scale}} := p_{\mathrm{ref}}/(\rho_{\mathrm{ref}} g)$ represents the height over which pressure changes of magnitude p_{ref} would occur in a constant density atmosphere in hydrostatic balance. It is a rough measure of the thickness of the atmosphere: $h_{\mathrm{scale}} = O(10)$ km. We consider h_{scale} as our basic scale and use two-scale asymptotics to resolve much smaller horizontal 'micro scales' in one *ansatz* and to cover the much larger synoptic scales in another. The first *ansatz* is designed to assess deep convective motions that may occur in cumulus clouds, while the second analyses short-time dynamics on the large scales of weather patterns.

If p_{ref}, ρ_{ref}, v_{ref} and $l_{\mathrm{ref}} := h_{\mathrm{scale}}$ are taken as reference variables, the conservation laws for mass, momentum and energy in a dry rotating atmosphere ([16], [17], [27]) read

$$\begin{aligned}
&\rho_t + \nabla_\| \cdot (\rho \boldsymbol{u}) + (\rho w)_z && = 0, \\
&(\rho \boldsymbol{u})_t + \nabla_\| \cdot (\rho \boldsymbol{u} \circ \boldsymbol{u}) + (\rho \boldsymbol{u} w)_z + \frac{1}{\mathrm{M}^2} \nabla_\| p + \frac{1}{\mathrm{Ro}} \rho \left(\boldsymbol{u}^\perp \mathcal{S} + \boldsymbol{i} w \mathcal{C} \right) && = \boldsymbol{D}_{\rho \boldsymbol{u}}, \\
&(\rho w)_t + \nabla_\| \cdot (\rho \boldsymbol{u} w) + (\rho w^2)_z + \frac{1}{\mathrm{M}^2} p_z + \frac{1}{\mathrm{Fr}^2} \rho g - \frac{1}{\mathrm{Ro}} \rho u \mathcal{C} && = D_{\rho w}, \\
&(\rho e)_t + \nabla_\| \cdot ((\rho e + p) \boldsymbol{u}) + ((\rho e + p) w)_z && = D_{\rho e},
\end{aligned} \tag{2.3.1}$$

with the equation of state

$$p = (\gamma - 1) \left(\rho e - \frac{1}{2} \mathrm{M}^2 \rho \boldsymbol{v} \cdot \boldsymbol{v} - \frac{\mathrm{M}^2}{\mathrm{Fr}^2} \rho g z \right). \tag{2.3.2}$$

We are considering a simplified geometrical setup in which the acceleration of gravity acts in the vertical direction $\boldsymbol{k}$ of a Cartesian frame of reference of coordinates x, y, z, unit vectors $\boldsymbol{i}$, $\boldsymbol{j}$ and $\boldsymbol{k}$ and velocity components u, v and w; $\boldsymbol{u} := \boldsymbol{i} u + \boldsymbol{j} v$ is called the horizontal wind and $\boldsymbol{u}^\perp := -\boldsymbol{i} v + \boldsymbol{j} u$. This frame of reference rotates with constant angular velocity Ω around the axis defined by the unit vector $\boldsymbol{\Omega} := \boldsymbol{j}\mathcal{C} + \boldsymbol{k}\mathcal{S}$, where $\mathcal{C}$ and $\mathcal{S}$ are the sinus and the cosine of the latitude. They depend on the space coordinate but not on time. The operator $\nabla_\|$ indicates differentiation in the horizontal directions and $f_z := \partial f/\partial z$. $\mathrm{M} := v_{\mathrm{ref}}/\sqrt{p_{\mathrm{ref}}/\rho_{\mathrm{ref}}}$, $\mathrm{Fr} := v_{\mathrm{ref}}/\sqrt{g_{\mathrm{ref}} l_{\mathrm{ref}}}$ and $\mathrm{Ro} := v_{\mathrm{ref}}/(2\Omega l_{\mathrm{ref}})$ are the Mach number, the Froude number and the Rossby number, respectively. The terms $\boldsymbol{D}_{\rho \boldsymbol{u}}$, $D_{\rho w}$ on the right-hand side represent the effects of microscopical (molecular and/or turbulent) transport of momentum. $D_{\rho e}$ accounts for both microscopical transport of energy and for diabatic heating (latent heat and radiation). These processes are usually parameterized by means of empirical relationships or models. The above system of conservation laws implies the following evolution equation for the entropy $s = p/\rho^{\gamma}$:

$$s_t + \boldsymbol{u} \cdot \nabla_{\|} s + w s_z = \frac{\gamma - 1}{\rho^{\gamma}} \left(D_{\rho e} - \mathrm{M}^2 (\boldsymbol{u} \cdot \boldsymbol{D}_{\rho \boldsymbol{u}} + w D_{\rho w}) \right). \tag{2.3.3}$$

2.3.2. *Asymptotic ansatz*

Let U be a shortcut for a solution component or a function of a solution of the governing equations (2.3.1), such as the pressure p, density ρ or the velocity field $\boldsymbol{v} = \boldsymbol{u} + w\boldsymbol{k}$. In general, U depends on the horizontal coordinate $\boldsymbol{x} := \boldsymbol{i}x + \boldsymbol{j}y$, the vertical coordinate z, the time t, the singular perturbation parameter M and on other dimensionless parameters like the Rossby and Reynolds numbers, Ro, Re, etc. We focus the attention on the behavior of U as $\mathrm{M} \to 0$, noticing that our choice of a reference length implies

$$\mathrm{Fr} \equiv \mathrm{M} \qquad \text{for} \qquad l_{\mathrm{ref}} := h_{\mathrm{scale}}. \tag{2.3.4}$$

Assuming that all dimensionless parameters other than the Froude and Mach numbers are fixed as $\mathrm{M} \to 0$, we write $U = U(\boldsymbol{x}, z, t; \mathrm{M})$ and consider an asymptotic expansion of the solution in terms of the Mach/Froude number:

$$U(\boldsymbol{x}, z, t; \mathrm{M}) := \sum_{i=0} \mathrm{M}^i U^{(i)}(\boldsymbol{\eta}, \boldsymbol{x}, \boldsymbol{\xi}, z, \tau, t). \tag{2.3.5}$$

Here

$$\boldsymbol{\eta} = \frac{1}{\mathrm{M}} \boldsymbol{x}, \qquad \tau = \frac{1}{\mathrm{M}} t \tag{2.3.6}$$

are new small-scale horizontal and time coordinates, while

$$\boldsymbol{\xi} = \mathrm{M} \boldsymbol{x} \tag{2.3.7}$$

represents a new large-scale horizontal coordinate in analogy with the multiple-scales *ansatz* pursued in Section 2.1.3.

In particular, we will consider $U^{(i)} = U^{(i)}(\boldsymbol{\eta}, z, \tau)$ for the analysis of deep convection on very small horizontal scales and $U^{(i)} = U^{(i)}(\boldsymbol{x}, z, t)$, $U^{(i)} = U^{(i)}(\boldsymbol{x}, \boldsymbol{\xi}, z, t)$ for the analysis on the meso and synoptic scales. In this paper we will not consider any multi-scale *ansatz* in the vertical direction.

2.3.3. *Results*

Leading-order pressure and density. Leading-order pressure and density are found to be in hydrostatic balance both at micro scale and at meso and synoptic scales:

$$p^{(0)} = p^{(0)}(z), \qquad \rho^{(0)} = \rho^{(0)}(z). \tag{2.3.8}$$

This result is not surprising. Indeed, it is common practice in numerical methods for atmospheric flows to solve for $p - p^{(0)}$, $\rho - \rho^{(0)}$ where the hydrostatic basic state $p^{(0)}$, $\rho^{(0)}$ is more or less arbitrarily assigned. (see [21, pp. 39–41].)

Meso and synoptic scales: Vertical velocity constraint. A less trivial result is obtained by averaging the continuity equation and the entropy equation (2.3.3) on the meso scales and taking into account (2.3.8). One gets

$$\begin{aligned} \partial_z\left(\rho^{(0)}\overline{w^{(0)}}^{x}\right) &= \rho^{(0)}\frac{1}{|\mathcal{D}_x|}\int\limits_{\partial\mathcal{D}_x} \boldsymbol{u}^{(0)}\cdot\boldsymbol{n}\,\mathrm{d}L, \\ \overline{w^{(0)}}^{x}\,\partial_z s^{(0)} &= \rho^{(0)-\gamma}(\gamma-1)\overline{D_{\rho e}^{(0)}}^{x}. \end{aligned} \tag{2.3.9}$$

In the above system $\overline{w^{(0)}}^{x}$ represents the average of $w^{(0)}$ on the $\boldsymbol{x}$-scales. The system shows that, if $\partial_z s^{(0)} \neq 0$, the leading-order mass flux through $\partial\mathcal{D}_x$ and the vertical rate of change of the leading-order average heating $\overline{D_{\rho e}^{(0)}}^{x}$ are coupled by a simple relationship. In the special case of zero mass flux the first equation requires $\partial_z(\rho^{(0)}\overline{w^{(0)}}^{x})$ to be zero. Since $\rho^{(0)} \to 0$ for $z \to \infty$ and $\overline{w}^{x}$ is bounded, $\overline{w^{(0)}}^{x}$ must be zero. In this case the source term of the energy equation, $D_{\rho e}^{(0)}$ must have a zero small-scale average. Thus we have found a constraint that parameterizations of turbulent and radiative heating must fulfill in the limit of vanishing Mach and Froude numbers.

For a stable stratification, $\partial_z s^{(0)} \neq 0$, the entropy evolution equation yields the leading-order (non-averaged) vertical velocity $w^{(0)}$ as a function of the heating term $D_{\rho e}^{(0)}$:

$$w^{(0)} = \frac{\rho^{(0)-\gamma}(\gamma-1)D_{\rho e}^{(0)}}{\partial_z s^{(0)}}. \tag{2.3.10}$$

Note that this is *not* a form or consequence of the continuity equation. In fact, insertion of (2.3.10) in the continuity equation yields a sequence of two-dimensional Poisson problems for the second-order pressure $p^{(2)}$, one for each constant z-level. This pressure, in turn, is needed to determine the leading-order horizontal wind.

Thus, the leading-order analysis yields the following picture of slow atmospheric motion on meso and synoptic scales: Pressure and density are in hydrostatic balance to leading order; the vertical wind has zero small-scale average or has an average that is coupled to the boundary conditions for the horizontal wind; for stable stratifications the vertical wind is a function of the source term of the energy equation and of the stability parameter represented by $\partial s^{(0)}/\partial z$. This functional dependence defines a divergence constraint for the horizontal wind and allows its computation by layer-wise numerical integration of a set of quasi-two-dimensional Navier-Stokes equations. This picture and, in particular, Equation (2.3.10) are consistent with the results obtained on the micro scales where the interest is focused on deep convection. On such scales the dynamics of vertical velocity and density perturbation are described by the following system of partial differential equations (see [26]):

$$\begin{aligned} \frac{c^{(0)2}}{\rho^{(0)\gamma}}\frac{\mathrm{D}\widetilde{\rho^{(1)}}^{\eta}}{\mathrm{D}\tau} - \partial_z s^{(0)}\widetilde{w^{(0)}}^{\eta} &= -\frac{\gamma-1}{\rho^{(0)\gamma}}\widetilde{D_{\rho e}^{(0)}}^{\eta}, \\ \frac{\mathrm{D}\widetilde{w^{(0)}}^{\eta}}{\mathrm{D}\tau} + \frac{\widetilde{\rho^{(1)}}^{\eta}}{\rho^{(0)}}g &= \frac{1}{\rho^{(0)}}\widetilde{D_{\rho w}^{(-1)}}^{\eta}. \end{aligned} \tag{2.3.11}$$

In this system $\widetilde{w^{(0)}}^{\eta}$ represents the difference between the leading-order vertical velocity and its small-scale average (*i.e.* the average over scales of order $O(\mathrm{M}h_{\mathrm{scale}})$). These equations describe the well known oscillations of density perturbations in a stably stratified atmosphere that are associated with the Brunt-Väisälä frequency and the additional driving of vertical motions due to net heat sources. Notice that stationary solutions of this system on the (very short) τ-time scale, with $\mathrm{D}/\mathrm{D}\tau \equiv 0$, reproduce the diagnostic vertical velocity equation from (2.3.10) which was obtained in the single space, single time scale regime.

Anelastic and Boussinesq approximation. A popular approximation to the full continuity equation $(2.3.1)_1$ is the *anelastic* approximation ([28], [29]): $\nabla \cdot (\overline{\rho}\boldsymbol{v}) = 0$. Citing [27], we observe that $\overline{\rho}$ is 'a steady reference-state density that varies only along the coordinate axis parallel to the gravitational restoring force' *i.e.* $\overline{\rho}$ is equal to $\rho^{(0)}$. In this case

$$\nabla \cdot (\overline{\rho}\boldsymbol{v}) = \nabla \cdot (\rho^{(0)}\boldsymbol{v}) = \nabla_x \cdot (\rho^{(0)}\boldsymbol{u}^{(0)}) + \partial_z(\rho^{(0)}w^{(0)}) + O(\mathrm{M}) = O(\mathrm{M}). \tag{2.3.12}$$

The last equality follows from (2.3.8) and shows that the anelastic equation approximates the full continuity equation up to terms of order $O(\mathrm{M})$ as $\mathrm{M} \to 0$. Notice also that the Boussinesq approximation, $\nabla \cdot \boldsymbol{v} = 0$, will generally introduce unacceptable leading-order errors, unless the vertical velocity w is very small everywhere.

Meso and synoptic scales: first order perturbation. Let $\zeta^{(0)}$ be the synoptic density-weighted vorticity of the average leading-order horizontal velocity

$$\zeta^{(0)} := \rho^{(0)}\boldsymbol{k} \cdot \nabla_{\xi}\overline{\boldsymbol{u}^{(0)}}^{x} = -\rho^{(0)}\nabla_{\xi} \cdot \overline{\boldsymbol{u}^{(0)\perp}}^{x}. \tag{2.3.13}$$

Remember that $\boldsymbol{\xi} := \mathrm{M}\boldsymbol{x}$. As pointed out in the previous paragraphs, pressure and density are in hydrostatic balance as $\mathrm{M} \to 0$. They only depend on the vertical coordinate z. Thus, pressure tendencies $\partial p/\partial t$ are small of order $O(\mathrm{M})$. These tendencies do not vary on the meso scales: they are functions of t, $\boldsymbol{\xi}$ and of the vertical coordinate but *not* of $\boldsymbol{x}$. They are coupled with $\zeta^{(0)}$ through the following system

$$\begin{aligned} \partial_{tt}p^{(1)} - {c^{(0)}}^2\Delta_{\xi}p^{(1)} &= -{c^{(0)}}^2\overline{f}^{x}\zeta^{(0)} + P, \\ \partial_t\zeta^{(0)} &= \frac{\overline{f}^{x}}{{c^{(0)}}^2}\partial_t p^{(1)} + Q. \end{aligned} \tag{2.3.14}$$

This system describes the interaction between long-wave acoustics and large-scale vorticity. The terms P and Q on the right-hand side depend on the $\boldsymbol{x}$-average of the first-order perturbation vertical velocity $\overline{w^{(1)}}^{x}$, on variations of the Coriolis parameters $\mathcal{f}$ and $\mathcal{C}$ on the large scales, on correlations between fluctuations of the leading-order velocity, on large-scale turbulent stresses and on the heating. In the special case $P = Q = 0$ the system (2.3.14) supports traveling waves of the form

$$(p^{(1)}, \zeta^{(0)}) = (p_0^{(1)}, \zeta_0^{(0)})e^{\mathrm{i}(kx+ly-\omega t)} \tag{2.3.15}$$

provided that $\kappa^2 := k^2 + l^2$ satisfies the dispersion relation

$$\omega^2 = {c^{(0)}}^2\kappa^2 + \overline{f}^{x2}. \tag{2.3.16}$$

This is the dispersion relation for perturbations of the *shallow water* approximation linearized about a *geostrophic* basic state, see [27] page 15.

The closure of system (2.3.14) for the first-order perturbation pressure $p^{(1)}$ and for the large-scale vorticity $\zeta^{(0)}$ requires an equation for the vertical velocity $\overline{w^{(1)}}^{x}$. For a stable stratification, $\partial_z s^{(0)} \neq 0$, one obtains:

$$
\begin{aligned}
\partial_{tt} p^{(1)} - c^{(0)^2} \Delta_{\xi} p^{(1)} = {} & -c^{(0)^2} \overline{f}^{x} \zeta^{(0)} + \rho^{(0)} \partial_t \left(g \overline{w^{(1)}}^{x} - c^{(0)^2} \partial_z \overline{w^{(1)}}^{x} \right) \\
& + \frac{1}{\mathrm{Ro}} c^{(0)^2} \rho^{(0)} \nabla_{\xi} \cdot \left(\overline{\overline{\widetilde{\boldsymbol{u}^{(0)\perp}}}^{x} \widetilde{\mathcal{S}}^{x}}^{x} + \boldsymbol{i} \overline{\overline{\widetilde{w^{(0)}}}^{x} \widetilde{\mathcal{C}}}^{x} \right) \\
& + c^{(0)^2} \partial_z \Big(\rho^{(0)} \overline{\overline{\widetilde{\boldsymbol{u}^{(0)}}}^{x} \overline{\widetilde{w^{(0)}}}^{x}}^{x} \Big) \\
& + (\gamma - 1) \partial_t \overline{D^{(1)}_{\rho e}}^{x} - c^{(0)^2} \nabla_{\xi} \cdot \overline{\boldsymbol{D}^{(0)}_{\rho \boldsymbol{u}}}^{x}, \\
\partial_t \zeta^{(0)} = {} & \frac{\overline{f}^{x}}{c^{(0)^2}} \partial_t p^{(1)} + \overline{f}^{x} \partial_z \Big(\rho^{(0)} \overline{w^{(1)}}^{x} \Big) \\
& - \frac{1}{\mathrm{Ro}} \rho^{(0)} \boldsymbol{k} \cdot \nabla_{\xi} \times \left(\overline{\overline{\widetilde{\boldsymbol{u}^{(0)\perp}}}^{x} \widetilde{\mathcal{S}}^{x}}^{x} + \boldsymbol{i} \overline{\overline{\widetilde{w^{(0)}}}^{x} \widetilde{\mathcal{C}}^{x}}^{x} \right) \\
& - \boldsymbol{k} \cdot \nabla_{\xi} \times \partial_z \left(\rho^{(0)} \overline{\overline{\widetilde{\boldsymbol{u}^{(0)}}}^{x} \overline{\widetilde{w^{(0)}}}^{x}}^{x} \right) \\
& - \frac{\overline{f}^{x}}{c^{(0)^2}} (\gamma - 1) \overline{D^{(1)}_{\rho e}}^{x} + \boldsymbol{k} \cdot \nabla_{\xi} \times \overline{\boldsymbol{D}^{(0)}_{\rho \boldsymbol{u}}}^{x}, \\
N^{(0)^2} c^{(0)^2} \overline{w^{(1)}}^{x} = {} & - \frac{1}{\rho^{(0)}} \partial_t \left(g p^{(1)} + c^{(0)^2} \partial_z p^{(1)} \right) \\
& + \frac{g}{\rho^{(0)}} (\gamma - 1) \overline{D^{(1)}_{\rho e}}^{x}.
\end{aligned}
\tag{2.3.17}
$$

where $N^{(0)}$ is the buoyancy frequency

$$
N^{(0)^2} := -g \left(\frac{\partial_z \rho^{(0)}}{\rho^{(0)}} + g \frac{\rho^{(0)}}{\gamma p^{(0)}} \right). \tag{2.3.18}
$$

System (2.3.17) supports internal gravity waves, acoustic waves and the Lamb wave and represents the link between our asymptotic framework and the classical theory of small perturbations of the state of rest for compressible stratified fluids (see [17, p. 171]). There are two major differences between the classical theory and the present case. The first one is in the equation for the vertical velocity $\overline{w^{(1)}}^{x}$. In the classical theory such an equation is obtained through combination of vertical momentum and continuity and contains a second-order term $\partial_{tt} \overline{w^{(1)}}^{x}$ on the left-hand side. The second difference lies in the right-hand side of (2.3.17) where $\boldsymbol{x}$-scale correlations of fluctuations of the leading-order solution, microscopical transport and radiative heating appear as forcing terms in the equations. In the classical theory these terms are absent due to linearization and the particular choice of the basic state (of rest).

The results outlined above have interesting implications for the issues of modeling and computing slow atmospheric flows. So far, we can draw the following conclusions:

1. Pressure perturbations of order $O(\mathrm{M}^2)$ affect the velocity field at leading order. Therefore, single-variable representations of the pressure field do not allow a meaningful computation of pressure gradients.

2. For stable or moderately stable stratifications, the vertical velocity w must satisfy a diagnostic constraint. This constraint takes the form of a simple relationship between vertical velocity, stratification and heating and is *not* a form of the continuity equation. This result has three implications:

   (a) Parameterizations of sub-scale physical processes involved in the energy budget (*e.g.* turbulent heat transfer) are not completely free and must satisfy an integral constraint.

   (b) The second-order perturbation pressure can be computed by solving, at each $z =$ const. level, a two-dimensional elliptic problem. The numerical solution of this problem involves a plain Laplace operator and can be computed efficiently with standard methods.

   (c) There is no truly three-dimensional motion at meso and synoptic scales. Whether numerical methods based on the unconstrained-constraint integration of the full three-dimensional equation of motion can predict vertical winds which are consistent with the asymptotic behavior of the true solution is an open question.

3. The analysis of deep convection on the micro scales reveals a vertical velocity-density perturbation dynamics which is perfectly consistent with the above-mentioned diagnostic constraint for w: in the quasi-steady limit Equation (2.3.11) simply reduces to Equation (2.3.10).

4. On meso and synoptic scales pressure tendencies can be computed by integration of a system of second-order partial differential equations which supports internal and barotropic gravity waves, acoustic waves and Lamb wave. This system is a generalization of the set of equations obtained in the classical theory of perturbations from the state of rest for compressible stratified flows.

3. Mathematical issues

In this section we review a rigorous mathematical framework for deriving the relationships commonly used in multi-scale asymptotics. For a detailed analysis we refer to [10]. For concreteness, attention is focused on the low-Mach-number limit of the Euler equations. The results, *e.g.* Lemma 3.1 and Lemma 3.2, however, are generally valid.

Motivated by observations of many systems of practical relevance, we require that each quantity, say $U(\boldsymbol{x}, t)$, is bounded in the sense that for all $\boldsymbol{x} \in \mathcal{G} \subset \mathbf{R}^d$ and for each $t \in \mathbf{R}_0^+$ there exists a time-dependent upper bound $c_U(t)$ such that

$$|U(\boldsymbol{x}, t)| \leq c_U(t). \tag{3.0.1}$$

3.1. Asymptotic expansions

The success of an asymptotic analysis depends crucially on the choice of an asymptotic sequence – the functions $\mathrm{M}^0, \mathrm{M}^1, \mathrm{M}^2 \ldots$ in, *e.g.*, Equation (2.1.22) – as well as on the choice of suitable time and space scales. These choices depend on both the governing equations and on the regime being considered. Throughout this paper, the following asymptotic sequence has been used in the analysis of low-Mach-number flows ($\epsilon = \mathrm{M}$):

$$\phi_n(\epsilon) = \epsilon^n, \; n \in \mathbf{N}_0. \tag{3.1.1}$$

The sequence satisfies

$$\phi_n(\epsilon) = o\left(\phi_{n-1}(\epsilon)\right), \; \epsilon \to 0. \tag{3.1.2}$$

The choice of this asymptotic sequence is not obvious! In the governing equation, *e.g.* in (2.1.15)–(2.1.20), the perturbation parameter is the square of the Mach number ϵ^2. In fact, in the literature one can find many asymptotic analyses which are based on the sequence $\phi_n(\epsilon) = \epsilon^{2n}$.

Using the decay property (3.1.2) one can prove the following Lemma:

Lemma 3.1 *Let $\{\phi_n(\epsilon)\}_{n\in\mathbf{N}_0}$ be an asymptotic sequence and L_n, $n = 0, \ldots, N$ arbitrary terms which are independent of ϵ. Then*

$$\sum_{n=0}^{N} \phi_n(\epsilon) L_n = o\left(\phi_N(\epsilon)\right), \; \epsilon \to 0 \tag{3.1.3}$$

holds if and only if $L_n = 0$, $n = 0, \ldots, N$.

Proof: Let us assume the existence of terms $L_0, \ldots, L_N$ with $\max\limits_{n=0,\ldots,N} |L_n| > 0$ for which Equation (3.1.3) holds and let $m = \min\limits_{n=0,\ldots,N, L_n \neq 0} n$. Using (3.1.2), which is supposed to hold, we obtain, for $m \leq N$:

$$\sum_{n=0}^{m} \phi_n(\epsilon) L_n = o\left(\phi_N(\epsilon)\right) - \sum_{n=m+1}^{N} \phi_n(\epsilon) L_n = o\left(\phi_m(\epsilon)\right).$$

The last equality is a consequence of (3.1.3). Dividing by $\phi_m(\epsilon)$ and taking the limit $\epsilon \to 0$ one gets

$$\lim_{\epsilon\to 0} \frac{\sum_{n=0}^{m} \phi_n(\epsilon) L_n}{\phi_m(\epsilon)} \equiv L_m = \lim_{\epsilon \to 0} \frac{o\left(\phi_m(\epsilon)\right)}{\phi_m(\epsilon)}$$

in contradiction to the definition of m. The opposite direction is trivial.

For further detailed discussions of asymptotic sequences, asymptotic series, order symbols, etc., one may consult the classical textbooks by van Dyke [30], Schneider [1], or Kevorkian and Cole [31, 32].

Simple examples show that, in general, an asymptotic single-scale expansion can not be successfully employed in the context of problems that contain phenomena on different scales [32, 33]. The following sections provide results for single as well as multiple-scales expansions in space. The exact definitions of the asymptotic expansions are given in each section separately.

3.1.1. *Single-scale asymptotic analysis for confined domains*

In this section the Euler equations are considered in a domain $\mathcal{G} \times \mathbf{R}_0^+$ with a bounded spatial part $\mathcal{G} \subset \mathbf{R}^d$. Due to the boundedness of $\mathcal{G}$ there is always a bound $\mathrm{M}_l > 0$ such that for each $\mathrm{M} \in (0, \mathrm{M}_l)$ no phenomena on a scale $\boldsymbol{\xi} = \mathrm{M}\boldsymbol{x}$ are present in the distribution of the solution vector $\boldsymbol{u} = (\rho, m_1, \ldots, m_d, \rho e)^T$, provided we assume that $\boldsymbol{u}_{\text{ref}}$ is fixed as $\mathrm{M} \to 0$. Hence, we define the asymptotic solution space $\mathcal{U}_s$ as the set of all functions $\boldsymbol{u} : \mathcal{G} \times \mathbf{R}_0^+ \times (0, \widetilde{\mathrm{M}}) \to \mathbf{R}^{d+2}$, $\widetilde{\mathrm{M}} > 0$ which satisfy the governing equations and for which each physical quantity U can be expressed by an asymptotic expansion

$$U(\boldsymbol{x}, t; \mathrm{M}) = \sum_{i=0}^{2} \mathrm{M}^i U^{(i)}(\boldsymbol{x}, t) + o(\mathrm{M}^2), \;\; \mathrm{M} \to 0,$$

which is valid in $\mathcal{G} \times \mathbf{R}_0^+$. To study the properties of the functions $U^{(i)}, i = 0, 1, 2$ we introduce the asymptotic expansion into the Euler equations and collect the terms which are multiplied by the same power of the Mach number M. These terms are functions which do not depend on M and, according to Lemma 3.1, must vanish in the limit $\mathrm{M} \to 0$. This leads to the following asymptotic system

$$\partial_t \rho^{(i)} + \nabla_x \cdot \boldsymbol{m}^{(i)} = 0, \quad i = 0, 1, 2, \tag{3.1.4}$$

$$\nabla_x p^{(0)} = \mathbf{0}, \tag{3.1.5}$$

$$\nabla_x p^{(1)} = \mathbf{0}, \tag{3.1.6}$$

$$\partial_t \boldsymbol{m}^{(0)} + \nabla_x \cdot \left(\boldsymbol{m}^{(0)} \circ \boldsymbol{v}^{(0)}\right) + \nabla_x p^{(2)} = \mathbf{0}, \tag{3.1.7}$$

$$\partial_t (\rho e)^{(i)} + \nabla_x \cdot ((\rho e + p)\boldsymbol{v})^{(i)} = 0, \quad i = 0, 1, 2 \tag{3.1.8}$$

in $\mathcal{G} \times \mathbf{R}_0^+$. Here we have introduced $\boldsymbol{m} := \rho \boldsymbol{v}$. The above hierarchy of equations is the starting point of the analysis presented in Section 2.1.2.

3.1.2. *Multiple-scale asymptotic analysis for unbounded domains*

In contrast to the previous section we consider a spatial domain $\mathcal{G} \subset \mathbf{R}^d$ which satisfies

$$\left(d_{\min}(\mathcal{G})\right)^{-1} = O\left(\frac{\mathrm{M}}{\ell_{\text{ref}}}\right), \qquad \mathrm{M} \to 0 \tag{3.1.9}$$

with

$$d_{\min}(\mathcal{G}) := \min_{k=1,\ldots,d} |x_k|.$$

This means that, for each M, the domain $\mathcal{G}$ is large enough to accommodate perturbations of wavelength $\ell_{\text{ref}}/\mathrm{M}$, *i.e.*, in dimensionless coordinates, perturbations on the large scale $\boldsymbol{\xi} = \mathrm{M}\boldsymbol{x}$. Note that this assumption implies that either $\mathcal{G} = \mathbf{R}^d$ or that $\mathcal{G}$ depends on M or that the physical dimensions of $\mathcal{G}$ are fixed, but the reference length ℓ_{ref} is proportional to M. Furthermore, without loss of generality we can require that there always exists a positive real number M' such that

$$\mathbf{B}_x\left(\frac{1}{\mathrm{M}}\right) := \left\{ x \in \mathbf{R}^d \middle| |x| < \frac{1}{\mathrm{M}} \right\} \subset \mathcal{G}$$

for all $\mathrm{M} \leq \mathrm{M}'$. To take into account multiple space scales consider the mapping

$$\begin{aligned} g : \mathbf{R}^d \times \mathbf{R}_0^+ \times (0, \widetilde{\mathrm{M}}) &\rightarrow \mathbf{R}^{2d} \times \mathbf{R}_0^+, \\ (x, t; \mathrm{M}) &\stackrel{g}{\longmapsto} (x, \mathrm{M}x, t) . \end{aligned} \tag{3.1.10}$$

This mapping allows us to define a particular space $\widetilde{\mathcal{U}}_m$ of functions which are solution of the Euler equations. These are those functions U that can be expressed in $\mathcal{G} \times \mathbf{R}_0^+$ by a multiple-scale asymptotic expansion

$$U(x, t; \mathrm{M}) = \sum_{i=0}^{j} \mathrm{M}^i\, U^{(i)}(g(x, t; \mathrm{M})) + o\left(\mathrm{M}^j\right), \;\; \mathrm{M} \rightarrow 0, \;\; j = 0, 1, 2,$$

which is uniformly valid outside an arbitrary large ball $\mathbf{B}_x(s)$, $s \in \mathbf{R}^+$. The asymptotic functions $U^{(i)}$ depend on two space coordinates, $\boldsymbol{\eta}$ and $\boldsymbol{\xi}$:

$$\begin{aligned} U^{(i)} : \mathbf{R}^d \times \mathbf{R}^d \times \mathbf{R}_0^+ &\rightarrow \mathbf{R}, \;\; i = 0, 1, 2 \\ (\boldsymbol{\eta}, \boldsymbol{\xi}, t) &\stackrel{U^{(i)}}{\longmapsto} U^{(i)}(\boldsymbol{\eta}, \boldsymbol{\xi}, t) , \end{aligned}$$

Note that one cannot derive the uniform validity of the asymptotic expansion for $j = 0, 1$ from the fact that the asymptotic expansion is uniformly valid for $j = 2$. Therefore, it is necessary to introduce this assumption in the multiple-scale case. First, let us consider the so-called sub-linear growth condition for asymptotic functions. This condition represents a basic tool throughout the analysis.

Lemma 3.2
Let $U : \mathbf{R}^d \times \mathbf{R}_0^+ \times (0, \widetilde{\mathrm{M}}) \rightarrow \mathbf{R}$ be uniformly bounded with respect to x and M, *let $s \in \mathbf{R}^+$, and let the asymptotic multiple-scale expansion*

$$U(x, t; \mathrm{M}) = \sum_{i=0}^{j} \mathrm{M}^i\, U^{(i)}(g(x, t; \mathrm{M})) + o\left(\mathrm{M}^j\right), \;\; \mathrm{M} \rightarrow 0$$

be valid in $\mathbf{R}^d \times \mathbf{R}_0^+$ and uniformly valid in $\mathbf{R}^d \setminus \mathbf{B}_x(s) \times \mathbf{R}_0^+$ for each $j = 0, 1, 2$. Then the leading-order asymptotic function $U^{(0)}$ satisfies

$$U^{(0)}(g(x, t; \mathrm{M})) = o\left(|x|^{\alpha}\right), \;\; x \in \partial\mathbf{B}_x\left(\frac{1}{\mathrm{M}}\right), \;\; \mathrm{M} \rightarrow 0 \tag{3.1.11}$$

for all $\alpha > 0$. Furthermore, the first-order and second-order asymptotic functions satisfy

$$U^{(i)}(g(x, t; \mathrm{M})) = o\left(|x|\right), \;\; x \in \partial\mathbf{B}_x\left(\frac{1}{\mathrm{M}}\right), \;\; \mathrm{M} \rightarrow 0, \;\; i = 1, 2, \tag{3.1.12}$$

respectively.

Proof: From the uniform validity of the asymptotic expansion we have

$$U^{(i)}(g(x, t; \mathrm{M})) = o\left(\mathrm{M}^{-1}\right), \;\; \mathrm{M} \rightarrow 0 \tag{3.1.13}$$

uniformly in $\mathbf{R}^d \setminus \mathbf{B}_{\boldsymbol{x}}(s) \times \mathbf{R}_0^+$ for $i = 1, 2$. Given a null sequence $\{\mathrm{M}_n\}_{n\in\mathbf{N}}$, we define an arbitrary vector sequence $\{\boldsymbol{x}_n\}_{n\in\mathbf{N}}$ with $\boldsymbol{x}_n \in \partial\mathbf{B}_{\boldsymbol{x}}\left(\frac{1}{\mathrm{M}_n}\right)$. Now, we define the null sequence $\{\widetilde{\mathrm{M}}_n\}_{n\in\mathbf{N}}$ as a subsequence of $\{\mathrm{M}_n\}_{n\in\mathbf{N}}$ by taking only those elements which satisfy $\widetilde{\mathrm{M}}_n < s^{-1}$. Furthermore, we define the vector sequence $\{\widetilde{\boldsymbol{x}}_n\}_{n\in\mathbf{N}}$ as a subsequence of $\{\boldsymbol{x}_n\}_{n\in\mathbf{N}}$ which is given by the elements satisfying $\widetilde{\boldsymbol{x}}_n \in \partial\mathbf{B}_{\boldsymbol{x}}\left(\frac{1}{\widetilde{\mathrm{M}}_n}\right)$. Using the uniform validity of (3.1.13), we obtain

$$0 = \lim_{n\to\infty} \frac{U^{(i)}(\boldsymbol{g}(\widetilde{\boldsymbol{x}}_n, t; \widetilde{\mathrm{M}}_n))}{\widetilde{\mathrm{M}}_n^{-1}} = \lim_{n\to\infty} \frac{U^{(i)}(\boldsymbol{g}(\widetilde{\boldsymbol{x}}_n, t; \widetilde{\mathrm{M}}_n))}{|\widetilde{\boldsymbol{x}}_n|} = \lim_{n\to\infty} \frac{U^{(i)}(\boldsymbol{g}(\boldsymbol{x}_n, t; \mathrm{M}_n))}{|\boldsymbol{x}_n|},$$

which proves (3.1.12). In order to derive Equation (3.1.11), we use the uniform boundedness of U. This yields

$$U(\boldsymbol{x}, t; \mathrm{M}) = o(r^\alpha), \ \boldsymbol{x} \in \partial\mathbf{B}_{\boldsymbol{x}}(r), \ r \to \infty \tag{3.1.14}$$

for all $\alpha > 0$. Using the asymptotic expansion, we obtain

$$U(\boldsymbol{x}, t; \mathrm{M}) = U^{(0)}(\boldsymbol{g}(\boldsymbol{x}, t; \mathrm{M})) + o(1), \ \mathrm{M} \to 0 \tag{3.1.15}$$

uniformly in $\mathbf{R}^d \setminus \mathbf{B}_{\boldsymbol{x}}(s) \times \mathbf{R}_0^+$. A simple combination of the Equations (3.1.14) and (3.1.15) leads to

$$\begin{aligned} U^{(0)}(\boldsymbol{g}(\boldsymbol{x}, t; \mathrm{M})) &= U(\boldsymbol{x}, t; \mathrm{M}) + o(1) \\ &= o\left(\mathrm{M}^{-\alpha}\right) + o(1) = o\left(|\boldsymbol{x}|^\alpha\right), \end{aligned}$$

where $\boldsymbol{x} \in \partial\mathbf{B}_{\boldsymbol{x}}\left(\frac{1}{\mathrm{M}}\right)$ and $\mathrm{M} \to 0$.

Notice that the validity of Theorem 3.2 does not depend on the special form of the mapping $\boldsymbol{g}$ and therefore the sub-linear growth condition is also valid in the case of a single-scale expansion if an unbounded domain is considered. The assumption that the physical quantity U is uniformly bounded is motivated by (3.0.1) and guarantees that we have chosen the reference values in a proper manner. Using the derivatives

$$\partial_t(U^{(i)} \circ \boldsymbol{g})(\boldsymbol{x}, t; \mathrm{M}) = \partial_t U^{(i)}(\boldsymbol{g}(\boldsymbol{x}, t; \mathrm{M})), \tag{3.1.16}$$

$$\partial_{x_j}(U^{(i)} \circ \boldsymbol{g})(\boldsymbol{x}, t; \mathrm{M}) = \partial_{\eta_j} U^{(i)}(\boldsymbol{g}(\boldsymbol{x}, t; \mathrm{M})) + \mathrm{M}\partial_{\xi_j} U^{(i)}(\boldsymbol{g}(\boldsymbol{x}, t; \mathrm{M})), \ j = 1, \ldots, d \tag{3.1.17}$$

and introducing the asymptotic expansion into the Euler equations, one obtains

$$\begin{aligned} &\partial_t \rho^{(0)} + \nabla_\eta \cdot \boldsymbol{m}^{(0)} \\ &\quad + \mathrm{M}\left(\partial_t \rho^{(1)} + \nabla_\eta \cdot \boldsymbol{m}^{(1)} + \nabla_\xi \cdot \boldsymbol{m}^{(0)}\right) + \mathrm{M}^2\left(\partial_t \rho^{(2)} + \nabla_\eta \cdot \boldsymbol{m}^{(2)} + \nabla_\xi \cdot \boldsymbol{m}^{(1)}\right) \\ &\quad = o\left(\mathrm{M}^2\right), \ \mathrm{M} \to 0 \ \text{in} \ \widetilde{\mathcal{D}}(\mathrm{M}), \end{aligned} \tag{3.1.18}$$

$$\begin{aligned} &\mathrm{M}^{-2}\nabla_\eta p^{(0)} + \mathrm{M}^{-1}\left(\nabla_\eta p^{(1)} + \nabla_\xi p^{(0)}\right) \\ &\quad + \left(\partial_t \boldsymbol{m}^{(0)} + \nabla_\eta \cdot \left(\boldsymbol{m}^{(0)} \circ \boldsymbol{v}^{(0)}\right) + \nabla_\eta p^{(2)} + \nabla_\xi p^{(1)}\right) \\ &\quad = o(1), \ \mathrm{M} \to 0 \ \text{in} \ \widetilde{\mathcal{D}}(\mathrm{M}), \end{aligned} \tag{3.1.19}$$

$$\begin{aligned}
&\partial_t(\rho e)^{(0)} + \nabla_{\eta} \cdot ((\rho e + p)\boldsymbol{v})^{(0)} \\
&\quad + \mathrm{M}\left(\partial_t(\rho e)^{(1)} + \nabla_{\eta} \cdot ((\rho e + p)\boldsymbol{v})^{(1)} + \nabla_{\xi} \cdot ((\rho e + p)\boldsymbol{v})^{(0)}\right) \\
&\quad + \mathrm{M}^2\left(\partial_t(\rho e)^{(2)} + \nabla_{\eta} \cdot ((\rho e + p)\boldsymbol{v})^{(2)} + \nabla_{\xi} \cdot ((\rho e + p)\boldsymbol{v})^{(1)}\right) \\
&\quad = o\left(\mathrm{M}^2\right), \ \mathrm{M} \to 0 \ \text{in} \ \widetilde{\mathcal{D}}(\mathrm{M}),
\end{aligned} \tag{3.1.20}$$

where the manifold $\widetilde{\mathcal{D}}(\mathrm{M})$ is defined to be

$$\widetilde{\mathcal{D}}(\mathrm{M}) := \left\{(\boldsymbol{\eta}, \boldsymbol{\xi}, t) \in \mathbf{R}^{2d} \times \mathbf{R}_0^+ \,\middle|\, (\boldsymbol{\eta}, \boldsymbol{\xi}, t) = \boldsymbol{g}(\boldsymbol{x}, t; \mathrm{M}),\ (\boldsymbol{x}, t) \in \mathcal{G} \times \mathbf{R}_0^+\right\}.$$

In contrast to the single-scale expansion, the manifold $\widetilde{\mathcal{D}}$ depends on the reference parameter M and therefore we are not able to decompose the above system (3.1.18)–(3.1.20) in the same manner as described in Section 3.1.1. In order to overcome this difficulty, it is convenient to introduce the space $\mathcal{U}_m \subset \widetilde{\mathcal{U}}_m$ of those asymptotic functions $U^{(0)}$, $U^{(1)}$, $U^{(2)}$ which satisfy the Equations (3.1.18)–(3.1.20) in

$$\mathcal{D} := \left\{(\boldsymbol{\eta}, \boldsymbol{\xi}, t) \in \mathbf{R}^{2d} \times \mathbf{R}_0^+ \,\middle|\, \boldsymbol{\xi}_j < \widetilde{\mathrm{M}}\eta_j,\ j = 1, \ldots, d\right\}$$

and fulfill the sub-linear growth condition in the sense that, for fixed $\boldsymbol{\xi}$, one has

$$U^{(0)}(\boldsymbol{\eta}, \boldsymbol{\xi}, t) = o\left(|\boldsymbol{\eta}|^{\alpha}\right), \ \boldsymbol{\eta} \in \partial\mathbf{B}_{\eta}\left(\frac{1}{\mathrm{M}}\right), \ \mathrm{M} \to 0$$

for all $\alpha > 0$ and

$$U^{(i)}(\boldsymbol{\eta}, \boldsymbol{\xi}, t) = o\left(|\boldsymbol{\eta}|\right), \ \boldsymbol{\eta} \in \partial\mathbf{B}_{\eta}\left(\frac{1}{\mathrm{M}}\right), \ \mathrm{M} \to 0, \ i = 1, 2.$$

Then, in analogy with the single-scale case, we can express Equations (3.1.18) – (3.1.20) in the equivalent form

$$\partial_t \rho^{(0)} + \nabla_{\eta} \cdot \boldsymbol{m}^{(0)} = 0, \tag{3.1.21}$$

$$\partial_t \rho^{(1)} + \nabla_{\eta} \cdot \boldsymbol{m}^{(1)} = -\nabla_{\xi} \cdot \boldsymbol{m}^{(0)}, \tag{3.1.22}$$

$$\partial_t \rho^{(2)} + \nabla_{\eta} \cdot \boldsymbol{m}^{(2)} = -\nabla_{\xi} \cdot \boldsymbol{m}^{(1)}, \tag{3.1.23}$$

$$\nabla_{\eta} p^{(0)} = \boldsymbol{0}, \tag{3.1.24}$$

$$\nabla_{\eta} p^{(1)} = -\nabla_{\xi} p^{(0)}, \tag{3.1.25}$$

$$\partial_t \boldsymbol{m}^{(0)} + \nabla_{\eta} \cdot \left(\boldsymbol{m}^{(0)} \circ \boldsymbol{v}^{(0)}\right) + \nabla_{\eta} p^{(2)} = -\nabla_{\xi} p^{(1)}, \tag{3.1.26}$$

$$\partial_t (\rho e)^{(0)} + \nabla_{\eta} \cdot ((\rho e + p)\boldsymbol{v})^{(0)} = 0, \tag{3.1.27}$$

$$\partial_t (\rho e)^{(1)} + \nabla_{\eta} \cdot ((\rho e + p)\boldsymbol{v})^{(1)} = -\nabla_{\xi} \cdot ((\rho e + p)\boldsymbol{v})^{(0)}, \tag{3.1.28}$$

$$\partial_t (\rho e)^{(2)} + \nabla_{\eta} \cdot ((\rho e + p)\boldsymbol{v})^{(2)} = -\nabla_{\xi} \cdot ((\rho e + p)\boldsymbol{v})^{(1)} \tag{3.1.29}$$

in $\mathcal{D}$. This hierarchy of equations is the starting point of the analysis presented in Section 2.1.3.

3.1.3. *Examples*

The sub-linear growth conditions stated in Lemma 3.2 provide the key to derive many of the results presented in previous sections. We discuss two examples. First, consider Equation (2.1.36) and let $p^{(0)} \in \mathcal{U}_m$. Then Equation (3.1.24) implies $p^{(0)} = p^{(0)}(\boldsymbol{\xi}, t)$. Equation (3.1.9) together with the mapping $\boldsymbol{g}$ allows the integration of (3.1.25) over the open ball $\mathbf{B}_\eta\left(\frac{1}{\mathrm{M}}\right)$. Using sub-linear growth condition, we have

$$\begin{aligned}\nabla_\xi p^{(0)} &= -\frac{1}{|\mathbf{B}_{\boldsymbol{\eta}}\left(\frac{1}{\mathrm{M}}\right)|}\int_{\mathbf{B}_\eta\left(\frac{1}{\mathrm{M}}\right)} \nabla_\eta p^{(1)}\,\mathrm{d}\boldsymbol{\eta} = -\frac{1}{|\mathbf{B}_\eta\left(\frac{1}{\mathrm{M}}\right)|}\int_{\partial\mathbf{B}_\eta\left(\frac{1}{\mathrm{M}}\right)} p^{(1)}\cdot\boldsymbol{n}\,\mathrm{d}s\\ &= O(\mathrm{M}^d)\,O(\mathrm{M}^{1-d})\,o(\mathrm{M}^{-1}) = o(1), \qquad (\mathrm{M}\to 0).\end{aligned}$$

Hence, we obtain $p^{(0)} = p^{(0)}(t)$ and the first-order momentum equation (3.1.25) gives $p^{(1)} = p^{(1)}(\boldsymbol{\xi}, t)$ as stated in (2.1.36). In the second example we emphasize the meaning of small-scale averaging in the derivation of Equation (2.1.41). Consider the small-scale average operator

$$\overline{U}(\boldsymbol{\xi}, t) := \lim_{\mathrm{M}\to 0}\frac{1}{\left|\mathbf{B}_\eta\left(\frac{1}{\mathrm{M}}\right)\right|}\int\limits_{\mathbf{B}_\eta\left(\frac{1}{\mathrm{M}}\right)} U(\boldsymbol{\eta}, \boldsymbol{\xi}, t)\,\mathrm{d}\boldsymbol{\eta}.$$

Again, using the sub-linear growth condition, one has

$$\frac{1}{\left|\mathbf{B}_\eta\left(\frac{1}{\mathrm{M}}\right)\right|}\int\limits_{\mathbf{B}_\eta\left(\frac{1}{\mathrm{M}}\right)} \nabla_\eta p^{(2)} + \nabla_\eta\cdot\left(\boldsymbol{m}^{(0)}\circ\boldsymbol{v}^{(0)}\right)\,\mathrm{d}\boldsymbol{\eta} = o(1), \qquad (\mathrm{M}\to 0).$$

As seen above, $p^{(1)} = \overline{p^{(1)}}$. Thus, averaging the momentum equation (3.1.26) leads to

$$\begin{aligned}\mathbf{0} &= \lim_{\mathrm{M}\to 0}\frac{1}{\left|\mathbf{B}_\eta\left(\frac{1}{\mathrm{M}}\right)\right|}\int\limits_{\mathbf{B}_\eta\left(\frac{1}{\mathrm{M}}\right)} \left(\partial_t\boldsymbol{m}^{(0)} + \nabla_\eta\cdot\left(\boldsymbol{m}^{(0)}\circ\boldsymbol{v}^{(0)}\right) + \nabla_\eta p^{(2)} + \nabla_\xi p^{(1)}\right)\mathrm{d}\boldsymbol{\eta}\\ &= \lim_{\mathrm{M}\to 0}\frac{1}{\left|\mathbf{B}_\eta\left(\frac{1}{\mathrm{M}}\right)\right|}\int\limits_{\mathbf{B}_\eta\left(\frac{1}{\mathrm{M}}\right)} \partial_t\boldsymbol{m}^{(0)}\,\mathrm{d}\boldsymbol{\eta} + \nabla_\xi p^{(1)}\\ &= \overline{\partial_t\boldsymbol{m}^{(0)}} + \nabla_\xi p^{(1)},\end{aligned}$$

which is the first equation of system (2.1.41).

4. Numerics

4.1. Discrete identification of a multiple-scale low-Mach-number flow regime

In the following sections, unless otherwise stated, all quantities are assumed to be in dimensional form. The superscript $\hat{}$ denotes the dimensional form. For the sake of simplicity we neglect the superscript whenever its omission does not lead to misunderstandings.

The identification of a low-Mach-number long-wave regime is closely related to the determination of a discrete equivalent $\check{M}$ of the Mach number $\check{M} \in (0, 1]$ and to the decomposition of the primitive variables $\hat{\mathbf{U}} = (\hat{\rho}, \hat{\boldsymbol{v}}, \hat{p})^T$ needed in the multiple scale *ansatz*. In particular, the density $r\hat{h}o$ and the velocity $\hat{\boldsymbol{v}}$ can be decomposed in the form

$$\hat{\rho} = \overline{\rho} + \rho', \quad \hat{\boldsymbol{v}} = \overline{\boldsymbol{v}} + \boldsymbol{v}',$$

where the superscript $'$ denotes short wave phenomena and the overbar characterizes long-wave parts, respectively. On the other hand, it is appropriate to decompose the pressure into three terms

$$\hat{p} = \hat{p}^{(0)} + \check{M}\hat{p}^{(1)} + \check{M}^2\hat{p}^{(2)}. \tag{4.1.1}$$

Consider, first, the pressure decomposition. Notice that no truncation error of order $o(\check{M}^2)$ appears in Equation (4.1.1). Here, in contrast to the *ansatz* introduced in the asymptotic analysis, the pressure splitting is exact and $\hat{p}^{(0)}$, $\hat{p}^{(1)}$, and $\hat{p}^{(2)}$ are not solutions of the asymptotic equations discussed in Section 2.1. The idea, however, is to try to construct functions $\hat{p}^{(0)}$, $\hat{p}^{(1)}$, and $\hat{p}^{(2)}$ that have the same asymptotic behavior as the corresponding pressure components of the asymptotic sequence. First, consider the leading-order pressure $\hat{p}^{(0)}$. If the reference parameter $\check{M}$ is sufficiently small, we require $\hat{p}^{(0)}$ to be spatially homogeneous and the first-order pressure $\hat{p}^{(1)}$ has to represent exclusively long-wave phenomena. Due to the exactness of the pressure decomposition (4.1.1) the second-order pressure $\hat{p}^{(2)}$ can then be computed from $\hat{p}$, $\hat{p}^{(0)}$ and $\hat{p}^{(1)}$ and represents the remaining long- and short-wave influences. If the data are in the low-Mach- number regime, we expect this construction to provide a discrete pressure decomposition that guarantees the boundedness constraints

$$\hat{p}^{(i)} \leq C\hat{p}_{\text{ref}}, i = \{1, 2, 3\}\,. \tag{4.1.2}$$

If not, we will assume the flow field not to fall into the single-time, multiple-space-scale regime considered here (notice, however, that this *ansatz* excludes multiple-time, single-space scale regimes as considered, *e.g.*, in [5], [3], [34], [35], [4] and references therein). Notice also that the constant C must be independent of $\check{M}$.

We propose discrete-filter operations that, given a pressure distribution, determine $\check{M}$ and the relevant underlying length scales and simultaneously provide a pressure decomposition that, in the low-Mach-number long-wave regime, satisfies the above boundedness constraint. It turns out that pressure decomposition (4.1.1), determination of an effective numerical value $\check{M}$ corresponding to the physical Mach number M and extraction of short and long wavelengths have to be synchronized in a suitable way to achieve a decomposition satisfying the boundedness constraint (4.1.2).

Notice that if $\check{M}$ is much smaller than the ratio between the smallest flow scale $\hat{l}_{\text{ref}}$ and the maximal diameter of the domain $\hat{\Omega}$, *i.e.*,

$$\frac{1}{\check{M}} \gg \frac{d_{\max}(\hat{\Omega})}{\hat{l}_{\text{ref}}} = d_{\max}(\Omega),$$

long-wave acoustic phenomena on a scale $\boldsymbol{\xi} = \check{M}\, \boldsymbol{x}$ cannot be accommodated and the multiple-scale expansion degenerates into a single-scale expansion. In this case computing the first-order pressure and the long-wave parts of density and velocity is inappropriate. The limit of $\check{M} = \mathrm{M} = 0$ will be discussed in more detail in Section 4.3. If we define

$$\hat{p}^{(0)}(t) := \hat{p}_{ref}(t) = \frac{1}{|\hat{\Omega}|} \int\limits_{\hat{\Omega}} \hat{p}(\boldsymbol{x}, t)\, \mathrm{d}\boldsymbol{x},$$

then the requirements mentioned above regarding the leading-order pressure $\hat{p}^{(0)}$ hold for all $\check{M} \in (0, 1]$. Furthermore, it is possible to rewrite the pressure decomposition (4.1.1) as

$$\phi = \overline{\phi} + \phi',$$

with $\phi := \hat{p} - \hat{p}^{(0)}$, $\overline{\phi} := \check{M}\hat{p}^{(1)}$ and $\phi' := \check{M}^2\hat{p}^{(2)}$. We can describe a decomposition algorithm in the following way:

Let I denote the index set of the discretization. Decompose every given signal $(f_i)_{i\in I}$ of a function f in the form

$$f_i = \overline{f}_i + f_i', \quad \forall i \in I,$$

where $(\overline{f}_i)_{i\in I}$ represents only long-wave phenomena and $(f_i')_{i\in I}$ contains the remaining long- and short-wave parts. Furthermore $(f_i')_{i\in I}$ has to be small compared to $(\overline{f}_i)_{i\in I}$ with regard to the maximum norm, i.e.

$$\|(\overline{f}_i)_{i\in I}\|_\infty \gg \|(f_i')_{i\in I}\|_\infty .$$

An optimal decomposition algorithm should satisfy the following requirements:

A. The decomposition as well as $\check{M}$ have to be calculated simultaneously to satisfy (4.1.1) and (4.1.2) in an optimal way.

B. The reference parameter $\check{M}$ should continuously depend on the data of the discretization.

C. The algorithm should be almost parameter-free and must terminate automatically.

D. The method should locally be self-adaptive with respect to the data of the discretization, *i.e.* the decomposition should be invariant in regions where the signal is sufficiently smooth.

E. Relevant features of the solution should be neither dislocated nor smoothed.

F. The algorithm must be independent of the type of discretization. It must also be robust to grid refinements and to numerical noise. In this context it seems to be worthwhile to prove the continuity of the decomposition operators used.

G. The method must be fast, since the physical quantities have to be decomposed at each time step. Thus, if n denotes the number of discretization points or control volumes, the algorithm has to be of order $O(n \log n)$ or even faster.

Due to the last requirement, averaging operators based on convolutions with a box function or a Gaussian kernel must be discarded. We also disregard filtering algorithms based on Fourier transformations. These would require certain manipulations of high frequencies associated with aliasing errors and non-periodicity of the analyzed signals. The net effects on the resulting signal components are ambiguous and hard to control, especially on unstructured grids. Besides, Fourier decomposition is a global operation.

Wavelets have local properties in space and time, but until now they are restricted to structured grids. This makes them too rigid in the context of remark (F). Furthermore, the reference parameter $\check{M}$ must be precisely determined for small values, but this is impossible by means of a single wavelet analysis. Sampling the high-frequency parts of the wavelet analysis again, it is possible for us to determine all frequencies at the cost of loosing the multiple-scale properties and the order of the algorithm.

A survey of scale-space evolution algorithms which are based on parabolic differential equations is given in [36]. Besides the advantages of a curve evolution based on linear diffusion equations there are serious problems arising in this approach (Weickert [36, p. 6]):

a. "Gaussian smoothing does not only reduce noise, but also blurs important features such as edges and, thus, makes them harder to identify. Since Gaussian scale-space is designed to be completely uncommitted, it cannot take into account any *a-priori* information on structures which are worth being preserved (or even enhanced).

b. Linear diffusion filtering dislocates edges when moving from finer to coarser scales. So structures which are identified at a coarse scale do not give the right location and have to be traced back to the original image[..]. In practice, relating dislocated information obtained at different scales is difficult and bifurcations may give rise to instabilities. These coarse-to-fine tracking difficulties are generally denoted as the correspondence problem."

To overcome these problems anisotropic and nonlinear diffusion processes or even reactive-diffusion filters are being considered, but these decomposition strategies are not parameter-free and controlling the termination of the resulting algorithms is still an open problem. Furthermore, the computational effort associated with this class of complex filter algorithms increases in comparison with the simpler linear diffusion approaches, which already have an operation count similar to convolution processes. Hence, such techniques are also unsuitable for the kind of problem considered.

4.1.1. *Long-wave short-wave filters based on polygonal curves*

In the area of discrete geometrical data analysis, [37], polygonal curves are simplified to extract the important visual parts neither changing the coarse structure of the borderline nor dislocating the relevant features. This is done via a step–by–step algorithm. During each curve-analysis step a certain number of successive edges are replaced by a straight line connecting the endpoints of such a set of edges. The key to success of this very simple algorithm is hidden in the order of these replacements. In each step of the evolution a relevance measure assigns a cost value to each pair of edges. If the cost value is small, that is, the structure embraced by the pair of edges is insignificant, it will be replaced. The sequence of replacements enforced by the relevance measure produces a hierarchy of structures.

In our case a suitable relevance measure has to be defined to classify those points of a discrete distribution $(f_i)_{i\in I}$ which are not represented by the function $\overline{f}$ which has to be calculated. Then, the values of these points have to be replaced to give $\overline{f}$ some kind of discrete smoothness.

First of all, we must derive an appropriate definition to distinguish long-wave and short-wave phenomena. In contrast to the Fourier *ansatz* we will call any function to be of long-wave type, if the distance between its turning points is sufficiently large. On the basis of this definition, a sine function belongs to the same class as an arbitrary function having equivalent minimal distances between its turning points.

This definition of 'long-waviness' suffers from the drawback that high-frequency parts in the sense of Fourier can conceal themselves in long-wave structures. An example is given by a wave steepening into a shock. Although the frequency increases rapidly with respect to the Fourier series, the distance between two turning points can still remain constant during the steepening process. Consequently, the wavelength of the function as defined by our 'turning point' measure is constant. Thus, it might be necessary to execute a Fourier analysis of the signal *after* it has been decomposed.

Generally, in areas with short-wave structures turning points are very frequent, whereas in areas with long wave structures only a few turning points can be found. We will exploit this property of the wavelength to define a cost function and finally derive the relevance measure.

4.1.2. *A discrete-wavelength decomposition algorithm*

For the sake of simplicity we confine ourselves to a single-space dimension. Let $(\sigma_i)_{i\in I}$ be a given disjunct decomposition of the domain Ω into control volumes, boxes for short. Let I be an index set and n its size: $n := \#(I)$. We associate the values $(f_i)_{i\in I}$ to the centers of the boxes. Connecting each value with its neighbors with a segment leads to a piecewise linear function comparable to the polygonal contours investigated in [38]. We denote an edge connecting the values f_i and f_{i+1} by $k_{i+\frac{1}{2}}$ and the turn angle (the angle enclosed by the edges $k_{i-\frac{1}{2}}$ and $k_{i+\frac{1}{2}}$) by α_i.

Using the sign of α_i, we can now subdivide the domain Ω into overlapping concave and convex subdomains Ω_j, $j \in J$, where J denotes the index set of subdomains. Each sub-domain consists of points with the same sign of the turning angle plus the left and right turning points. Therefore, each point on the boundary of a sub-domain is called a discrete turning point of the function f. Note that discrete turning points always appear in pairs. Between a pair of turning points there might be points with vanishing turning angles.

The discrete-curve decomposition of a function f produces a sequence of functions $\{f^0 = f, \dots, f^m\}$ and a sequence of index sets $\{I^0 = I, \dots, I^m\}$ with $\#I^{l+1} < \#I^l$ and $\#I^m \leq 2$. In this description the index set I^l denotes those points which are still unchanged at the beginning of the l-th decomposition step. To determine those points which are to be removed during the l-th step, every point f_i^l, $i \in I^l$, is allocated a cost value $k(i, I^l) \in \mathbf{R}^+$ via the relevance measure to be determined in the next section. Let

$$k_{\min}(I^l) := \min\{k(i, I^l) : i \in I^l\}.$$

The set $I_{\min}(I^l)$ denotes the set of points with minimal costs at the l-th step,

$$I_{\min}(I^l) := \{i \in I^l : k(i, I^l) = k_{\min}(I^l)\}.$$

Now the algorithm removes the index set $I_{\min}(I^l)$ from I^l,

$$I^{l+1} = I^l \setminus I_{\min}(I^l).$$

4.1.3. *The relevance measure and the singular reference parameter*

The relevance measure represents the significance of each individual structure compared to all others and therefore controls the sequence of the decomposition steps.

In our case the relevance measure should be characterized by the wavelength of the function. The wavelength itself was fixed by the distance of the turning points from each other. So the cost function of a cell average f_i should primarily be a function of the sum of the distances from its neighboring turning points with different signs of the turning angles compared to the angle at index i. The cost function should only secondarily depend on other properties like its relative position between its turning points or its curvature. The sum of the distances to the so-called neighboring turning points are equal to the length of the concave or convex subdomains $|\Omega_j|$ projected onto the x-axis. Note that σ_i can be a member of up to three different subdomains. Because the number of subdomains changes during the evolution, we also introduce a sequence of index sets $\{J^0 = J, \dots, J^m\}$ of subdomains Ω_j.

Now, define the cost function $k(i, I^l) \in \mathbf{R}^+$ to be

$$k(i, I^l) := \Big| \min_{j \in \{J^l,\ \sigma_i \in \Omega_j\}} |\Omega_j| - \min_{j \in J^l} |\Omega_j| \Big| * |\alpha_i|, \quad \forall i \in I^l. \tag{4.1.3}$$

Obviously, the cost function represents a non-negative function which is equal to zero for all points of the smallest sub-domain. Hence, in addition to the points with collinearly arranged function values, all points of the smallest sub-domain are removed during the next step.

If we define the minimal wavelength of the function f_l as the minimal distance of two neighboring subdomains Ω_j, Ω_{j+1}, *viz.*,

$$\lambda_{\min}(f^l) := \min_{j \in J^l} \left(|\Omega_j| + |\Omega_{j+1}| - |\Omega_j \cap \Omega_{j+1}| \right)$$

with $\Omega_j := \emptyset$, $\forall j \notin J^l$, then the inequality

$$\lambda_{\min}(f^l) \le \lambda_{\min}(f^{l+1})$$

does not hold for all $l \in \{0, \dots, (m-1)\}$, in general. Nevertheless, for the stopping criterion, it follows

$$\lambda_{\min}(f^0) \le \lambda_{\min}(f^m) = |\Omega|.$$

In the case $\lambda_{\min}(f^0) \ne |\Omega|$ the inequality is sharp. Thus, the minimal wavelength increases during the evolution, although it is theoretically possible that the progress is not monotone.

Now we can easily define the reference parameter $\check{M}$ as

$$\check{M}(f^l) := \min \left\{ \frac{1}{\lambda_{\min}(f^l)}, 1 \right\}. \tag{4.1.4}$$

4.1.4. *Properties of the scale-decomposition scheme*

A summary of all steps of the algorithm described in the previous chapters is now given:

1. Set $\check{M} = 1,\ l = 0,\ \overline{f}_i = f_i,\ f_i' = 0,\ \forall i \in I^0 = I$;

2. Subdivide Ω into overlapping concave and convex subdomains Ω_j of the function $\overline{f}_i$, $\forall i \in I^0$;

3. Compute $\check{M}$ by means of Equation (4.1.4) and check the validity of the estimates in (4.1.2) as well as $\#I^l > 2$. If one of these conditions is not satisfied, the evolution is stopped;

4. Assign a relevance measure $k(i, I^l)$ to each box σ_i, $i \in I^l$, using the cost function (4.1.3);

5. Compute $I_{\min}(I^l) = \{i \in I^l : k(i, I^l) = 0\}$;

6. Reduce the index set of the maintained function values,
   $I^{l+1} = I^l \setminus I_{min}(I^l)$;
7. Compute $\overline{f}_i$, $i \in I^0 \setminus I^{l+1}$ taking the maintained points f_i, $i \in I^{l+1}$ into account;

8. Compute $f_i' = f_i - \overline{f}_i$, $i \in I^0 \setminus I^{l+1}$;

9. Increase the index of the evolution, $l = l + 1$;

10. Go to (2.).

We add some remarks regarding the smoothness of the discrete function $\overline{f}$. The use of a simple linear interpolation in order to define discrete values for the smooth function $\overline{f}$ at those points which are actually removed does not lead to a suitable recovery procedure, since it is possible that all points in larger parts of the domain have been removed. The tempting alternative of cubic splines has the considerable disadvantage that it may or may not generate new turning points. This would necessitate a recursive decomposition procedure with an undetermined, possibly infinite number of steps until termination.

We have decided to employ the following *ansatz*. First of all, we add some of the removed points in such a way that no additional turning points are introduced. Only then we apply one of the recovery techniques mentioned above. The resulting scheme yields good results already in the case that the recovery step is given in the form of a linear interpolation documented in Section (4.1.5) below.

In [38] those properties which are independent of the relevance measure chosen are jointly employed in the numerical framework. Now, we want to investigate in addtion the properties that are dependent on the underlying relevance measure (4.1.3) in order to introduce them in a proper manner.

a. The decomposition of the pressure is enforced simultaneously with the computation of the reference parameter. During each step of the decomposition procedure the wavelength of the smoothed function increases, while the reference parameter will decrease.

b. The accuracy of the value of $\check{M}$ depends only on the quality of the discretization.

c. The algorithm is parameter-free and terminates automatically. At the end of the evolution process a straight line is obtained, if $\#I^{ab} \leq 2$.

d. The method is self-adaptive to the data. Due to the chosen relevance measure, only subdomains of Ω including short-wave phenomena will be processed during one decomposition step. All other points remain unchanged.

e. The scheme does not introduce any shape-rounding effects and there is no dislocation of relevant features, because the remaining points do not change their values.

f. The cost function is not continuous and passes this property to the algorithm employed to decompose the physical quantities. Nevertheless, owing to the fact that parts of the domain Ω are treated as a whole in the evolution, numerical experiments emphasize that the splitting does not depend on the discretization. Furthermore, noise is a short-wave phenomenon and therefore noise elimination takes place in the early stages of the evolution process.

g. The algorithm requires $O(n \log n)$ operations.

4.1.5. *Numerical results*

We have chosen two one-dimensional test cases to demonstrate the numerical accuracy of the decomposition algorithm. In both cases we use a regular discretization of the domain $\Omega = [-51, 51]$ into control volumes σ_i with $|\sigma_i| = \frac{1}{10}$, $\forall i \in I$, $I = \{0, \ldots, 1019\}$.

First, we considered a long sinusoidal pressure wave ($\check{M} = \frac{1}{102}$) which is disturbed by a regular short-wave noise function,

$$\hat{p} = \hat{p}^{(0)} + \check{M}\hat{p}^{(1)} + \check{M}^2\hat{p}^{(2)}$$

with the reference of the pressure $\hat{p}_{\text{ref}}$ set to $\hat{p}_{\text{ref}} = \frac{4}{\check{M}^2}$ and

$$p^{(0)} = 1,\ p^{(1)} = 2\gamma(1 + \cos(2\pi\check{M}x),\ p^{(2)} = 2\gamma\sin(80\pi\check{M}x).$$

Figures 1 and 2 show the analytical long- and short-wave components of the pressure wave $\hat{p}$. The total pressure distribution defined the by adding of long- and short-wave components is shown in Figure 3. In Figures 4 and 5 the decomposed long- and short-wave components are plotted. The long-wave part of the pressure is reproduced very well and it is separated from the short-wave perturbation. There are only two pairs of turning points left, so the reference parameter $\check{M}$ has taken the value of the frequency of the analytical pressure wave in a natural way.

The other example is concerned with a density distribution moving in a long-wave acoustic field. The initial conditions are

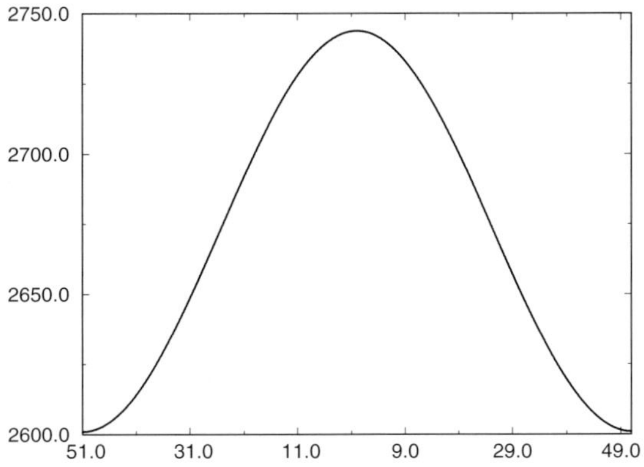

Figure 1. Long wave pressure part.

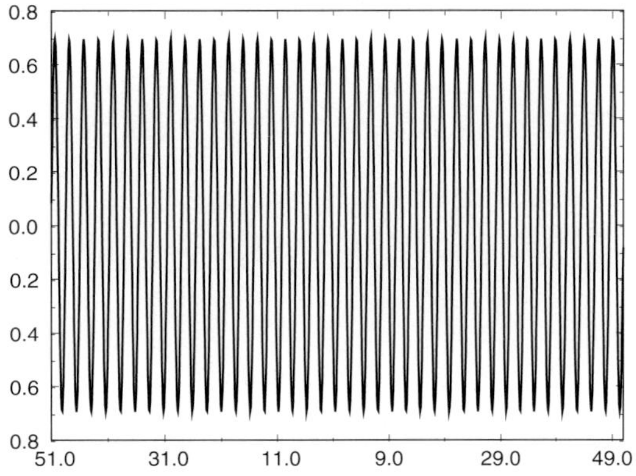

Figure 2. Pressure perturbation.

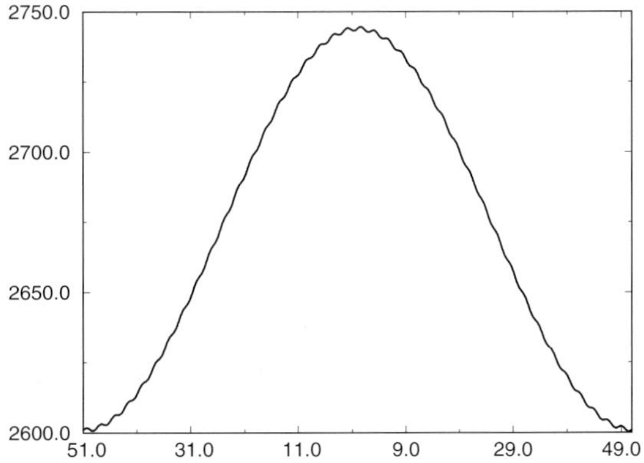

Figure 3. Total pressure distribution $\hat{p}$.

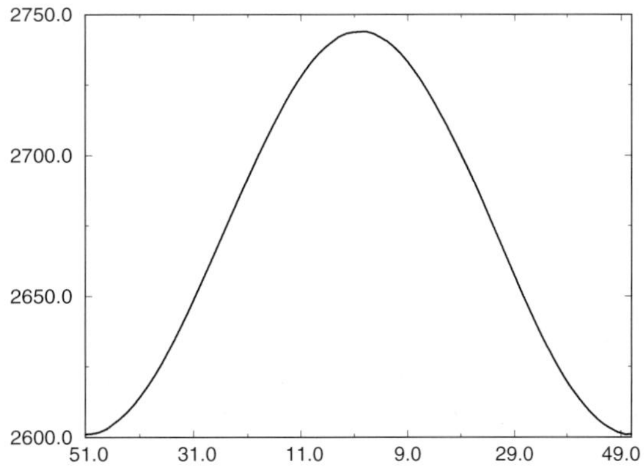

Figure 4. Filtered long wave part.

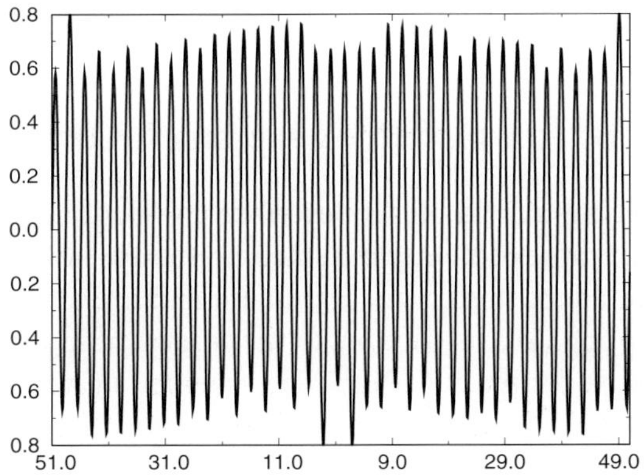

Figure 5. Remaining noise.

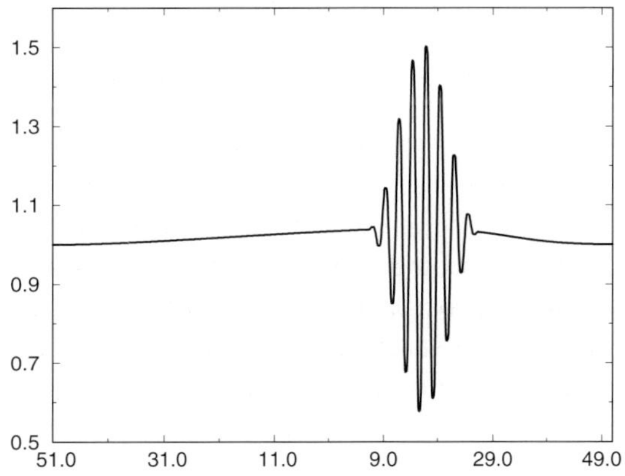

Figure 6. Density profile.

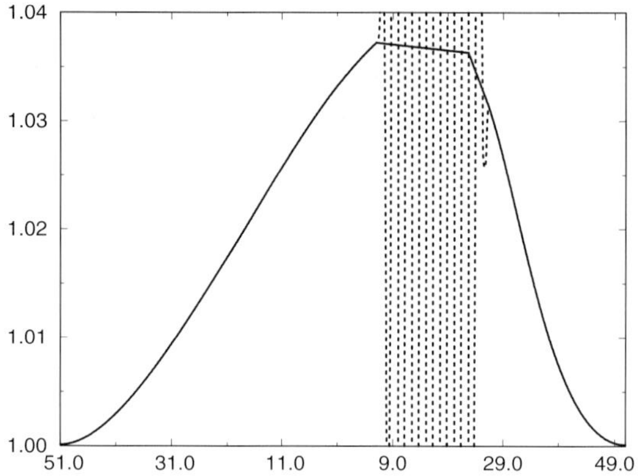

Figure 7. Filtered long wave signal.

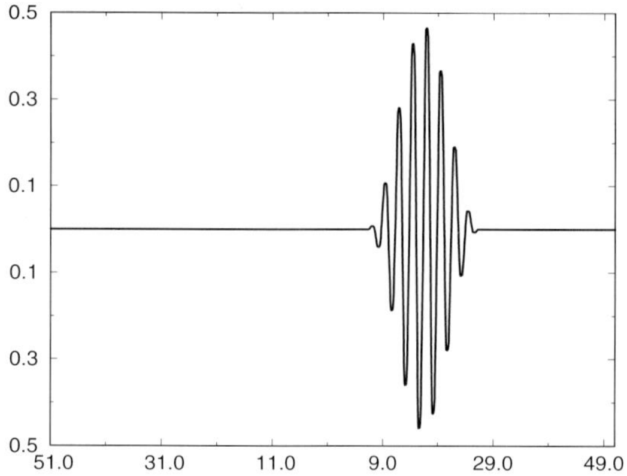

Figure 8. Short wave oscillations.

$$\begin{aligned}
\hat{\rho}(x, t = 0) &= 1 + \frac{1}{51}\left(1 + \cos\left(\frac{2\pi x}{102}\right)\right) + \Phi(x)\frac{1}{2}\sin\left(\frac{80\pi x}{102}\right),\\
\hat{v}(x, t = 0) &= \sqrt{\gamma}\left(1 + \cos\left(\frac{2\pi x}{102}\right)\right),\\
\hat{p}(x, t = 0) &= 51^2\left(1 + \frac{\gamma}{51}\left(1 + \cos\left(\frac{2\pi x}{102}\right)\right)\right),
\end{aligned}$$

and

$$\Phi(x) = \begin{cases} \frac{1}{2}\left(1 - \cos(\frac{10\pi x}{102})\right), & \text{if } 0 \le x \le \frac{102}{5},\\ 0 & \text{otherwise.}\end{cases}$$

The signal of the density is filtered at time $T = 5,071$.

This example clearly shows the advantages of the proposed decomposition algorithm. The decomposition is almost completely restricted to the sub-domain where the short-wave oscillations of the density distribution are found. In the other parts of Ω the filtered function and the original one are almost the same, neglecting numerical noise of size 10^{-6}. Therefore, no shape-rounding effects have taken place.

4.2. Extension of incompressible methods to weakly compressible flows

In the following we will show how to use the asymptotic analysis described in Section 2 to extend an incompressible method to the weakly compressible regime. Incompressible solvers are usually based on the primitives variables. We adopt this formulation here and use a staggered grid to stabilize the pressure-velocity coupling. We restrict ourselves to describing the extension of a projection method. The modification for the SIMPLE-type scheme is given in [39].

4.2.1. *A compressible projection method*

Projection methods for incompressible flows have been introduced by Chorin [40, 41]. In such methods discrete approximate solutions to the incompressible Navier-Stokes equations are advanced in time through two steps. First, the new-time-level density and an intermediate velocity field are obtained by a discrete integration of the convection-diffusion system

$$\begin{aligned} &\rho_t + \boldsymbol{v} \cdot \nabla \rho = 0, \\ &\boldsymbol{v}_t + (\boldsymbol{v} \circ \nabla)\, \boldsymbol{v} = -\frac{1}{\rho \mathrm{Re}} \nabla \cdot \tau, \end{aligned} \tag{4.2.1}$$

over $t^n < t \le t^{n+1}$ with the old-time-level distributions $\boldsymbol{v}^n$, ρ^n as initial data

$$\boldsymbol{v}(t^n) = \boldsymbol{v}^n \qquad \rho(t^n) = \rho^n\,. \tag{4.2.2}$$

(Note that the original method was designed for constant density flow, for which (4.2.1a) is void. The density-advection equation is ignored.)

The auxiliary velocity field $\boldsymbol{v}^{*,n+\frac{1}{2}}$ obtained from this step is not divergence-free due to the absence of the pressure gradient in the momentum equation. A correction is constructed through discrete integration over $t^n < t \le t^{n+1}$ of the projection equation

$$\boldsymbol{v}_t + \frac{1}{\rho} \nabla p^{(2)} = 0, \tag{4.2.3}$$

with the intermediate field $\boldsymbol{v}^{*,n+\frac{1}{2}}$ as initial data, and with the divergence constraint for the new-time-level velocity field,

$$\nabla \cdot \boldsymbol{v}^{n+1} = 0\,. \tag{4.2.4}$$

In the weakly compressible regime, we split the full compressible Navier-Stokes equations into two subsystems in a similar way. The convection-diffusion system now reads as

$$\begin{aligned} \rho_t + \boldsymbol{v} \cdot \nabla \rho \quad &= 0, \\ \boldsymbol{v}_t + (\boldsymbol{v} \circ \nabla)\, \boldsymbol{v} \ &= -\frac{1}{\rho \mathrm{Re}} \nabla \cdot \tau, \\ p_t + \boldsymbol{v} \cdot \nabla p \quad &= 0, \end{aligned} \tag{4.2.5}$$

and the 'sonic' system as

$$\begin{aligned} \rho_t + \rho \nabla \cdot \boldsymbol{v} \quad &= 0, \\ \boldsymbol{v}_t + \frac{1}{\mathrm{M}^2 \rho} \nabla p \ &= 0, \\ p_t + \gamma p \nabla \cdot \boldsymbol{v} \ &= \frac{\gamma}{\mathrm{Pr}\,\mathrm{Re}} \nabla \cdot q, \end{aligned} \tag{4.2.6}$$

which contains all effects that spread with sound velocity and degenerate to an elliptic constraint as $\mathrm{M} \to 0$. Heat conduction with heat flux q is retained here, since it influences the divergence constraint as seen in Equation (2.1.30) whereas diffusion does not. Neglecting the viscous and heat-conduction terms, we may call this decomposition a hyperbolic-elliptic splitting that is motivated by the wave speed as mentioned above, but also by the asymptotic analysis, (see [2, 39]). In the system (4.2.5) all the terms which are associated with elliptic expressions in the zero-Mach-number limit are neglected.

The convection-diffusion system is discretized by means of an explicit MUSCL-type upwind scheme for the convective terms and implicit second-order central differences for the diffusion part. Thus, the CFL-time step condition contains only the finite flow velocity.

The sonic system is discretized implicitly to avoid time-step restrictions due to the sound velocity tending to infinity in the limit. Here, the pressure expansion $p\left(\mathbf{x}, t; \hat{\mathrm{M}}\right) = p^{(0)}(t) + \hat{\mathrm{M}} p^{(1)}(\xi, t) + \hat{\mathrm{M}}^2 p^{(2)}(\bar{\mathbf{x}}, \xi, t)$ comes into play. Notice that the numerical parameter $\hat{\mathrm{M}}$ is now identified with the Mach number M, as we consider a case where compressibility effects do not change dramatically as the solution evolves in time. First, the total pressure has to be decomposed into its leading, first- and second-order contributions. Using the fact the leading-order pressure $p^{(0)}$ becomes constant on both space scales and the first-order term $p^{(1)}$ on the small space scale, they may be defined by averaging procedures,

$$p^{(0)} := \frac{1}{|V|} \int_V p \, \mathrm{d}V, \tag{4.2.7}$$

and

$$p^{(1)} := \frac{1}{\hat{\mathrm{M}} |V_{\mathrm{ac}}|} \int_{V_{\mathrm{ac}}} \left(p - p^{(0)}\right) \mathrm{d}V, \tag{4.2.8}$$

where V_{ac} is the acoustic domain. On a two-dimensional Cartesian grid of constant grid spacing h and coordinates x and y, V_{ac} is the rectangle $[x-h/(2\mathrm{M}), x+h/(2\mathrm{M})] \times [y-h/(2\mathrm{M}), y+h/(2\mathrm{M})]$. The second-order pressure $p^{(2)}$ then becomes

$$p^{(2)} := \frac{1}{\hat{\mathrm{M}}^2} \left(p - p^{(0)} - \hat{\mathrm{M}} p^{(1)}\right). \tag{4.2.9}$$

for consistency. For $\hat{\mathrm{M}} \to 0$, the acoustic domain exceeds the computational domain and $p^{(1)}$ becomes zero and drops out of the equations. As $\hat{\mathrm{M}}$ becomes very small, but non-zero, the influence of the pressure term $p^{(2)}$ on the thermodynamics of the system becomes negligible. It is thus regarded as a new variable independent of the total thermodynamic pressure, which is consistent with the earlier asymptotic results.

The temporal evolution of $p^{(0)}$ and $p^{(1)}$ can be determined from (2.1.32) and (2.1.44). From Equation (2.1.32), we obtain $p_t^{(0)}$ by applying a suitable ODE-integrator. To determine $p_t^{(1)}$, the acoustic system (2.1.44) has to be solved. To do so, we have two possibilities. The first is that the large-scale derivatives ∇_ξ are replaced by $\nabla_x / \hat{\mathrm{M}}$ using the chain rule and the equations are solved on the same grid as used to resolve the flow structures. This requires an implicit scheme, since the time step for an explicit scheme would be restricted by a CFL-condition including the speed of sound. Thus, the time step would tend to zero for $\hat{\mathrm{M}} \to 0$. On the other hand, Equations (2.1.44) include only large scale effects. No resolution of short-scale phenomena is necessary and a much coarser grid can be used. The coarsening factor is known to be $1/\hat{\mathrm{M}}$, which is the factor between the two space scales $\boldsymbol{x}$ and $\boldsymbol{\xi}$. In a discretization based on such an adapted coarse grid the Mach number cancels from the CFL-condition for the acoustic equations and they can be solved explicitly with low computational effort. The only potentially expensive computational step remains the extraction of the appropriate long-wave solution components needed to initiate the coarse-grid time step. The naïve integral filters from

(4.2.7), (4.2.8) must be replaced by more elaborate techniques as discussed in Section 4.1 to overcome this obstacle.

System (4.2.6) is solved in the following way. Introducing the pressure decomposition and rewriting it as a system for density, velocity and the incompressible pressure $p^{(2)}$, we obtain

$$\begin{aligned} \rho_t + \rho\nabla\cdot\boldsymbol{v} &= 0, \\ \boldsymbol{v}_t + \frac{1}{\rho}\nabla p^{(2)} &= -\frac{1}{\hat{\mathrm{M}}\rho}\nabla p^{(1)}, \\ \hat{\mathrm{M}}^2 p_t^{(2)} + \gamma p\nabla\cdot\boldsymbol{v} &= \frac{\gamma}{\mathrm{Pr}\,\mathrm{Re}}\nabla\cdot q - p_t^{(0)} - \hat{\mathrm{M}}p_t^{(1)}. \end{aligned} \tag{4.2.10}$$

For $q = 0$, this system can be viewed as the second step of the incompressible projection method since the right-hand sides of Equations (4.2.10b) and (4.2.10c) vanish for $\hat{\mathrm{M}} \to 0$ and the equations coincide with system (4.2.6).

Equations (4.2.10) are solved by nested iterations. All equations are discretized implicitly. Then, the density is fixed at the value obtained from the convection-diffusion step solving Equations (4.2.5). The leading- and first-order pressure terms at the new time level are already determined, $p^{(0),n+1} = p^{(0),n} + \Delta t p_t^{(0)}$ and $p^{(1),n+1} = p^{(1),n} + \Delta t p_t^{(1)}$, where $p^{(0),n}$ and $p^{(1),n}$ are obtained from the discretized Equations (4.2.7), (4.2.8). The old-time-level pressure $p^{(2),n}$ was given by Equation (4.2.9), the one at the new time level has to be guessed. Usually, it is either assumed to be $p^{(2),*} = p^{(2),n}$ or $p^{(2),*} = 0$. Then, the estimate for the total pressure at the next time level is set according the consistency condition $p^* = p^{(0),n+1} + \hat{\mathrm{M}}p^{(1),n+1} + \hat{\mathrm{M}}^2 p^{(2),*}$. With these, the new time level velocity $\boldsymbol{v}^*$ can be guessed from the discretized velocity equation (4.2.10b). The new-time-level pressure can then be set to be the guess plus a correction,

$$p^{(2),n+1} = p^{(2),*} + \delta p^{(2)}, \qquad p^{n+1} = p^* + \hat{\mathrm{M}}^2\delta p^{(2)}. \tag{4.2.11}$$

Inserted into the discretized pressure equation (4.2.10c), a Poisson-equation is obtained for the pressure-correction term $\delta p^{(2)}$ which reads as

$$\begin{aligned} &\frac{\hat{\mathrm{M}}^2}{\Delta t}\delta p^{(2)} - \gamma p^*\nabla\cdot\left(\tfrac{\Delta t}{\rho}\nabla\delta p^{(2)}\right) = \\ &\qquad -\gamma p^*\nabla\cdot\boldsymbol{v}^* + \frac{\gamma}{\mathrm{Re}\,\mathrm{Pr}}\nabla\cdot q - p_t^{(0)} - \hat{\mathrm{M}}p_t^{(1)} \\ &\qquad\quad -\frac{\hat{\mathrm{M}}^2}{\Delta t}\left(p^{(2),*} - p^{(2),n}\right). \end{aligned} \tag{4.2.12}$$

The estimate p^* for the total pressure is used as the linearisation term here. With the solution $\delta p^{(2)}$ of this equation, the estimates $p^{(2),*}$ and p^* are improved according to (4.2.11), the corresponding velocity estimate is calculated and the process is repeated up to convergence. With the velocity obtained from this inner iteration, a new density estimate is obtained and the inner loop is started again. Most of the computing time for the simulation is spent in solving the linear system resulting from discretizing the pressure-correction equation (4.2.12). For this we use preconditioned Krylov-Subspace schemes or multi-grid techniques.

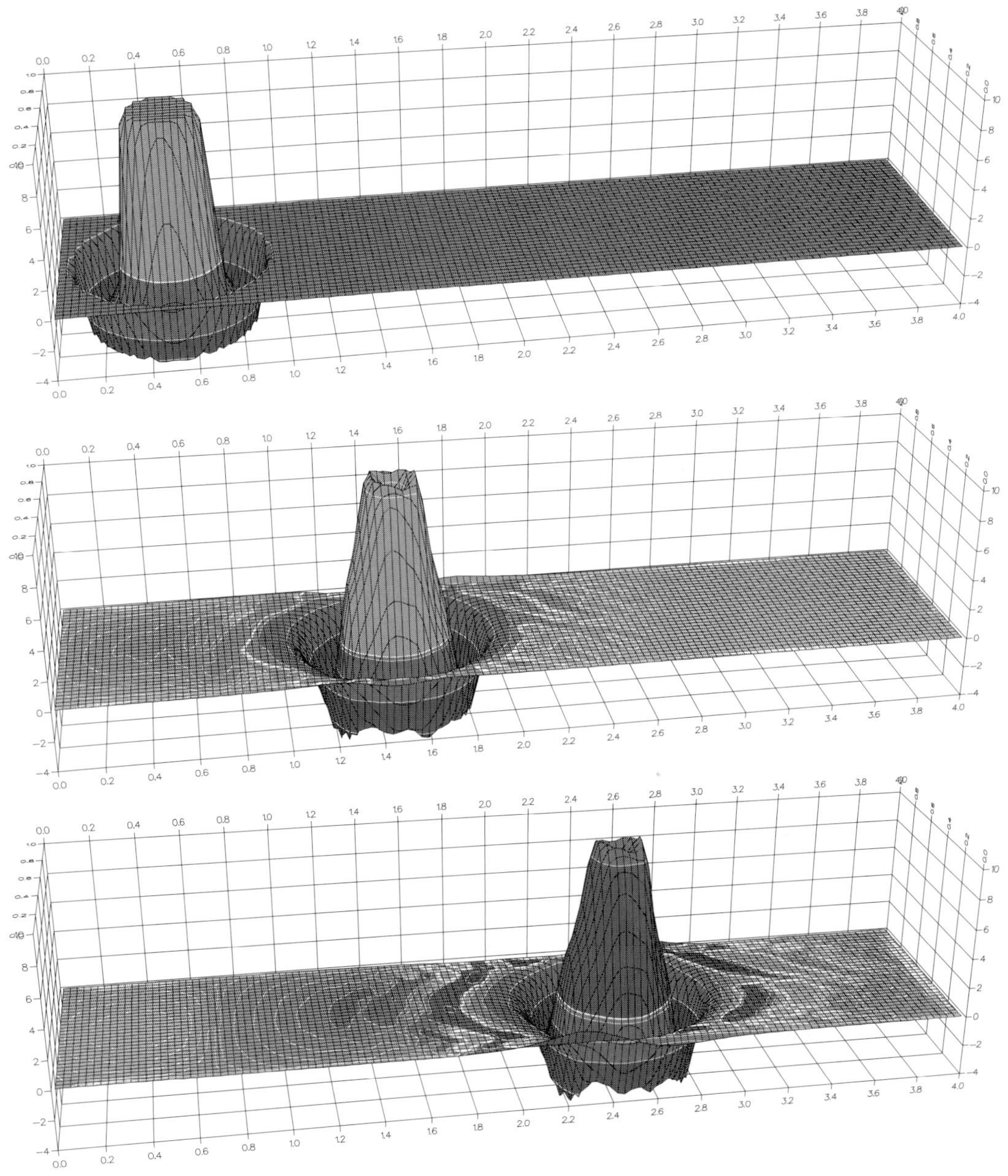

Figure 9 Vorticity at times t=0,1,2.

4.2.2. *Numerical results*

The algorithm given above does not depend on any special spatial discretization. For the numerical examples to be given now, a Cartesian staggered-grid arrangement was applied defining the scalar variables at the cell centers, the horizontal vector components on the vertical cell interfaces and the vertical components on the horizontal interfaces. As already mentioned, the convection-diffusion system (4.2.5) uses an explicit MUSCL-type scheme for the convective terms, and implicit central differences for the diffusive part. The sonic system

Figure 10. Streamline visualization of the lid driven cavity flow at Re = 1000.

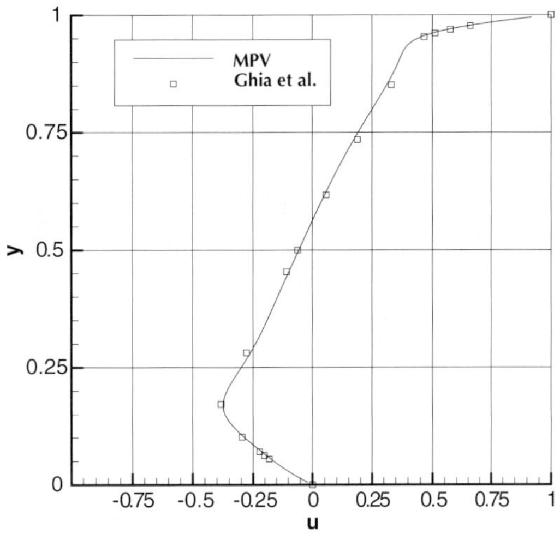

Figure 11. Horizontal velocity at the vertical cross section through the center of the cavity.

(4.2.10) is discretized with implicit central differences for all terms.

Gresho and Chan's transported vortex. The first test case is an incompressible Euler calculation proposed by Gresho [43, 44] as a test case with known, but non-trivial solution. A triangular vortex is convected through a channel while revolving around its center. It should be transported without any damping of the vorticity. Figure 9a shows the vorticity distribution at the initial stage, Figures 9b and 9c the vorticity after one and two turnarounds. The vortex is transported one spatial unit to the right during one revolution.

It can be clearly seen, that the vorticity is well reproduced. There is only limited damping of the magnitude; the distortions are due to the fact that a rotational symmetric vortex has to be approximated on a Cartesian grid. These asymmetries are reduced with grid refinement. The calculations shown here have been carried out on a 160×40 mesh for a 4×1-units domain.

The standard driven cavity test. This second test case involves an incompressible viscous flow at Reynolds number Re = 1000. It is characterized by a large vortex in the center of the cavity and two smaller ones in the lower left and right corners, see Figure 10.
The computational domain is the unit square, discretized by 100×100 grid points. At first, the standard incompressible test case is compared to results in the literature. Figure 11 shows the horizontal velocity at the vertical cross section at $x = 0{\cdot}5$, Figure 12 the vertical velocity at $y{=}0{\cdot}5$. The solid lines show the results obtained with the MPV code. They are in very good agreement with the benchmark solutions obtained by Ghia *et al.* [45] on a 129×129 Cartesian grid.
Driven cavity with differentially heated side-walls. The driven-cavity test case can be extended to the weakly compressible regime by differential heating of the vertical walls. The left wall was kept at constant temperature of $30°C$ $(293{\cdot}6K)$. The right wall was heated to $293{\cdot}6K + \Delta T$. As long as the temperature difference ΔT is small, the Boussinesq approximation can

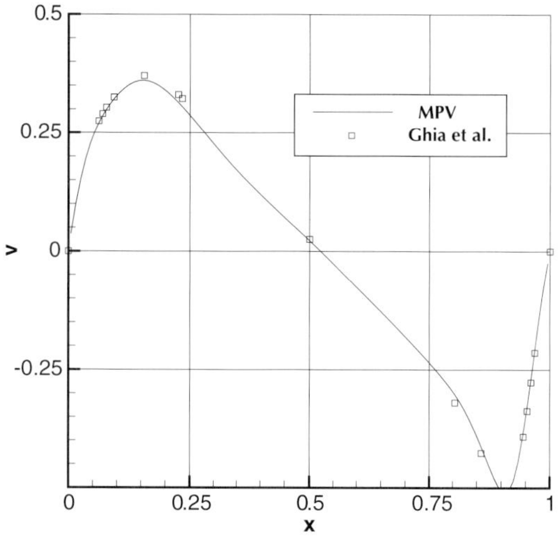

Figure 12. Vertical velocity at the horizontal cross section through the center of the cavity.

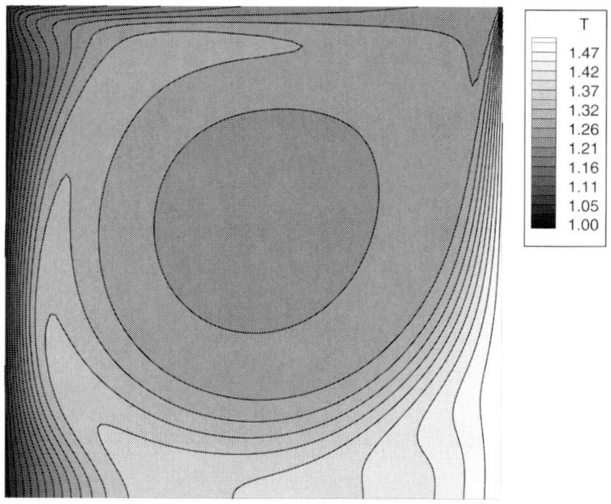

Figure 13. Isotherms for a driven cavity with differentially heated vertical walls with temperature difference $\Delta T = 150$ K..

be applied. Results of calculations with the full MPV-scheme and incompressible calculations involving the Boussinesq approximation were quite similar in this case. For ΔT larger than a few degrees, the Boussinesq approximation is no longer valid. To show this discrepancy, we applied a temperature difference $\Delta T = 150K$ and compared the results of the MPV scheme with those obtained by an incompressible Boussinesq approximation. The Mach number for this test case is $M = 0{\cdot}0005$, the Prandtl number $Pr = 0{\cdot}7$ and the Reynolds number Re = 1000 as above. Figure 13 shows the temperature distribution of the steady state.

Figures 14 and 15 show the temperature profiles at the left and right walls. The solid lines give the results of the MPV scheme, the dashed lines are the results obtained by means of an incompressible solver with Boussinesq approximation. It can be clearly seen that the incompressible code shows a much thicker thermal boundary layer than the MPV scheme. This is due to the fact that it cannot reproduce the thermal density changes.

Baroclinic vorticity generation by long-wave acoustics. The next test problem considered is again a weakly compressible one and shows the approximation of acoustic waves. Here, we want to show the interaction of a long-wave-length acoustic wave with small entropy fluctuations on the local length scale. The initial data,

$$\begin{aligned}
\rho(x, y, 0) &= 1{\cdot}0 + 0{\cdot}2M(1{\cdot}0 + \cos(\pi x/L)) + \Phi(y)\\
p(x, y, 0) &= 1{\cdot}0 + M\gamma(1{\cdot}0 + \cos(\pi x/L))\\
u(x, y, 0) &= \sqrt{\gamma}(1 \cdot 0 + \cos(\pi x/L)),\\
v(x, y, 0) &= 0{\cdot}0,
\end{aligned}$$

with

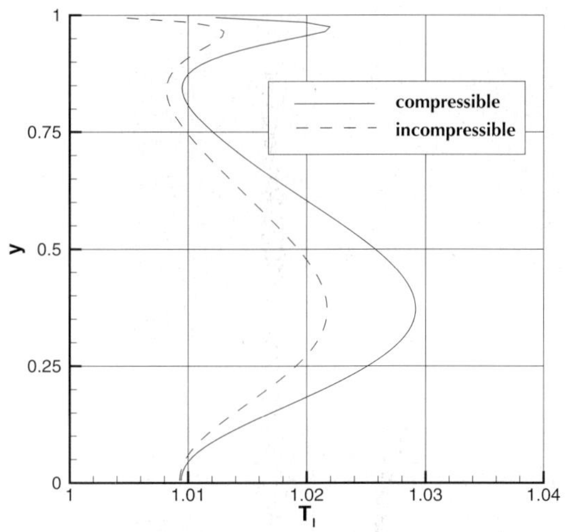

Figure 14. Horizontal velocity at the vertical cross section through the center of the cavity.

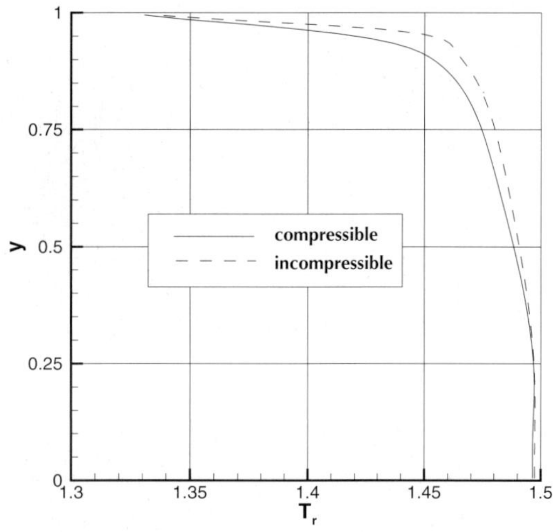

Figure 15. Vertical velocity at the horizontal cross section through the center of the cavity.

$$\Phi(y) = \begin{cases} \frac{0{\cdot}8}{L_y} y & \text{for } 0 \le y \le \frac{1}{2} L_y, \\ \frac{0{\cdot}8}{L_y} \left(y - \frac{1}{2} L_y)\right) - 0{\cdot}4 & \text{for } \frac{1}{2} L_y \le y \le L_y, \end{cases}$$

represent a saw-tooth-like density stratification in the vertical direction set into motion by a right-running acoustic pulse in the horizontal direction. At $\hat{\mathrm{M}} = 1/20$, the computational domain is a double-periodic domain of $[-L : L] \times [0 : L_y]$ with $L = \frac{1}{\hat{\mathrm{M}}} = 20$ and $L_y = \frac{2L}{5} = 8$, just long enough to let one period of the acoustic wave take place. Due to the non-homogeneous density, the fluid is subject to a higher acceleration for lower density values and a lower acceleration at higher densities. This leads to the well-known phenomenon of baroclinic instability. A shear layer of sinusoidal shape is generated and moved with the acoustic wave. Figure 16 shows the density at different times. The first plot shows the initial data, the second one the first forming of the sinusoidal shape at time $t = 6$. The interface starts rolling up and small vortices are formed. This can be clearly seen at the later times $t = 9, 11, 14$.

The smallest structures resolved have a thickness of a few grid zones length, smaller ones are damped out by numerical viscosity. In Figure 17 vorticity contours are given at the same times. It can be seen that in the initially vortex-free flow two thin regions of vorticity are created along the density interface. As the interface starts rolling up, counter-rotating vortices are formed. These computations demonstrate that the interaction of the long-wave-length acoustic wave with the local flow structures creates small-scale vortices.

A model problem for the flow in a disk brake. Here we discuss some simulations of the cooling flow in a simplified disk-brake facility. The key issue in this problem is the increase of the rate of heat transport due to the stripping of the boundary layer by the brake block. To study this problem, the heat-transport coefficient is evaluated qualitatively and the physical phenomena

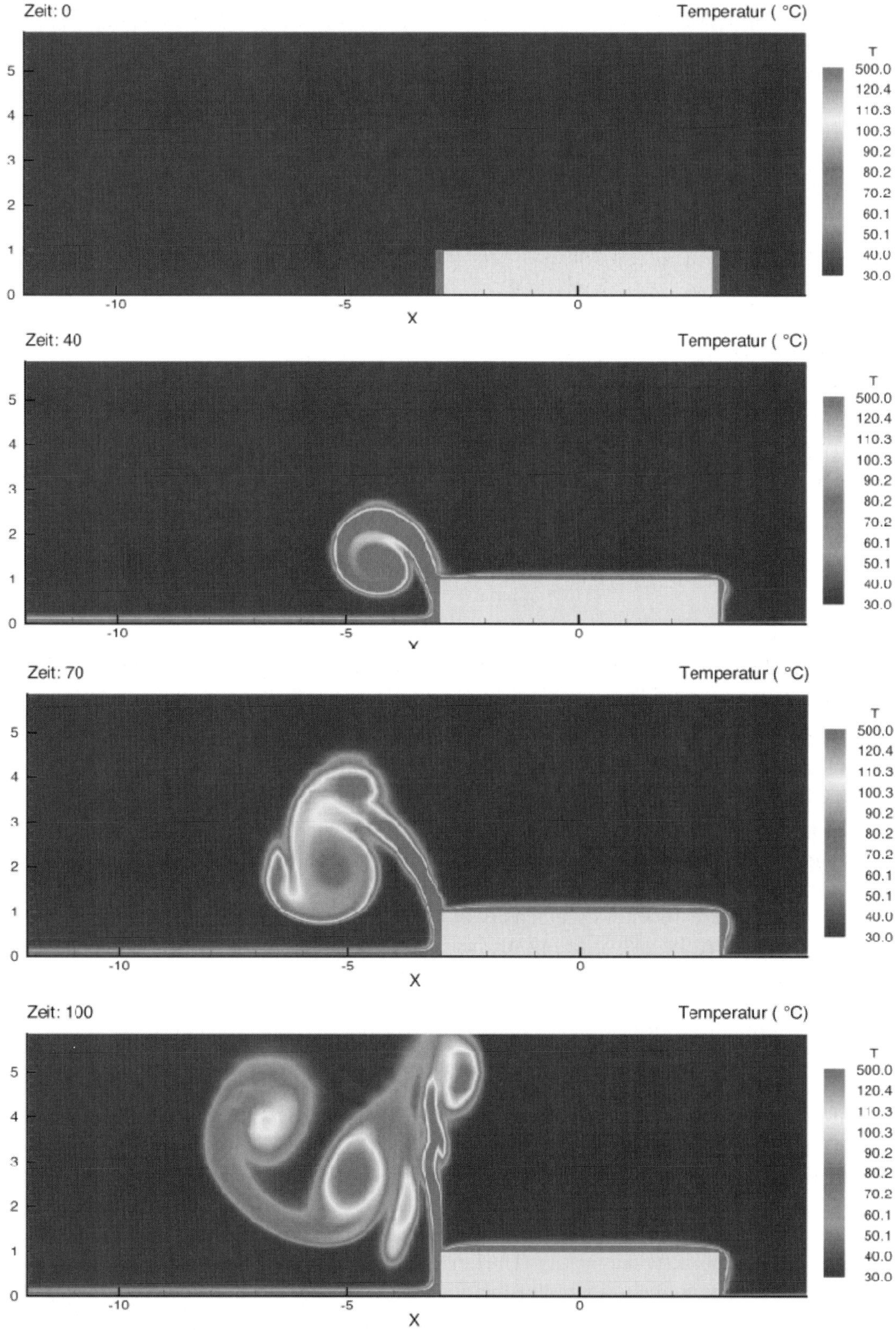

Figure 18 Temporal development of the temperature distribution in the brake-facility.

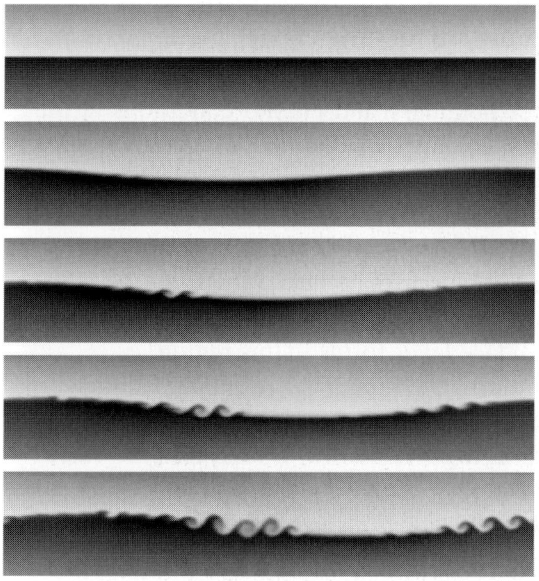

Figure 16. Density at times $t = 0, 6, 9, 11, 14$.

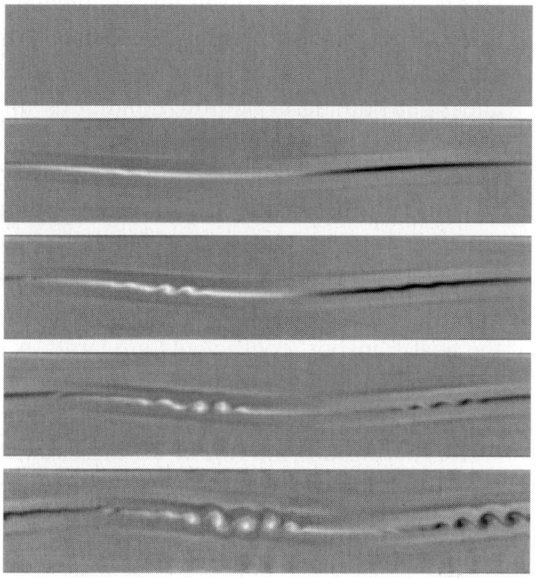

Figure 17. Vorticity at times $t = 0, 6, 9, 11, 14$.

are investigated. For the sake of simplicity the brake block is modeled as a fixed standing rectangular obstacle. The brake disk moves at constant velocity. The starting point is the dimensionless compressible Navier-Stokes equations. With following reference values

$$\begin{aligned}
\rho_{\text{ref}} &= 1{\cdot}205 && [\text{kg/m}^3]\\
T_{\text{ref}} &= 303{\cdot}15 && [\text{K}]\\
p_{\text{ref}} &= 1{\cdot}0499 \times 10^5 && [\text{N/m}^2]\\
l_{\text{ref}} &= 2 \times 10^{-2} && [\text{m}]\\
c_p &= 1{\cdot}\,0006 \times 10^3 && [\text{J/(kg K)}]\\
\lambda &= 2.637 \times 10^{-2} && [\text{W/(m K)}]\\
U_{\text{ref}} &= 21{\cdot}8 && [\text{m/sec}^2]\\
\nu(T = 500^\circ C) &= 7{\cdot}890 \times 10^{-5} && [\text{m}^2/\text{sec}]\\
\mu(= \rho_{\text{ref}} \cdot \nu) &= 9{\cdot}50 \times 10^{-5} && [\text{kg/(m sec)}],
\end{aligned}$$

the three dimensionless numbers, Reynolds-, Prandtl- and Mach number, important for the flow become

$$\begin{aligned}
\text{Re} &= 5525{\cdot}98\\
\text{Pr} &= 3{\cdot}627\\
\text{M} &= 0{\cdot}0738.
\end{aligned}$$

The total computing area covers the range from $-18l_{\text{ref}}$ to $18l_{\text{ref}}$ in the x-direction and from $0l_{\text{ref}}$ to $6l_{\text{ref}}$ in the y-direction. The brake block is fixed at the origin and has length $6l_{\text{ref}}$ and height $1l_{\text{ref}}$. We have used an equidistant grid in the x-direction ($\Delta x = 0.1 l_{\text{ref}}$). In the y-direction we adopted a non-equidistant grid ($\Delta y_{\min} = 0 \cdot 03 l_{\text{ref}}$) for a better resolution in the boundary layer. For the simulation we assume periodic conditions for all variables at the left and right boundary, outflow conditions at the upper boundary and no-slip conditions at the brake block and brake disk. As initial condition the state at rest is assumed. The temperatures at the brake block and brake disk are given as $500°C$ and at the outflow $30°C$. In Figure 18 the time-development of the hot temperature boundary layer is shown. During the so-called 'stripping-process' of the boundary layer at the brake block a strong left-rotating vortex develops rapidly and interacts with the outer flow. This process transports heat away from the brake disk very efficiently. Behind the brake block cold air is in contact with the hot brake disk and the heat-transport coefficients are large. A new boundary layer forms in this region.

For small brake-disk velocities and low temperatures, the numerical results are similar to those of an incompressible simulation. For higher temperatures (about $500°C$ when braking while driving down a hill or during sudden braking at high velocities) a compressible calculation is necessary. Future work will examine the stability of the boundary layer. By use of the linear stability analysis based on the Orr-Sommerfeld equations appropriate perturbations will be overlaid to the flow field.

4.2.3. *Summary*

The numerical results for the test problems clearly demonstrate that the MPV concept works very well in the incompressible as well as in the low-Mach- number regime. No modifications are necessary for viscous and inviscid fluid flows. The basics are the results of a multi-scale asymptotic analysis and these results are used to motivate a pressure decomposition where every term of the asymptotic series plays a different physical role. The asymptotic equations themselves are only used to obtain estimates which have to be corrected. Therefore, the scheme always solves the full compressible Euler or Navier-Stokes equations and is not restricted to very small Mach numbers, as it would be when solving the asymptotic equations only. The MPV concept proposed may be used to extend any incompressible solver to the low-Mach-number regime.

4.3. A FULLY CONSERVATIVE ZERO-MACH-NUMBER VARIABLE-DENSITY FLOW SOLVER

The conservation laws governing the evolution of a compressible calorically perfect gas are given by (2.1.15). Neglecting viscous effects, chemical reactions and taking into account a gravitational force field, we have these equations:

$$\begin{aligned}
\rho_t + \nabla \cdot (\rho \overline{\boldsymbol{v}}) &= 0\,, \\
(\rho \overline{\boldsymbol{v}})_t + \nabla \cdot (\rho \overline{\boldsymbol{v}} \circ \overline{\boldsymbol{v}}) + \frac{1}{\mathrm{M}^2} \nabla p &= \frac{1}{\mathrm{Fr}^2} \rho \overline{\boldsymbol{g}}, \\
(\rho e)_t + \nabla \cdot ((\rho e + p) \boldsymbol{v}) &= \frac{\mathrm{M}^2}{\mathrm{Fr}} \rho \boldsymbol{v} \cdot \overline{\boldsymbol{g}},
\end{aligned} \tag{4.3.1}$$

with the equation of state

$$p = (\gamma - 1)(\rho e - \tfrac{1}{2} \mathrm{M}^2 \rho \boldsymbol{v} \cdot \boldsymbol{v}). \tag{4.3.2}$$

After scaling, the acceleration $\boldsymbol{g}$ is a constant unit vector. All variables are dimensionless and $O(1)$. Beside the Mach number M, a new characteristic number appears in (4.3.1). This is the

Froude number $\mathrm{Fr} := v_{\mathrm{ref}}/\sqrt{g l_{\mathrm{ref}}}$. In the limit of $\mathrm{M} \to 0$, but finite Fr, solutions of (4.3.1) exhibit the singular behavior analyzed in Section 2.1 (see also [2], [46], [6]). Let us stress the issues which are particularly relevant for the computation of numerical approximations of (4.3.1) in the $\mathrm{M} \to 0$ limit:

1. ∇p tends to zero, but $\nabla p/\mathrm{M}^2$ tends to some non-trivial function $\nabla p^{(2)}$.

2. The total energy per unit volume ρe tends to $p/(\gamma - 1)$ and the right-hand side of the energy equation tends to zero. Thus, the energy fluxes $\nabla \cdot ((\rho e + p)\boldsymbol{v})$ must satisfy a constraint for the time derivative of ρe to have zero gradients. This is a constraint for the divergence of the velocity field $\boldsymbol{v}$.

3. Assume this constraint is $\nabla \cdot \boldsymbol{v} = 0$ (this is a special, but important case). Then the first equation of (4.3.1) requires the density ρ to be simply advected along particle paths. Thus, if ρ is constant at the initial time (and density variations are not entering the computational domain through the boundary), it should remain constant at any time.

4. The eigenvalues of the Jacobian of the flux function associated with the homogeneous part of (4.3.1) degenerate: These eigenvalues are $\boldsymbol{v} \cdot \boldsymbol{n}$ and $\boldsymbol{v} \cdot \boldsymbol{n} \pm c/\mathrm{M}$ with $c^2 = \gamma p/\rho$ and $\boldsymbol{n}$ any unit vector.

Due to this singular behavior, when attempting to compute numerical solutions of (4.3.1) for $\mathrm{M} \to 0$, using a standard finite-volume method for compressible flow, one faces at least the following difficulties:

1. A finite arithmetic single variable representation of the pressure p does not allow a meaningful computation of $\nabla p/\mathrm{M}^2$.

2. The energy fluxes do not satisfy the correct divergence constraint. As a consequence the total energy does not tend to $p/(\gamma - 1)$ and has non-zero gradients.

3. The velocity field does not satisfy the correct divergence constraint.

4. The numerical method fails to compute the correct rate of change of density along particle paths. For the special case in which the divergence constraint for the velocity is simply $\nabla \cdot \boldsymbol{v} = 0$, the numerical method fails to preserve an initially constant density distribution.

5. Explicit methods suffer from a Courant-Friedrichs-Lewy [47] time step stability restriction of the kind $\delta t < O(\mathrm{M})$.

The first and the last difficulties can be overcome by replacing (4.3.1) with its asymptotic limit:

$$
\begin{aligned}
\rho_t + \nabla \cdot (\rho \overline{\boldsymbol{v}}) &= 0, \\
(\rho \boldsymbol{v})_t + \nabla \cdot (\rho \boldsymbol{v} \circ \boldsymbol{v}) + \nabla p^{(2)} &= \frac{1}{\mathrm{Fr}^2} \rho \boldsymbol{g}, \\
(\rho e)_t + \nabla \cdot ((\rho e + p)\boldsymbol{v}) &= 0, \\
p &= (\gamma - 1)\rho e\,, \\
p &= p^{(0)}(t)\,.
\end{aligned}
\tag{4.3.3}
$$

Let us comment on the above equations. According to the results discussed in Section 2.1, the pressure field p has been decomposed into a zero-gradient time dependent thermodynamic component $p^{(0)}$ plus a second-order perturbation $\mathrm{M}^2 p^{(2)}$. The thermodynamic pressure $p^{(0)}$ is given by (2.1.38). For $(\gamma - 1)(\rho e)_t$ to be equal to $\mathrm{d}p^{(0)}/\mathrm{d}t$, the energy fluxes through the boundary ∂V of any arbitrary control volume $V \subset \Omega$ must satisfy the following constraint

$$
\frac{1}{|V|} \oint_{\partial V} (\rho e + p)\boldsymbol{v} \cdot \boldsymbol{n} \mathrm{d}s = -\frac{1}{\gamma - 1} \frac{\mathrm{d}p^{(0)}}{\mathrm{d}t}. \tag{4.3.4}
$$

Since $\rho e + p$ has no variations in space, Equation (4.3.4) is a divergence constraint for the velocity field $\boldsymbol{v}$. It implicitly defines the perturbation pressure $p^{(2)}$: At any time the gradients $\nabla p^{(2)}$ must guarantee that the acceleration $\boldsymbol{v}_t$ has a well-defined spatially-constant time-dependent divergence. We can compute this divergence by taking the time derivative of Equation (4.3.4). In the special case $\mathrm{d}p^{(0)}/\mathrm{d}t = 0$ (closed vessel, periodic boundary conditions, etc.) Equation (4.3.4) becomes the well-known $\nabla \cdot \boldsymbol{v} = 0$ constraint of the 'incompressible' Euler equations and $\nabla p^{(2)}$ a Lagrangian multiplier projecting convective and gravitational accelerations onto the space of divergence-free vector fields.

System (4.3.3) enjoys a time-step stability restriction of the kind $\delta t < O(1)$ and explicitly introduces a two-variable representation of the pressure field which allows a meaningful finite arithmetic computation of pressure gradients in the $M = 0$ limit. Notice that, because of the implicitness of $p^{(2)}$, (4.3.3) is *not* a hyperbolic system. Our aim is to solve (4.3.3) numerically by means of a finite-volume method in conservation form:

$$
\mathbf{U}_V^{n+1} = \mathbf{U}_V^n - \frac{\delta t}{|V|} \sum_{I \in \mathcal{I}_{\partial V}} |I| \mathbf{F}_I + \delta t \mathbf{W}_V \qquad \forall\, V \in \mathcal{V}. \tag{4.3.5}
$$

Here $V \in \mathcal{V}$ is a cell of a conformal space discretization of the flow domain Ω and $|V|$ is the volume of V. $I \in \mathcal{I}$ is an interface between two adjacent cells and $|I|$ is the area of I. By $\mathcal{V}$, $\mathcal{I}$ we denote the set of all cells and of all interfaces, respectively. $\mathcal{I}_{\partial V} \subset \mathcal{I}$ represents the set of those interfaces which lay on the boundary ∂V of V. $\mathbf{U}_V^n$ is a numerical approximation to the average $\mathbf{u}_V^n$ of the solution $\mathbf{u}$ of (4.3.3) over the cell V at time t^n. $\mathbf{F}_I$ and $\mathbf{W}_V$ are numerical approximations to the averages $\mathbf{f}_I$ and $\mathbf{w}_V$ of the flux function $\mathbf{f}$ and of the right-hand side $\mathbf{w}$ of (4.3.3). These averages are taken over the time interval $[t^n, t^{n+1} := t^n + \delta t]$ and over the interface I and the cell V for $\mathbf{f}_I$ and $\mathbf{w}_V$, respectively:

$$
\mathbf{u} := \begin{pmatrix} \rho \\ \rho \boldsymbol{v} \\ \rho e \end{pmatrix}, \qquad \mathrm{f} := \begin{pmatrix} \rho \boldsymbol{v} \cdot \boldsymbol{n} \\ \boldsymbol{v} \rho \boldsymbol{v} \cdot \boldsymbol{n} + p^{(2)} \boldsymbol{n} \\ h \rho \boldsymbol{v} \cdot \boldsymbol{n} \end{pmatrix}, \qquad \mathbf{w} := \begin{pmatrix} 0 \\ \rho \boldsymbol{g} / \mathrm{Fr}^2 \\ 0 \end{pmatrix}.
$$

In the above expressions h is the enthalpy per unit mass of the gas: $\rho h := \rho e + p$. Let us focus the attention on the numerical flux $\mathbf{F}_I$:

$$\mathbf{F}_I := \mathbf{f}(\mathbf{U}_I, \boldsymbol{n}_I) = \begin{pmatrix} (\rho\overline{\boldsymbol{v}})_I \cdot \boldsymbol{n}_I \\ \boldsymbol{v}_I(\rho\boldsymbol{v})_I \cdot \boldsymbol{n}_I \\ h_I(\rho\boldsymbol{v})_I \cdot \boldsymbol{n}_I \end{pmatrix} + \begin{pmatrix} 0 \\ p_I^{(2)}\boldsymbol{n}_I \\ 0 \end{pmatrix}. \tag{4.3.6}$$

We want to construct numerical fluxes $\mathbf{F}_I$ or, equivalently, interface averages $\mathbf{U}_I$, that fulfill the following requirements:

1. On smooth solutions $\mathbf{F}_I$ approximates the average flux $\mathbf{f}_I$ up to errors of order $O(\delta t^2)$.

2. The energy fluxes satisfy the divergence constraint

$$\frac{1}{|V|}\sum_{I\in\mathcal{I}_{\partial V}} |I| h_I(\rho\boldsymbol{v})_I \cdot \boldsymbol{n}_I = -\frac{1}{\gamma-1}\frac{p_V^{(0),n+1} - p_V^{(0),n}}{\delta t}, \qquad \forall\, V \in \mathcal{V}. \tag{4.3.7}$$

   This is a discrete form of the divergence constraint (4.3.4) which is consistent with the finite-volume method (4.3.5).

3. The mass fluxes guarantee the correct rate of change of density along particle paths. Notice that, in the special case $\mathrm{d}p^{(0)}/\mathrm{d}t = 0$, this can be achieved if (and only if) the interface velocities $\boldsymbol{v}_{\mathcal{I}}$ which are responsible for the advection of the density are divergence-free in the sense that

$$\frac{1}{|V|}\sum_{I\in\mathcal{I}}^{\partial V} |I|\, \boldsymbol{v}_I \cdot \boldsymbol{n}_I = 0\,.$$

4. The discrete velocity field $\boldsymbol{v}_V^{n+1}$ satisfies some discrete form of the divergence constraint (4.3.4).

In the next subsection we describe how to construct numerical fluxes which fulfill the requirements listed above. This is done via a semi-implicit procedure. First, we compute an explicit approximation to the fluxes of an auxiliary hyperbolic system. In our implementation this is done in a predictor stage in which the influence of pressure gradients on the convective fluxes is neglected over a half time step. Secondly, a Poisson-type equation for cell-centered pressures $p_V^{(2)}$ is solved. These pressures allow the computation of convective fluxes of mass and energy that fulfill requirement 1) to 3) and the update of density and total energy cell averages. At this point the grid-cell interface pressure-induced momentum fluxes $p_I^{(2)}\boldsymbol{n}_I$ are yet unknown. This pressure is obtained by solving another suitable discrete form of the energy conservation law. This yields a second elliptic problem. The solution of this problem leads to a new cell-centered velocity field $\boldsymbol{v}_V^{n+1}$ which *exactly* satisfies a discrete divergence constraint that is consistent with energy conservation. For a detailed description of the flux construction algorithm and of the discrete operators involved in such construction we refer to [7]. In Section 4.3.2 we discuss some numerical results.

4.3.1. *Semi-implicit construction of numerical fluxes*

The auxiliary system

$$
\begin{aligned}
\rho_t + \nabla \cdot (\rho \overline{\boldsymbol{v}}) &= 0\,, \\
(\rho \boldsymbol{v})_t + \nabla \cdot (\rho \boldsymbol{v} \circ \boldsymbol{v}) + \nabla p &= \frac{1}{\mathrm{Fr}^2} \rho \boldsymbol{g}\,, \\
(\rho e)_t + \nabla \cdot ((\rho e + p^{(0)}) \boldsymbol{v}) &= 0\,,
\end{aligned}
\tag{4.3.8}
$$

with equation of state

$$
p = (\gamma - 1)\rho e \tag{4.3.9}
$$

and flux function

$$
\mathbf{f}^* := \begin{pmatrix} & \rho \boldsymbol{v} \cdot \boldsymbol{n} \\ \boldsymbol{v} & \rho \boldsymbol{v} \cdot \boldsymbol{n} + p\boldsymbol{n} \\ h^{(0)} & \rho \boldsymbol{v} \cdot \boldsymbol{n} \end{pmatrix}, \qquad \rho h^{(0)} := \rho e + p^{(0)}
$$

enjoys the following properties:

1. The system has the same convective fluxes as the zero-Mach-number governing equations (4.3.3).

2. The system is hyperbolic.

3. The eigenvalues of the Jacobian of the flux function $\mathbf{f}^*$ are $\boldsymbol{v} \cdot \boldsymbol{n}$ and $\boldsymbol{v} \cdot \boldsymbol{n} \pm c$ with $c^2 := (\gamma - 1)h^{(0)}$ (see [6]).

4. Solutions of (4.3.8) satisfy, for homogeneous pressure p and zero flow divergence at time $t = 0$, the following estimates at time $t > 0$ (see [7]):

$$
\nabla \cdot \boldsymbol{v} = O(t), \qquad \nabla p = O(t^2).
$$

Let $\mathbf{F}^*_{\jmath}$ be numerical fluxes obtained with an explicit high-resolution upwind method for the auxiliary system (4.3.8). In our implementation, for instance, the method is a MUSCL scheme (see [48], [49], [50], [51], [52]) based on slope limiting of characteristic variables and the numerical flux proposed by Einfeldt [53] which has been extended for system (4.3.8) according to the characteristic analysis presented in [6].

Due to the first and the last of the above items, the difference between the rate of change of the conserved variables $\mathbf{u}$ as given by (4.3.3) and the rate of change of $\mathbf{u}$ in the auxiliary system (4.3.8) is, over short times t and up to terms $O(t^2)$:

$$
\mathbf{u}_t|_{(4.3.3)} - \mathbf{u}_t|_{(4.3.8)} = - \begin{pmatrix} 0 \\ \nabla p^{(2)} \\ 0 \end{pmatrix}. \tag{4.3.10}
$$

On each interface $I \in \mathcal{I}$ the numerical flux $\mathbf{F}_I$ can therefore be obtained from the auxiliary numerical flux $\mathbf{F}_I^*$ by correcting the interface average momenta $(\rho \boldsymbol{v})_I$ and velocities $(\boldsymbol{v})_I$ as follows:

$$(\rho \boldsymbol{v})_I = (\rho \boldsymbol{v})_I^* - \frac{\delta t}{2} (\nabla p^{(2)})_I, \tag{4.3.11}$$

$$(\boldsymbol{v})_I = (\boldsymbol{v})_I^* - \frac{\delta t}{2} \frac{1}{\rho_I} (\nabla p^{(2)})_I\,. \tag{4.3.12}$$

The term $(\nabla p^{(2)})_I$ represents the interface average gradients of the unknown pressure $p^{(2)}$. Let us approximate $p^{(2)}$ with a piecewise linear function between cell-centered values $p_{\mathcal{V}}^{(2)}$. Inserting (4.3.11) into the discrete divergence constraint (4.3.7), one obtains a discrete Poisson-type problem for the cell-centered values $p_{\mathcal{V}}^{(2)}$. The associated discrete Laplace operator has an enthalpy-weighted five points (in two space dimensions) compact stencil.

Once $(\rho \boldsymbol{v})_I$, $(\boldsymbol{v})_I$ have been computed ($\rho_I = \rho_I^*$ and $h_I = h_I^*$ because no correction in density or energy is needed, see (4.3.10)) the convective part of $\mathbf{F}_I$ is known and one can update both density and energy. Notice that, per construction

$$\begin{aligned}(\rho e)_V^{n+1} &= (\rho e)_V^n - \delta t \frac{1}{|V|} \sum_{I \in \mathcal{I}_{\partial V}} |I| h_I (\rho \boldsymbol{v})_I \cdot \boldsymbol{n}_I \\ &= (\rho e)_V^n + \delta t \frac{1}{\gamma - 1} \frac{p_V^{(0),n+1} - p_V^{(0),n}}{\delta t} = \frac{1}{\gamma - 1} p_V^{(0),n+1}\,.\end{aligned} \tag{4.3.13}$$

For the computation of $(\rho \boldsymbol{v})_V^{n+1}$, however, we still need the pressure component of the momentum flux $p^{(2)} \boldsymbol{n}$ at the interfaces. This is the last term of the numerical flux (4.3.6). This pressure is computed by imposing requirement number 4), *i.e.*, the discrete velocity field $\boldsymbol{v}_V^{n+1}$ shall satisfy some discrete form of the divergence constraint (4.3.4). Let $(\rho \boldsymbol{v})_V^{**}$ be the intermediate cell averages

$$(\rho \boldsymbol{v})_V^{**} := (\rho \boldsymbol{v})_V^n - \frac{\delta t}{|V|} \sum_{I \in \mathcal{I}_{\partial V}} |I| \boldsymbol{v}_I (\rho \boldsymbol{v})_I \cdot \boldsymbol{n}_I + \delta t \frac{\rho \boldsymbol{g}}{\mathrm{Fr}^2}, \qquad \forall\, V \in \mathcal{V}\,. \tag{4.3.14}$$

The finite-volume method (4.3.5) for the final cell averages $(\rho \boldsymbol{v})_V^{n+1}$ can be expressed in terms of $(\rho \boldsymbol{v})_V^{**}$ and of the unknown interface pressures $p_{\mathcal{I}}^{(2)}$ as:

$$(\rho \boldsymbol{v})_V^{n+1} = (\rho \boldsymbol{v})_V^{**} - \frac{\delta t}{|V|} \sum_{I \in \mathcal{I}_{\partial V}} |I|\, p_I^{(2)} \boldsymbol{n}_I, \qquad \forall\, V \in \mathcal{V}\,. \tag{4.3.15}$$

Let $\overline{\mathcal{V}}$ be a *dual* discretization of Ω. $\overline{\mathcal{V}}$ consists of control volumes $\overline{V}$ centered around the nodes of the original grid. The interfaces between the cells of $\overline{\mathcal{V}}$ are denoted by $\overline{I}$. As usual $\overline{\mathcal{I}}$ is the set of all such interfaces.

In Figure 19 a cell-centered and a node-centered control volume, V and $\overline{V}$, are drawn for a two-dimensional Cartesian grid. We compute $p_{\mathcal{I}}^{(2)}$ by linear interpolation on the set of nodal values $p_{\overline{\mathcal{V}}}^{(2)}$. These nodal values are computed following an idea originally proposed by Geratz [54]. Consider a second-order finite-volume method for the total energy ρe on the dual discretization:

$$(\rho e)_{\overline{V}}^{n+1} := (\rho e)_{\overline{V}}^n - \frac{\delta t}{|\overline{V}|} \sum_{\overline{I} \in \overline{\mathcal{I}}_{\partial \overline{V}}} \frac{|\overline{I}|}{2} \left((\rho h \boldsymbol{v})_{\overline{I}}^n + (\rho h \boldsymbol{v})_{\overline{I}}^{n+1} \right) \cdot \boldsymbol{n}_{\overline{I}}, \qquad \forall\, \overline{V} \in \overline{\mathcal{V}} \tag{4.3.16}$$

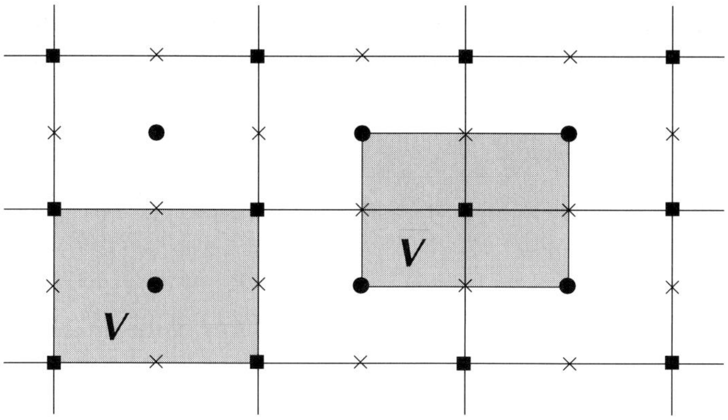

Figure 19 Cell (V) and node centered ($\overline{V}$) control volumes; cell centers, nodes and the midpoints of the interfaces are marked by circles, squares and crosses, respectively

and require the energy fluxes through the interfaces of the dual discretization to satisfy the divergence constraint

$$\frac{\delta t}{|\overline{V}|}\sum_{\overline{I}\in\overline{\mathcal{I}}_{\partial\overline{V}}} \frac{|\overline{I}|}{2}\left((\rho h \boldsymbol{v})^n_{\overline{I}} + (\rho h\boldsymbol{v})^{n+1}_{\overline{I}}\right)\cdot \boldsymbol{n}_{\overline{I}} = -\frac{1}{\gamma-1}\,\frac{p^{(0),n+1}_{\overline{V}} - p^{(0),n}_{\overline{V}}}{\delta t}, \qquad \forall\, \overline{V}\in\overline{\mathcal{V}}\,. \tag{4.3.17}$$

Equation (4.3.17) is a discrete form of the divergence constraint (4.3.4) which is consistent with the finite-volume method (4.3.16). $p^{(0),n+1}$ is the same as was computed and used in the first projection step. The dual interface averages can be expressed to the desired-order accuracy by means of cell averages:

$$(\rho h\boldsymbol{v})^{(\cdot)}_{\overline{\mathcal{I}}} = L^{\mathcal{V}}_{\overline{\mathcal{I}}}\left((\rho h\boldsymbol{v})^{(\cdot)}_{\mathcal{V}}\right)\,. \tag{4.3.18}$$

$L^{\mathcal{V}}_{\overline{\mathcal{I}}}$ is a (linear) operator mapping cell averages into interface averages. Together with (4.3.15) and (4.3.18), Equation (4.3.17) defines a discrete Poisson-type problem for the nodal values $p^{(2)}_{\overline{\mathcal{V}}}$. This problem is the equivalent of (4.3.7), (4.3.11) on the dual grid. Its associated discrete Laplace operator has an enthalpy-weighted nine points (in two space dimensions) compact stencil. In the special case $\mathrm{d}p^{(0)}/\mathrm{d}t = 0$ the nodal values $p^{(2)}_{\overline{\mathcal{V}}}$ guarantee that the discrete velocity field $\boldsymbol{v}^{n+1}_{\mathcal{V}}$ satisfies the divergence constraint

$$\frac{1}{|\overline{V}|}\sum_{\overline{I}\in\overline{\mathcal{I}}_{\partial\overline{V}}} |\overline{I}|L^{\mathcal{V}}_{\overline{I}}(\boldsymbol{v}^n_{\mathcal{V}})\cdot\boldsymbol{n}_{\overline{I}} = 0 \quad\Rightarrow\quad \frac{1}{|\overline{V}|}\sum_{\overline{I}\in\overline{\mathcal{I}}_{\partial\overline{V}}} |\overline{I}|L^{\mathcal{V}}_{\overline{I}}(\boldsymbol{v}^{n+1}_{\mathcal{V}})\cdot\boldsymbol{n}_{\overline{I}} = 0\,. \tag{4.3.19}$$

We have fulfilled requirements 1) – 4) and the construction of the numerical fluxes $\mathbf{F}_I$ is completed.

4.3.2. *Numerical results*

We discuss the numerical results obtained on five test problems. The first four problems are chosen to assess the accuracy and the efficiency of the method and test its capability to cope with large density variations and small-scale gravity-driven flows. For these tests either the exact solution or at least some properties of the exact solution are known. This allows a meaningful validation of the method and provides a flavor of the difficulties that must be faced in the numerical simulation of more realistic flows. Problem number five is included to show that the proposed numerical method can be extended to cope with boundary-driven compression/expansion, viscous forces and heat transfer. All test problems can be run with trivial geometries and boundary conditions.

The computations have been performed on regular Cartesian grids. The discrete operators and the linear systems for the cell-centered and for the node-centered pressures are those explicitly given in [7]. These two linear systems must be solved at each time step. This has been done by means of a multi-grid preconditioned conjugate-gradient method. The difference with respect to the standard conjugate-gradient solver is that, in each iteration, the new residual vector is computed by applying a multi-grid cycle to the previous residual vector. There are several ways of visiting the grid levels during the multi-grid procedure, such as a V-cycle, W-cycle, F-cycle [55] and nested cycle. In our case, the F-cycle turned out to provide the best contraction rate. As smoother a Gauss-Seidel method was used with two pre- and post- smoothing steps on each grid level. In two space dimensions a standard nine-point prolongation operator was used. This operator is defined through bilinear interpolation. In three dimensions trilinear interpolation provides a 27-point prolongation operator. The adjoint prolongation operator served as restriction operator. In the presence of large density variations, the coefficients of both linear systems can change by orders of magnitude. In this case the linear coarse-grid operators need to be constructed by Galerkin's approximation [55].

As expected, the computations show that the CPU time needed to solve the systems depends linearly on the number of unknowns. The solution of the linear systems accounts for about 95% of the time required for a computation and demands a memory allocation of roughly one K-byte per computational point. In each solution the residuals

$$r_2(p^{(2)}_{\mathcal{V}}) := \left\| D^{\mathcal{I}}_{\mathcal{V}}\left((\rho h^0 \boldsymbol{v})^*_{\mathcal{I}}\right) - \frac{\delta t}{2} D^{\mathcal{I}}_{\mathcal{V}}\left(h^{0,*}_{\mathcal{I}} G^{\mathcal{V}}_{\mathcal{I}}\left(p^{(2)}_{\mathcal{V}}\right)\right) \right\|_2,$$

$$r_2(p^{(2)}_{\overline{\mathcal{V}}}) := \left\| D^{\mathcal{V}}_{\overline{\mathcal{V}}}\left(h^{n+1}_{\mathcal{V}} (\rho \boldsymbol{v})^{**}_{\mathcal{V}}\right) + D^{\mathcal{V}}_{\overline{\mathcal{V}}}\left((\rho h \boldsymbol{v})^{n}_{\mathcal{V}}\right) - \delta t\, D^{\mathcal{V}}_{\overline{\mathcal{V}}}\left(h^{n+1}_{\mathcal{V}} G^{\overline{\mathcal{V}}}_{\mathcal{V}}\left(p^{(2)}_{\overline{\mathcal{V}}}\right)\right) \right\|_2$$

have been driven down to 10^{-7}. In the above definitions $\|a_{\mathcal{V}}\|_2$ represents the Euclidean norm of a vector whose components are the values $a_{\mathcal{V}}$, *i.e.*,

$$\|a_{\mathcal{V}}\|_2 := \left(\sum_{V \in \mathcal{V}} a^2_V\right)^{1/2}$$

and similarly for $\|a_{\overline{\mathcal{V}}}\|_2$.

Convergence studies. This test problem was originally proposed in Almgren *et al.* [56]. It has been designed to assess the accuracy of the method on constant-density flows. For any time t and $0 < x < 1$, $0 < y < 1$, the velocity field

$$u(x, y, t) := 1 - 2\cos(2\pi(x - t))\sin(2\pi(y - t)),$$

$$v(x, y, t) := 1 + 2\sin(2\pi(x - t))\cos(2\pi(y - t)),$$

together with the pressure $p^{(2)}(x, y, t)$

$$p^{(2)}(x, y, t) := -\cos(4\pi(x - t)) - \cos(4\pi(y - t))$$

Table 1. Constant density: errors and convergence rates in the 2-norm and in the maximum norm.

	32×32	Rate	64×64	Rate	128×128
2-norm	0·193646	2·07	0·0458949	2·10	0·010705
max-norm	0·236456	2·09	0·0553504	2·11	0·012821

is an exact solution of the zero-Mach-number governing equations (4.3.3) with constant pressure $p(x, y, t)$, constant density $\rho(x, y, t)$ and periodic boundary conditions on the unit square. Starting from $t = 0$, we have computed numerical approximations $u_{i,j}^N$ to the cell-averages $u(x_i, y_j, t^N)$ of the exact velocity u at time $t^N = 3$. Similarly $v_{i,j}^N$, $\rho_{i,j}^N$ are numerical approximations to the cell-averages $v(x_i, y_j, t^N)$ and $\rho(x_i, y_j, t^N)$ of the exact v, ρ at time $t^N = 3$.

Three equally spaced regular Cartesian grids of spacings $h = 1/32$, $h = 1/64$ and $h = 1/128$ have been used on the unit square. On each grid the time step was chosen according to a fixed Courant number of 0·8. The initial cell averages $(\rho \boldsymbol{v})_{i,j}^0$ have been computed for $\boldsymbol{v}_{i,j}^0$ to be discretely divergence free

$$(\rho \boldsymbol{v})_{i,j}^0 = (\rho \boldsymbol{v})(x_i, y_j, 0) - G_{i,j}^{\overline{\mathcal{V}}}\left(p_{\overline{\mathcal{V}}}^{(2),0}\right),$$

i.e., the initial pressure $p_{\overline{\mathcal{V}}}^{(2),0}$ is a solution of the Poisson problem

$$D_{\overline{\mathcal{V}}}^{\mathcal{V}}\left(\frac{1}{\rho_{i,j}^0} G_{i,j}^{\overline{\mathcal{V}}}\left(p_{\overline{\mathcal{V}}}^{(2),0}\right)\right) = D_{\overline{\mathcal{V}}}^{\mathcal{V}}\left(\frac{(\rho \boldsymbol{v})(x_i, y_j, 0)}{\rho_{i,j}^0}\right)$$

with $\rho_{i,j}^0 = \rho(x_i, y_j, 0) = 1$. In the MUSCL scheme for the computation of the auxiliary numerical fluxes $\mathbf{F}_{\imath}^*$ unlimited slopes have been used. For each grid we have measured the 2-norm e_2 and the maximum norm e_∞ of the cell-error $e_{i,j}$ at time $t^N = 3$:

$$e_{i,j} := \left|\rho(x_i, y_j, t^N) - \rho_{i,j}^N\right| + \left|u(x_i, y_j, t^N) - u_{i,j}^N\right| + \left|v(x_i, y_j, t^N) - v_{i,j}^N\right|,$$

$$e_2 := \left(\sum_{i,j} \left(e_{i,j} h\right)^2\right)^{1/2}, \qquad e_\infty := \max_{i,j} \left\{e_{i,j}\right\}.$$

Notice that this is essentially a measure of the velocity error: due to the exact projection of the interface velocity, the density error in the 2-norm is of the same order as $r_2(p_{\mathcal{V}}^{(2)})$, *i.e.*, 10^{-7}. Table 1 shows e_2, e_∞ on the three grids, together with the corresponding convergence rates. These have been computed as follows: Given, *e.g.*, coarse and fine grid 2-norm errors $e_{2,c}$, $e_{2,f}$ and the corresponding grid spacings h_c, h_f the convergence rate is

$$\frac{\log(e_{2,c}/e_{2,f})}{\log(h_c/h_f)}.$$

We have constructed the exact velocity field (4.3.2) by differentiating the streamline function

Table 2. Variable density: errors and convergence rates in the 2-norm and in the maximum norm.

	32 × 32	Rate	64 × 64	Rate	128 × 128
2-norm	0·229332	2·02	0·0563924	2·16	0·0125899
max-norm	0·263492	1·98	0·0664518	1·68	0·0207160

$$\phi(x, y, t) := y - x + \frac{1}{\pi}\cos(2\pi(x - t))\cos(2\pi(y - t))$$

and taking $u := \partial\phi/\partial y$, $v := -\partial\phi/\partial x$. The function ϕ represents a vortical motion $\varphi := \phi - y + x$ superimposed on a translation. The vortical motion is simply advected by the velocity field $\boldsymbol{v}$, *i.e.*,

$$\frac{\mathrm{D}\varphi}{\mathrm{D}t} := \frac{\partial\varphi}{\partial t} + \boldsymbol{v}\cdot\nabla\varphi = 0\,,$$

as one can verify by inspection. Thus, variable-density exact solutions to the governing equations (4.3.3) can be constructed if we take

$$\rho(x, y, t) := f(\varphi)$$

with some smooth function f. We used

$$f(\varphi) := 2 + (\pi\varphi)^2\,. \tag{4.3.20}$$

The constant on the right-hand side is taken to avoid negative densities. The square ensures that densities monotonically increase from the center to the outer boundary of each vortex: a density distribution with local maxima in vortex cores would undergo Rayleigh-Taylor instability. With (4.3.20) an exact solution for the density of (4.3.3) is

$$\rho(x, y, t) := 2 + 0{\cdot}5\cos^2(2\pi(x - t))\cos^2(2\pi(y - t)).$$

In Table 2 the error norms for the variable-density computations are shown. As for the constant-density case we obtain second-order accuracy both in the 2-norm and in the maximum norm.

Advection of a vortex. We consider the advection of a vortex in a channel. The computational domain is the rectangle $[0, 4] \times [0, 1]$. The upper and lower boundaries are walls; periodic boundary conditions are imposed at the left and right boundaries. The grid consists of 80×20 cells. The initial condition is:

$$\begin{aligned} \rho(x, y, 0) &= 1, \qquad & u(x, y, 0) &= 1 - v_\theta(r)\sin\theta, \\ p(x, y, 0) &= 1, \qquad & v(x, y, 0) &= v_\theta(r)\cos\theta \end{aligned}$$

with

$$v_\theta(r) = \begin{cases} r/0{\cdot}2 & \text{if}\, 0 < r < 0.2 \\ 2 - r/0{\cdot}2 & \text{if}\, R < r < 0.4 \\ 0 & \text{if}\, r > 0{\cdot}4 \end{cases} \quad \text{and} \quad r = \sqrt{(x - 0{\cdot}5)^2 + (y - 0{\cdot}5)^2}$$

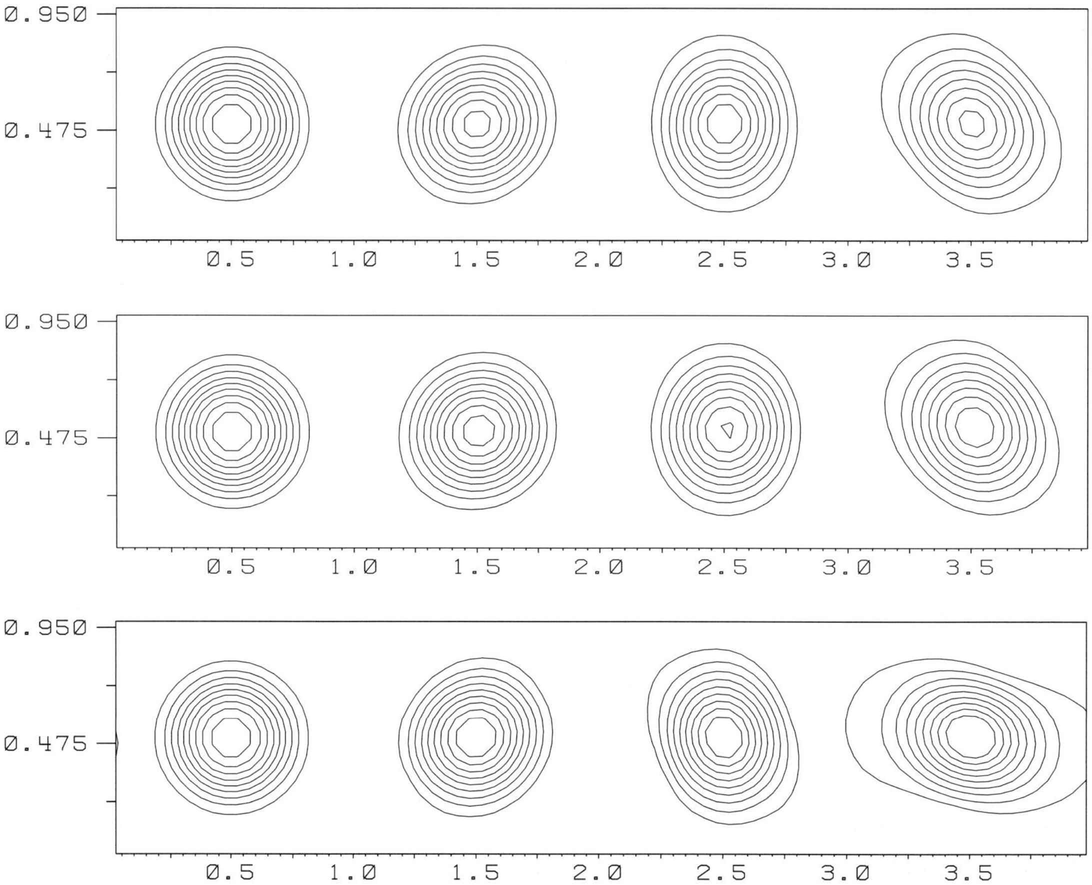

Figure 20. Advection of a vortex at times t =0·0, 1·0, 2·0, 3·0: 9 contour lines of the stream-function in [0·02,0·18]. Unlimited slopes (top), monotonized central-difference (middle) and Sweby's limiter with $k := 1{\cdot}8$ (bottom).

For the above initial data the exact velocity for $t > 0$ can be computed: $u(x, y, t) = u(x - u_\infty t, y, 0)$ and $v(x, y, t) = v(x - u_\infty t, y, 0)$, *i.e.*, the initial data are simply advected by the background velocity u_∞. This problem was originally proposed by Gresho *et al.* [57].

In Figure 20 we show contour lines of the stream function for three computations. They have been done using different slope limiters in the MUSCL step of the Godunov-type method for the computation of the auxiliary fluxes. Due to the rough discretization the results exhibit a significant deformation of the vortex. In contrast to the results shown in [57] Figure 13, however, the core of the vortex is advected along the axis of the channel in agreement with the exact solution. The first computation (unlimited slopes) shows a loss of vorticity comparable with [57] by exhibiting a stronger deformation of the vortex. The second and the third computations (monotonized central-difference and Sweby's limiter with $k := 1{\cdot}8$, see, *e.g.*, Schulz-Rinne [58]) show a slightly stronger deformation of the initial vorticity distribution, but a much better conservation of the maximum level of vorticity.

Driven cavity flows. The driven-cavity test problems proposed in [59] have been the subject of many numerical computations (see, *e.g.*, [60], [61]), and the Section 4.2 above. For Reynolds numbers (Re) up to 1000 most computations seem to converge towards a steady state and there is an excellent agreement between stationary solutions obtained with different numerical schemes. Thus, these problems are very well suited to validate new numerical methods. Here driven cavity flows at Reynolds numbers 100 and 1000 have been computed. Our main goals are:

- Show that the method can be easily extended to cope with viscous flows;
- Investigate the behavior of the method with respect to the coupling between pressure and velocity fields;
- Investigate the behavior of the method with respect to convergence towards stationary solutions;
- Compare our numerical results with established reference solutions.

In agreement with [59], we consider a viscous zero-Mach-number flow with no heat conduction. Viscous effects only enter in the momentum equation through a viscous stress and are accounted for via operator-splitting techniques. A delicate issue in the numerical computation of incompressible flows is that of the coupling between pressure and velocity fields. For finite discretizations this problem (often referred to as 'local grid decoupling' or 'checkerboard instability') can be described as follows. Assume that the null space of the discrete gradient operator, $\ker(G_{\nu}^{\overline{\nu}})$, contains highly oscillating fields. Since $G_{\nu}^{\overline{\nu}}$ has a local stencil, this is usually the case whenever $\dim(\ker(G_{\nu}^{\overline{\nu}})) > 1$. If the solution $p_{\overline{\nu}}$ of the Poisson-type problem has components in $\ker(G_{\nu}^{\overline{\nu}})$, one obtains pressure oscillations which do not influence the velocity field: pressure and velocity field decouple.

For two-dimensional equally spaced Cartesian grids and the implementation described in [7] one finds that $\dim(\ker(G_{\nu}^{\overline{\nu}})) = 2$ and $\ker(G_{\nu}^{\overline{\nu}})$ contains, besides constant pressures $p_{\overline{\nu}}^{c}$, a non-trivial highly oscillating mode $p_{\overline{\nu}}^{o}$. Therefore, we expect to observe pressure-velocity decoupling whenever the iterative linear system solver converges towards solutions $p_{\overline{\nu}}$ with components in $\ker(G_{\nu}^{\overline{\nu}})$. The method of conjugate gradients preserves, by exact arithmetics, the components of $p_{\overline{\nu}}^{k}$ in $\ker(G_{\nu}^{\overline{\nu}})$. Since we always start our iteration with $p_{\overline{\nu}}^{0} := 0$ we expect a numerical solution obtained in a reasonable number of iteration steps to be oscillation free. This is confirmed by our numerical results. On the other hand, numerical solutions obtained through a random choice of $p_{\overline{\nu}}^{0}$ may exhibit pressure-velocity decoupling.

The understanding of the pressure-velocity decoupling in the limit of vanishing grid size requires a deeper analysis. We have investigated numerically the effects 1) of grid refinement at constant convection-based Courant number CFL of 0·8 and 2) of time-step refinement for a fixed grid size. Some results are shown in Figures 21 and 22. Neither in the first nor in the second case do we observe the onset of pressure-velocity decoupling. Figure 23 shows the time history of the residual

$$r_2^n := \sum_{\nu \in V} h^2 \|\mathbf{U}_{\nu}^n - \mathbf{U}_{\nu}^{n-1}\|_2$$

for a $\mathrm{Re} = 1000$ computation on several grids. The residual is plotted versus the number of computational steps. The cost of a single step on a 64×64 grid is of about 1·3 seconds on a DEC Alpha 21164 CPU running at 500 MHz. For the 128×128 grid-cells computation pres-

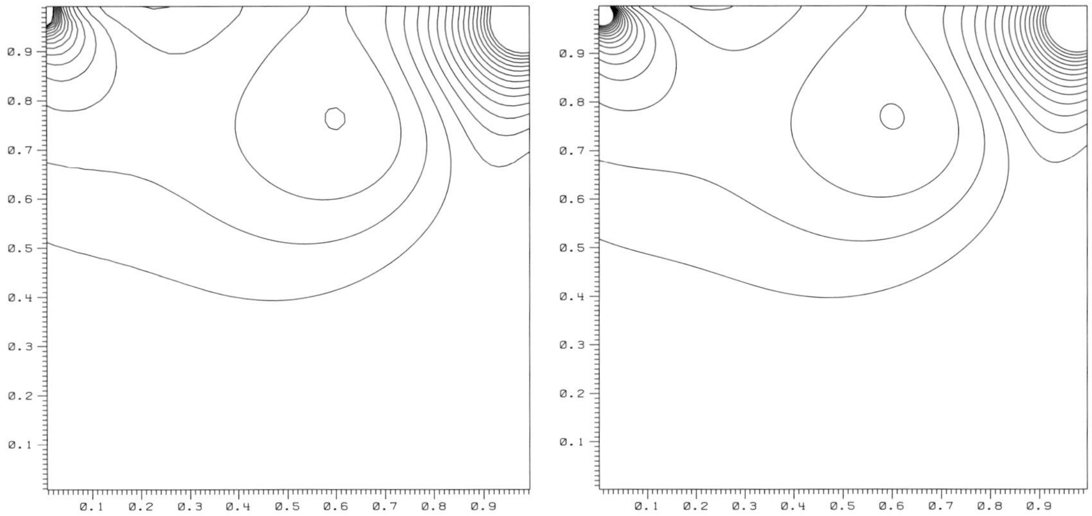

Figure 21. Driven cavity at Re= 100, C =0·8: 30 contour lines of the nodal pressure $p_{\mathcal{V}}^{(2)}$ in [−0·4,0·4]. 64 × 64 (left) and 256 × 256 (right) grid cells.

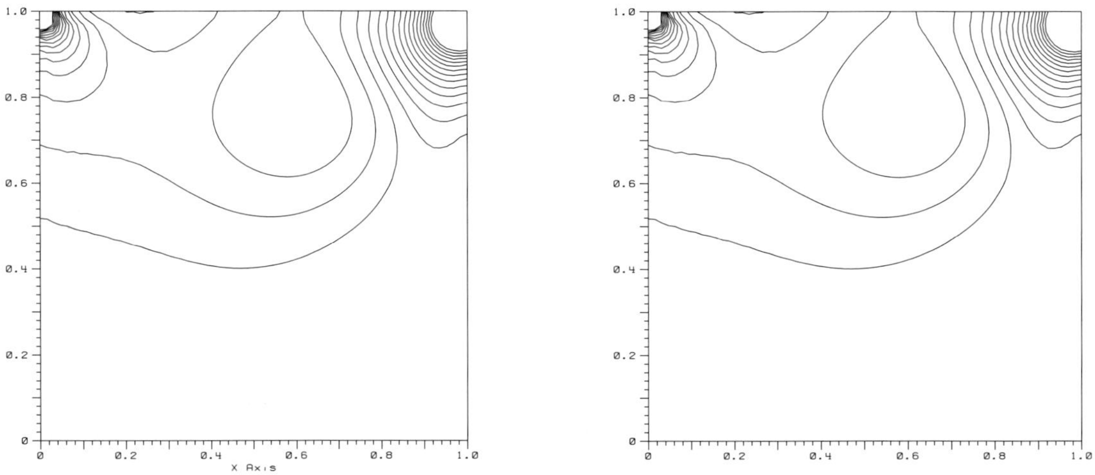

Figure 22. Driven cavity at Re= 100, 64 × 64 grid cells: 30 contour lines of the nodal pressure $p_{\mathcal{V}}^{(2)}$ in [−0·4,0·4]. C =0·08 (left) and C =0·008 (right).

sure and streamlines of the numerical solution after 5000 time steps are shown in Figures 24 and 25. These results are in a good qualitative agreement with the ones presented by Ghia *et al.* [59, p. 400]. For a more quantitative comparison the horizontal (vertical) component of the velocity along the vertical (horizontal) line through the geometric center of the cavity have been drawn in Figure 26. The solid line represents the numerical solution obtained with the present method. The dots are values of a reference solution taken from [59]. The accuracy of this solution has been confirmed by many independent computations.

Falling droplet. A heavy ‘droplet’ falls through a light fluid atmosphere and impacts into the surface of the heavy fluid in a cavity. The density ratio is 1000:1 and the Froude number equal to one. The flow is assumed to be inviscid and there is no account for surface tension or for a

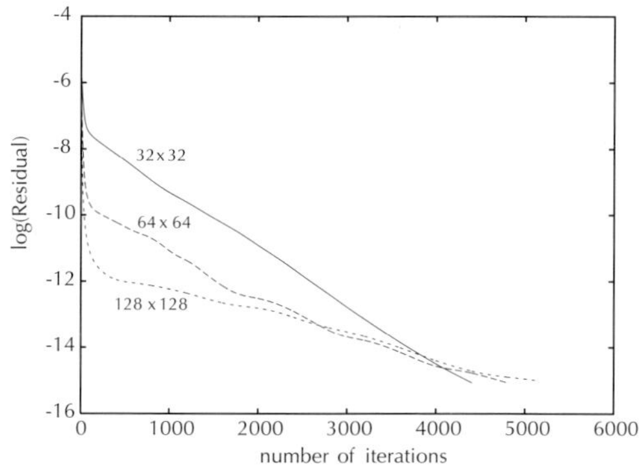

Figure 23. Driven cavity at Re= 1000: Residual versus number of iterations for 32 × 32, 64 × 64 and 128 × 128 grid cells computations; coarser grid solutions have been taken as initial data for finer grid solutions.

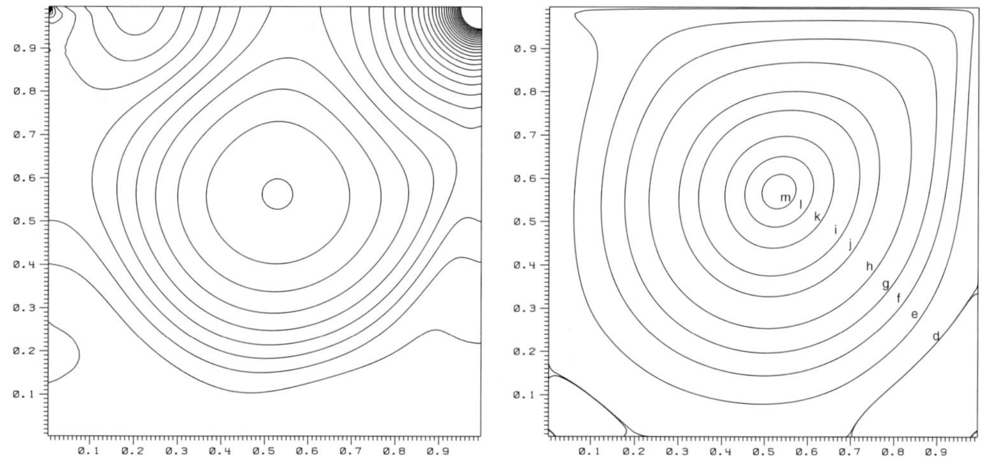

Figure 24. Driven cavity at Re= 1000, 128 × 128 grid cells: 30 contour lines of the nodal pressure $p^{(2)}_{\overline{\nu}}$ in [-0·4;0·4] (left) and streamlines (right). Streamline values and labels from [41].

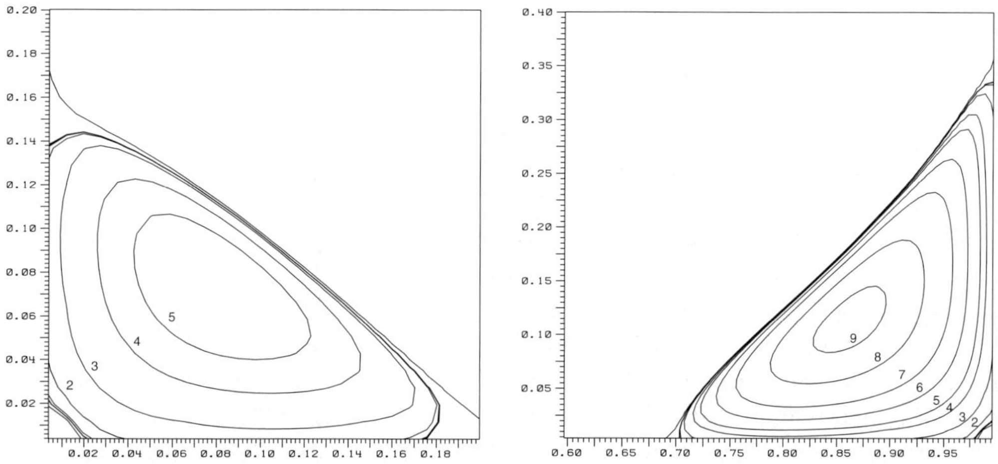

Figure 25. Driven cavity at Re=1000, 128 × 128 grid cells: streamlines in the left and right bottom secondary vortices. Values and labels from [41].

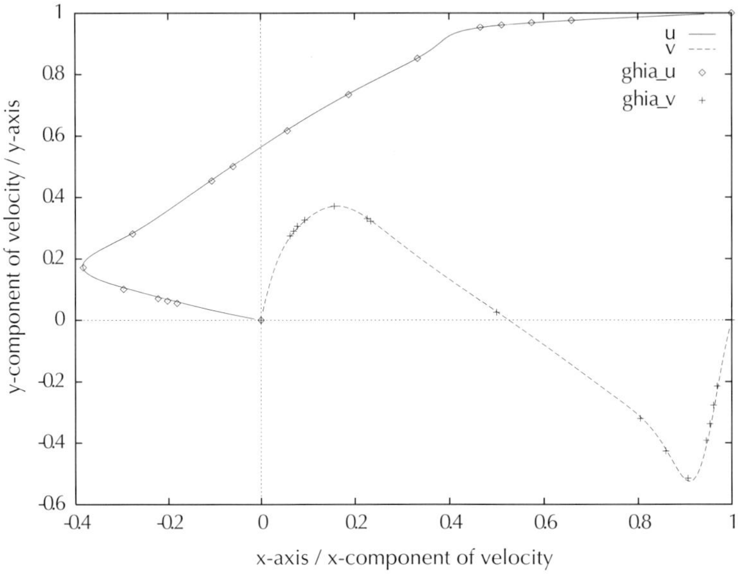

Figure 26. Driven cavity at Re=1000, 128 × 128 grid cells: horizontal (vertical) component of the velocity along the vertical (horizontal) line through the geometric center of the cavity; present results (solid line) and reference solution from [41].

change of the equation of state (hence, the quotes on 'droplet'!). The computational domain is the rectangle $[0, 1] \times [0, 2]$. We present both two- and three-dimensional computations. The goal is to investigate the capability of the method to cope with large density variations.

From a numerical point of view the effect of density variations is to increase the condition number of the discrete Poisson-type operators associated with the numerical computation of the pressure $p^{(2)}$. One might expect poor convergence in the iterative solution of the linear systems and, in the worst case, oscillations in the pressure field $p_{\jmath}^{(2)}$. Since our interface pressures $p_{\jmath}^{(2)}$ are computed via a discrete Poisson-type operator which, for two-dimensional equally spaced Cartesian grids, exhibits local grid decoupling, we are thus particularly interested in the behavior of $p_{\jmath}^{(2)}$ in the two dimensional case.

Two-dimensional case. This problem was originally proposed in Puckett *et al.* [62] to test a tracking method for incompressible variable-density flows. Here the interface between light and heavy fluid is captured, but we still expect our second-order method to describe properly the main features of the flow. The computational grid consists of 64 × 128 cells. The initial data are:

$$\rho(x, y, 0) = \begin{cases} 1000{\cdot}0 & \text{if } 0{\cdot}0 \le y \le 1{\cdot}0 \text{ or } 0{\cdot}0 \le r \le 0{\cdot}2, \\ 1{\cdot}0 & \text{if } 1{\cdot}0 < y \le 2{\cdot}0 \text{ or } 0{\cdot}2 < r, \end{cases}$$

$$p(x, y, 0) = 1, \quad \boldsymbol{v}(x, y, 0) = 0 \quad \text{and} \quad r = \sqrt{(x - 0{\cdot}5)^2 + (y - 1{\cdot}75)^2}\,.$$

Figure 27 shows density contours at a sequence of output times. After the impact of the droplet some areas of lighter fluid appear within the heavy fluid (last three frames). This is consistent with the results shown in [62] where this effect was referred to as 'trapped air bubbles'. For this sequence we monitored the cell interface pressure $p_{\jmath}^{(2)}$ without noticing any spurious oscillations or local grid decoupling effects. The multi-grid preconditioned conjugate-gradients

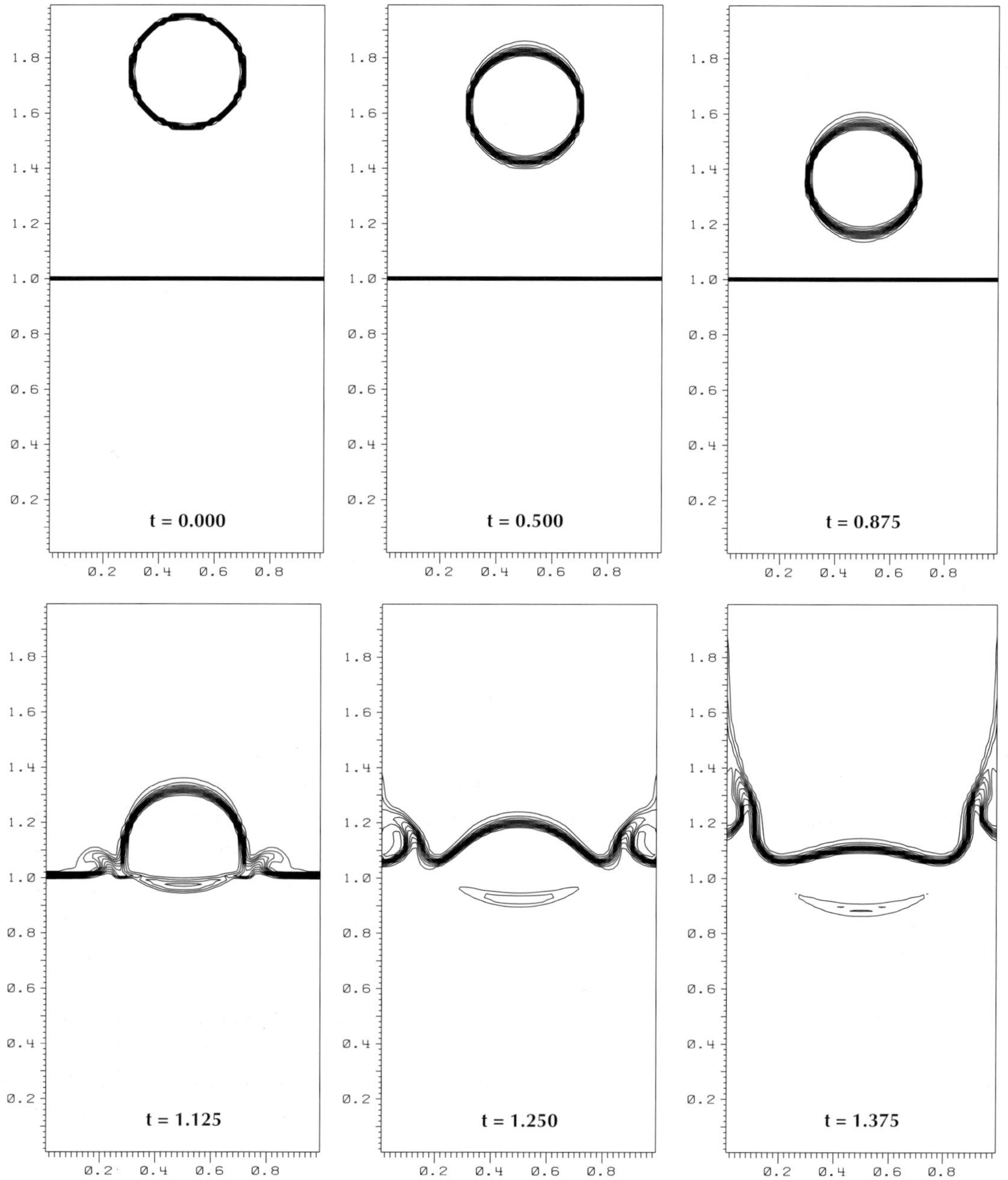

Figure 27. Two-dimensional falling 'droplet' at $Fr = 1$ and density ratio 1000: Contour lines of density in [1,1000].

technique allows the iterative solution of the linear systems for the pressure in about the same number of iterations as for the constant-density case.

Three-dimensional case. This is a simple extension of the previous case to three space dimensions. The grid consists of $64 \times 64 \times 128$ cells. The initial data are:

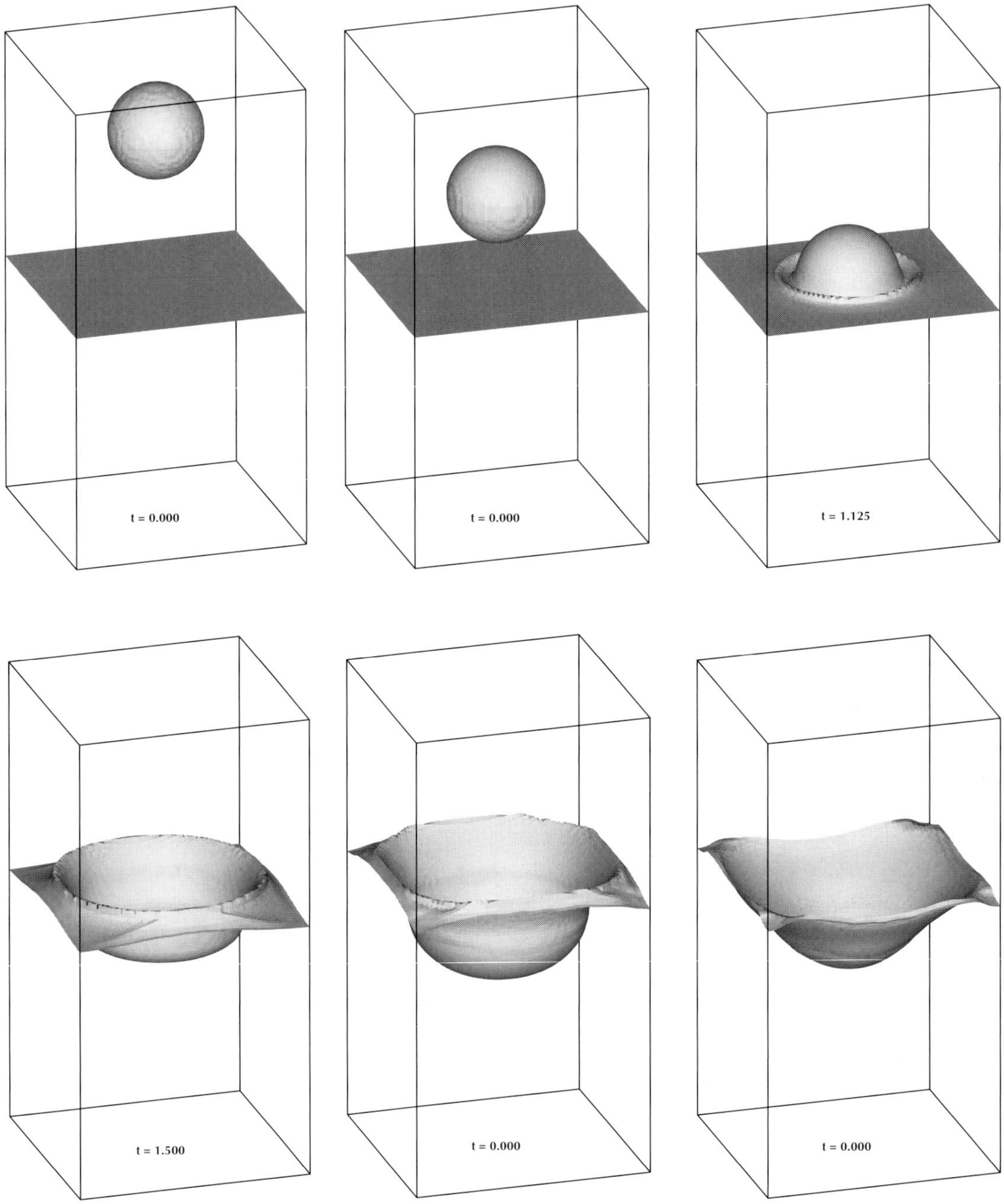

Figure 28. Three-dimensional falling 'droplet' at Fr= 1 and density ratio 1000: Iso-surface $\rho = 500$ of density.

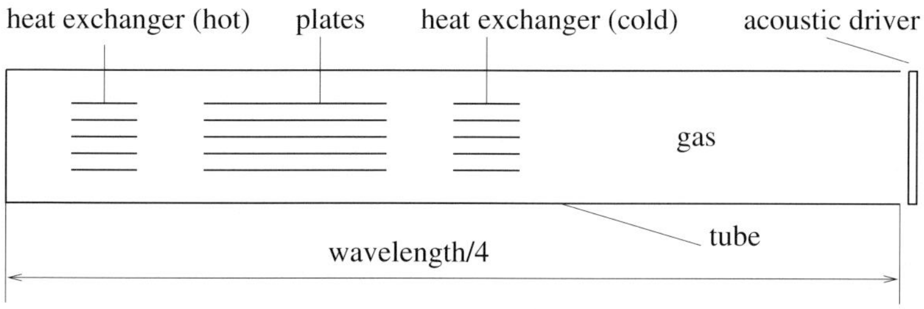

Figure 29. Sketch of a simplified thermo-acoustic refrigerator.

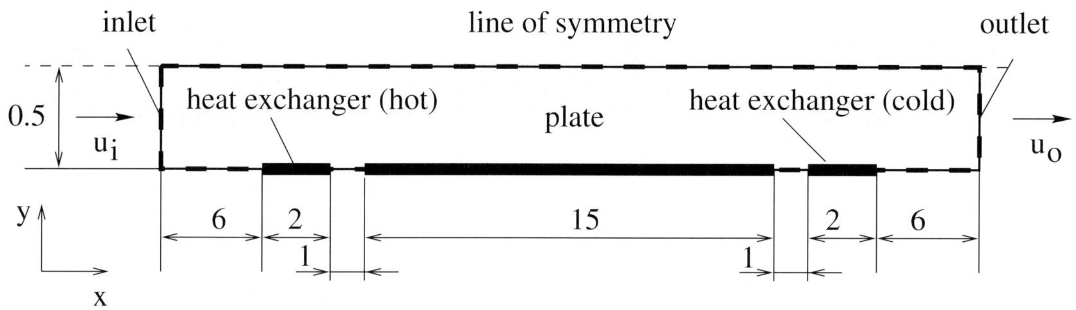

Figure 30. Computational domain (dotted line).

$$\rho(x, y, 0) = \begin{cases} 1000{\cdot}0 & \text{if } 0{\cdot}0 \leq z \leq 1{\cdot}0 \text{ or } 0{\cdot}0 \leq r \leq 0{\cdot}2, \\ 1{\cdot}0 & \text{if } 1{\cdot}0 < z \leq 2{\cdot}0 \text{ or } 0{\cdot}2 < r, \end{cases}$$

$$p(x, y, 0) = 1, \quad \boldsymbol{v}(x, y, 0) = 0 \quad \text{and} \quad r = \sqrt{(x - 0{\cdot}5)^2 + (y - 0{\cdot}5)^2 + (z - 1{\cdot}75)^2}\,.$$

Figure 28 shows the density iso-surface 500 as the droplet falls and impacts into the surface of the heavy fluid in the closed cavity.

Thermo-acoustic refrigerator. This example shows that our method can treat compressible zero-Mach-number flows with heat conduction. A thermo-acoustic refrigerator basically consists of a resonance tube, a stack of plates, two heat exchangers and an acoustic driver (usually a loudspeaker), (see, [63, 64, 65]). The basic components of a thermo-acoustic refrigerator are sketched in Figure 29: The flow within the tube is characterized by two length scales, namely the short hydrodynamic and the long acoustic scale. The Mach number in the tube is very small, typically $O(10^{-3})$. Thus, the flow between the plates, which are much shorter than the tube, can be assumed to be incompressible with a prescribed velocity field imposed on the inlet and outlet boundaries. The calculation focuses on the flow along the plate and the heat exchangers. The plate is modeled as a zero thickness plate with finite thermal mass. The thickness of the heat exchangers is zero as well. The geometry of the domain for this simplified simulation is shown in Figure 30.

The problem is defined in terms of several characteristic numbers: the Prandtl number Pr, the Reynolds number Re and the ratio of specific heats γ. The temperatures of the heat exchangers T_{hot}, T_{cold} are kept constant. The temperature distribution within the plate is governed by a heat conduction equation:

Table 3. Specific numbers.

Pr	0·68	u_{inlet}	$0{\cdot}7711\cos(t)$
Re	200	v_{inlet}	0·0
γ	5/3	u_{outlet}	$1{\cdot}0267\cos(t)$
T_{hot}	1·0267	v_{outlet}	0·0
T_{cold}	0·9733	λ	0·05
Pe_s	300	κ	41·14

$$\frac{\partial T}{\partial t} = \frac{1}{\mathrm{Pe}_s}\left(\frac{\partial^2}{\partial x^2}T + \frac{2\kappa}{\lambda}\frac{\partial}{\partial y}T\bigg|_{\text{gas}}\right). \tag{4.3.21}$$

Where Pe_s denotes the Peclet number of the solid, κ represents the ratio of the thermal conductivities and λ is the thermal penetration depth. The specific values are listed below. After 200 acoustic cycles a periodical solution is reached. Figure 31 shows the temperature at different times during the 201-th acoustic cycle. The heat fluxes through the surface of the exchangers during an acoustic cycle are shown in Figure 32.

4.3.3. *Summary*

The results discussed above show that a finite-volume compressible-flow solver can be extended to handle incompressible, zero-Mach-number flows. Our approach is general enough to include a wide variety of underlying compressible-flow schemes. The major ingredients of the required extensions are two pressure Poisson solutions. These allow us to enforce zero-Mach-number elliptic divergence constraints for the convective numerical fluxes, as well as for the final cell-centered velocity fields.

The design of the scheme directly draws on the low-Mach-number asymptotic analysis of the governing equations in conservation form presented in Section 2.1. This analysis shows how the well-known velocity- divergence constraint of incompressible flows emerges in a natural way from an associated divergence constraint on the energy flux as the Mach number vanishes. The insight gained in this way is used to construct numerical fluxes of mass, momentum and energy that are consistent with the zero-Mach-number limit. The scheme thus represents a discretization of the full conservation equations rather than one of an asymptotic limit system which would explicitly introduce a velocity-divergence constraint! The computational examples given are chosen to demonstrate various features of the proposed method. Thus, we show second-order accuracy for a test problem proposed by Almgren *et al.*, [56], and we obtain competitive results on the test problem of an advected zero circulation vortex as proposed by Gresho and Chen, [57]. After adding a first-order-in-time extension to viscous incompressible flow, we find very close agreement with published results in the literature for standard driven-cavity test problems (see Ghia, Ghia and Shin, [59]). Notably, grid refinement at constant convection-based CFL number of 0·8 as well as decreasing time steps at constant spatial resolution do not affect the results. This suggests stability and convergence of the method, even though we cannot provide rigorous proofs at this stage. Excellent behavior of the scheme is found for variable density flows. A ‘falling droplet’ with a density ratio of 1000, simulated by a suitable choice of an initial entropy distribution in an ideal gas, is handled without evidence of pressure, velocity or density oscillations.

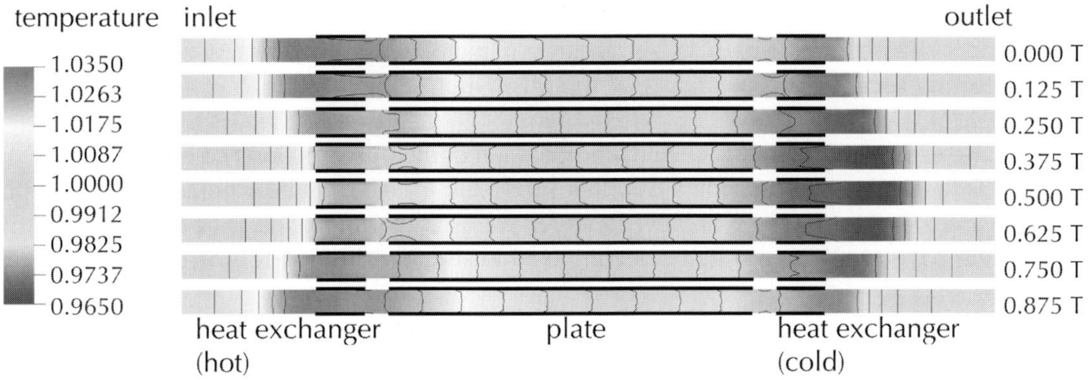

Figure 31. Temperature field during different times of an acoustic cycle; $T := 2\pi$.

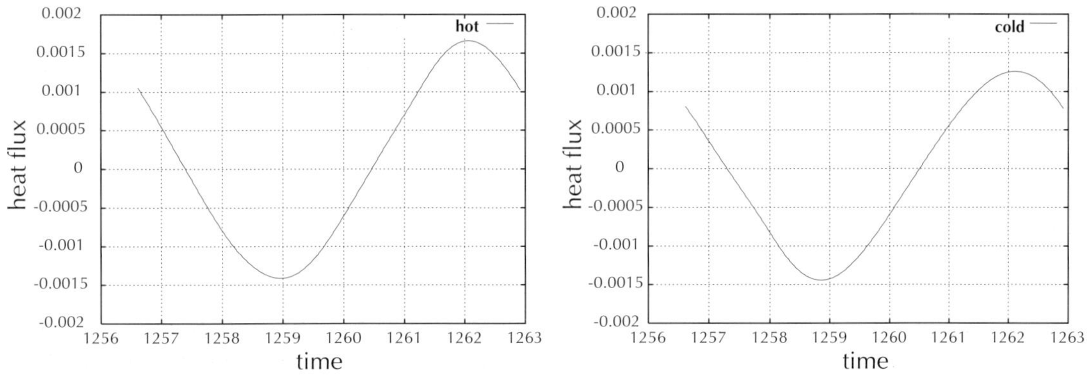

Figure 32. Heat fluxes through the surface of the hot (left) and cold (right) heat exchanger during an acoustic cycle.

4.4. A GODUNOV-TYPE SCHEME FOR WEAKLY COMPRESSIBLE FLOWS

This section is devoted to the extension of a standard conservative finite-volume method designed for the compressible Euler equations to unsteady low-Mach-number flows. Neglecting the gravitational force field, we can rewrite the conservation laws (4.3.1) in one space dimension in the form

$$\partial_t \mathbf{u}(x,t) + \partial_x \mathbf{f}(\mathbf{u}(x,t)) = \mathbf{0} \quad \text{in } \Omega \times \mathbf{R}_0^+, \tag{4.4.1}$$

where $\mathbf{u} = (\rho, m = \rho v, \rho e)^T$ denotes the vector of conserved variables and $\mathbf{f}(\mathbf{u}) = (m, mv + p/\check{M}^2, Hm)^T$ represents the convective-flux function. As seen in Section 4.3, the eigenvalues of the Jacobian of the flux function are v and $v \pm c/\check{M}^2$ with the speed of sound $c = \sqrt{\gamma \frac{p}{\rho}}$. Our aim is to construct a finite-volume (FV) Godunov-type method for (4.4.1) which is adaptive in the sense that 1) for $M \to 0$ it reduces to a FV method for zero-Mach-number flows of the kind described in Section 4.3, 2) for $M \to 1$ it becomes a standard compressible flow solver. We proceed in three steps:

- Decompose the flux function in the form
  $$\mathbf{f}(\mathbf{u}) = \mathbf{h}(\mathbf{u}) + \mathbf{a}(\mathbf{u}); \tag{4.4.2}$$
- Employ a modified Riemann solver to approximate the fluxes of the system

$$\partial_t \mathbf{u} + \partial_x \mathbf{h}(\mathbf{u}) = \mathbf{0}; \tag{4.4.3}$$

- Correct the obtained fluxes to get an approximate solution of system (4.4.1) which is uniformly accurate independently of the Mach number.

In the first step we define a suitable decomposition of the flux function **f** of Equation (4.4.2). The aim is to decompose **f** in such a way that subsystem (4.4.3) can be efficiently solved with some standard Godunov-type scheme. Moreover, we want the correction to be applied in step 3 to vanish as $\check{\mathrm{M}} \to 1$. Thus, a suitable decomposition should fulfill the following requirements

- System (4.4.3) is hyperbolic and the eigenvalues of the Jacobian of the flux function **h** are $O(1)$ as $\check{\mathrm{M}} \to 0$.

- The flux function **h** has to reproduce **f** in the case that $\check{\mathrm{M}}$ tends to one, *i.e.*, $\lim_{\check{\mathrm{M}} \to 1} \mathbf{a}(\mathbf{u}) = \mathbf{0}$.

In agreement with [6] we define the flux function **h** as follows

$$\mathbf{h}(\mathbf{u}) = \begin{pmatrix} m \\ mv + p \\ H^* m \end{pmatrix}. \tag{4.4.4}$$

Here the 'enthalpy' H^* is defined in terms of a non-local pressure p_{NL}:

$$H^* = \frac{\rho e + p_{\mathrm{NL}} + \check{\mathrm{M}}^2 p}{\rho}.$$

This pressure is computed by use of the averaging operator

$$\varphi_\sigma(t) = (\mathcal{M}_\sigma \varphi)(t) = \frac{1}{|\sigma|} \int_\sigma \varphi(x,t)\, \mathrm{d}x, \quad \forall \sigma \subset \Omega$$

as

$$p_{\mathrm{NL}} = \left(1 - \check{\mathrm{M}}^2\right) p_\Omega(0). \tag{4.4.5}$$

A straightforward calculation shows that the eigenvalues of the Jacobian of the flux function **h** are v and

$$v \pm \underbrace{\sqrt{\frac{p + (\gamma - 1)(p_{\mathrm{NL}} + \check{\mathrm{M}}^2 p)}{\rho}}}_{=:\, c^*}.$$

Moreover, the Jacobian of **h** has a complete set of eigenvectors. Furthermore, one has

$$\mathbf{a}(\mathbf{u}) = \mathbf{f}(\mathbf{u}) - \mathbf{h}(\mathbf{u}) = \begin{pmatrix} 0 \\ -\dfrac{\check{\mathrm{M}}^2 - 1}{\check{\mathrm{M}}^2} p \\ -(\check{\mathrm{M}}^2 - 1) p v - p_{\mathrm{NL}} v \end{pmatrix}. \tag{4.4.6}$$

Taking into account the definition of the non-local pressure (4.4.5) we easily see that $\mathbf{a}(\mathbf{u})$ satisfies the condition

$$\lim_{\check{M}\to 1} \mathbf{a}(\mathbf{u}) = \mathbf{0}.$$

Therefore, the decomposition $\mathbf{f}(\mathbf{u}) = \mathbf{h}(\mathbf{u}) + \mathbf{a}(\mathbf{u})$ satisfies both requirements (4.4.4) and (4.4.6). For the numerical computation of unsteady flows it is advisable to use a second-order time stepping method. Let $\mathbf{u}^n(x)$ denote an approximation of the solution vector $\mathbf{u}(x, t^n)$ of

$$\partial_t \mathbf{u} + \partial_x \mathbf{g}(\mathbf{u}) = \mathbf{0}. \tag{4.4.7}$$

Then

$$\begin{aligned} \mathbf{u}_{\mathbf{g}}^0(x) &= \mathbf{u}(x, t^0) \\ \mathbf{u}_{\mathbf{g}}^{n+1}(x) &= \mathbf{u}_{\mathbf{g}}^n(x) - \delta t \Phi_{\mathbf{g}}(\mathbf{u}, \delta t), \quad n = 0, 1, \ldots \end{aligned} \tag{4.4.8}$$

with

$$\Phi_{\mathbf{g}}(\mathbf{u}, \delta t) = -\partial_x \mathbf{g}\left(\mathbf{u}_g^n(x) - \frac{\delta t}{2} \partial_x \mathbf{g}\left(\mathbf{u}_{\mathbf{g}}^n(x)\right)\right)$$

represents a consistent second order predictor-corrector time discretization. Let $\Phi = \left(\Phi_{(\rho)}, \Phi_{(m)}, \Phi_{(\rho e)}\right)^T$ be a second-order method for the Euler equations (4.4.1) and $\Phi_{\mathbf{h}}$ a second-order method for the auxiliary system (4.4.3). The difference between the intermediate momenta of the Euler equations and those of the auxiliary system can be written as

$$m^{n+\frac{1}{2}} - m_h^{n+\frac{1}{2}} = \frac{\delta t}{2} \frac{\check{M}^2 - 1}{\check{M}^2} \partial_x p^{n+\frac{1}{2}} + O(\delta t^2). \tag{4.4.9}$$

This influence yields the asymptotic correction of the mass fluxes in the form

$$\begin{aligned} \rho^{n+1} &= \rho^n - \delta t \Phi_{(\rho)}(\mathbf{u}, \delta t) + O(\delta t^3) \\ &\overset{(4.4.9)}{=} \rho^n - \delta t \partial_x \left(m_{\mathbf{h}}^{n+\frac{1}{2}} + \frac{\delta t}{2} \frac{\check{M}^2 - 1}{\check{M}^2} \partial_x p^{n+\frac{1}{2}} \right) + O(\delta t^3) \\ &= \rho^n - \delta t \Phi_{\mathbf{h},(\rho)}(\mathbf{u}, \delta t) - \delta t \Phi_{\mathbf{a},(\rho)}(\mathbf{u}, \delta t) + O(\delta t^3) \end{aligned}$$

with

$$\Phi_{\mathbf{a},(\rho)}(\mathbf{u}, \delta t) = \frac{\delta t}{2} \frac{\check{M}^2 - 1}{\check{M}^2} \partial_x^2 p^{n+\frac{1}{2}}. \tag{4.4.10}$$

The same analysis applied to the intermediate velocity distribution gives ($\rho^{n+\frac{1}{2}} = \rho_{\mathbf{h}}^{n+\frac{1}{2}}$)

$$v^{n+\frac{1}{2}} = v_{\mathbf{h}}^{n+\frac{1}{2}} + \frac{\delta t}{2\rho^{n+\frac{1}{2}}} \frac{\check{M}^2 - 1}{\check{M}^2} \partial_x p^{n+\frac{1}{2}} + O(\delta t^2).$$

Consequently, the entire momentum flux including the second-order asymptotic correction terms reads

$$m^{n+1} = m^n - \delta t \Phi_{\mathbf{h},(m)}(\mathbf{u}, \delta t) - \delta t \Phi_{\mathbf{a},(m)}(\mathbf{u}, \delta t) + O(\delta t^3)$$

where

$$\Phi_{\mathbf{a},(m)}(\mathbf{u},\delta t) = -\frac{\check{\mathrm{M}}^2-1}{\check{\mathrm{M}}^2}\partial_x p^{n+\frac{1}{2}} + \partial_x(p-p_{\mathbf{h}})^{n+\frac{1}{2}} \\ +\delta t\frac{\check{\mathrm{M}}^2-1}{\check{\mathrm{M}}^2}\partial_x\left(v_{\mathbf{h}}^{n+\frac{1}{2}}\partial_x p^{n+\frac{1}{2}}\right). \tag{4.4.11}$$

Analogously, we achieve the energy flux in the form

$$(\rho e)^{n+1} = (\rho e)^n - \delta t\Phi_{\mathbf{h},(\rho e)}(\mathbf{u},\delta t) - \delta t\Phi_{\mathbf{a},(\rho e)}(\mathbf{u},\delta t) + O(\delta t^3) \tag{4.4.12}$$

where

$$\Phi_{\mathbf{a},(\rho e)}(\mathbf{u},\delta t) \\ = -(\check{\mathrm{M}}^2-1)\partial_x(pv)^{n+\frac{1}{2}}\partial_x(p_{\mathrm{NL}}v)^{n+\frac{1}{2}} \\ +\check{\mathrm{M}}^2\partial_x\left((p^{n+\frac{1}{2}}-p_{\mathbf{h}}^{n+\frac{1}{2}})v^{n+\frac{1}{2}}\right) + \frac{\delta t}{2}\frac{\check{\mathrm{M}}^2-1}{\check{\mathrm{M}}^2}\partial_x(H_{\mathbf{h}}^{*,n+\frac{1}{2}}\partial_x p^{n+\frac{1}{2}}) \\ +\frac{\delta t}{2}\partial_x\left(\left((\check{\mathrm{M}}^2-1)\partial_x(pv)^{n+\frac{1}{2}} + \partial_x(p_{\mathrm{NL}}v)^{n+\frac{1}{2}}\right)v^{n+\frac{1}{2}}\right). \tag{4.4.13}$$

Two facts are remarkable. First, the additional terms representing the differences between $\Phi_{\mathbf{a}}$ and a straightforward discretization of the asymptotic fluxes $\mathbf{a}(\mathbf{u})$ are of order $O(\delta t^2)$ and hence negligible in the case of a first-order time-stepping procedure. Second, the above-mentioned additional terms can be substituted by an arbitrary first-order approximation without affecting the order of accuracy. For example, one can evaluate the pressure derivative within equation (4.4.9) at any time level $t^m \in [t^n, t^{n+1}]$.

4.4.1. *Evolution of the spatially homogeneous leading-order pressure*

According to the results of the asymptotic single- as well as multiple-scale analysis, the evolution of the spatial homogeneous leading-order pressure $p^{(0)}$ is due to compression from the boundary $\partial\Omega$ or expansion of the reacting gas itself. In agreement with Equation (2.1.38) we employ

$$p^{(0),n+1} = p^{(0),n} - \delta t\gamma p^{(0),n}\left(v_r^n - v_\ell^n\right), \tag{4.4.14}$$

where v_r and v_ℓ represent the velocity at the right and left boundary of the computational domain, respectively.

4.4.2. *Evolution of the acoustic first-order pressure*

The evolution of the acoustic pressure $p^{(1)}$ is governed by the system of linearized acoustics (2.1.41), which can be rewritten in the form

$$\partial_t\overline{m} + \frac{1}{\check{\mathrm{M}}}\partial_x p^{(1)} = 0, \quad \partial_t p^{(1)} + \frac{1}{\check{\mathrm{M}}}\partial_x(\overline{c}\,\overline{m}) = 0, \tag{4.4.15}$$

where

$$\bar{c} = \sqrt{\gamma \frac{p^{(0)}}{\bar{\rho}}}.$$

The system (4.4.15) is hyperbolic with characteristic signal speeds $\pm\frac{\bar{c}}{\check{M}} = O\left(\frac{1}{\check{M}}\right)$. Since the time-step size with explicit discretization of the auxiliary system satisfies $\delta t = O(\delta x)$, it is necessary to employ a suitable discretization technique for the solution of system (4.4.15) to overcome the restrictive Courant-Friedrichs-Lewy condition for an explicit forward Euler time discretization. Implicit time-stepping schemes have the advantage that the numerical domain of dependence always covers the physical one and hence, these methods inherently fulfill the CFL-condition. Due to the fortunate property that the Jacobian of the flux function within the system (4.4.15) is very simple, we decided to employ an implicit discretization. The linear system arising from the linearization of the flux function is solved by means of the BiCGSTAB method [66] preconditioned by an incomplete LU factorization technique. The initialization of the long-wave momentum and density distribution is performed by the decomposition algorithm described in Section 4.1.

4.4.3. *Energy-Poisson-Equation*

Introducing the equation of state written as

$$(\rho e)^{n+1} = \frac{p^{(0),n+1}}{\gamma - 1} + \frac{\check{M} p^{(1),n+1}}{\gamma - 1} + \frac{\check{M}^2 p^{(2),n+1}}{\gamma - 1} + \frac{\check{M}^2}{2} \rho^{n+1} \left(v^{n+1}\right)^2$$

into the energy equation (4.4.12) and omitting terms of order $O(\delta t^3)$ we have

$$\begin{aligned} \frac{\check{M}^2}{\gamma - 1} p^{(2),n+1} = {} & (\rho e)^n - \frac{p^{(0),n+1}}{\gamma - 1} - \frac{\check{M} p^{(1),n+1}}{\gamma - 1} - \frac{\check{M}^2}{2} \rho^{n+1} \left(v^{n+1}\right)^2 \\ & - \delta t \Phi_{\mathbf{h},(\rho e)} (\mathbf{u}, \delta t) - \delta t \Phi_{\mathbf{a},(\rho e)} (\mathbf{u}, \delta t) . \end{aligned}$$

Note that the leading-order pressure $p^{(0),n+1}$ and the acoustic pressure $p^{(1),n+1}$ can be computed by means of (4.4.14) and by solving the system of linearized acoustics (4.4.15), respectively. We use a particular implicit discretization of the asymptotic part (4.4.13):

$$\begin{aligned} & \Phi_{\mathbf{a},(\rho e)} (\mathbf{u}, \delta t) \\ & = -(\check{M}^2 - 1)\partial_x (p v)^{n+\frac{1}{2}} - \partial_x (p_{\mathrm{NL}} v)^{n+\frac{1}{2}} \\ & \quad + \check{M}^2 \partial_x \left((p^{n+\frac{1}{2}} - p_{\mathbf{h}}^{n+\frac{1}{2}}) v^{n+1} \right) + \frac{\delta t}{2} \frac{\check{M}^2 - 1}{\check{M}^2} \partial_x (H_{\mathbf{h}}^{*,n+\frac{1}{2}} \partial_x p^{n+1}) \\ & \quad + \frac{\delta t}{2} \partial_x \left(\left((\check{M}^2 - 1)\partial_x (p v)^{n+1} + \partial_x (p_{\mathrm{NL}} v)^{n+1} \right) v^{n+1} \right) . \end{aligned} \tag{4.4.16}$$

Using the pressure decomposition in the asymptotic flux function (4.4.16) and assuming that the pressure terms $p^{n+\frac{1}{2}}$ and $p_{\mathbf{h}}^{n+\frac{1}{2}}$ as well as the enthalpy form $H_{\mathbf{h}}^{*,n+\frac{1}{2}}$ are available, we can discretize the derivatives within the asymptotic fluxes by means of central differences to achieve a linear system of equations

$$\mathbf{A}\mathbf{p}^{(2),n+1} = \mathbf{b}, \quad \mathbf{p}^{(2),n+1} = \left(p_1^{(2),n+1}, \ldots, p_k^{(2),n+1}\right)^T \tag{4.4.17}$$

for the calculation of the second-order pressure. The matrix and the right- hand side of Equation (4.4.17) contain density and velocity contributions at the time level t^{n+1}. Due to this fact we employ a fix-point iteration where the asymptotic flux corrections $\Phi_{\mathbf{a}}$ are computed simultaneously. Therefore, we take $v_{\mathbf{h}}^{n+1}$ and $\rho_{\mathbf{h}}^{n+1}$, provided by the discretization of the auxiliary system as the initial guess for the velocity and density (*i.e.* $v^{n+1,0} = v_{\mathbf{h}}^{n+1}$, $\rho^{n+1,0} = \rho_{\mathbf{h}}^{n+1}$), respectively. Now, the matrix and the right-hand side are fixed and one can solve the system (4.4.17) to obtain $p^{(2),n+1,0}$. Using the variables $\rho^{n+1,0}$, $v^{n+1,0}$, and $p^{(2),n+1,0}$, we can evaluate the asymptotic fluxes completely by means of a Crank-Nicolson approach. Clearly, a prescription for the computation of $p^{n+\frac{1}{2}}$, $p_{\mathbf{h}}^{n+\frac{1}{2}}$, and $H_{\mathbf{h}}^{*,n+\frac{1}{2}}$ is required to close the system. The pressure $p_{\mathbf{h}}^{n+\frac{1}{2}}$ and the enthalpy form $H_{\mathbf{h}}^{*,n+\frac{1}{2}}$ are given by the discretization of the auxiliary system (4.4.3) using $\tilde{\delta t} = \frac{\delta t}{2}$. In order to compute the pressure $p^{n+\frac{1}{2}}$, we use the same procedure as mentioned above for the time interval $[t^n, t^{n+\frac{1}{2}} = t^n + \tilde{\delta t}]$ where the approximation

$$
\begin{aligned}
&\Phi_{\mathbf{a},(\rho e)}(\mathbf{u}, \delta t) \\
&\quad = -(\check{\mathrm{M}}^2 - 1)\partial_x (pv)^{n+\frac{1}{2}} - \partial_x (p_{\mathrm{NL}} v)^{n+\frac{1}{2}} \\
&\qquad + \check{\mathrm{M}}^2 \partial_x \left((p^{n+\frac{1}{2}} - p_{\mathbf{h}}^{n+\frac{1}{2}}) v^{n+\frac{1}{2}} \right) + \tilde{\delta t} \frac{\check{\mathrm{M}}^2 - 1}{\check{\mathrm{M}}^2} \partial_x \left(H_{\mathbf{h}}^{*,n+\frac{1}{2}} \partial_x p^{n+\frac{1}{2}} \right) \\
&\qquad + \tilde{\delta t} \partial_x \left(\left((\check{\mathrm{M}}^2 - 1)\partial_x (pv)^{n+\frac{1}{2}} + \partial_x (p_{\mathrm{NL}} v)^{n+\frac{1}{2}} \right) v^{n+\frac{1}{2}} \right)
\end{aligned}
$$

is employed and the Crank-Nicolson time-stepping scheme for the asymptotic fluxes is replaced by a backward Euler approach.

4.4.4. *Spatial discretization*

We discretize the Euler equations (4.4.1) on a bounded domain $\Omega \subset \mathbf{R}$. The computational domain is subdivided into control volumes $\sigma_1, \ldots, \sigma_k$. First, consider the auxiliary hyperbolic system (4.4.3). We call $\mathbf{u}$ a weak solution of system (4.4.3) on Ω, if

$$\frac{\mathrm{d}}{\mathrm{d}t} \int_\sigma \mathbf{u} \, \mathrm{d}x + \mathbf{h}(\mathbf{u})\Big|_{\sigma^\ell}^{\sigma^r} = \mathbf{0}$$

holds for all control volumes $\sigma \subset \Omega$ with σ^r and σ^ℓ representing the right and left boundary of σ, respectively. To compute numerical fluxes for the cell averages

$$\mathbf{u}_i(t) = \left(\mathcal{M}_{\sigma_i} \mathbf{u}\right)(t), \tag{4.4.18}$$

we used the modified Riemann-problem solver described in [6]. In the limit $\check{\mathrm{M}} \to 1$ this flux is consistent with a standard numerical flux for the Euler equations. To achieve a second-order scheme in space and time, we use linear reconstruction of the characteristic variables in spatial directions as well as a MUSCL approach in time. A detailed description of this Godunov-type algorithm is given in [67].

The approximation of the asymptotic fluxes is done via finite differences. Each physical quantity $\varphi_{i+\frac{1}{2}}$ at the interface between two adjacent boxes σ_i and σ_{i+1} is computed as a mean value of the corresponding cell averages. Similarly, we approximate the derivatives $(\partial_x \varphi)_{i+\frac{1}{2}}$ in the form of central differences.

4.4.5. *Time-dependent scaling*

In order to gain a suitable scaling of the physical quantities one has to calculate the required reference values from a given flow field. In a standard approach the reference values are set at the initial time and kept constant throughout the computation. Here, however, we want to consider flows which may evolve through different regimes. Almost incompressible initial conditions may turn into weakly compressible flows or *vice versa*. Thus, it is necessary to implement some kind of dynamical time-dependent scaling and the reference values must adapt to the evolving numerical solution.

As seen in Section 2.1, a scaling is defined in terms of $\hat{v}_{\text{ref}}$, $\hat{p}_{\text{ref}}$ and $\hat{\rho}_{\text{ref}}$ via the Mach number $\check{\mathrm{M}}$

$$\check{\mathrm{M}} = \frac{\hat{v}_{\text{ref}}}{\sqrt{\frac{\hat{p}_{\text{ref}}}{\hat{\rho}_{\text{ref}}}}}, \tag{4.4.19}$$

which is a measure of the compressibility of the flow. As seen in Section 4.1, $\check{\mathrm{M}}$ is also a measure of the wave length of acoustic waves. We assume that a nearly incompressible flow field is characterized by the following two properties of the dimensional quantities. First, we expect that the divergence of the velocity filed is very small, and second, the flow field shows only tiny fluctuations within the pressure distribution.

The time-dependent reference values for density and pressure are determined by averaging each quantity over the whole domain Ω,

$$\hat{\rho}_{\text{ref}} = \hat{\rho}_{\text{ref}}(t) = \frac{1}{|\Omega|} \int_{\Omega} \hat{\rho}(\boldsymbol{x}, t)\, \mathrm{d}\boldsymbol{x}$$

and

$$\hat{p}_{\text{ref}} = \hat{p}_{\text{ref}}(t) = \frac{1}{|\Omega|} \int_{\Omega} \hat{p}(\boldsymbol{x}, t)\, \mathrm{d}\boldsymbol{x},$$

respectively. Using these values, we can determine the global reference of the speed of sound as

$$\hat{c}_{\text{ref}} = \sqrt{\gamma \frac{\hat{p}_{\text{ref}}}{\hat{\rho}_{\text{ref}}}}.$$

We define three indicators to measure the different properties of weakly compressible fluid flows. One of these is coupled with the pressure distribution as stated in Section 4.1. To prevent misunderstandings, the parameter which is calculated within the decomposition algorithm is called $\check{\mathrm{M}}_l$. The other two characterize the velocity flow field and are given by

$$\check{M}_v = C_v \frac{\max_{\boldsymbol{x}\in\Omega} |\hat{\boldsymbol{v}}|}{\hat{c}_{\text{ref}}}$$

and

$$\check{M}_{\text{div}} = C_{\text{div}} \sqrt{\frac{\max_{\boldsymbol{x}\in\Omega} |\partial_x \hat{\boldsymbol{v}}|}{\hat{c}_{\text{ref}}}}.$$

Now, we can formally define the reference parameter $\check{\mathrm{M}}$. In order to ensure that the standard Godunov-type solver is used in the case that one indicator is greater or equal than a critical value α, $\alpha \leq 1$, we define

$$\tilde{\check{\mathrm{M}}} = \min \left\{ \alpha, \max(\check{M}_v, \check{M}_{\mathrm{div}}, \check{\mathrm{M}}_l) \right\},$$

and a function

$$g(\tilde{\check{\mathrm{M}}}) = 1 + \frac{\frac{1}{\alpha} - 1}{\alpha^2} \tilde{\check{\mathrm{M}}}^2 .$$

The reference parameter is then given by

$$\check{\mathrm{M}} = g(\tilde{\check{\mathrm{M}}})\tilde{\check{\mathrm{M}}}, \quad \check{\mathrm{M}} \in [0, 1],$$

where Equation (4.4.19) now yields the velocity reference in the form

$$\hat{v}_{\mathrm{ref}} = \frac{\check{\mathrm{M}} \hat{c}_{\mathrm{ref}}}{\sqrt{\gamma}} .$$

We complete the scaling process by taking the reference time equal to the advective time scale:

$$\hat{t}_{\mathrm{ref}} = \frac{\hat{l}_{\mathrm{ref}}}{\hat{v}_{\mathrm{ref}}} .$$

Due to the indicators $\check{M}_v$ and $\check{M}_{\mathrm{div}}$ it is guaranteed that the dimensionless velocity field is limited,

$$\max_{x \in \Omega} |\boldsymbol{v}| = \frac{\max_{\boldsymbol{x} \in \Omega} |\hat{\boldsymbol{v}}|}{\hat{v}_{\mathrm{ref}}} = \sqrt{\gamma} \frac{\check{M}_v}{\check{\mathrm{M}}},$$

and that the divergence-free condition in the limit of a vanishing Mach number is fulfilled,

$$\max_{x \in \Omega} |\partial_x \hat{\boldsymbol{v}}| = \frac{\max_{\boldsymbol{x} \in \Omega} |\partial_x \boldsymbol{v}|}{\hat{v}_{\mathrm{ref}}} = \sqrt{\gamma} \frac{\check{M}^2_{\mathrm{div}}}{\check{\mathrm{M}}} .$$

We want to add some remarks regarding the adaptive concept. As seen before, a time-dependent adaption can easily be achieved by a simple rescaling of each physical quantity after each time step. Because the decomposition of the physical quantities has to be carried out at each time step, the computation of the indicators, as well as of the average values, is not very expensive in comparison to the costs of the decomposition. But it is possible to decrease the computational effort and avoid the rescaling of density and pressure, if their average values do not rapidly change in time. Then, only the velocity field has to be rescaled after each time step to employ the influences of a varying reference parameter $\check{\mathrm{M}}$ to the flow field.

4.4.6. *Numerical results*

In order to test the novel decomposition algorithm described in Section 4.1 we study 1) a weakly compressible flow with large density fluctuations and 2) the transition of a weakly incompressible flow into the compressible regime. These problems were originally proposed in [6].

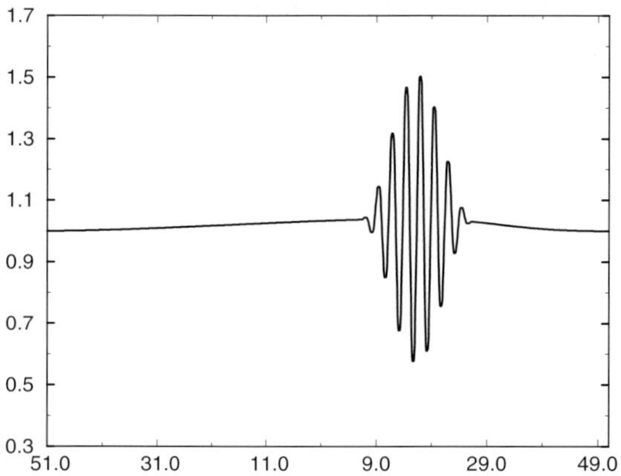

Figure 33. Density distribution at time $t = 5,071$ s.

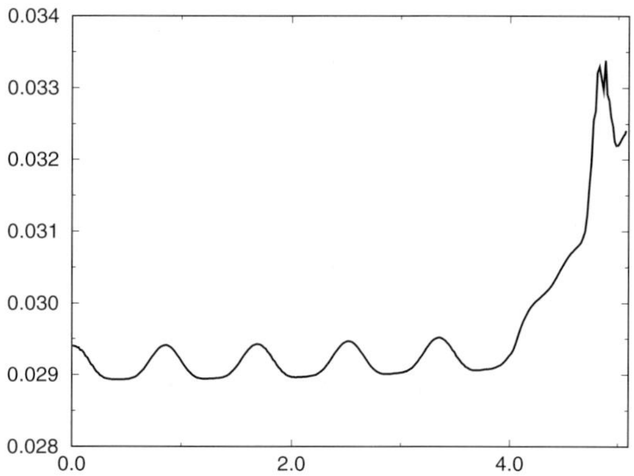

Figure 34. Time dependent adaptation of the Mach number parameter $\hat{M}$.

First, consider a long-wave acoustic signal periodically passing through a region in which the density has some short-wave-length large-amplitude fluctuations. Every period the density distribution will be displaced by the acoustic signal in one direction. The computation is carried over on $\hat{\Omega} = [-51, 51]$ which is subdivided into 1020 regular cells. The initial conditions are given by

$$\hat{\rho}(x, t = 0) = \left(1 + 2\check{\mathrm{M}}\left(1 + \cos(2\check{\mathrm{M}}\pi x)\right) + \Phi(x)\tfrac{1}{2}\sin(80\check{\mathrm{M}}\pi x)\right),$$

$$\hat{v}(x, t = 0) = \sqrt{\gamma}\left(1 + \cos(2\check{\mathrm{M}}\pi x)\right),$$

$$\hat{p}(x, t = 0) = \frac{4}{\check{\mathrm{M}}^2}\left(1 + 2\gamma\check{\mathrm{M}}\left(1 + \cos(2\pi\check{\mathrm{M}}x)\right)\right)$$

and

$$\Phi(x) = \begin{cases} \frac{1}{2}\left(1 - \cos(10\check{\mathrm{M}}\pi x)\right), & \text{if } 0 \le x \le \dfrac{2}{10\check{\mathrm{M}}}, \\ 0, & \text{otherwise.} \end{cases}$$

Periodic boundary conditions are imposed on the left and on the right end of $\hat{\Omega}$. The parameter $\check{\mathrm{M}}$ and the reference length $\hat{l}_{\mathrm{ref}}$ are set to $\frac{1}{102}$ and 1, respectively.

In Figure 33 the density profile at the end of the computation ($T = 5,071$) is plotted. The amplitudes of the short-wave density perturbation are almost preserved. The time evolution of the reference parameter $\check{\mathrm{M}}$ is shown in Figure 34. The small oscillations of $\check{\mathrm{M}}$ are due to the fact that the acoustic wave moves faster than the density distribution. Therefore, density fluctuations change their position relative to the extrema of the acoustic wave. At the beginning of the computation the indicator $\check{M}_v$ controls the progress of the reference parameter. Then, the acoustic wave slowly steepens and at the end the indicator $\check{M}_{\mathrm{div}}$ takes over.

In the second example we compute the steepening of an acoustic wave into a shock. We use a regular discretization of the domain $\hat{\Omega} = [-22, 44]$ into 330 boxes with periodic boundary conditions. The initial conditions are

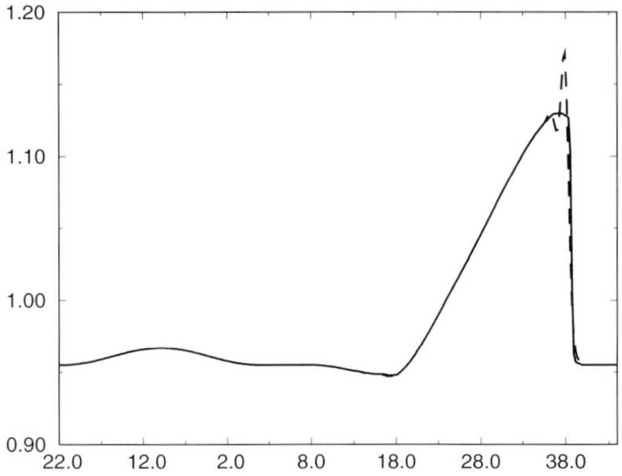

Figure 35. Density distribution at time $t = 33$ s.

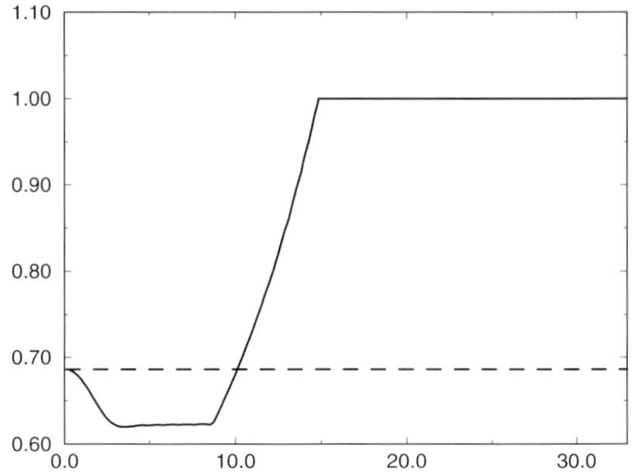

Figure 36. Evolution of the reference parameter.

$$\hat{\rho}(x, t = 0) = \left(0{\cdot}955 + \frac{1}{11}\left(1 - \cos\left(\frac{\pi x}{11}\right)\right)\right),$$

$$\hat{v}(x, t = 0) = \frac{\sqrt{\gamma}}{11}\left(1 - \cos\left(\frac{\pi x}{11}\right)\right),$$

$$\hat{p}(x, t = 0) = \left(2\gamma + \frac{\gamma}{11}\left(1 - \cos\left(\frac{\pi x}{11}\right)\right)\right),$$

for $x \leq 0$ and

$$\hat{\rho}(x, t = 0) = 0{\cdot}955, \quad \hat{v}(x, t = 0) = 0, \quad \hat{p}(x, t = 0) = 2\gamma,$$

otherwise. The reference length is $\hat{l}_{\text{ref}} = 1$. Figure 35 shows the density distribution at time $t = 33$ with and without adaptation of $\check{\mathrm{M}}$. The time evolution of $\check{\mathrm{M}}$ is given in Figure 36. The comparison between the two computations indicates the necessity of dynamically computing $\check{\mathrm{M}}$ and adapting the computational method in the case of transitions from weakly compressible to compressible flow regimes.

4.5. Hyperbolic elliptic splitting for MHD equations

In this section the aim is to use results of Section 2.2 in distinguishing between terms related with pure convection and terms related with the fast wave speeds. As in the case of Euler equations the idea behind this is to get a splitting of the system that enables us to treat the convection terms that remain hyperbolic also in the limit, and the fast-wave-speed terms, that become elliptic in the limit, with different numerical methods.

Especially for fusion plasma, the time-step stability constraint of explicit schemes may often be too restrictive and several implicit or semi-implicit schemes have been proposed (see [68], [69]) and references therein. For the case of small Mach numbers, the implicitness of our approach reduces to a simple pressure-correction equation. The convection system for the MHD equations takes the following form:

$$\rho_t + \rho_x u = 0\,, \tag{4.5.1}$$

$$u_t + uu_x + \frac{1}{2Av^2\rho}(B^2)_x = 0\,, \tag{4.5.2}$$

$$v_t + uv_x - \frac{1}{\mathrm{Av}^2\rho} B_1 B_{2x} = 0\,, \tag{4.5.3}$$

$$w_t + uw_x - \frac{1}{\mathrm{Av}^2\rho} B_1 B_{3x} = 0\,, \tag{4.5.4}$$

$$B_{2t} + B_2 u_x + u B_{2x} - B_1 v_x = 0\,, \tag{4.5.5}$$

$$B_{3t} + B_3 u_x + u B_{3x} - B_1 w_x = 0\,, \tag{4.5.6}$$

$$p_t + p_x u = 0\,. \tag{4.5.7}$$

As usual we can write Equations (4.5.1) – (4.5.7) as a system

$$q_t + Aq_x = 0 \tag{4.5.8}$$

of evolution equations with the vector

$$q = (\rho, u, v, w, B_2, B_3, p)^T \tag{4.5.9}$$

of the physical variables. The Matrix A has the eigenvalues

$$\lambda_1 = u - \frac{1}{\mathrm{Av}}\sqrt{\frac{B^2}{\rho}}\,, \qquad \lambda_2 = u - \frac{1}{\mathrm{Av}} c_A\,, \tag{4.5.10}$$

$$\lambda_3 = \lambda_4 = \lambda_5 = u\,, \tag{4.5.11}$$

$$\lambda_6 = u + \frac{1}{\mathrm{Av}} c_A\,, \qquad \lambda_7 = u + \frac{1}{\mathrm{Av}}\sqrt{\frac{B^2}{\rho}} \tag{4.5.12}$$

(c_A is the Alfvén velocity) and a complete set of eigenvectors. Therefore, system (4.5.1) – (4.5.7) is strictly hyperbolic. The eigenvalues of A correspond to the different wave speeds associated with fluid convection and the magneto-sonic waves. The remaining terms of (2.2.2) reduce to the system of the sonic waves (4.2.6). Therefore, we have all terms that cause fast wave speeds in this system. We call this system the elliptic system. The extension to multiple space dimensions is straightforward.

While the hyperbolic system (4.5.1) – (4.5.7) can be approximated with an explicit scheme, the system of the sonic waves should be discretized implicitly in time. This can be done as described in Section 4.2 for the equations of gas dynamics.

In the case of small Alfvén number the terms $B_2 u_x - B_1 v_x$ and $B_3 u_x - B_1 w_x$ become constant in space. Their time behavior is determined by the boundary values, so the information propagates with infinite velocity and may be considered to be elliptic. The leading-order terms of the derivatives of the magnetic terms in the velocity equations now vanish and B becomes constant in space. In this case the splitting of the fast and slow waves may not lead to an efficient approximation in comparison with a fully implicit scheme.

5. Conclusions

In this paper we have introduced the notion of asymptotically adaptive numerical methods. This research was motivated by the fact that singularly perturbed systems typically require very different numerical solution methods, depending on whether they are far from or very close to the singular limit. This issue becomes critical when, in some general application, the solution dynamically evolves towards or away from an asymptotic limit regime. In that case, standard numerical methods will fail one way or the other. Here we have proposed a new strategy for the construction of uniformly applicable discretizations. The approach requires a close interplay between application-oriented asymptotics and numerical analysis. Typically, one must first detect the origin of some failure of an available standard numerical scheme. This failure is then traced back to an asymptotic singular limit. Next, an asymptotic analysis is pursued, yielding valuable insight into the changes of the problem's mathematical structure as this limit regime is approached. This knowledge is then exploited to construct an extended numerical scheme that operates with uniform efficiency and accuracy for arbitrary values of the asymptotic parameter. Finally, for problems in which the asymptotic parameter itself is part of the solution, one needs additional control strategies which determine the actual instantaneous or local parameter value and trigger an automatic adjustment of the numerical discretization.

In this paper we have described a number of typical applications that call for an implementation of the above strategy. These include variable-density low-Mach-number flows, atmospheric flows under a distinguished limit of low-Froude and low-Mach numbers, and Magneto-Hydrodynamics at low Mach or Alfvén numbers. For the 'pure' low-Mach-number case, we have almost completed the program sketched above. A numerical method for fully compressible, multi-dimensional flows has been modified to also 'survive' the zero-Mach-number limit. We have proposed a numerical detection strategy that provides quantitative indicators of whether a flow is in the appropriate asymptotic regime or not. This detection scheme, currently implemented only for a single space dimension, has then be used to design an asymptotically adaptive numerical method that, in fact, operates uniformly accurately and efficiently for arbitrary Mach number. A second application of these asymptotic results has been the extension of an existing standard zero-Mach-number flow solver to the weakly compressible regime. This extension addresses, in particular, the problem of long-wavelength acoustics, which is typically ignored in competing approaches. Our *ansatz* has led to a natural kind of 'multi-gridding': In the regime considered acoustic perturbations live on large length scales only, and are thus amenable to accurate representation on relatively coarse computational grids. The quasi-incompressible, vortical component of the flow field is active on much smaller length scales and is thus represented on a finer grid.

For the atmospheric flow and magneto-hydrodynamics problems we have described asymptotic analyses that provide the necessary insight for understanding characteristic failures of standard numerical schemes. Our current efforts are directed at exploiting these results in the construction of new numerical techniques along the lines described above.

Acknowledgments

The asymptotic analysis of slow atmospheric motions has been developed together with Dr. A. S. Almgren of Lawrence Berkeley National Lab. The numerical simulation of the flow in the stack of a thermo-acoustic refrigerator is part of a collaboration with Prof. Dr. O. Knio of the Department of Mechanical Engineering of the Johns Hopkins University. We thank the Deutsche Forschungsgemeinschaft and the Potsdam Institut für Klimafolgenforschung for generously supporting our work.

References

1. W. Schneider, *Mathematische Methoden in der Strömungsmechanik*. Braunschweig: Vieweg (1979) 261 pp.
2. S. Klainerman and A. J. Majda, Compressible and incompressible fluids. *Comm. Pure Appl. Math.* 35 (1982) 629–651.
3. J. K. Hunter, A. Majda and R. Rosales, Resonantly interacting weakly nonlinear hyperbolic waves II. Several space variables. *Stud. Appl. Math.* 75 (1986) 187–226.
4. A. Majda, R. R. Rosales and M. Schönbek, A canonical system of integro-differential equations arising in resonant nonlinear acoustics. *Stud. Appl. Math.* 79 (1988) 205–262.
5. R. Klein and N. Peters, Cummulative effects of weak pressure waves during the induction period of a thermal explosion in a closed cylinder. *J. Fluid Mech.* 187 (1988) 197–230.
6. R. Klein, Semi-implicit extension of a Godunov-type scheme based on low Mach number asymptotics I: One-dimensional Flow. *J. Comp. Phys.* 121 (1995) 213–237.
7. Th. Schneider, N. Botta, R. Klein and K. J. Geratz, Extension of finite volume compressible flow sovers to multi-dimensional, variable density zero Mach number flow. *J. Comp. Phys.* 155 (1999) 248–286.
8. F. A. Williams, *Combustion Theory (2nd ed.)*. Menlo Park, CA: Benjamins Cummins (1985) 680 pp.
9. A. Majda and J. Sethian, The derivation and numerical solution of the equations for zero Mach number combustion. *Com. Sci. Techn.* 42 (1985) 185–205.
10. A. Meister, Asymptotic single and multiple scale expansions in the low Mach number limit. *SIAM J. Appl. Math.* 60 (1999) 256–271.
11. M. Oevermann, M. Berger, J. Goodman and R. Klein, A numerical method for two-phase incompressible flow with large density ratio and sharp interface resolution. *J. Comp. Phys.* (2000) submitted.
12. C. A. J. Fletcher, *Computational techniques for fluid-dynamics–Specific techniques for different flow categories*, Vol. 2. Berlin, Heidelberg: Springer (1997) 493 pp.
13. R. I. Issa, Solution of the implicily discretised fluid flow equations by operator-splitting. *J. Comp. Phys.* 62 (1985) 40–65.
14. J. Kim and P. Moin, Application of a fractional step method to incompressible Navier-Stokes equations. *J. Comp. Phys.* 59 (1985) 308–323.
15. S. V. Patankar, *Numerical Heat Transfer and Fluid Flow*. Hemisphere Publishing Corporation (1980) 197 pp.
16. H. Pichler, *Dynamik der Atmosphäre.* Mannheim, Wien, Zürich: bibliografisches Institut (1986) 459 pp.
17. A. E. Gill, *Atmosphere-Ocean Dynamics*. London: Academic Press (1982) 662 pp.
18. J. Pedlosky, *Geophysical Fluid Dynamics* (2nd ed.). Berlin, Heidelberg, New York: Springer (1987) 710 pp.
19. R. K. Zeytounian, *Asymptotic Modeling of Atmospheric Flows*. Heidelberg, Berlin, New York: Springer (1990) 396 pp.
20. R. K. Zeytounian, *Meteorological Fluid Dynamics. Lecture Notes in Physics* m5. Heidelberg, Berlin, New York: Springer (1991) 346 pp.
21. G. Doms and U. Schättler, *Nonhydrostatic Limited-Area LM (Lokal-Modell) of DWD: Part I, Scientific Documentation*. Offenbach: Deutscher Wetterdienst (1997) 169 pp.
22. G. Volpe, On the use and accuracy of cimpressible flow codes at low Mach numbers. AIAA Paper 91-1662 (1991).
23. J. L. Sesterhenn, *Zur numerischen Berechnung kompressibler Strömungen bei kleinen Mach-Zahlen*. PhD thesis Swiss Federal Institue of Technology, Diss. ETH No. 11334 (1995) 71 pp.
24. C. Viozat, Implicit upwind schemes for low Mach number compressible flows. INRIA 3084 (1997).
25. Th. Schneider, N. Botta, K. J. Geratz and R. Klein, Extension of finite volume compressible flow solvers to multi-dimensional, variable density zero Mach number flow. ZIB, Konrad-Zuse-Zentrum für Informationstechnik Berlin SC 98-31 (1998) and J. Comp. Phys. 155 (1999) 248–286.
26. N. Botta, R. Klein and A. S. Almgren, Dry atmosphere asymptotics. PIK Potsdam Institute for Climate Impact Research 55 Potsdam (1999).
27. D. R. Durran, *Numerical Methods for Wave Equations in Geophysical Fluid Dynamics*. New York, Berlin, Heidelberg: Springer (1999) 465 pp.

28. Y. Ogura and N. Phillips, Scale analysis for deep and shallow convection in the atmosphere. *J. Atmos. Sci.* 19 (1962) 173–179.
29. F. Lipps and R. Hemler, A scale analysis of deep moist convection and some related numerical calculations. *J. Atmos. Sci.* 29 (1982) 2192–2210.
30. M. van Dyke, *Perturbation Methods in Fluid Dynamics*. Stanford, CA: Parabolic Press (1975) 271 pp.
31. J. Kevorkian and J. D. Cole, *Perturbation Methods in Applied Mathematics*. New York, Heidelberg, Berlin: Springer (1980) 558 pp.
32. J. Kevorkian and J. D. Cole, *Multiple Scale and Singular Petrubation Methods*. New York, Berlin: Springer (1996) 632 pp.
33. A. Meister, Zur mathematischen Fundierung einer Mehrskalenanalyse der Euler-Gleichungen. *Hamburger Beiträge zur Angewandten Mathematik, Reihe F, Computational Fluid Dynamics and Data Analysis 1* (1997).
34. S. Schochet, Resonant nonlinear geometric optics for weak solutions of conversation laws. *J. Diff. Eqs.* 113 (1994) 106–231.
35. J. L. Joly, G. Métivier and J. Rauch, Resonant one-dimensional nonlinear geometric optics. *J. Funct. Anal.* 114 (1993) 106–231.
36. J. Weickert, A review of nonlinear diffusion filtering. *Scale-Space Theory in Computer Visions, Lecture Notes in Comp. Sciences* 1252 (1997) 3–28.
37. J. L. Latecki and R. Lakämper, Convexity rule for shape decomposition based on discrete contour evolution. *Comp. Vision Image Understanding* 73 (1999) 441–454.
38. J. L. Latecki and R. Lakämper, Polygon evolution by vertex deletion (1999). In preparation.
39. C.-D. Munz, S. Roller, R. Klein and K. J. Geratz, The extension of incompressible flow solvers to the weakly compressible regime. submitted to Comp. and Fluids (2000).
40. A. J. Chorin, A numerical method for solving incompressible viscous flow problems. *J. Comp. Phys.* 2 (1967) 12–26.
41. A. J. Chorin, Numerical solution of the Navier-Stokes equations. *J. Comp. Phys.* 22 (1968) 745–762.
42. R. Temam, Sur l'approximation de la solution des equations de Navier-Stokes par la methode des pas fractionaires. *Arch. Rat. Mech. Anal.* 32 (1969) 135–153.
43. P. M. Gresho, On the theory of semi-implicit projection methods for viscous incompressible flows and its implementation via a finite element method that also introduces a nearly consistent mass matrix. Part 1: Theory. *Int. J. Num. Meth. Fluids* 11 (1990) 587–620.
44. P. M. Gresho, On the theory of semi-implicit projection methods for viscous incompressible flows and its implementation via a finite element method that also introduces a nearly consistent mass matrix. Part 2: Implementation. *Int. J. Num. Meth. Fluids* 11 (1990) 621–659.
45. U. Ghia, K. Ghia and C. Shin, High-Re solutions for incompressible flow using the Navier-Stokes equations and a multigrid method. *J. Comp. Physics* 48 (1982) 387–411.
46. S. Schochet, Asymptotics for symmetric hyperbolic systems with a large parameter. *J. Diff. Eqs.* 75 (1988) 1–27.
47. R. Courant, K. O. Friedrichs and H. Lewy, Über die partiellen Differenzengleichungen der Physik. *Math. Ann.* 100 (1928) 32.
48. B. van Leer, Towards the ultimate conservative difference scheme. I. The quest of monotonicity. *Lect. Notes in Phys.* 18 (1973) 163–168.
49. B. van Leer, Towards the ultimate conservative difference scheme. II. Monotonicity and conservation combined in a second-order scheme. *J. Comp. Phys.* 14 (1974) 361–370.
50. B. van Leer, Towards the ultimate conservative difference scheme. III. Upstream-centered finite-difference schemes for ideal compressible flow. *J. Comp. Phys.* 23 (1977) 263–275.
51. B. van Leer, Towards the ultimate conservative difference scheme. IV. A new approach to nummerical convection. *J. Comp. Phys.* 23 (1977) 276–299.
52. B. van Leer, Towards the ultimate conservative difference scheme. V. A second-order sequel to Godunov's method. *J. Comp. Phys.* 32 (1979) 101–136.
53. B. Einfeldt, On Godunov type methods for gas dynamics. Inst. f. Geometrie u. Praktische Mathematik, RWTH Aachen, Bericht Nr. 41 (1986).
54. K. J. Geratz, *Erweiterung eines Godunov-Typ-Verfahrens für zwei-dimensionale kompressible Strömungen auf die Fälle kleiner und verschwindender Machzahl*. PhD thesis Rheinisch-Westfälische Technische Hochschule Aachen, D 82 (1997) 101 pp.

55. P. Wesseling, *An Introduction to multigrid methods*. Chichester: John Wiley & Sons (1991) 284 pp.
56. A. S. Almgren, J. B. Bell, P. Colella, L. H. Howell and M. L. Welcome, A conservative adaptive projection method for the variable density incompressible Navier-Stokes equations. Ernest Orlando Lawrence Berkeley National Laboratory LBNL-39075 (1996).
57. P. M. Gresho and S. T. Chan, On the theory of semi-implicit projection methods for viscous incompressible flow and its implementation via a finite element method that also introduces a nearly consistent mass matrix. Part 2: Implementation. *Int. J. Num. Meth. Fluids* 11 (1990) 621–659.
58. C. W. Schulz-Rinne, *The Riemann problem for two-dimensional gas dynamics and new limiters for high-order schemes*. PhD thesis Swiss Federal Institute of Technology, Diss. ETH No. 10297 (1993) 122 pp.
59. K. N. Ghia, U. Ghia and C. T. Shin, High-Resolutions for incompressible flow using the Navier-Stokes equations and a multigrid method. *J. Comp. Phys.* 48 (1982) 387–411.
60. W. Y. Soh and J. W. Goodrich, Unsteady solution of the incompressible Navier-Stokes equations. *J. Comp. Phys.* 79 (1988) 113–134.
61. C. H. Bruneau and C. Jouron, an efficient scheme for solving steady incompressible Navier-Stokes equations. *J. Comp. Phys.* 89 (1990) 389–413.
62. E. G. Puckett, A. S. Almgren, J. B. Bell, D. L. Marcus and W. J. Rider, A high-order projection method for tracking fluid interfaces in variable density incompressible flow. *J. Comp. Phys.* 130 (1997) 269–282.
63. A. S. Worlikar, O. M. Knio and R. Klein, Numerical simulation of a thermoacoustic refigerator. II stratified flow around the stack. *J. Comp. Phys.* 144 (1998) 299–324.
64. A. S. Worlikar and O. M. Knio, Numerical study of oscillatory flow and heat transfer in a loaded thermoacoustic stack. *Num. Heat Transf.* 35 (1999) 49–65.
65. Th. Schneider, R. Klein, E. Besnoin and O. Knio, Computational analysis of a thermoacoustic refrigerator. In: *Proceedings of the joint EAA/ASA meeting* March 1999.
66. H. A. van der Vorst, BI-CGSTAB: A fast and smoothly converging variant of BI-CG for the solution of nonsymmetric linear systems. *SIAM J. Sci. Stat. Comp.* 13 (1992) 631–644.
67. E. F. Toro, *Riemann Solvers and Numerical Methods for Fluid Dynamics*. Berlin, Heidelberg, New York: Springer (1997) 624 pp.
68. K. Lerbinger and J. F. Luciani, a new semi-implicit method for MHD computations. *J. Comp. Phys.* 97 (1991) 444–459.
69. O. S. Jones, U. Shumlak and D. S. Eberhardt, An implicit scheme for nonideal magnetohydrodynamics. *J. Comp. Physics* 130 (1997) 231–242.

Journal of Engineering Mathematics **39:** 345–366, 2001.

Asymptotic analysis of the flow of shear-thinning foodstuffs in annular scraped heat exchangers

A. D. FITT and C. P. PLEASE
Faculty of Mathematical Studies, University of Southampton, Southampton SO17 1BJ, UK (e-mail: adf@maths.soton.ac.uk; cpp@maths.soton.ac.uk)

Received 22 February 2000; accepted in revised form 10 August 2000

Abstract. The problem of isothermal flow of a shear-thinning (pseudoplastic) fluid in the gap between two concentric cylinders is considered. A pump provides an axial pressure gradient which causes flow down the device. The outer cylinder is fixed and has 'scrapers' attached to it to cause flow mixing, whilst the inner cylinder rotates about its axis to provide shear and thus thin the fluid. The goal is to determine the optimal distribution of power between rotation and pumping. Although ostensibly the flow is nonlinear and three-dimensional we show that judicious use of fairly straightforward asymptotic methods can yield a great deal of information about the device, including cross-sectional flow predictions and throughput results. Furthermore, these results are derived for a variety of different flow conditions. Some numerical calculations are carried out using a commercial CFD code. These show good agreement with the asymptotic analysis.

Key words: asymptotic analysis, food industry, lubrication theory, shear-thinning fluids, slow flow.

1. Introduction

In this study we discuss a problem connected with a process commonly employed in the food manufacturing industry. Since the problem ostensibly concerns a non-Newtonian fluid flowing in a complex, moving geometrical region in three dimensions, one might expect that the only way of deriving results of any practical use would be to proceed along purely numerical lines. We shall show, however, that by judicious use of asymptotic analysis the problem *as originally posed* may be virtually completely solved. It is the nature of real-world problems that when questions about an industrial process are posed and answered, inevitably they lead only to additional more detailed and difficult questions; we shall indicate how further aspects of the problem may be examined and what mathematical and numerical tools might be utilised.

It is worth emphasising that one's expectations of what asymptotic analysis might be able to achieve for practical problems must be constantly tempered with experience and realism. It would be absurd to suggest that all problems may be completely solved using asymptotic analysis alone. In most industrial settings it will eventually be necessary to employ numerical computation to fully analyse a process. This computation may be carried out using purpose-designed software or 'black box' CFD packages. This study should therefore be thought of as in no way belittling the role that numerical analysis has to play; rather it puts forward the views that (i) a great deal of insight can be gained from carrying out carefully considered asymptotic analysis and (ii) that in an ideal world *a combination* of asymptotic and numerical analysis should be used to attack a given problem.

In order to illustrate how effective asymptotic analysis can be in tackling complicated problems it is tempting to "invent" a superficially practical problem where all of the parameter values are chosen so that asymptotic analysis succeeds in every subproblem. We have resisted this temptation however and considered a real industrial problem as presented by a real industrial company. Although the fluid mechanics, rheology and asymptotic analysis that are required to make progress are not in themselves too testing, the problem provides a graphic illustration of the power of asymptotics.

1.1. Industrial Background

The problem discussed below was brought to our attention during a meeting between representatives from Crown Chemtech Ltd. (a small company based in Reading, Berkshire, UK which makes machinery for food processing), members of the Food Sciences Group at the University of Reading and members of the Departments of Mathematics at the Universities of Reading and Southampton.

In its simplest form, the problem may be stated thus: a fatty oil flows in the annular gap between two concentric circular cylinders, driven by a known axial pressure gradient: predict the throughput for given device parameters. In the current setting, however, a number of complicating factors arise. Firstly, the fatty fluid is to be turned (for example) into margarine. Essentially, this is done by cooling the fluid as it progresses down the annulus, which therefore acts as a heat exchanger. At the upstream end of the device, the product may thus truly be regarded as a fluid, but at exit it more resembles a paste. Secondly, the fluid is non-Newtonian and has a temperature-dependent viscosity. In particular, it is known that a good model for the rheology of the fluid is to assume that it is a power-law shear-thinning fluid with zero yield stress and a known shear-thinning exponent. Thirdly, the inner cylinder rotates with a given angular velocity Ω and the outer cylinder has a number of 'scrapers' (knife blades) that are attached normally to it.

Clearly there are many different aspects of this problem that could be studied. In the first instance, however, our industrial partner wished to neglect altogether the impact of the heat exchange process on the flow, and consider the whole process to be isothermal. (Details of the thermal problem may be found in [1].) Also, it was clear that, although the present device was concerned with the manufacture of margarine, it could also be used for a wide range of other food products, including molten chocolate, jam, jelly and spreads such as peanut butter (all of which may be considered to be power-law shear-thinning fluids with a zero yield stress ([2])).

The shear-thinning nature of liquid food products has an important effect on the prediction and optimisation of device throughput, since it may be exploited to increase production. Indeed, if the flow in such devices were to be generated *only* by a pressure gradient, it is known that unacceptably large pressure gradients would be required to achieve the desired throughput. The general idea therefore is to divert some energy away from the pumps producing the pressure gradient in the annulus and use it instead to rotate the inner cylinder. Evidently by doing this the shear in the fluid will be increased, leading to a thinning of the fluid that will greatly ease its passage down the device.

In the absence of heat transfer, we shall therefore suppose that a given amount of power is available to run the device and assume that the point at issue is one of optimisation, namely how much of the power should be used for rotation of the inner cylinder and how much should be invested in producing a pressure gradient from the pump. (Evidently an optimum use of energy must exist, for whilst *some* rotation of the inner cylinder is known to increase the throughput it is clear that if all of the energy were to be used in cylinder rotation then the throughput would be zero.)

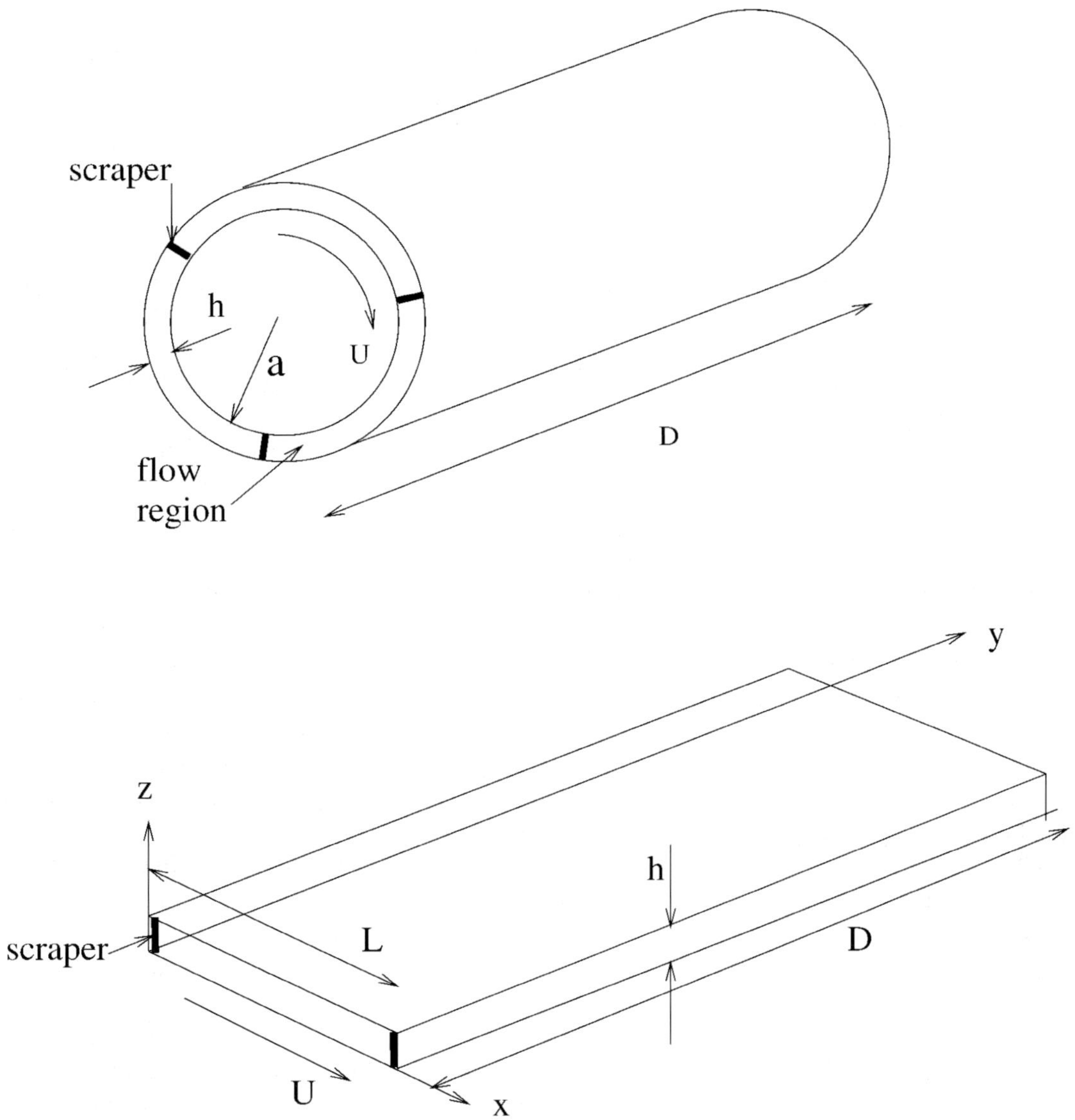

Figure 1. (Top) Schematic view of device (bottom) schematic view of 'unwrapped' device.

2. Mathematical modelling

Schematic diagrams of the device and the coordinate system that will be employed are shown in Figure 1. The radius of the inner cylinder is denoted by a, the (linear) rotation speed of the inner cylinder by $U = \Omega a$, the length of the device by D, and the annular gap width by h. We also denote the 'unwrapped' distance between two scrapers by L. Kinematic viscosity and density are denoted by ν and ρ, respectively, and t_R and V denote a typical residence time in the machine and a typical speed in the y-direction, respectively.

2.1. NUMERICAL PROBLEM

We begin by stating the full problem. This would have to be solved numerically if we decided to analyse the device without using any form of theoretical or asymptotic analysis.

The flow is governed by the three-dimensional unsteady Navier-Stokes equations

$$\rho \frac{\mathrm{d}\boldsymbol{q}}{\mathrm{d}t} = \operatorname{div} T, \tag{1}$$

$$\operatorname{div} \boldsymbol{q} = 0, \tag{2}$$

where the fluid velocity is denoted by $\boldsymbol{q}$, T is the stress tensor and we have assumed that there are no body forces. Following the information given, we treat the fluid as an isothermal shear-thinning fluid with zero yield stress and use the constitutive law (see, for example [3, p. 55]).

$$T = -p\delta_{ij} + \Lambda \mid (d_{kl}d_{kl})^{\frac{m-1}{2}} \mid 2d_{ij}, \tag{3}$$

where p denotes pressure, d_{ij} is the infinitesimal rate-of-strain tensor given by

$$d_{ij} = \tfrac{1}{2}(\nabla \boldsymbol{q} + \nabla \boldsymbol{q}^T),$$

$m \in (0, 1]$ is the 'shear-thinning index' (the model is still valid for shear *thickening* fluids such as corn starch where $m > 1$, but we do not pursue this further here) and Λ is a constant that is usually referred to as the 'consistency' of the fluid (which has dimensions s^{m-2} kg m^{-1}).

The full numerical problem thus consists of solving the Equations (1) and (2) with the constitutive law (3), subject to the boundary conditions

$$\boldsymbol{q} = 0 \quad \text{at} \quad r = a + h; \quad \boldsymbol{q} = U\boldsymbol{e}_\theta \quad \text{at} \quad r = a,$$

(where $\boldsymbol{e}_\theta$ is a unit vector in the θ-direction),

$$\boldsymbol{q} = 0 \quad \text{on the surface of each scraper.}$$

To complete the problem specification, it is also necessary to prescribe suitable initial conditions and to specify conditions at $y = 0$ and $y = D$.

Evidently the numerical problem is demanding. Even if we avoid consideration of start-up and spin-down processes and examine only steady flow, the problem is essentially three-dimensional and contains moving boundaries. The boundary conditions that must be imposed at the ends $y = 0$ and $y = D$ of the device are awkward, for though it might not be too hard to enforce a zero stress condition at $y = D$, the correct condition at the inlet to the device seems to be that the normal stress is given (in practice this is probably what would be measured as the 'pressure') and the other two components of the stress are zero. At the inlet, therefore, no velocity profiles are known. The numerical problem also involves all the usual complications due to the convective derivative terms and the fact that the pressure occurs in the equations only in the guise of a Lagrange multiplier. Furthermore, the nonlinearity of the fluid must be taken into consideration. There are a number of 'black box' commercial codes that claim to be able to successfully carry out calculations of this sort, but even with the power of today's PCs, it is doubtful whether three-dimensional runs would be able to be carried out in the sort of execution times that would make parametric studies possible.

As an alternative to large-scale computing, we therefore examine the problem from an asymptotic-analysis point of view. It is worth observing that, though some previous authors have concerned themselves with two-dimensional lubrication theory for slider bearings (see, for example [4–7]), not much attention seems to have been given to three-dimensional problems involving shear-thinning fluid flow. The flow of a Bingham fluid in eccentric annular geometries was studied in [8] in a case where the inner cylinder was at rest. The resulting simple shear profile allowed mathematical progress to be made. A modification of this study was presented in [9], where a system with concentric cylinders was examined but the inner cylinder was allowed to rotate. Analytical progress could be made for a paradigm problem with a one-dimensional shear dependence, but for the general problem a numerical solution was required. Both [8] and [9] addressed the problem of the existence or otherwise of plug flow regions.

3. Asymptotic analysis

We begin by examining the sizes of the typical problem parameters and non-dimensionalizing (1) and (2). According to the device parameters supplied by Crown Chemtech. Ltd.,

$$U \sim 2\ \mathrm{m\,s^{-1}}, \quad h \sim 0{\cdot}015\ \mathrm{m}, \quad D \sim 2\ \mathrm{m}, \quad a \sim 0{\cdot}075\ \mathrm{m},$$

$$\nu \sim 0{\cdot}1\ \mathrm{m^2\,s^{-1}}, \quad \rho \sim 800\ \mathrm{kg\,m^{-3}},$$

so that the cylinder circumference is about 0·47 m. Assuming (since this is the standard configuration) that there are three scrapers, we thus have

$$L \sim 0{\cdot}15\ \mathrm{m}.$$

A typical residence time t_R is about 10s and hence

$$V \sim 0{\cdot}2\ \mathrm{m\,s^{-1}}.$$

We scale according to $t = (L/U)\bar{t}$, $x = L\bar{x}$, $y = (L/\epsilon)\bar{y}$, $z = \epsilon L\bar{z}$, $u = U\bar{u}$, $v = \epsilon U\bar{v}$, $w = \epsilon U\bar{w}$ where $\epsilon = h/L$ and the bars denote non-dimensional quantities. The scalings for t, u, w, x and z are straightforward, and those used for v and y (which are motivated by the fact that the pipe is long compared to its radius) lead to a distinguished limit that is suggested by the values of the physical parameters in the problem. We also anticipate that lubrication theory-type analysis will be appropriate and set $p = (U/h)^{m-1}(\mu_0 U/L\epsilon^2)\bar{p}$ where for convenience we have defined $\mu_o = \Lambda 2^{(1-m)/2}$.

The next task is to insert the above scalings into (1) and (2) and determine the dominant terms in the equations. This procedure should be carried out on the full shear-thinning Navier–Stokes equations in cylindrical coordinates. We omit the details of this as they are well-known and simply note the result, which is that the (dimensional) leading-order problem is

$$0 = -p_x + (\mu_0(u_z^2)^{\frac{m-1}{2}} u_z)_z, \tag{4}$$

$$0 = -p_y + (\mu_0(u_z^2)^{\frac{m-1}{2}} v_z)_z, \tag{5}$$

$$0 = -p_z, \tag{6}$$

$$0 = u_x + w_z, \tag{7}$$

these equations being valid under the assumptions that

$$\epsilon \ll 1, \quad \epsilon^2 \mathrm{Re} \ll 1,$$

where the equivalent of the Reynolds number for shear-thinning flow is defined by

$$\mathrm{Re} = \frac{LU\rho}{\mu_0}\left(\frac{h}{U}\right)^{m-1}.$$

In the interests of clarity, all subsequent equations will be presented in dimensional form.

With the parameter values suggested above, (assuming Newtonian flow with $m = 1$ and $\mu_0/\rho = \nu$) we find that

$$\epsilon \sim 10^{-1}, \quad \mathrm{Re} \sim 3 \quad \epsilon^2\mathrm{Re} \sim 0{\cdot}03,$$

thus justifying the neglect of the other terms in the full equations. Inertia is absent from (4–7) (which apply for either steady or unsteady flow, though we shall only consider steady flow), the leading-order balance being between viscous forces and the pressure gradient. All centrifugal forces are absent. One other point concerning (4–7) is worthy of note: although for convenience we have omitted the modulus signs in the viscosity terms, it is understood that the viscosity must be positive, so that some care must be taken whenever terms such as $\sqrt{u_z^2}$ are simplified.

We now proceed to solve (4–7) with appropriate boundary conditions. The fact that p is independent of z allows (4) to be solved with little difficulty; we divide the region $0 \le z \le h$ into two subregions $[0, z_c]$ and $[z_c, h]$ where z_c (which is to be determined) is defined so that $u_z \le 0$ for $z \in [0, z_c]$ and $u_z \ge 0$ for $z \in [z_c, h]$ (see Figure 2 for a typical velocity profile). Solving the x-momentum equation (4) by integration, we impose the boundary conditions

$$u_z(x, y, z_c) = 0, \quad u(x, y, 0) = U, \quad u(x, y, h) = 0.$$

This shows that

$$u = \begin{cases} \left(\dfrac{m}{1+m}\right)\left(\dfrac{p_x}{\mu_0}\right)^{\frac{1}{m}}[(z_c - z)^{\frac{1+m}{m}} - z_c^{\frac{1+m}{m}}] + U & (z \le z_c) \\ \left(\dfrac{m}{1+m}\right)\left(\dfrac{p_x}{\mu_0}\right)^{\frac{1}{m}}[(z - z_c)^{\frac{1+m}{m}} - (h - z_c)^{\frac{1+m}{m}}] & (z \ge z_c) \end{cases}. \tag{8}$$

We must also ensure that u is continuous at $z = z_c$; this gives

$$\left(\frac{m}{1+m}\right)\left(\frac{p_x}{\mu_0}\right)^{\frac{1}{m}}(z_c^{\frac{1+m}{m}} - (h - z_c)^{\frac{1+m}{m}}) = U, \tag{9}$$

which we shall return to presently.

At this point, two elements are missing from the model, for p_x is yet to be determined and the effect of the scrapers has not yet been considered. The pressure gradient is easily determined in the normal way by using the (continuity) equation (7). By integrating from $z = 0$ to h, we find that, since w is zero on both $z = 0$ and $z = h$, we have

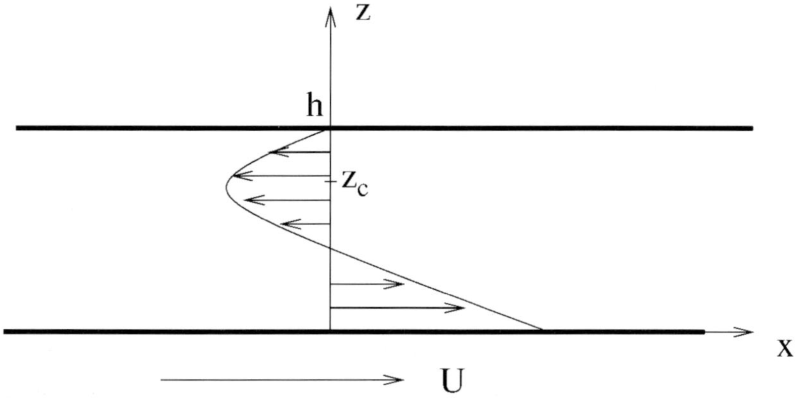

Figure 2. Typical velocity profile for flow in annular gap between scrapers.

$$\int_0^h u_x \, dz = 0 \tag{10}$$

and thus (using (8))

$$\left(\frac{1}{1+m}\right)\left(\frac{1}{\mu_0}\right)^{\frac{1}{m}} p_x^{\frac{1}{m}-1} p_{xx} I_0 = 0,$$

where I_0 is an integral which is not zero. Thus $p_{xx} = 0$ and so p_x is independent of x.

Now we consider the effect of the scrapers: we assume that each scraper does not allow flow across it (though see Section 5) and restricts the fluid between scrapers to that region. To model the scrapers in a form suitable for lubrication theory we therefore impose the weaker condition that

$$\int_0^h u \, dz = 0 \tag{11}$$

at $x = 0$ and $x = L$. (Of course, lubrication theory is not valid in regions immediately adjacent to scrapers; here, where the flow changes direction, the full Navier Stokes equations with the appropriate boundary conditions must be solved. This will make no appreciable difference to the throughput qualities of the device, however. Lubrication theory may also not be valid near to the input and output regions of the device.) Applying (11) to (8) now gives

$$\left(\frac{m}{1+2m}\right)\left(\frac{p_x}{\mu_0}\right)^{\frac{1}{m}} [-z_c^{\frac{1+2m}{m}} - (h - z_c)^{\frac{1+2m}{m}}] + U z_c = 0 \tag{12}$$

and when (9) is combined with (12), we find that

$$1 = \frac{z_c(1+2m)[z_c^{\frac{1+m}{m}} - (h - z_c)^{\frac{1+m}{m}}]}{(1+m)[z_c^{\frac{1+2m}{m}} + (h - z_c)^{\frac{1+2m}{m}}]}.$$

The position z_c is thus determined *only* by the shear-thinning index m and satisfies

$$1 = \frac{\lambda(1+2m)[\lambda^{\frac{1+m}{m}} - (1 - \lambda)^{\frac{1+m}{m}}]}{(1+m)[\lambda^{\frac{1+2m}{m}} + (1 - \lambda)^{\frac{1+2m}{m}}]}, \tag{13}$$

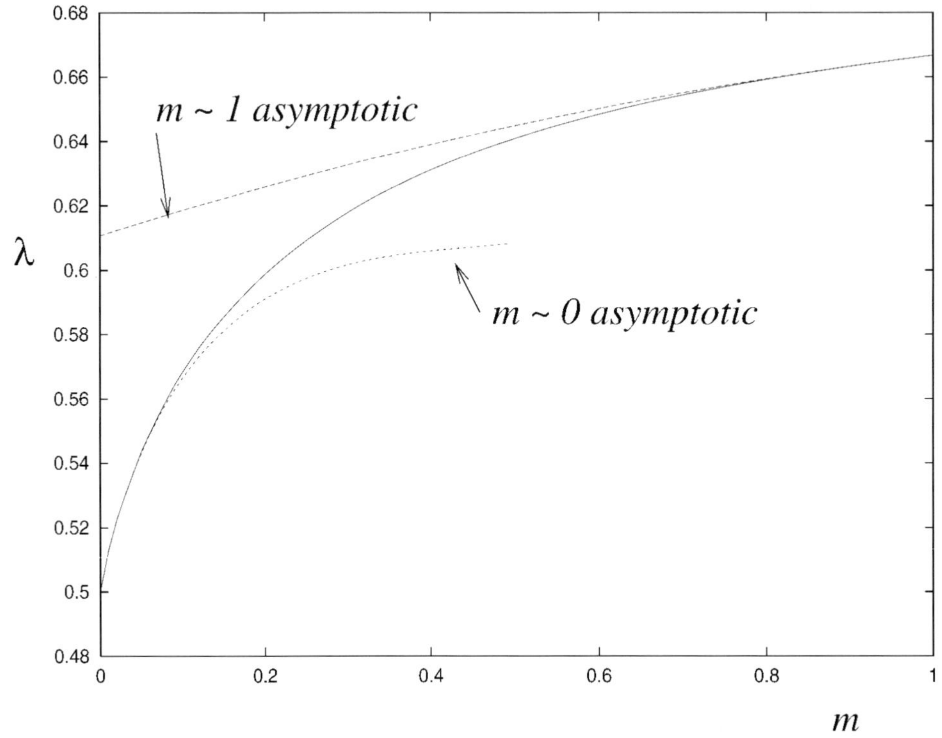

Figure 3. Dependence of λ upon m (solid line) and asymptotic estimates (14) and (15) (broken lines).

where $z_c = \lambda h$. It may easily be confirmed that (13) has a unique solution $1/2 < \lambda \leq 2/3$ for all m with $0 < m \leq 1$, and this solution may easily be located using any standard numerical method for solving a single nonlinear equation. It is also a simple matter to determine the behaviour of λ when m is close to 0 or 1 ('asymptotics on the asymptotics'). We find that

$$\begin{aligned}\lambda = {} & \tfrac{2}{3} + \tfrac{1}{27}(1 - \tfrac{8}{3}\log 2)(1-m) + \\ & + \tfrac{1}{81}(1 - 4\log 2 - \tfrac{4}{9}(\log 2)^2)(1-m)^2 + O((1-m)^3) \quad (m \sim 1),\end{aligned} \tag{14}$$

$$\lambda = \frac{1}{2} - \frac{m}{4}\log m + \frac{\log 2}{4}m + \frac{3}{8}m^2 \log m + O(m^2) \quad (m \sim 0). \tag{15}$$

The parameter λ is shown as a function of m along with the asymptotic estimates (14) and (15) in Figure 3. It may also be confirmed from (8) and (9) that for a given m, $u = 0$ at $z = (2\lambda - 1)h$.

Once λ is known, p_x is determined for a given U by (12) (and is a constant). We find that

$$p_x = \frac{\mu_0}{h^{1+m}} \left(\frac{(1+2m)U\lambda}{m[\lambda^{\frac{1+2m}{m}} + (1-\lambda)^{\frac{1+2m}{m}}]} \right)^m . \tag{16}$$

Having determined u, we may now turn to the y-momentum equation (5). We must solve

$$p_y = (\mu_0 (u_z^2)^{\frac{m-1}{2}} v_z)_z$$

with $v = 0$ at $z = 0$ and $z = h$, and once again this may be accomplished by simple integration. Again we solve separately for the two regions $0 \leq z \leq z_c$ and $z_c \leq z \leq h$ and use the boundary conditions $v_z = 0$, v continuous at $z = z_c$. This gives

$$v = \begin{cases} \dfrac{m p_y p_x^{\frac{1-m}{m}} [(z_c - z)^{\frac{1}{m}} (-(z-h)(h-z_c)^{\frac{1}{m}} - z z_c^{\frac{1}{m}}) - h z_c^{\frac{1}{m}} (h - z_c)^{\frac{1}{m}}]}{(1+m)\mu_o^{\frac{1}{m}} h^{\frac{1}{m}} [\lambda^{\frac{1}{m}} + (1-\lambda)^{\frac{1}{m}}]} & (z \leq z_c) \\ \dfrac{m p_y p_x^{\frac{1-m}{m}} [(z - z_c)^{\frac{1}{m}} ((z-h)(h-z_c)^{\frac{1}{m}} + z z_c^{\frac{1}{m}}) - h z_c^{\frac{1}{m}} (h - z_c)^{\frac{1}{m}}]}{(1+m)\mu_o^{\frac{1}{m}} h^{\frac{1}{m}} [\lambda^{\frac{1}{m}} + (1-\lambda)^{\frac{1}{m}}]} & (z \geq z_c) \end{cases}.$$

The throughput $\mathcal{T}_1$ (measured in kg m^{-1} s^{-1}) may now be found. (Here and henceforth we work, unless otherwise stated, in terms of the throughput due to the portion of the device *between two scrapers*. This value should be multiplied by the number of scrapers and the distance between adjacent scrapers to determine the total device mass flux (kg s^{-1}).) We have

$$\mathcal{T}_1 = \rho \int_0^h v \, \mathrm{d}z$$

and, after some calculation, we find that

$$\mathcal{T}_1 = \frac{-\rho p_y h^{2+m} U^{1-m}}{\mu_0} f_1(\lambda, m), \tag{17}$$

where f_1 (which depends only on m and λ) is defined by

$$f_1(\lambda, m) = \frac{m^m \lambda^{1-m} [m^2 \Gamma (\lambda^{2+1/m} + (1-\lambda)^{2+1/m}) + (1+2m)\lambda^{1/m}(1-\lambda)^{1/m}]}{(1+2m)^m (1+m)^2 \Gamma [\lambda^{2+1/m} + (1-\lambda)^{2+1/m}]^{1-m}}$$

and

$$\Gamma = \lambda^{1/m} + (1-\lambda)^{1/m}.$$

The quantity $\mathcal{T}_1$ depends on the problem parameters in a revealing manner; the dependence on U becomes more greatly pronounced as m decreases (and so the fluid 'shear-thins' by a greater amount) but otherwise $\mathcal{T}_1$ depends linearly on ρ and the pressure gradient.

3.1. Newtonian case ($m = 1$)

A partial check of the results derived so far is provided by setting $m = 1$ in the results derived so far to retrieve the Newtonian version of the flow. We find that

$$u = \frac{U(h-3z)(h-z)}{h^2}, \quad v = \frac{p_y z(z-h)}{2\mu_0},$$

$$z_c = \frac{2h}{3}, \quad p_x = \frac{6\mu_0 U}{h^2}, \quad \mathcal{T}_1 = \frac{-p_y \rho h^3}{12\mu_0}.$$

Naturally in this limit the throughput is independent of U since now rotating the inner cylinder cannot change the fluid viscosity.

3.2. Optimization problem

We may now address (at least, in principle) the optimization problem that was the original focus of this study. We know that

$$\mathcal{T}_1 \propto (-p_y) U^{1-m}$$

and we must choose U and $-p_y = \triangle P$ say to maximize this quantity subject to the total power expenditure being fixed. There are many alternative, but essentially equivalent ways of formulating the optimization problem; we adopt the simplest and assume that the total power expended (measured in W = kg m^2 s^{-3}) is composed of (a) a component due to the rotation of the inner cylinder and (b) a component due to the maintenance of a pressure gradient via a pump. Using elementary engineering correlations (other more accurate formulae may be found in the literature; see, for example [10]), we now assert that

P_p = pump power = (Volume flow rate)(Pressure difference)

and

P_r = rotation power = (Torque)(Angular velocity).

The total volume flow rate (for a device with three scrapers) is given by $3L\mathcal{T}_1/\rho$, whilst the pressure difference down the device is $D\triangle P$. Thus

$$P_p = \frac{3L(\triangle P)^2 h^{2+m} U^{1-m} D}{\mu_0} f_1(\lambda, m).$$

As far as the torque is concerned, the shear stress τ may easily be calculated to be given by

$$\tau = \mu_0 (u_z^m)_{z=0} = p_x z_c.$$

With a three-scraper device the force is thus $3Lp_x z_c D$ and the torque is therefore $3Lap_x z_c D$ where a is is the radius of the inner cylinder. The power is thus given by $3LUp_x z_c D$, and using (16) now gives

$$P_r = 3LU^{1+m} Dh^{-m} \lambda \mu_0 g(\lambda, m),$$

where

$$g(\lambda, m) = \frac{\lambda^m (1+2m)^m}{m^m [\lambda^{2+1/m} + (1-\lambda)^{2+1/m}]^m}.$$

We now seek to maximise $\triangle P U^{1-m}$ subject to

$$P_r + P_p = P_t, \tag{18}$$

where P_t is the prescribed total available power. Perfect power transmission has been assumed in (18); in reality, for most devices it is reasonable to assume that the ratio of actual power delivered to the power input is a constant, usually known as the efficiency (which may be as low as 0·1 or 0·2; see, for example [11] or [12]). We do not include efficiencies here, but obviously they could easily be incorporated if they were known.

To maximise $\triangle P U^{1-m}$ subject to

$$K_1 U^{1+m} + K_2 (\triangle P)^2 U^{1-m} - P_t = 0,$$

where

$$K_1 = 3LDh^{-m} \mu_o \lambda g(\lambda, m), \qquad K_2 = \frac{3Lh^{2+m} D f_1(\lambda, m)}{\mu_0};$$

we introduce a Lagrange multiplier ϕ and instead maximize

$$\triangle P U^{1-m} - \phi(K_1 U^{1+m} + K_2 (\triangle P)^2 U^{1-m} - P_t).$$

The three equations to be solved are thus

$$0 = U^{1-m} - 2\phi K_2 \triangle P U^{1-m}, \tag{19}$$

$$0 = (1-m) \triangle P U^{-m} - \phi[(1+m)K_1 U^m + (1-m)K_2(\triangle P)^2 U^{-m}], \tag{20}$$

$$0 = K_1 U^{1+m} + K_2(\triangle P)^2 U^{1-m} - P_t. \tag{21}$$

Solving (19) and (20), we find that

$$\triangle P = \frac{1}{2K_2\phi}, \quad U = \left(\frac{1-m}{4\phi^2(1+m)K_1K_2}\right)^{\frac{1}{2m}}.$$

From (21) we therefore find that ϕ satisfies the nonlinear algebraic equation

$$K_1\left(\frac{1-m}{4\phi^2(1+m)K_1K_2}\right)^{\frac{1+m}{2m}} + \frac{1}{4\phi^2K_2}\left(\frac{1-m}{4\phi^2(1+m)K_1K_2}\right)^{\frac{1-m}{2m}} = P_t \tag{22}$$

and the optimization problem is, in principle solved, once (22) is solved.

In a real industrial device the key parameters may take a range of values; here we analyse the predictions of (22) to calculate the optimum values of U and $\triangle P$ for a typical industrial device specification. Assuming that $m = \frac{2}{3}$ (a value typical for margarine, jam and peanut butter) we first solve (13) to give $\lambda \sim 0{\cdot}65250$ whence $f_1(\lambda, m) \sim 0{\cdot}09793$ and $g(\lambda, m) \sim 4{\cdot}37993$. Now we take typical values $\rho = 800$ kg m^{-3}, $h = 0{\cdot}015$ m, $L = 0{\cdot}15$ m, $D = 2$ m and also assume that $\mu_0 = 10{\cdot}0$ (this value may vary for different food products, but certainly the order of magnitude of this figure is correct). We now have to give a value for the total power P_t; we take $P_t = 2{\cdot}5$ kW (a typical value for a medium-sized device that might be used in a factory for food processing). Solving (22), we find that $\phi \sim 31{\cdot}4953$ and the throughput is thus maximised when $U \sim 0{\cdot}9911$ m s^{-1} and $\triangle P = 131620{\cdot}4$ Pa, the associated volume flow rate being $3L\mathcal{T}_1/\rho = 0{\cdot}007914$ m^3 s^{-1}. Although detailed information for a range of devices and food products is not available, it may be confirmed that the figures for U and $\triangle P$ are really rather close to the values that have been found by trial and error in factories.

Of course, the real advantage of the current asymptotic formulation is the amount of parameter-dependence information that may be gleaned from (17) and (22). For example, Figure 4 shows the optimal volume flow rate (in m^3 s^{-1}) as a function of m for the parameters $\rho = 800$ kg m^{-3}, $P_t = 2{\cdot}5$ kW, $h = 0{\cdot}015$ m, $L = 0{\cdot}15$ m, $D = 2$ m and $\mu_0 = 10{\cdot}0$. As m decreases and the fluid shear-thins to a greater extent, it is clear that the optimum volume flow rate increases dramatically. It may also be confirmed that, as expected, decreasing m also decreases the optimal pressure gradient and increases the value of the optimal rotation speed as it becomes more worthwhile to expend greater amounts of energy in thinning the fluid.

4. Analysis for other orders of magnitude of U

Though the analysis of the previous section has covered the case that we believe is of most relevance as far as the currently-operated industrial process is concerned, the framework that has been set up to examine the problem allows some other limits to be considered. These are now briefly analyzed for the sake of completeness.

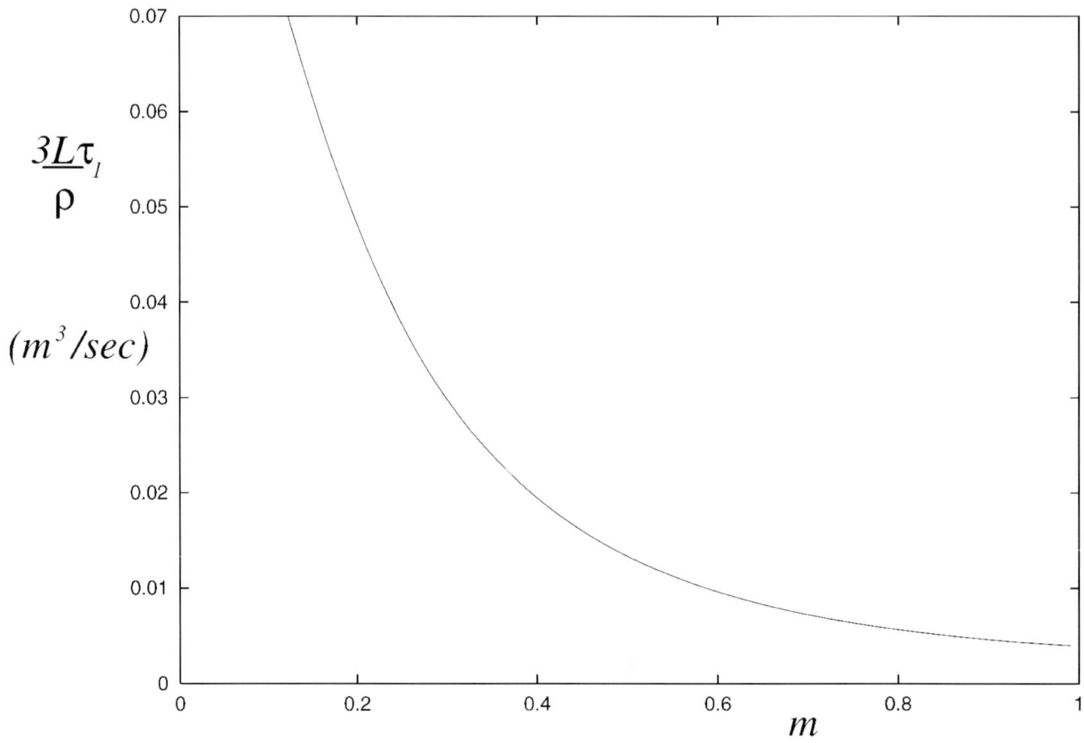

Figure 4. Optimum volume flow ($m^3\ s^{-1}$) as a function of shear-thinning exponent m.

4.1. Analysis for small rotation rates

When the rotation rate is very small so that U is not large enough to contribute to the leading order flow, we scale in an identical manner to that discussed at the beginning of Section 3, save for the fact that we set $v = (U/\epsilon)\bar{v}$ and

$$p = \frac{\mu_0 U}{L\epsilon^4}\left(\frac{U}{\epsilon h}\right)^{m-1}\bar{p}.$$

Essentially this scaling may be thought of as one that preserves the previous orders of magnitude of v and p_y for small U. The (dimensional) leading-order equations now become

$$0 = -p_x, \quad 0 = -p_y + (\mu_0(v_z^2)^{\frac{m-1}{2}} v_z)_z,$$

$$0 = -p_z, \quad u_x + v_y + w_z = 0$$

and hence the problem reduces to that of determining Poiseuille flow for a shear-thinning fluid. The quantity p_y is therefore a function of y alone (and for the sake of definiteness we will assume here and henceforth that p_y is negative so that flow takes place in a positive direction along the y-axis) and we must solve

$$p_y = (\mu_0(v_z^2)^{\frac{m-1}{2}} v_z)_z$$

with $v = 0$ at $z = 0$ and $z = h$. The flow is evidently symmetrical in the regions $0 \leq z \leq h/2$ and $h/2 \leq z \leq h$ and may be easily confirmed to be given by

$$v = \begin{cases} \left(\frac{m}{1+m}\right)\left(\frac{-p_y}{\mu_0}\right)^{\frac{1}{m}}\left[\left(\frac{h}{2}\right)^{\frac{1+m}{m}} - \left(\frac{h}{2} - z\right)^{\frac{1+m}{m}}\right] & (0 \le z \le h/2) \\ \left(\frac{m}{1+m}\right)\left(\frac{-p_y}{\mu_0}\right)^{\frac{1}{m}}\left[\left(\frac{h}{2}\right)^{\frac{1+m}{m}} - \left(z - \frac{h}{2}\right)^{\frac{1+m}{m}}\right] & (h/2 \le z \le h) \end{cases}.$$

A 'scraper condition' may be imposed as before to show that the quantity p_y is a constant, however to fully determine u and w it is necessary to consider a higher-order problem. Notwithstanding this, the throughput $\mathcal{T}_2$ may be established; we find by simple integration that

$$\mathcal{T}_2 = \frac{2\rho m(-p_y)^{1/m} h^{2+1/m}}{(1+2m)\mu_0^{1/m} 2^{2+1/m}}$$

and we note that, in contrast to the case considered in the previous section, the throughput is now independent of U, and depends on the $(1/m)$th power of the quantity $-p_y/\mu_0$.

Finally for this case, we note that setting $m = 1$ to retrieve the Newtonian case once again gives the familiar Poiseuille flow results

$$v = \frac{-p_y}{2\mu_0} z(h - z) \quad (0 \le z \le h),$$

$$\mathcal{T}_2 = \frac{-p_y \rho h^3}{12\mu_0}.$$

4.2. Analysis for larger small rotation rates

When U is too small to contribute to the leading-order flow, but large enough to play a role in the shear, an "intermediate case" arises which may be analyzed (and is the distinguished limit containing both of the previous cases). For this case, the relevant scalings are identical to those used at the beginning of Section 3, save for the fact that we set $v = U\bar{v}$ and

$$p = \frac{\mu_0 U}{L\epsilon^3}\left(\frac{U}{h}\right)^{m-1} \bar{p}.$$

As usual, to leading order p cannot depend upon z, but in this case we also find that p is independent of x. Hence the correct expansion for p is $p = p_0(y) + \epsilon p_1(x, y) + O(\epsilon^2)$. The dimensional pressure gradients in the x and y directions are thus comparable for this intermediate case. If this expansion for the pressure is used the (dimensional) equations governing the leading-order velocities are

$$0 = -p_x + (\mu_0(u_z^2 + v_z^2)^{\frac{m-1}{2}} u_z)_z, \tag{23}$$

$$0 = -p_y + (\mu_0(u_z^2 + v_z^2)^{\frac{m-1}{2}} v_z)_z, \tag{24}$$

$$0 = -p_z, \tag{25}$$

$$0 = u_x + w_z. \tag{26}$$

The major difference between this case and the previous cases that have been discussed is that it is now easy to see that u_z and v_z change sign at distinct positions, which we shall denote by $z = z_c$ and $z = z_b$, respectively. We shall also assume, as usual, that $p_x > 0$ and $p_y < 0$. Integration of (23) and (24) now gives

$$u_z \mu_0 (u_z^2 + v_z^2)^{\frac{m-1}{2}} = (z - z_c) p_x, \tag{27}$$

$$v_z \mu_0 (u_z^2 + v_z^2)^{\frac{m-1}{2}} = (z - z_b) p_y \tag{28}$$

and thus

$$v_z = \frac{u_z}{g} \left(\frac{z - z_b}{z - z_c} \right),$$

where

$$g = \frac{p_x}{p_y} \quad (< 0).$$

This expression for v_z may now be used in (27) to yield an equation for u_z. It is simplest to solve this equation separately in the regions $z \geq z_c$ and $z \leq z_c$. We find that

$$u = \begin{cases} \left(\dfrac{p_x}{\mu_0} \right)^{\frac{1}{m}} (-g)^{\frac{m-1}{m}} [f_c(z) - f_c(0)] + U & (z \leq z_c), \\[2ex] \left(\dfrac{p_x}{\mu_0} \right)^{\frac{1}{m}} (-g)^{\frac{m-1}{m}} f_c(z) & (z \geq z_c), \end{cases}$$

where

$$f_c(z) = \int_h^z \frac{s - z_c}{[g^2 (s - z_c)^2 + (s - z_b)^2]^{\frac{m-1}{2m}}} \, \mathrm{d}s.$$

The condition that u is continuous at $z = z_c$ now yields

$$U = f_c(0) \left(\frac{p_x}{\mu_0} \right)^{\frac{1}{m}} (-g)^{\frac{m-1}{m}},$$

and the mass flow condition

$$\int_0^h u_x \, \mathrm{d}z = 0$$

imposed by the scrapers once again confirms the hypothesis that p_x is constant, and is given by

$$p_x = \frac{\mu_0 (U z_c)^m}{(-g)^{m-1} \left(z_c f_c(0) - \int_0^h f_c(z) \, \mathrm{d}z \right)^m}.$$

Now that u has been determined, v may be found in a similar fashion. We find that

$$v = \begin{cases} \left(\dfrac{-p_y}{\mu_0} \right)^{\frac{1}{m}} f_b(z) & (z \leq z_b) \\[2ex] \left(\dfrac{-p_y}{\mu_0} \right)^{\frac{1}{m}} [f_b(z) - f_b(h)] & (z \geq z_b) \end{cases},$$

where

$$f_b(z) = \int_0^z \frac{z_b - s}{[g^2(s - z_c)^2 + (s - z_b)^2]^{\frac{m-1}{2m}}}\, ds$$

and the condition that v is continuous at $z = z_b$ requires that $f_b(h) = 0$. The problem is now, in principle, solved. For a given imposed pressure gradient p_y we must solve

$$U = \left(\int_h^0 \frac{s - z_c}{[g^2(s - z_c)^2 + (s - z_b)^2]^{\frac{m-1}{2m}}}\, ds \right) \left(\frac{p_x}{\mu_0} \right)^{\frac{1}{m}} \left(\frac{p_x}{-p_y} \right)^{\frac{m-1}{m}}, \tag{29}$$

$$p_x = \frac{\mu_0 (U z_c)^m}{(-g)^{m-1} \left(z_c f_c(0) - \int_0^h f_c(z)\, dz \right)^m}, \tag{30}$$

$$0 = \int_0^h \frac{z_b - s}{[g^2(s - z_c)^2 + (s - z_b)^2]^{\frac{m-1}{2m}}}\, ds \tag{31}$$

to determine p_x, z_b and z_c; the throughput $\mathcal{T}_3$ is then given by

$$\mathcal{T}_3 = \rho \int_0^h v\, dz = \rho \left(\frac{-p_y}{\mu_0} \right)^{\frac{1}{m}} \left[\int_0^h f_b(z)\, dz \right]. \tag{32}$$

Each of (29–31) and (32) may be simplified: when we set $z_b = \lambda_b h$ and $z_c = \lambda_c h$, change the order of integration in (30) and (32) and use some elementary properties of the functions f_b and f_c, some straightforward algebra reveals that in the most convenient formulation of the problem the throughput $\mathcal{T}_3$ may be expressed as

$$\mathcal{T}_3 = \frac{\rho \lambda_b U h}{(-g)},$$

where g, λ_b and λ_c satisfy the nonlinear equations

$$\int_0^1 \frac{s(s - \lambda_c)}{[g^2(s - \lambda_c)^2 + (s - \lambda_b)^2]^{\frac{m-1}{2m}}}\, ds = 0, \tag{33}$$

$$\int_0^1 \frac{\lambda_b - s}{[g^2(s - \lambda_c)^2 + (s - \lambda_b)^2]^{\frac{m-1}{2m}}}\, ds = 0, \tag{34}$$

$$\int_0^1 \frac{\lambda_c - s}{[g^2(s - \lambda_c)^2 + (s - \lambda_b)^2]^{\frac{m-1}{2m}}}\, ds = \frac{\alpha}{(-g)} \tag{35}$$

and

$$\alpha = \left(\frac{\mu_0}{-p_y} \right)^{\frac{1}{m}} \frac{U}{h^{\frac{1+m}{m}}}. \tag{36}$$

Unlike the previous two cases that were considered, the parameters λ_b and λ_c are not now determined solely by m; the intimate coupling of u and v now means that even for fixed m the flow will be different for each value of α.

Equations (33–35) may be presented in various different forms (for example, judicious combination of (34) and (35) allows a transcendental equation for λ_b, λ_c and g to be derived) but in general it is impossible to avoid the need to solve nonlinear equations that involve integrals. This may be accomplished in an efficient manner by using library routines. For the results presented below, the NAG quadrature routine D01AHF (Patterson's method) was used to evaluate the integrals and the nonlinear equation solver C05NBF (which uses the MINPACK routine HYBRD1; for details see [13]) was used to solve the equations. In all cases solutions to (33–35) proved to be quick to determine, though for values of m near to zero a good initial guess (best provided from computed results for nearby small m-values) is required.

4.3. Unified Throughput Law

Although the optimization problem that is most relevant to the the real industrial problem has already been solved in Section 3.2, it is of some interest to gather together the throughput laws for each of three asymptotic cases that have been considered. The most informative way to plot throughput against pressure gradient and rotation speed is to plot the normalised throughput $\bar{\mathcal{T}}$, defined by

$$\bar{\mathcal{T}} = \frac{\mathcal{T}(\mu_0)^{\frac{1}{m}}}{\rho(-p_y)^{\frac{1}{m}} h^{\frac{1+2m}{m}}}$$

against α as defined in (36). In Figure 5, the quantities

$$\bar{\mathcal{T}}_1 = f_1(\lambda, m)\alpha^{1-m}, \quad \bar{\mathcal{T}}_2 = \left(\frac{m}{1+2m}\right)2^{-1-1/m}, \quad \bar{\mathcal{T}}_3 = \frac{\lambda_b \alpha}{(-g)}$$

are plotted against α for $m = 0{\cdot}3$ (broken line) and $m = 0{\cdot}6$ (dot-dashed line). In each case the horizontal line denotes $\bar{\mathcal{T}}_2$ which is independent of α. (The solid line shows the throughput for the Newtonian case for purposes of comparison; of course, when $m = 1$, $\bar{\mathcal{T}}_1$, $\bar{\mathcal{T}}_2$ and $\bar{\mathcal{T}}_3$ are identical.) For both $m = 0{\cdot}3$ and $m = 0{\cdot}6$ the higher of the two non-constant throughput curves corresponds to $\bar{\mathcal{T}}_3$ and the lower to $\bar{\mathcal{T}}_1$; when $\alpha \ll 1$ the quantity $\bar{\mathcal{T}}_1$ tends to zero, rather than the correct value $\bar{\mathcal{T}}_2$. The asymptotics give us a great deal of information about the throughput of the device; the parametric dependence is rather subtle, and may be too involved to be detected by all but the most careful experimental or numerical studies.

5. Enhanced modelling of scrapers

Thus far we have assumed that the scrapers located on the outer cylinder allow no fluid to pass through them. Though it is true that such scrapers really *are* designed to literally scrape the inner cylinder clean of any material (and thus contact with the inner cylinder is constantly maintained), for design reasons it is sometimes necessary to manufacture scrapers that contain holes. If we wished to analyse such a device, the models presented above could easily be modified by altering the scraper condition (10). The simplest way of modelling perforated scrapers would be to use a scraper condition of the form

$$\int_0^h u_x \, dz = Q,$$

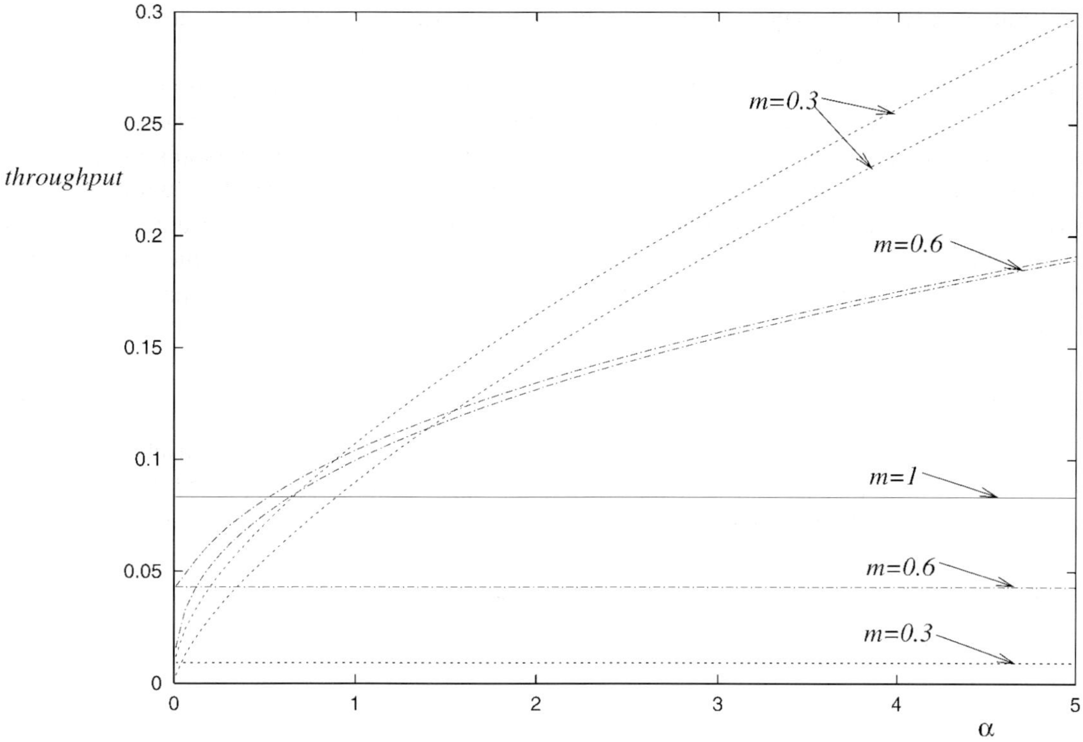

Figure 5. Non-dimensional throughputs $\mathcal{T}_1$, $\mathcal{T}_2$ and $\mathcal{T}_3$ for $m = 0{\cdot}3$ (broken line) and $m = 0{\cdot}6$ (dot-dashed line) (see text).

where Q is the mass loss through a scraper, which we assume in known. More elaborate models might also be posed in which Q depended on the pressure drop across a scraper. The main change to the analysis presented above would be that p_x would now no longer be constant; apart from this, however, our previous analysis would carry through in more or less the same form.

As far as asymptotic analysis is concerned, there is also scope for further treatment of the scrapers. Near to a scraper, it is clear that the flow adjacent to the inner cylinder amounts to the classical 'driven flow in a corner' problem where flow is produced by moving a wall parallel to itself. It has long been known that for Newtonian flow ($m = 1$) a simple exact solution is available (see, for example [14]). Recently, it was shown ([15]) that the problem could be solved for a shear-thinning fluid in the limit $m \to 0$. Moreover, the resulting predictions remained remarkably accurate even for values of m quite close to 1. (See also [14].) It could therefore reasonably be observed that asymptotic methods have succeeded for this part of the problem also.

In the region near to a scraper but adjacent to the wall of the outer cylinder, we may expect to observe recirculating eddies. For a Newtonian flow, this problem was solved completely in [16]; at present, however, little progress appears to have been made in solving this problem for shear-thinning flows (though see, for example, [14] and [17]).

6. Numerical calculations

We have already seen that some of the details of the parametric dependence of the throughput etc. are rather subtle and involved. Is it therefore possible that the results of all the asymptotic work described above could be predicted solely by a numerical attack on the problem? Let us consider what sort of numerical work could be undertaken. First, we could produce made-to-measure codes to examine various aspects of the problem. This would allow us to carry out careful and accurate numerical calculations to determine the throughput, find what the flow in a cross-section looks like, examine the flow near a scraper, etc. The disadvantage of an investigation of this sort is that many different codes will have to be written, as different numerical schemes may have to be used in different parts of the flow. If the company is question is NASA and the product is the space shuttle, then detailed investigation of this sort may be feasible. For a small industrial concern, however, a second option is more viable: to use a commercial code.

To remain true to the industrial background of this problem, we have therefore used an off-the-shelf package to perform numerical calculations. We chose to use FASTFLO^{TM} (see, for example [18]) because rather than being a 'black box' CFD code it is essentially a finite–element partial differential equation solver. This allows the user to take more control of the calculations by specifying the mesh, method and tolerances that are to be used. It is worth pointing out that although we decided to solve the full problem (*i.e.* with inertia included) some numerical simplifications may result if only the slow flow problem is tackled.

For all the calculations below, an Augmented Lagrangian Method (ALM) was coded into FASTFLO^{TM}. The ALM method (for a full explanation see, for example [19] or standard CFD texts) solves the (Newtonian) non-dimensional Navier–Stokes equations

$$\mathrm{Re}(\bar{\boldsymbol{q}}.\bar{\nabla})\bar{\boldsymbol{q}} = -\bar{\nabla}\bar{p} + \bar{\nabla}^2\bar{\boldsymbol{q}}, \quad \bar{\nabla}.\bar{\boldsymbol{q}} = 0$$

by using the iterative procedure

$$-\mathrm{Pen}\bar{\nabla}(\bar{\nabla}.\bar{\boldsymbol{q}}_n) + \bar{\nabla}\bar{p}_{n-1} - \bar{\nabla}^2\bar{\boldsymbol{q}}_n + \mathrm{Re}\bar{\boldsymbol{q}}_{n-1}.\bar{\nabla}\bar{\boldsymbol{q}}_n = 0, \tag{37}$$

$$\bar{p}_n = \bar{p}_{n-1} - \mathrm{Pen}\bar{\nabla}.\bar{\boldsymbol{q}}_n$$

for $n = 0, 1, \ldots$ where Pen is a 'Penalty parameter' and we denote non-dimensional variables as usual using an overbar. The method is more accurate and reliable than the popular 'penalty' method as there are no restrictions on the size of the penalty parameter. Also, in contrast to the penalty method, the algorithm (if it converges) introduces no further errors into the calculation and provides a solution for the pressure. The alterations that must be made to the algorithm for shear-thinning flow are of a fairly minor nature: in addition to the inclusion of the nonlinear terms in (37) an initial estimate for the viscosity must be provided. All of the calculations detailed below were carried out on a P200 microcomputer running the LINUX version of FASTFLO^{TM}.

6.1. Cross-sectional flow comparisons

We first used FASTFLO^{TM} to make numerical/asymptotic comparisons for the cross-sectional flow between two scrapers. Computations were performed in a box of aspect ratio 1:10, the bottom wall moving with a speed of 1 parallel to itself. The finite-element mesh was composed of six-noded triangles (quadratic element) and had 500 corner nodes. The penalty parameter

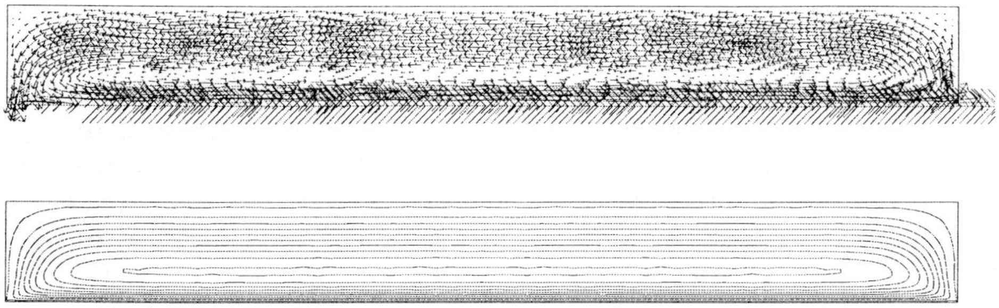

Figure 6. Numerical results (arrow plot, streamline plot) for $m = \frac{2}{3}$.

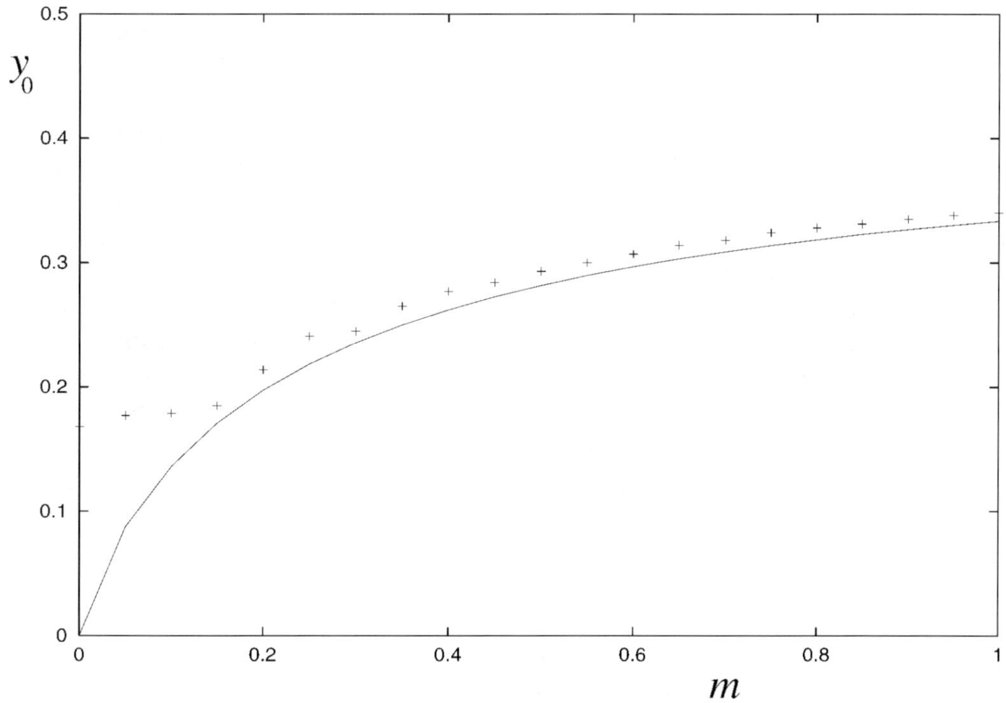

Figure 7. Ordinate y_0 where horizontal velocity u is zero. Numerical results (symbols) and asymptotic predictions (solid line) *vs.* m.

was taken to be Pen $= 200$. Computations typically converged in 10–20 iterations (using about a minute of CPU time) and the usual convergence checks were made by ensuring that as the number of mesh elements increased the converged solutions became indistinguishable.

Figure 6 shows typical numerical results (in the form of a velocity arrow plot and a streamline plot) for $m = \frac{2}{3}$. Although there are some inherent problems in presenting results for a flow region with a 10:1 aspect ratio and individual arrows are hard to distinguish, we note that the flow seems to be independent of x except for regions near to the edges of the box. For the vast majority of the flow, therefore the asymptotic results should be accurate. Both of the plots in Figure 6 also clearly indicate the region where the horizontal velocity changes sign. A quantitative comparison between the numerical and asymptotic results may be carried out by comparing the positions y_0 where $u = 0$ for different m.

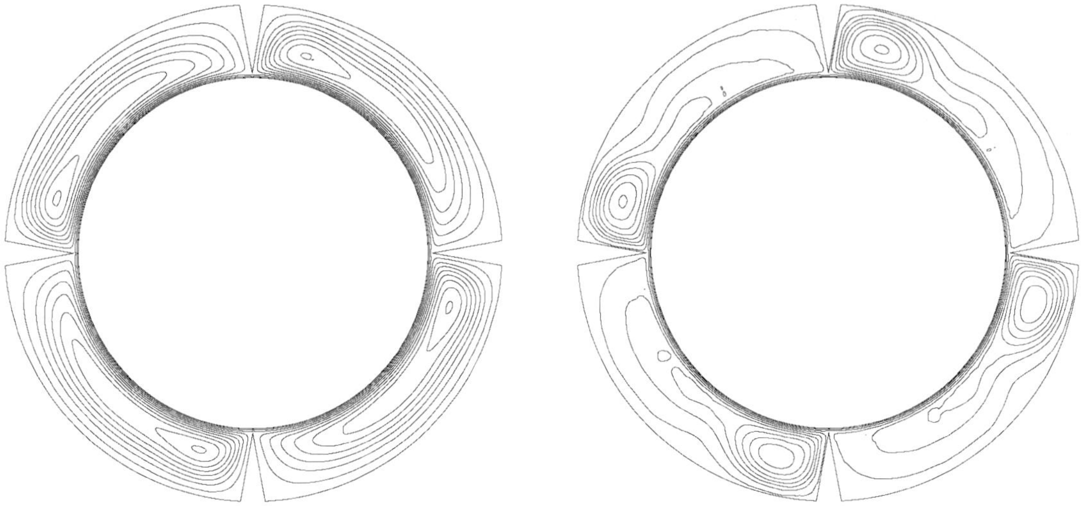

Figure 8. Numerical results (streamlines) for four-scraper device with inner cylinder rotating anti-clockwise with non-dimensional speed $\bar{u}$ =100. (Left-hand: Newtonian, right-hand: $m = \frac{2}{3}$).

In Figure 7 y_0 is plotted for various m for the same 10:1 aspect ratio box and the same numerical parameters that were used to produce Figure 6. The symbols denote numerically calculated points, and the solid line is the asymptotic theory developed in Section 3. We note that for most values of m the agreement between the symbols and the asymptotic results is excellent. It is no surprise that for small values of m, the numerical results seem to be rather unreliable; the computation of shear-thinning flow for small values of the shear-thinning parameter is a notoriously difficult problem. In any case, when $m \ll 1$ there is no guarantee that the asymptotic results are correct anyway, as m is now an additional small parameter in the problem and there are a number of different distinguished limits that may be relevant. Some further numerical experiments (not shown here) are revealing. The slightly larger values of y_0 that are predicted by the numerical code do not seem to be a result of the discretization, but are rather a consequence of the regions near to the scrapers where the full Navier-Stokes equations apply. Although it is virtually impossible to tell from Figure 6, the effect of these end regions seems to be to slightly increase the value of y_0 near to the centre of the flow region, whilst slightly decreasing it closer to the edges of the flow region.

6.2. NUMERICAL RESULTS FOR MORE COMPLICATED FLOW CONDITIONS

Although we have tackled many aspects of the problem successfully using asymptotic analysis, there are inevitably some cases where asymptotic analysis does not seem appropriate. As discussed in the introduction, we have to be realistic about what asymptotics can accomplish for us, and there are some device configurations where numerical analysis is our only hope. Figure 8 shows numerical results (streamlines) for a (symmetrical, two-dimensional) four-scraper device where there is a small gap beneath each scraper. The inner cylinder rotates with a non-dimensional speed of 100, so that inertia is no longer negligible. Clearly in this case the shear-thinning nature of the fluid is crucial and changes the flow structure a good deal.

7. Conclusions and further work

In this study we have examined only the simplest possible incarnation of the problem of determining the flow of shear-thinning foodstuffs in annular scraped heat exchangers. The geometry of the device clearly allows many simplifications if an asymptotic approach is used, and though the parametric dependence of the throughputs in the three different regimes could no doubt have been determined by more *ad hoc* methods, the detailed description of the flow that an asymptotic approach provides is likely to prove invaluable for further studies.

Of course, not all of the asymptotic possibilities of the problem have been exploited. For example, a regular perturbation scheme could easily be employed to consider the case $m = 1 - \epsilon$ ($\epsilon \ll 1$) to determine the details of 'nearly Newtonian' flow. Although from a mathematical point of view this leads only to some very standard analysis, the results would be of practical importance to the device manufacturers as they would allow corrections to be made to flow rate predictions for foodstuffs with $\epsilon \ll 1$. As discussed above, corner solutions could also be examined to determine the local details of the flow near a scraper, and many other asymptotic limits might usefully be pursued.

In particular, we note that, although for the case of the flow of margarine the shear-thinning index is normally greater than 0·5, in previous work we have been interested in the flow of materials where the shear-thinning index tends to zero (see, for example [20] and [15]). The cases studied above could also be looked upon from an $m \to 0$ perspective. As usual, we see that in this limit the flow details are very sensitive to the value of m; in particular the throughput can change by an order of magnitude when other flow parameters change only by an order one amount. There is clearly scope for further work here.

The main simplifying assumption in the work presented above, namely isothermality, constitutes a serious shortcoming and needs to be remedied when further work is carried out. A local analysis will be required, and this will eventually have to be coupled in to the full model. It should be noted, however, that a 'rough and ready' approximation to the case where temperature dependence is included may be obtained by allowing μ_0 to be a given function of y, thereby assuming that the temperature in the heat exchanger is known and depends only only the distance down the apparatus.

Acknowledgements

The authors are grateful to Nick Hall-Taylor of Crown Chemtech Ltd. for bringing the problem to their attention and to Mike Baines and Leo Pyle of Reading University and Jon Chapman of Oxford University for helpful discussions.

References

1. N. Hall-Taylor, A model for the behaviour of scraped surface heat exchangers. Internal Report, Crown Chemtech Ltd., July (1995).
2. N. Hall-Taylor, Experimental results for the rheology pastes and spreads. Internal Report, Crown Chemtech Ltd., January (1998).
3. G. Astarita and G. Marrucci, *Principles of Non-Newtonian Fluid Mechanics*. London: McGraw-Hill (1974) 289pp.
4. M.W. Johnson and S. Mangkoesoebroto, Analysis of lubrication theory for the power law fluid. *J. Tribology, Trans. ASME* 115 (1993) 71–77.

5. T.R. Lin, Thermal elastohydrodynamic lubrication of rolling and sliding contacts with a power law fluid. *Wear* 154 (1992) 77–93.
6. D. Prasad and R.P. Chhabra, Thermal and normal squeezing effects in lubrication of rollers by a power-law fluid. *Wear* 145 (1991) 61–76.
7. J.M. Wang and G.B. Jin, The optimum design of the Rayleigh slider bearing with a power law fluid. *Wear* 129 (1989) 1–11.
8. P. Szabo and O. Hassager, Flow of viscoplastic fluids in eccentric annular geometries. *J. Non-Newtonian Fluid Mech.* 45 (1992) 149–169.
9. S.H. Bittleston and O. Hassager, Flow of viscoplastic fluids in a rotating concentric annulus. *J. Non-Newtonian Fluid Mech.* 42 (1992) 19–36.
10. R.H. Warring, *Pumps: Selection, Systems and Applications, 2nd Ed.* Morden, Surrey: Trade & Technical Press Ltd. (1984) 271pp.
11. V.S. Lobanoff and R.R. Ross, *Centrifugal Pumps, Design and Application.* Houston, Texas: Gulf Publications (1985) 374pp.
12. J. Davidson and O. von Bertele, *Process Pump Selection.* UK: Professional Engineering Publishing (2000) 202pp.
13. M.J.D. Powell, A hybrid method for nonlinear algebraic equations, In: P. Rabinowitz, (ed.) *Numerical Methods for Nonlinear Algebraic Equations*, Gordon & Breach (1970) pp. 45–48.
14. R.T. Fenner, On local solutions to non-Newtonian viscous flows. *Int. J. Non-Linear Mechanics* 10 (1975) 207–214.
15. S.J. Chapman, A.D. Fitt and C.P. Please, Extrusion of power-law shear-thinning fluids with small exponent. *Int. J. Non-Linear Mechanics* 32 (1997) 187–199.
16. H.K. Moffatt, Viscous and resistive eddies near a sharp corner. *J. Fluid Mech.* 18 (1964) 1–18.
17. P. Henriksen and O. Hassager, Corner flow of power law fluids. *J. Rheology* 33 (1989) 865–879.
18. NAG Ltd., *FASTFLO User Manual.* Oxford, England: Numerical Algorithms Group (1999).
19. X. Nicolas, P. Traore, A. Mojtabi and J.P. Caltagirone, Augmented Lagrangian method and open boundary conditions in 2D simulation of Poiseuille–Bernard channel flow. *Int. J. Num. Meth. Fluid Mech.* 25 (1997) 265–283.
20. M.E. Brewster, S.J. Chapman, A.D. Fitt and C.P. Please, Asymptotics of very small exponent power-law shear-thinning fluids in a wedge. *Euro. J. Appl. Math.* 6 (1995) 559–571.

Journal of Engineering Mathematics **39:** 367–385, 2001.

The evolution of travelling waves from chemical-clock reactions

S. J. PREECE, J. BILLINGHAM AND A. C. KING
School of Mathematics and Statistics, University of Birmingham, Edgbaston, Birmingham B15 2TT, U.K.

Received 6 March 2000; accepted in revised form 22 August 2000

Abstract. A clock-reaction is a chemical reaction which gives rise to an initial induction period before a significant change in concentration of one of the chemical species occurs. In this paper the development of travelling waves from a particular class of inhibited autocatalytic clock-reactions is analysed. The numerical solutions show that, after the induction period, a propagating reaction-diffusion front is initiated. This front is seen to accelerate initially and then to become a constant-speed travelling wave. An asymptotic analysis of the large-time travelling-wave behaviour is given and from this it is possible to fix a minimum wave speed. The asymptotic predictions of the wave speed are found to agree well with those of the numerical solution.

Key words: clock-reaction, autocatalysis, inhibition, reaction-diffusion, travelling wave

1. Introduction

Propagating reaction-diffusion fronts were first considered by Luther [1] over 80 years ago and a mathematical formulation was first carried out by Fisher [2] and analysed in more detail by Kolmogorov *et al.* [3]. Since the work of Merkin and Needham [4] on a simple autocatalytic chemical system much research has been published relating to the development of travelling waves. In this paper we analyse the development, in an unstirred medium, of a class of reaction schemes which is known to give rise to clock-reaction behaviour and are able to demonstrate the existence of travelling waves.

A reaction is regarded as displaying clock characteristics if, after the initial mixing, a significant induction period is observed before a rapid change in one of the product or reactant concentrations occurs. Such reactions often give rise to observable phenomena at the end of the induction period. Examples of clock-reactions include the arsenic(III) sulfide clock-reaction [5], the formaldehyde clock-reaction [6], the iodine bisulphate clock [7] and the hydration of carbon dioxide [8]. In this paper we consider two mechanisms which can give rise to clock-reaction behaviour. The first is simple autocatalysis, which can be described as an induction reaction. An example of such a scheme is,

$$A + nB \longrightarrow (n+1)B, \qquad \text{rate } kab^n, \tag{1}$$

where lower-case letters denote the respective chemical concentrations and k is a reaction-rate constant. Examples of solution-phase reactions which are thought to be well modelled by cubic autocatalysis are the iodate-arsenous acid reaction [9] and the iodine-bisulphate clock-reaction [10, pp. 54–56]. Another mechanism which can lead to clock-reaction behaviour is,

$$\begin{aligned} P &\longrightarrow B, && \text{rate } k_0 p, \\ B + C &\longrightarrow D, && \text{rate } k_1 bc, \end{aligned}$$

which can be described as an inhibition reaction. The clock chemical, B, is supplied to the system via the decay of the precursor chemical P. An inhibitor chemical, C, reacts with the clock chemical limiting its concentration. Once the inhibitor chemical is consumed the concentration of the clock chemical starts to increase. This system has been analysed mathematically in a well stirred situation by Billingham and Needham [11]. Examples of reactions which are thought to be well modelled by this mechanism are the photosynthesis of hydrogen chloride inhibited by ammonia and the polymerisation of vinyl acetate by benzoquinone [12].

In this paper we analyse the behaviour of the system,

$$P \longrightarrow A, \qquad \text{rate } k_0 p, \tag{2}$$

$$A + nB \longrightarrow (n+1)B, \qquad \text{rate } k_1 a b^n, \tag{3}$$

$$mB + C \longrightarrow D, \qquad \text{rate } k_2 b^m c, \tag{4}$$

for the cases $n = m = 1, 2$. This system was first analysed by Billingham and Needham [13] who assumed the decay of the precursor to be negligible. Using asymptotic methods, they were able to obtain expressions for the length of the induction period in each case. Preece *et al.* [14] extended this work to allow for the consumption of the precursor chemical. They obtained modified expressions for the induction period and showed that clock-reaction behaviour was only observed within certain parameter limits.

In the first part of this paper numerical solutions are presented which show that, when a clock-reaction is initiated with localised initial conditions, the rapid growth in concentration of autocatalyst sets up a propagating reaction-diffusion front. After an initial transient phase in which the reaction front accelerates, a constant-speed travelling wave is seen to develop. Using asymptotic methods, we construct a large-time travelling-wave solution for the cases $n = m = 1, 2$ and are able to fix a minimum wave speed in each case. Our asymptotic predictions of wave speed are found to agree well with those of the numerical solution.

2. Mathematical formulation

We formulate the problem for $n = m = 1$ or 2 setting $m = n$ and choose to study the case of one-dimensional slab geometry with the diffusion coefficients of the chemical species equal. From reaction scheme (2), (3), (4) we obtain the partial differential equations,

$$\frac{\partial p}{\partial t} = D\frac{\partial^2 p}{\partial \bar{x}^2} - k_0 p, \tag{5}$$

$$\frac{\partial a}{\partial t} = D\frac{\partial^2 a}{\partial \bar{x}^2} + k_0 p - k_1 a b^n, \tag{6}$$

$$\frac{\partial b}{\partial t} = D\frac{\partial^2 b}{\partial \bar{x}^2} + k_1 a b^n - n k_2 b^n c, \tag{7}$$

$$\frac{\partial c}{\partial t} = D\frac{\partial^2 c}{\partial \bar{x}^2} - k_2 b^n c, \tag{8}$$

which are to be solved subject to the initial conditions,

$$\begin{aligned} p(\bar{x},0) &= p_0, \\ a(\bar{x},0) &= 0, \\ b(\bar{x},0) &= \begin{cases} b_0 & \bar{x} \le |l| \\ 0 & \bar{x} > |l|, \end{cases} \end{aligned} \tag{9}$$

$$c(\bar{x}, 0) = c_0.$$

These conditions describe an initially uniform concentration of precursor and inhibitor chemical and an initial top hat distribution of the autocatalyst of half width l. We impose no flux boundary conditions on all chemical species at infinity. As no spatial variation will occur in the precursor concentration, Equation (5) can be integrated to give

$$p = p_0 \mathrm{e}^{-k_0 t}. \tag{10}$$

We define dimensionless variables as

$$\alpha = a/b_0, \quad \beta = b/b_0, \quad \gamma = c/b_0, \quad \bar{x} = \left(\frac{k_1 b_0^n}{D}\right)^{1/2} x, \quad \tau = k_1 b_0^n t, \tag{11}$$

and Equations (6), (7) and (8) reduce to

$$\frac{\partial \alpha}{d\tau} = \frac{\partial^2 \alpha}{\partial x^2} + \mu \mathrm{e}^{-\epsilon_0 \delta \tau} - \alpha \beta^n, \tag{12}$$

$$\frac{\partial \beta}{d\tau} = \frac{\partial^2 \beta}{\partial x^2} + \alpha \beta^n - \frac{n}{\delta} \beta^n \gamma, \tag{13}$$

$$\frac{\partial \gamma}{d\tau} = \frac{\partial^2 \gamma}{\partial x^2} - \frac{1}{\delta} \beta^n \gamma, \tag{14}$$

where the parameters ϵ_0, δ and μ are defined as

$$\epsilon = \frac{k_0}{k_1 b_0^n}, \quad \delta = \frac{k_1}{k_2 b_0}, \quad \mu = \epsilon P_0, \quad \text{with} \quad P_0 = \frac{p_0}{b_0}. \tag{15}$$

These equations must be solved subject to the initial conditions

$$\begin{aligned} \alpha(x, 0) &= 0, \\ \beta(x, 0) &= \begin{cases} 1 & x < |\xi| \\ 0 & x > |\xi|, \end{cases} \\ \gamma(x, 0) &= \lambda, \end{aligned} \tag{16}$$

where,

$$\lambda = \frac{c_0}{b_0}, \quad \xi = \left(\frac{D}{k_1 b_0^n}\right)^{\frac{1}{2}} l. \tag{17}$$

P_0 and λ are the dimensionless initial concentrations of P and C, respectively. The parameters ϵ and $1/\delta$ are measures of the reaction rate of steps (2) and (4), respectively, relative to the rate of the autocatalytic step (3). The diffusion coefficient has been scaled out of the governing equations and we have retained the imposed length scale in the initial conditions via the parameter ξ. It has been shown by Billingham and Needham [13] that in a well stirred environment the length of the induction period is of $O(\delta^{-1})$ for the cases $n = 1, 2$. We choose to consider the small δ situation, in which

$$\epsilon = \epsilon_0 \delta \quad \epsilon_0 = O(1) \quad \text{as} \quad \delta \to 0 \quad \text{with} \quad \mu = O(1). \tag{18}$$

This ensures that the effect of precursor consumption is significant at the time of initiation of the reaction-diffusion front and produces interesting phenomena which are described in the next section.

3. Numerical solution

We now present a numerical solution of Equations (12), (13) and (14) for the cases $n = 1$ and $n = 2$. We need consider only the region $x \geq 0$ as we have imposed initial conditions which are symmetric about $x = 0$. To solve the equations we apply the boundary conditions $\partial\alpha/\partial x = \partial\beta/\partial x = \partial\gamma/\partial x = 0$ at $x = 0$ and we note that there will be another symmetrically disposed solution in the region $x \leq 0$. We apply no flux boundary conditions at some large value of x, typically $x = 500$, as the system will tend to its unreacted spatially uniform state at large distances from the origin. The numerical scheme used is the method of lines. We discretise in x by replacing the spatial derivatives with second order central-difference approximations. This gives the three sets of ordinary differential equations,

$$\frac{\partial\alpha_i}{\partial\tau} = \frac{1}{(\Delta x)^2}\left[\alpha_{i+1} - 2\alpha_i + \alpha_{i-1}\right] + \mu e^{-\epsilon_0\delta\tau} - \alpha_i\beta_i^n, \tag{19}$$

$$\frac{\partial\beta_i}{\partial\tau} = \frac{1}{(\Delta x)^2}\left[\beta_{i+1} - 2\beta_i + \beta_{i-1}\right] + \alpha_i\beta_i^n - \frac{m}{\delta}\beta_i^m\gamma_i, \tag{20}$$

$$\frac{\partial\gamma_i}{\partial\tau} = \frac{1}{(\Delta x)^2}\left[\gamma_{i+1} - 2\gamma_i + \gamma_{i-1}\right] - \frac{1}{\delta}\beta_i^m\gamma_i, \tag{21}$$

for $i = 0, 1, 2, 3, \ldots, N$ and where α_i, β_i and γ_i represent the values of α, β and γ at time τ and distance $i\Delta x$ from the origin. The outer boundary conditions are given as $\alpha_{N-1} = \alpha_{N+1}$, $\beta_{N-1} = \beta_{N+1}$, $\gamma_{N-1} = \gamma_{N+1}$ and similarly the boundary conditions at the origin are $\alpha_{-1} = \alpha_1$, $\beta_{-1} = \beta_1$, $\gamma_{-1} = \gamma_1$. To solve the above set of coupled nonlinear ordinary differential equations we used the NAG DO2BBF fourth-order Runge Kutta scheme. A spatial step size of $\Delta x = 0{\cdot}025$ was used and the Runge-Kutta scheme had a variable step size. It was necessary to use a scheme which was fourth-order accurate in time to capture the rapid growth in concentration, characteristic of clock-reaction behaviour. When the spatial step size was halved the results remained consistent to within the excepted accuracy of the scheme. The outer boundary conditions were applied at large distances from the origin so that the propagating wave profile could be given sufficient time to develop.

The solutions for the cases $n = 1$ and $n = 2$ display very similar characteristics and so we present the two sets of results together. Figure 1 shows how the concentration of the autocatalyst varies during the early stages of the reaction. At $\tau = 0$ the concentration of the autocatalyst, β, is zero everywhere apart from a unit step function of half width ξ about the origin. As the reaction proceeds, diffusion smooths out the initial step function in the neighbourhood of $x = \xi$ and a spatially uniform decay occurs in the vicinity of the origin. The autocatalyst is consumed because it reacts with the inhibitor chemical. At later times the distinction is lost between the region of spreading and that of spatially uniform decay. The autocatalyst continues to decrease in concentration and remains monotonically decreasing from its maximum at the origin to zero at infinity.

We now enter the induction period where the autocatalyst concentration remains small and inhibitor from outside the region $|x| < \xi$ diffuses inwards to replace consumed inhibitor. The concentration α continues to increase everywhere because of the decay of the precursor chemical. At some time later the autocatalyst concentration stops decreasing and starts to grow. Inhibitor chemical is consumed in the region around the origin and there is then a very rapid growth in the concentration β and a corresponding decrease in the concentration α. The rapid growth in autocatalyst concentration is illustrated in Figure 2.

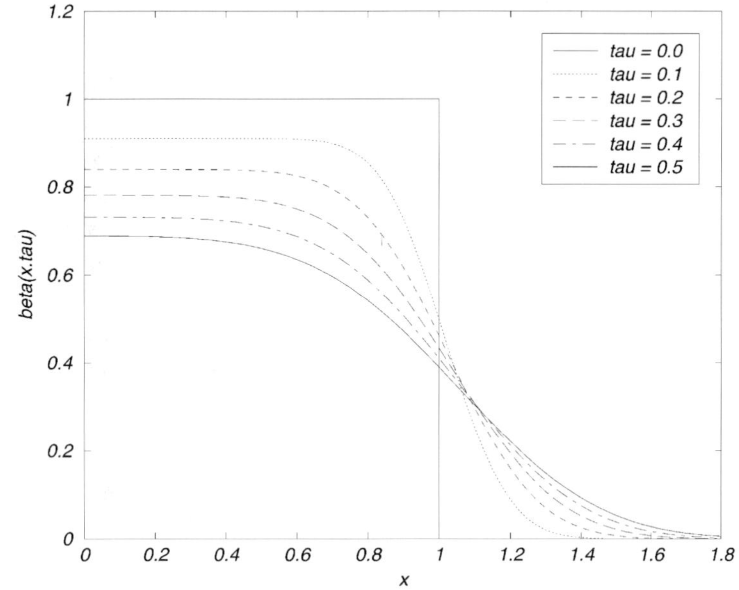

Figure 1. Initial decay of the autocatalyst for the case $n = 2$ with $\mu = 5$, $\lambda = 1$, $\epsilon_0 = 1$, $\xi = 1$ and $\delta = 0{\cdot}2$.

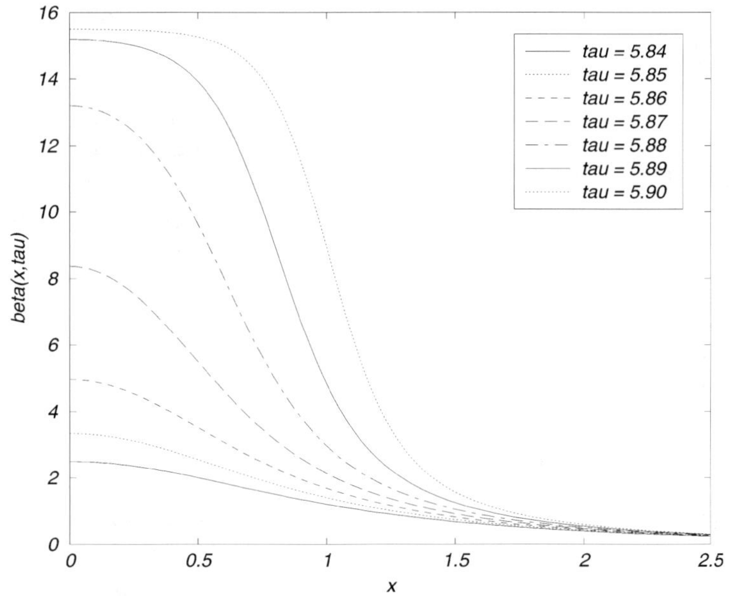

Figure 2. Rapid growth of the autocatalyst for the case $n = 2$ with $\mu = 5$, $\lambda = 1$, $\epsilon_0 = 1$, $\xi = 1$ and $\delta = 0{\cdot}2$.

At approximately $\tau = 5{\cdot}9$ we begin to observe the development of a wave indicated by a spatially uniform region in the vicinity of the origin followed by a sharp decay in β. Behind the profile we find that the concentration of the inhibitor chemical and chemical A have decayed to become very small.

The evolution of the wave profile for the case $n = 2$ is shown in Figure 3. For reasons of clarity we have shown only the autocatalyst concentrations. The wave profile is observed to both propagate away from the origin and grow as time increases. The growth corresponds to an increase in the concentration of the autocatalyst behind the wave and an increase in the

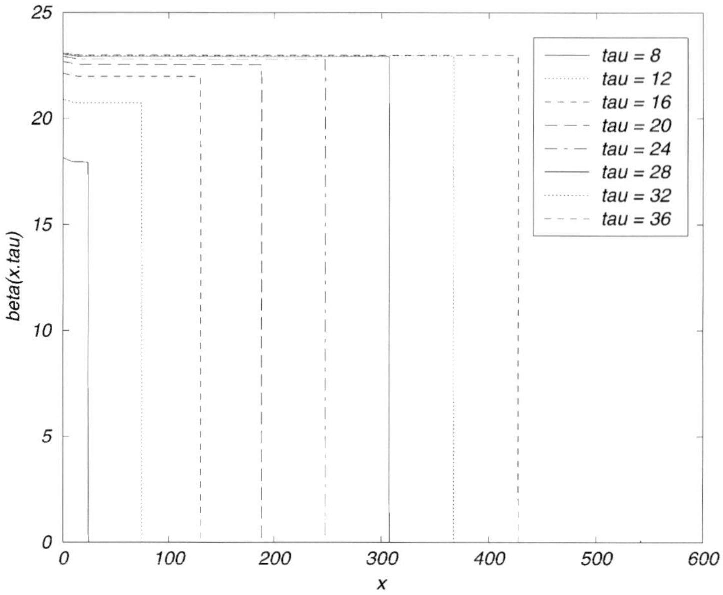

Figure 3. The evolution of the reaction-diffusion front for the case $n = 2$ with $\mu = 5$, $\lambda = 1$, $\epsilon_0 = 1$, $\xi = 1$ and $\delta = 0{\cdot}2$.

concentration of A in front of the wave. The increase in α in front of the wave is due to the decay of the precursor and behind the wave A reacts with B via autocatalysis to produce more B, hence β also increases. Once the precursor chemical has decayed fully, the growth of the wave stops and we find that the wavefront travels the same distance during each period of four time units, indicating that it is travelling with constant speed.

Figures 4 and 5 show the structure of fully developed travelling-wave profiles for the cases $n = 1$ and $n = 2$, respectively. Behind the wavefront all the inhibitor has been consumed and the system is in a fully reacted state. Ahead of the wavefront the solution is in an unreacted state with $\beta = 0$, $\gamma = \lambda$ and α being a spatially uniform constant which tends towards $\mu/(\epsilon_0\delta)$ as the decay of the precursor continues. All the reaction processes are taking place in the centred region whose boundary is propagating away from the origin. We note that the width of the region for the case $n = 1$ is considerably larger than for the case $n = 2$.

To investigate the speed of the wave profile it was necessary to determine the position of the front at each time step. This was defined to be the value of x at which $\partial\beta/\partial x$ took its minimum value. By considering the position of the wavefront at time τ and the position two time increments earlier, $\tau - 2\Delta\tau$, the wave speed was calculated via a central-difference approximation. A value of $\Delta\tau = 0{\cdot}25$ was used throughout. The results for the cases $n = 1$ and $n = 2$ are shown in Figures 6 and 7, respectively.

The intersection of the curve with the horizontal axis marks the end of the induction period and there is then a sharp jump associated with the initiation of the reaction-diffusion front. The wave speed is then seen to increase to its maximum value.

For the case $n = 1$ the numerical results suggest that the final steady wave speed varies like $\delta^{-1/2}$ but for the case $n = 2$ the they suggest the final wave speed to vary like δ^{-1}. In the asymptotic analysis we show that these assumptions must be made for us to construct asymptotic solutions.

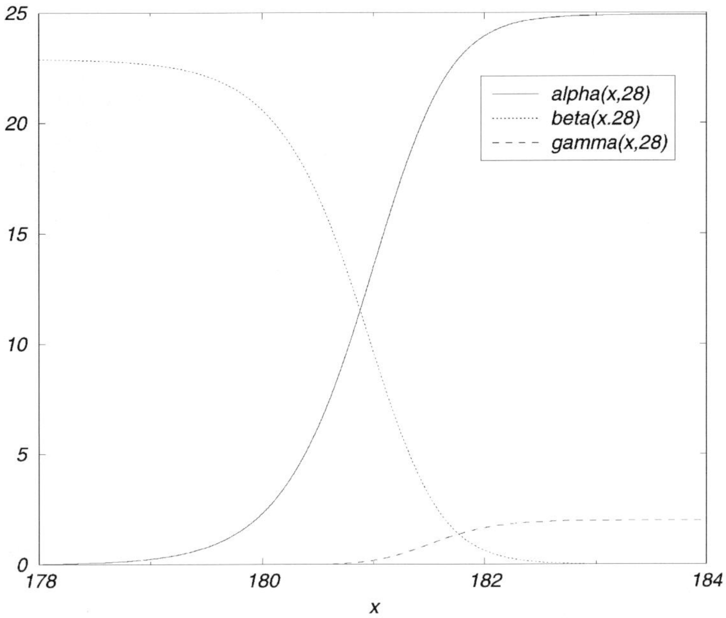

Figure 4. Travelling-wave profile for the case $n = 1$ with $\mu = 5$, $\lambda = 2$, $\epsilon_0 = 1$, $\xi = 1$ and $\delta = 0{\cdot}2$.

We note that the numerical results indicate that a constant-speed travelling-wave is observed only once the precursor has fully decayed. This implies that the concentration of chemical A ahead of the wave is the factor which controls the speed of the wave. Such results have also been found by Merkin and Needham [15] who studied a similar system which had a decay step instead of an inhibition step.

4. Travelling-wave analysis

After initiation of the reaction-diffusion front the numerical solution suggests that the governing equations will admit a solution of the form,

$$y = x - \sigma(\tau). \tag{22}$$

For small times after the end of the induction period, Figure 3 indicates that $\sigma = O(\tau^p)$, where $n > p$. Indeed, it has been shown by Merkin and Needham [15] that, for a cubic autocatalytic system which has a decay step rather than an inhibition step, the wavefront is an $O(\tau^2)$ distance away from the origin when the precursor chemical is assumed to decay slowly. Further investigation shows that the governing equations have a constant-speed travelling-wave solution when the precursor concentration become small, that is that $\sigma = O(\tau)$ when $e^{-\epsilon_0\delta\tau} \ll 1$. In this limit both the concentration of chemical A ahead of the front and the concentration of the autocatalyst behind the front become constant, as shown in Figures 4 and 5. We now analyse the behaviour of the system in this limit and fix the similarity variable y as,

$$y = x - c\tau, \tag{23}$$

where c is the constant speed of the propagating reaction-diffusion front.

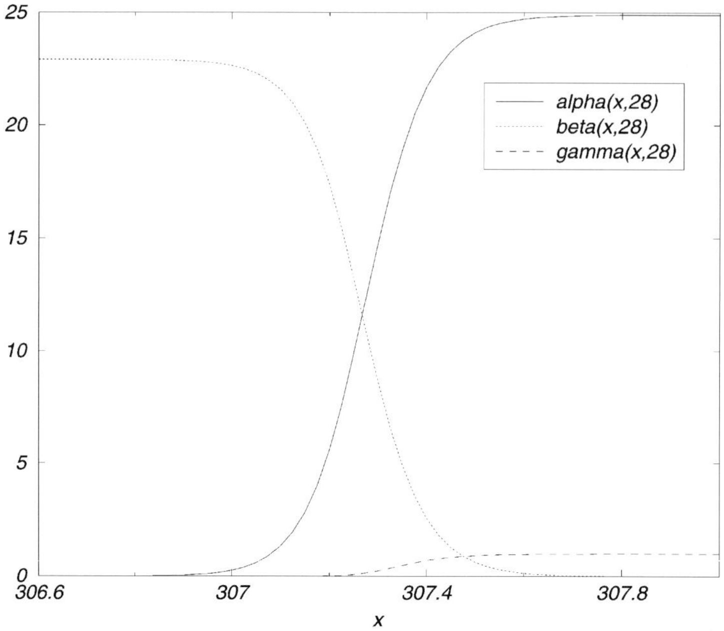

Figure 5. Travelling-wave profile for the case $n = 2$ with $\mu = 5$, $\lambda = 1$, $\epsilon_0 = 1$, $\xi = 1$ and $\delta = 0{\cdot}2$.

4.1. Formulation of the Travelling-Wave Problem

If we neglect the input of A from the precursor chemical and apply the change of variables (23) then Equations (12), (13) and (14) become,

$$\alpha_{yy} + c\alpha_y - \alpha\beta^n = 0, \tag{24}$$

$$\beta_{yy} + c\beta_y + \alpha\beta^n - \frac{1}{\delta}\beta^n\gamma = 0, \tag{25}$$

$$\gamma_{yy} + c\gamma_y - \frac{n}{\delta}\beta^n\gamma = 0, \tag{26}$$

where subscript y denotes differentiation with respect to y, the travelling-wave coordinate. For the purpose of this analysis we assume that the parameters λ and μ are of $O(1)$. The boundary conditions ahead of the wave are those of the unreacted state: there is no autocatalyst present, the concentration of the inhibitor is constant and α has grown to its maximum value via precursor decay so that,

$$\alpha \to \frac{\mu}{\epsilon_0\delta}, \quad \beta \to 0, \quad \gamma \to \lambda, \tag{27}$$

as $y \to \infty$. Figures 4 and 5 show fully developed constant-speed travelling waves. It can be seen that ahead of the travelling waves, condition (27) is satisfied and that behind the wavefronts both the concentrations α and γ are small, possibly zero, and β takes a constant value. If we now consider the variable ϕ defined as,

$$\phi = \alpha + \beta - n\gamma, \tag{28}$$

then addition of Equations (24), (25) and (26) shows ϕ to satisfy the ordinary differential equation,

$$\phi_{yy} + c\phi_y = 0. \tag{29}$$

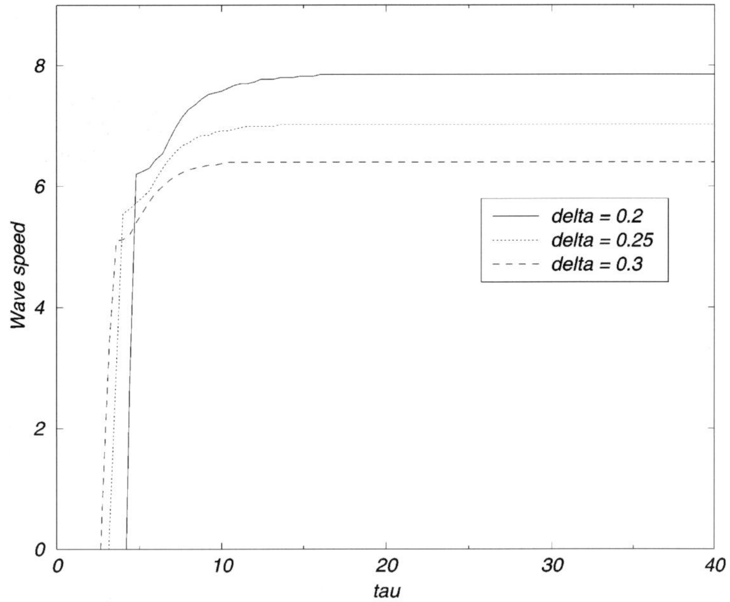

Figure 6. Wave Speed as a function of time for the case $n = 1$ with $\mu = 5$, $\lambda = 2$, $\epsilon_0 = 1$, $\xi = 1$ and $\delta = 0{\cdot}2,\ 0{\cdot}25,\ 0{\cdot}3$

Applying the boundary conditions as $y \to \infty$ we obtain,

$$\phi = \frac{\mu}{\epsilon_0 \delta} - n\lambda. \tag{30}$$

Equations (24), (25) and (26) show that for steady-state conditions both α and γ must be zero for β non-zero, hence from (28) and (30) we deduce the conditions behind the wave to be,

$$\alpha \to 0, \quad \beta \to \frac{\mu}{\epsilon_0 \delta} - n\lambda, \quad \gamma \to 0, \tag{31}$$

as $y \to -\infty$. We are now able to eliminate γ from Equation (25) and thus obtain the fourth-order system,

$$\alpha_{yy} + c\alpha_y - \alpha\beta^n = 0, \tag{32}$$

$$\beta_{yy} + c\beta_y + \alpha\beta^n - \frac{n}{\delta}\beta^n \left(\alpha + \beta - \frac{k}{\delta}\right) = 0, \tag{33}$$

subject to the boundary conditions,

$$\alpha \to \frac{k}{\delta} + n\lambda, \quad \beta \to 0, \qquad y \to \infty, \tag{34}$$

$$\alpha \to 0, \quad \beta \to \frac{k}{\delta}, \qquad y \to -\infty, \tag{35}$$

where

$$k = \frac{\mu}{\epsilon_0} - \delta n\lambda. \tag{36}$$

As this system is four dimensional it is difficult to use phase-space techniques. In the following sections we present asymptotic solutions valid for $\delta \ll 1$ for both cases. To construct

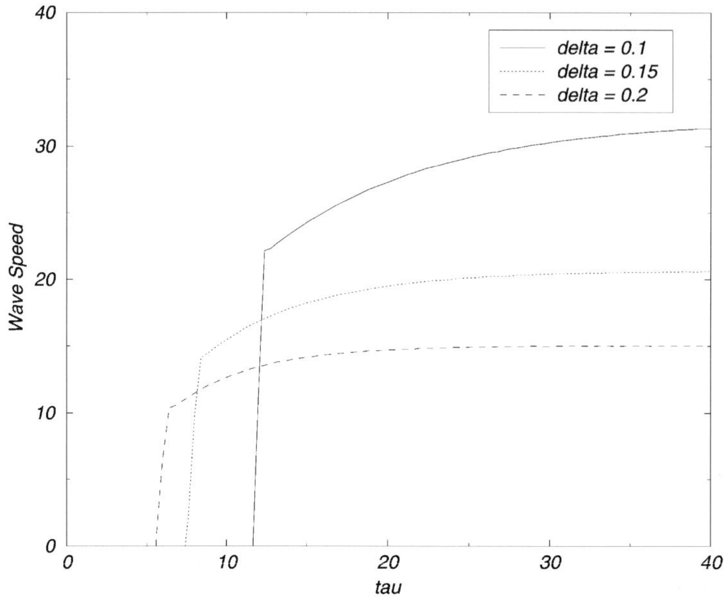

Figure 7. Wave speed as a function of time for the case $n = 2$ with $\mu = 5$, $\lambda = 1$, $\epsilon_0 = 1$, $\xi = 1$ and $\delta = 0{\cdot}1,\ 0{\cdot}15,\ 0{\cdot}2$.

these solutions we use information from the numerical solution of the initial-boundary-value problem to pose suitable expansions for the wave speed.

4.2. Quadratic Autocatalysis with Linear Inhibition ($n = 1$)

Solutions are required to Equations (32) and (33) with $n = 1$ subject to boundary conditions (34) and (35), where k is defined as,

$$k = \frac{\mu}{\epsilon_0} - \delta\lambda. \tag{37}$$

The results in Figure 6 suggest that the final wave speed is of $O(\delta^{-\frac{1}{2}})$, hence we pose an expansion of the form,

$$c = \frac{c_0}{\delta^{1/2}} + c_1 + o(1). \tag{38}$$

The structure of the small-δ asymptotic solution is shown schematically in Figure 8 and can been seen to consist of four asymptotic regions. Behind the wavefront, exponentially small corrections to the boundary condition are constructed. When developed, these solutions show the need for a further asymptotic region in which both α and β are of $O(\delta^{-1})$. Consideration of the system ahead of the wave also shows corrections to the boundary condition to be exponentially small. A further asymptotic region is then required to link these solutions into the central asymptotic region, region II, in which both α and β are of $O(\delta^{-1})$. This central region has width of $O(\delta^{\frac{1}{2}})$ and contains the equation for the travelling-wave solution the Fisher problem at leading order. We now give a description of the asymptotic structure of the full solution.

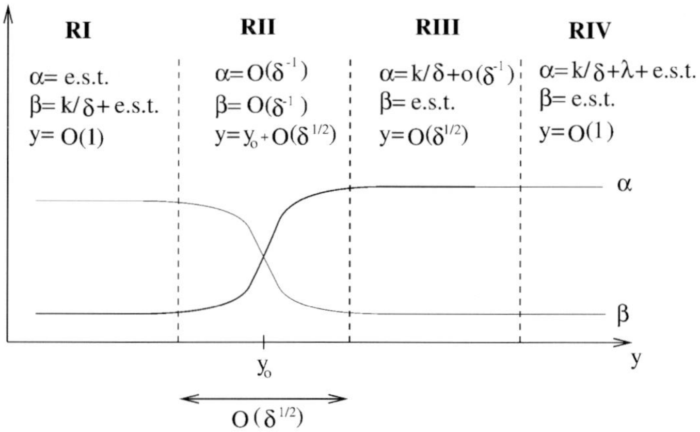

Figure 8. Schematic diagram showing the asymptotic structure of the travelling-wave solution for the case $n = 1$. Exponentially small terms have been denoted by e.s.t.

Region I:
For consistent asymptotic expansions to be developed we require exponentially small corrections to boundary conditions (35). Appropriate scalings are thus,

$$\bar{\alpha} = \delta^{\frac{1}{2}} \log \alpha = O(1), \quad \bar{\beta} = \delta^{\frac{1}{2}} \log \left(\frac{k}{\delta} - \beta \right) = O(1), \quad y = \bar{y} = O(1), \tag{39}$$

and by assuming appropriate expansions we obtain, in terms of the unscaled variables, the solutions

$$\alpha = A(\delta) \exp \left(\frac{\nu y}{\delta^{\frac{1}{2}}} - \frac{c_1 \nu y}{2\nu + c_0} \right) \left(1 + o(1) \right), \tag{40}$$

$$\beta = \frac{k}{\delta} - A(\delta) \exp \left(\frac{\nu y}{\delta^{\frac{1}{2}}} - \frac{c_1 \nu y}{2\nu + c_0} \right) \left(1 + o(1) \right), \tag{41}$$

where ν is given by,

$$\nu = -\frac{c_0}{2} + \frac{\sqrt{c_0^2 + 4k}}{2}, \tag{42}$$

and $A(\delta)$ is a constant of integration. Expansion (41) becomes non-uniform when the exponentially small term grows to become of $O(\delta^{-1})$. This occurs when,

$$y = \frac{\delta^{1/2}}{\nu} \log \left(\frac{1}{\delta A(\delta)} \right), \tag{43}$$

and suggests that a new asymptotic region is required in which both α and β are of $O(\delta^{-1})$.

We now construct a solution in the region ahead of the reaction front.

Region IV:
It is not immediately obvious what form the scalings should take for this region so we pose an asymptotic expansion of the form,

$$\alpha = \frac{k}{\delta} + \lambda + \tilde{\alpha} + \cdots, \quad \beta = \tilde{\beta} + \cdots, \tag{44}$$

with $y = \tilde{y} = O(1)$ and where $\tilde{\alpha} \ll 1$ and $\tilde{\beta} \ll 1$. If we now substitute in Equations (32) and (33) and linearise we obtain,

$$\tilde{\alpha}_{\tilde{y}\tilde{y}} + \left(\frac{c_0}{\delta^{1/2}} + c_1\right)\tilde{\alpha}_{\tilde{y}} - \left(\frac{k}{\delta} + \lambda\right)\tilde{\beta} \sim 0, \tag{45}$$

$$\tilde{\beta}_{\tilde{y}\tilde{y}} + \left(\frac{c_0}{\delta^{1/2}} + c_1\right)\tilde{\beta}_{\tilde{y}} + \left(\frac{k}{\delta} + \lambda\right)\tilde{\beta} - \frac{\lambda}{\delta}\tilde{\beta} \sim 0. \tag{46}$$

Equation (46) has a solution of the form,

$$\tilde{\beta} \sim \tilde{b}_1(\delta)\exp\left(\frac{\theta_+\tilde{y}}{\delta^{1/2}}\right) + \tilde{b}_2(\delta)\exp\left(\frac{\theta_-\tilde{y}}{\delta^{1/2}}\right), \tag{47}$$

where $\tilde{b}_1$ and $\tilde{b}_2$ are gauge functions and,

$$\theta_\pm = \frac{1}{2}\left(-c_0 \pm \sqrt{c_0^2 - 4(k-\lambda)}\right). \tag{48}$$

Substituting this result in Equation (45) and solving we have,

$$\begin{aligned}\tilde{\alpha} \sim{}& \tilde{a}_1(\delta) + \tilde{a}_2(\delta)\exp\left(\frac{-c_0\tilde{y}}{\delta^{1/2}}\right)\\ &+\frac{\left(\frac{k}{\delta}+\lambda\right)\tilde{b}_1(\delta)\exp\left(\frac{\theta_+\tilde{y}}{\delta^{1/2}}\right)}{\left(\frac{\theta_+^2}{\delta}+\frac{c_0\theta_+}{\delta}\right)} + \frac{\left(\frac{k}{\delta}+\lambda\right)\tilde{b}_2(\delta)\exp\left(\frac{\theta_-\tilde{y}}{\delta^{1/2}}\right)}{\left(\frac{\theta_-^2}{\delta}+\frac{c_0\theta_-0}{\delta}\right)}.\end{aligned} \tag{49}$$

We note that the gauge functions $\tilde{a}_1(\delta)$, $\tilde{a}_2(\delta)$, $\tilde{b}_1(\delta)$ and $\tilde{b}_2(\delta)$ are completely undetermined at present. Consideration of Equation (49) shows that when $\tilde{y} = O(\delta^{1/2})$ expansions (44) may become non-uniform. We now choose to examine the behaviour of the solution in this limit by considering the scaled variable,

$$\tilde{y} = \delta^{1/2}\breve{y}. \tag{50}$$

Rewriting the governing equations in terms of the variable $\breve{y}$ we obtain,

$$\frac{\alpha_{\breve{y}\breve{y}}}{\delta} + \left(\frac{c_0}{\delta^{1/2}} + c_1\right)\frac{\alpha_{\breve{y}}}{\delta^{1/2}} - \alpha\beta = 0, \tag{51}$$

$$\frac{\beta_{\breve{y}\breve{y}}}{\delta} + \left(\frac{c_0}{\delta^{1/2}} + c_1\right)\frac{\beta_{\breve{y}}}{\delta^{1/2}} + \alpha\beta - \frac{\beta}{\delta}\left(\alpha + \beta - \frac{k}{\delta}\right) = 0. \tag{52}$$

If we now pick all the gauge functions, $\tilde{a}_1(\delta)$, $\tilde{a}_2(\delta)$, $\tilde{b}_1(\delta)$ and $\tilde{b}_2(\delta)$, to be $O(1)$, then we find that appropriate expansions are,

$$\alpha = \frac{k}{\delta} + \breve{\alpha} + o(1), \quad \beta = \breve{\beta} + o(1). \tag{53}$$

To examine these new scalings we introduce a further asymptotic region.

Region III

Again the exact form of the scalings is not clear at present so we pose expansions (53) and scale y as,

$$y = \delta^{1/2}\breve{y}. \tag{54}$$

Equations (32) and (33) become, at leading order,

$$\breve{\alpha}_{\breve{y}\breve{y}} + c_0\breve{\alpha}_{\breve{y}} - k\breve{\beta} = 0, \tag{55}$$
$$\breve{\beta}_{\breve{y}\breve{y}} + c_0\breve{\beta}_{\breve{y}} + k\breve{\beta} - \breve{\beta}(\breve{\alpha} + \breve{\beta}) = 0. \tag{56}$$

These equations admit solutions of the form (47) and (49) and hence match with the solutions in region IV as $\breve{y} \to \infty$. It is not possible to obtain fully analytical solutions of Equations (55) and (56) valid for all $\breve{y}$, so we try to obtain the behaviour of the system as $\breve{y} \to -\infty$. Equations (55) and (56) linearise if we choose $\breve{\alpha} = -\breve{\beta}$ and the resulting system is easily solved to give,

$$\breve{\alpha} = -\breve{b}_1 e^{\zeta_+\breve{y}} - \breve{b}_2 e^{\zeta_-\breve{y}}, \tag{57}$$
$$\breve{\beta} = \breve{b}_1 e^{\zeta_+\breve{y}} + \breve{b}_2 e^{\zeta_-\breve{y}}, \tag{58}$$

where,

$$\zeta_\pm = -\frac{1}{2}\left(c_0 \pm \sqrt{c_0^2 - 4k}\right). \tag{59}$$

We now show, via a numerical integration, that this is the behaviour of Equations (55) and (56) as $\breve{y} \to -\infty$. We accomplished the integration by discretising the two equations with respect to $\breve{y}$ and then solving the resulting set of nonlinear equations using the NAG routine C05NBF. To obtain boundary conditions as $\breve{y} \to \infty$ we use the matching conditions from region IV and fix,

$$\breve{\alpha} = \lambda, \quad \breve{\beta} = 0, \tag{60}$$

hence we neglect terms which are exponentially small. As $\breve{y} \to -\infty$ we ensure that the solution is of the linearised form discussed above and fix,

$$\breve{\alpha} = -e^{\zeta_+\breve{y}} - \breve{b}_2 e^{\zeta_-\breve{y}}, \tag{61}$$
$$\breve{\beta} = e^{\zeta_+\breve{y}} + \breve{b}_2 e^{\zeta_-\breve{y}}. \tag{62}$$

The travelling-wave solution is invariant under a shift in the travelling-wave coordinate, y, so we can set $\breve{b}_1 = 1$ for the purpose of our numerical solution. Different values of the parameters were used, within the parameter regime $c_0 > 2\sqrt{k}$ which ensures that $\zeta_\pm$ is real, and the scheme was found to converge each time. Changing the point at which the solution was truncated had no effect on the overall solution. Figure 9 shows such a solution of Equations (55) and (56) subject to the boundary conditions described above. It can be observed that $\breve{\alpha}$ and $\breve{\beta}$ decay from being exponentially large as $\breve{y} \to -\infty$ to become λ and 0, respectively, as $\breve{y} \to \infty$. We conclude that the linearised behaviour of Equations (55) and (56) is the behaviour as $y \to -\infty$. In terms of the original variables these linearised solutions are written as,

$$\alpha \sim k - \breve{b}_1 e^{\zeta_+ y/\delta^{1/2}} - \breve{b}_2 e^{\zeta_- y/\delta^{1/2}}, \tag{63}$$
$$\beta \sim \breve{b}_1 e^{\zeta_+ y/\delta^{1/2}} + \breve{b}_2 e^{\zeta_- y/\delta^{1/2}}. \tag{64}$$

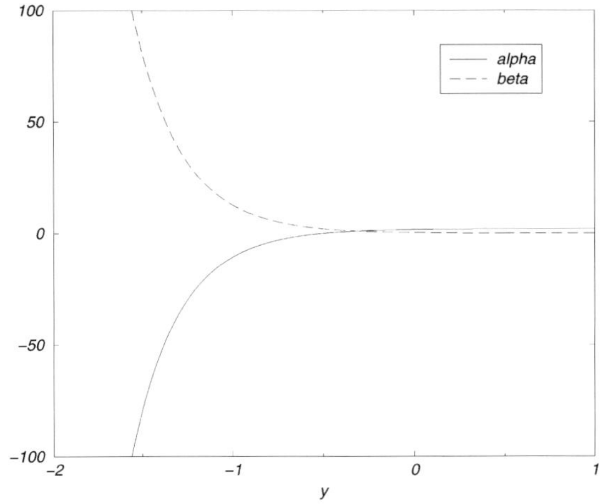

Figure 9. Numerical solution of Equations (55) and (56).

Expansion (63) becomes non-uniform as $y \to -\infty$ in particular when,

$$\breve{y} = -\frac{\delta^{1/2}}{\zeta_+} \log \delta. \tag{65}$$

This suggests a new region in which α and β are $O(\delta^{-1})$.

Region II:
This region is the central region and matches to region III as $\hat{y} \to \infty$ and region I as $\hat{y} \to -\infty$. Appropriate scalings are,

$$\hat{\alpha} = \delta\alpha = O(1), \quad \hat{\beta} = \delta\beta = O(1), \quad \hat{y} = \frac{y - y_0(\delta)}{\delta^{1/2}} = O(1). \tag{66}$$

For consistent matching $y_0(\delta)$ must be the same when evaluated in region I and in region III, giving the relation,

$$y_0(\delta) = -\frac{\delta^{1/2}}{\nu} \log\left[\delta A(\delta)\right] = -\frac{\delta^{1/2}}{\zeta_+} \log \delta, \tag{67}$$

must be satisfied. Under scalings (66) Equations (32) and (33) become,

$$\hat{\alpha}_{\hat{y}\hat{y}} + \left(c_0 + \delta^{1/2} c_1\right) \hat{\alpha}_{\hat{y}} - \hat{\alpha}\hat{\beta} = 0, \tag{68}$$

$$\hat{\beta}_{\hat{y}\hat{y}} + \left(c_0 + \delta^{1/2} c_1\right) \hat{\beta}_{\hat{y}} + \hat{\alpha}\hat{\beta} - \hat{\beta}\left(\frac{\hat{\alpha}}{\delta} + \frac{\hat{\beta}}{\delta} - \frac{k}{\delta}\right) = 0, \tag{69}$$

which at leading order take the form,

$$\hat{\alpha}_{0_{\hat{y}\hat{y}}} + c_0 \hat{\alpha}_{0_{\hat{y}}} - \hat{\alpha}_0 \hat{\beta}_0 = 0, \tag{70}$$

$$\hat{\alpha}_0 + \hat{\beta}_0 = k. \tag{71}$$

Elimination of $\hat{\alpha}_0$ gives the two-dimensional system,

$$\hat{\beta}_{0_{\hat{y}\hat{y}}} + c_0 \hat{\beta}_{0_{\hat{y}}} - \hat{\beta}_0\left(\hat{\beta}_0 - k\right) = 0, \tag{72}$$

Table 1. Comparison of the analytical minimum wave speed and the numerical estimate for the case $n = 1$ with $\mu = 5$, $\lambda = 2$, $\epsilon_0 = 1$ and $\xi = 1$.

δ	Numerical estimate of the wave speed	Analytical minimum wave speed
0·2	7·8	9·6
0·25	7·0	8·5
0·3	6·4	7·7

subject to the boundary conditions,

$$\hat{\beta}_0 \to k, \quad \hat{y} \to -\infty, \qquad \hat{\beta}_0 \to 0, \quad \hat{y} \to +\infty. \tag{73}$$

The parameter k can be removed from Equation (72) and boundary conditions (73) by appropriately scaling $\hat{\beta}_0$, c_0 and $\hat{y}$ Dropping hats and subscripts we obtain the equation,

$$\beta_{yy} + c_0\beta_y + \beta(1 - \beta) = 0, \tag{74}$$

subject to the conditions,

$$\beta \to 1, \quad y \to -\infty, \qquad \beta \to 0, \quad y \to +\infty. \tag{75}$$

Equation (74) is the equation for the travelling-wave solution of Fisher's Equation [2] which has been rigorously analysed by Kolmogorov *et al.* [3]. It is well known that a minimum-wave-speed solution exists and this has been shown by Larson [16] to be the solution which will develop from compact-support initial conditions. In terms of our original variables the minimum wave speed is given as,

$$C_{\min} = 2\left(\frac{\mu}{\epsilon_0\delta} - \lambda\right)^{1/2} + O(1). \tag{76}$$

Table 1 shows that agreement is observed to within the expected $O(1)$ accuracy between the numerical and asymptotic solution.

This completes the asymptotic solution for $\delta \ll 1$. We have shown that solutions which consist of exponentially small corrections to the boundary conditions can be developed to give a central region in which both $\alpha = O(\delta^{-1})$ and $\beta = O(\delta^{-1})$. Within this region we obtain the equation for the travelling-wave solution of the standard Fisher problem at leading order which fixes a minimum wave speed.

4.3. Cubic autocatalysis with quadratic inhibition ($n = 2$)

Solutions are required to Equations (32) and (33) with $n = 2$ subject to boundary conditions (34) and (35), where k is defined as,

$$k = \frac{\mu}{\epsilon_0} - 2\delta\lambda. \tag{77}$$

Figure 7 shows a plot of wave speed vs time for three different values of δ. These results indicate that the final wave speed is of $O(\delta^{-1})$ hence we pose an expansion of the form,

$$c = \frac{c_0}{\delta} + c_1 + o(1). \tag{78}$$

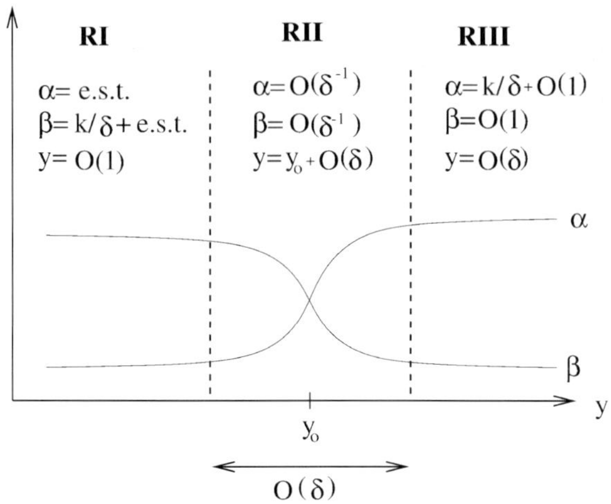

Figure 10. Schematic diagram showing the asymptotic structure of the travelling-wave solution for the case $n = 2$. Exponentially small terms have been denoted by e.s.t.

The structure of the small-δ asymptotic solution is shown schematically in Figure 10 and can been seen to consist of three asymptotic regions. Behind the wavefront exponentially small corrections to the boundary conditions are constructed. When developed these solutions show the need for a further asymptotic region in which both α and β are $O(\delta^{-1})$. Ahead of the wavefront, exponential corrections are again developed which again suggest a central region of width $O(\delta)$ in which α and β are of $O(\delta^{-1})$. The central region contains the equation for the travelling-wave solution of the cubic Fisher problem at leading order. We now give a brief description of the asymptotic structure.

Region I:
Appropriate scalings require exponentially small corrections to the boundary conditions behind the wave. These are,

$$\bar{\alpha} = \delta \log \alpha = O(1), \quad \bar{\beta} = \log\left(\frac{k}{\delta} - \beta\right) = O(1), \quad y = \bar{y} = O(1). \tag{79}$$

By posing suitable expansions we obtain, in terms of the unscaled variables, the expansions,

$$\alpha = A(\delta) \exp\left(\frac{\nu y}{\delta} - \frac{c_1 \nu y}{(2\nu + c_0)}\right)\left(1 + o(1)\right), \tag{80}$$

$$\beta = \frac{k}{\delta} - A(\delta) \exp\left(\frac{\nu y}{\delta} - \frac{c_1 \nu y}{(2\nu + c_0)} +\right)\left(1 + o(1)\right), \tag{81}$$

where

$$\nu = -\frac{c_0}{2} + \frac{\sqrt{c_0^2 + 4k^2}}{2}, \tag{82}$$

and $A(\delta)$ is a constant of integration. Non-uniformities occur when,

$$y = \frac{\delta}{\nu} \log\left(\frac{1}{\delta A(\delta)}\right). \tag{83}$$

Region III:
We now construct a solution in the region ahead of the reaction front. To obtain balances at leading order we choose the scalings,

$$\tilde{\alpha} = \alpha - \frac{k}{\delta} = O(1), \quad \tilde{\beta} = \beta = O(1), \quad \tilde{y} = \frac{y}{\delta} = O(1). \tag{84}$$

In terms of unscaled variables we obtain,

$$\alpha = \frac{k}{\delta} + 2\lambda - a(\delta)e^{-c_0 y/\delta}, \tag{85}$$

$$\beta = a(\delta)e^{-c_0 y/\delta}. \tag{86}$$

From these expansions we observe that non-uniformities will occur as $y \to -\infty$ in particular when,

$$y = -\frac{\delta}{c}\log\left[\frac{1}{a(\delta)}\right]. \tag{87}$$

Again, this suggests a central region of width $O(\delta)$ in which both α and β are of $O(\delta^{-1})$.

Region II:
This central region matches to region I as $y \to -\infty$ and to region III as $y \to \infty$. Appropriate scalings are,

$$\hat{\alpha} = \delta\alpha = O(1), \quad \hat{\beta} = \delta\beta = O(1) \quad \text{and} \quad \hat{y} = \frac{y - y_0}{\delta} = O(1), \tag{88}$$

which when substituted in the governing equations give the leading-order balances,

$$\hat{\alpha}_0 + \hat{\beta}_0 = k, \tag{89}$$

$$\hat{\alpha}_{0\hat{y}\hat{y}} + c_0\hat{\alpha}_{0\hat{y}} - \hat{\alpha}_0\hat{\beta}_0^{\,2} = 0. \tag{90}$$

Elimination of $\hat{\alpha}_0$ gives the second-order equation,

$$\hat{\beta}_{0\hat{y}\hat{y}} + c_0\hat{\beta}_{0\hat{y}} + \hat{\beta}_0^2(k - \hat{\beta}_0) = 0 \tag{91}$$

subject to the boundary conditions,

$$\hat{\beta}_0 \to k, \quad \hat{y} \to -\infty \qquad \hat{\beta}_0 \to 0, \quad \hat{y} \to \infty. \tag{92}$$

The parameter k can be removed from Equation (91) and boundary conditions (92) by an appropriate scaling and, dropping hats and subscripts, we obtain the equation,

$$\beta_{yy} + c_0\beta_y + \beta^2(1 - \beta) = 0 \tag{93}$$

subject to the boundary conditions,

$$\beta \to 1, \quad y \to -\infty, \quad \beta \to 0, \quad y \to \infty. \tag{94}$$

Equation (93) is the equation for the travelling-wave solution of the cubic Fisher problem and has been studied extensively by Gray, Showalter and Scott [17], Britton [18, pp. 102–108] and Billingham and Needham [19].

Table 2. Comparison of the analytical minimum wave speed and the numerical estimate for the case $n = 2$ with $\mu = 5$, $\lambda = 1$, $\epsilon_0 = 1$ and $\xi = 1$.

δ	Numerical estimate of the wave speed	Analytical minimum wave speed
0·1	31·3	33·9
0·15	20·6	22·2
0·2	15·0	16·3

As in the last section we would expect the numerical solution of the full initial-boundary-value problem to agree with the analytical prediction for the minimum wave speed. In terms of the original variables this is given as,

$$c_{\min} = \frac{1}{\sqrt{2}}\left(\frac{\mu}{\epsilon_0\delta} - 2\lambda\right) + O(1). \tag{95}$$

Table 2 shows that agreement between the numerical and asymptotic solution is observed to within the expected $O(1)$ accuracy.

At this point we note that an alternative region III can be constructed in which $y = O(1)$. In this case we obtain solutions for α and β which grow like $1/(y - y_0)$ as $y \to y_0$, where y_0 is a constant which can be fixed by matching. This form of solution matches to the non-minimum wave speed cubic Fisher solution in region II and illustrates an important difference between the solutions in the cases of quadratic and cubic autocatalysis. In the quadratic case all travelling wave solutions decay exponentially ahead of the wave. In contrast, for the case of cubic autocatalysis, only the minimum wave speed decays exponentially, all other solutions decaying algebraically as $y \to \infty$. The details of this alternative region have been omitted as we expect the large-time solution to have the minimum wave speed.

This completes the asymptotic solution for $\delta \ll 1$. We have constructed a three-region solution by developing correction terms to the boundary conditions. The central region has width of $O(\delta)$ and both $\alpha = O(\delta^{-1})$ and $\beta = O(\delta^{-1})$. Within this region we obtain the cubic Fisher problem at leading order and this fixes a minimum wave speed.

5. Conclusion

We have studied the model reaction scheme for the cases $n = 1$ and $n = 2$ allowing all the chemical species to diffuse in one-dimensional slab geometry. Fully numerical solutions have been given for both cases which are found to display similar features. Clock-reaction behaviour was found to be characterised by a very rapid growth of the autocatalyst in a thin region centred about the origin. At the end of the induction period a growing accelerating reaction-diffusion front was evolved. This was then seen to develop into a travelling-wave at constant speed.

Large-time constant-speed travelling-wave solutions have been constructed for both the cases $n = 1$ and $n = 2$ by means of small δ asymptotics. Four regions were required for the case $n = 1$ to describe the full travelling-wave solution. In the central asymptotic region it was found that the leading-order problem reduced to the Fisher equation, hence showing the existence of a family of travelling-wave solutions above a minimum wave speed. For the case $n = 2$ only three asymptotic regions were required. This time the central region was found to contain the cubic Fisher problem at leading order, again fixing a minimum wave speed. In both cases the asymptotic prediction of the wave speed agreed well with that of the numerical solution.

References

1. R. Luther, Discovery and analysis of chemical waves. *Elektrochem.* 12 (1906) 596–610.
2. R. A. Fisher, The wave of advance of advantageous genes. *Ann. Eugenics* 7 (1937) 355–369.
3. A. N. Kolmogorov, I. Petrovsky and N. Piscounoff, Study of the diffusion equation with increase in the quantity of matter, with application to a biological problem. *Bull. Univ. Moscow Ser. Int. Sec A.* 1 (1937) 1–17.
4. J. H. Merkin, D. J. Needham and S. K. Scott, The development of travelling waves in a simple isothermal chemical system I. Quadratic autocatalysis with linear decay. *Proc. R. Soc. London* A 424 (1989) 187–209.
5. K. W. Watkins and R. Distefano, The arsenic(iii) sulfide clock reaction. *J. Chem. Education* 64 (1987) 255–257.
6. P. Jones K. B. and Oldham, The theory of the formaldehyde clock reaction. *J. Chem. Education* 40 (1963) 366–368.
7. J. L. Lambert and G. T. Fina, Iodine clock reaction-mechanisms. *J. Chem. Education* 61 (1984) 1037–1038.
8. P. Jones, M. L. Haggett and J. L. Longridge, The hydration of carbon dioxide - A double clock experiment. *J. Chem. Education* 41 (1964) 610–612.
9. A. Hanna, A. Saul, and K. Showalter, Detailed studies of propagating fronts in the iodate oxidation of arsenous acid. *J. Amer. Chem. Soc.* 104 (1982) 3838–3844.
10. D. O. Cooke, *Inorganic Reaction Mechanisms.* London: The Chemical Society (1979).
11. J. Billingham, and D. J. Needham, Mathematical-modeling of chemical clock reactions. I. Induction, inhibition and the iodate arsenous-acid reaction. *Proc. R. Soc. London* A 340 (1992) 669–591.
12. G. W. Burnet and H. W. Melville, Determination of velocity coefficients for polymerisation processes. *Proc. R. Soc. London* A 189 (1947) 456–480.
13. J. Billingham and D. J. Needham, Mathematical modeling of chemical clock reactions II. A class of autocatalytic clock reaction schemes. *J. Eng. Math.* 27 (1993) 113–145.
14. S. J. Preece. A. C. King and J. Billingham, Chemical clock reactions: The effect of precursor consumption. *J. Mat. Chem.* 26 (1999) 47–73.
15. J. H. Merkin and D. J. Needham, Reaction-diffusion in a simple pooled chemical system. *Dynam. Stab. Sys.* 4 (1989) 141–167.
16. D. A. Larson, Transient bounds and time-asymptotic behaviour of solutions to nonlinear equations of Fisher type. *SIAM J. Appl. Mat.* 34 (1978) 93–103.
17. P. Gray, K. Showalter and S. K. Scott, Propagating reaction-diffusion fronts with cubic autocatalysis: The effects of reversibility. *J. Chimie Physique* 84 (1987) 1329–1333.
18. N. F. Britton, *Reaction Diffusion Equations and their Application to Biology.* London: Academic Press (1986).
19. J. Billingham and D. J. Needham, A note on the properties of a family of travelling-wave solutions arising in cubic autocatalysis. *Dynam. Stab. Sys.* 6 (1991) 33–39.